Bioactive Lipid Mediators

Takehiko Yokomizo • Makoto Murakami
Editors

Bioactive Lipid Mediators

Current Reviews and Protocols

Springer

Editors
Takehiko Yokomizo
Department of Biochemistry
Juntendo University Graduate School of Medicine
Tokyo, Japan

Makoto Murakami
Lipid Metabolism Project
Tokyo Metropolitan Institute of Medical Science
Tokyo, Japan

ISBN 978-4-431-55668-8 ISBN 978-4-431-55669-5 (eBook)
DOI 10.1007/978-4-431-55669-5

Library of Congress Control Number: 2015951082

Springer Tokyo Heidelberg New York Dordrecht London

Printed on acid-free paper

Springer Japan KK is part of Springer Science+Business Media (www.springer.com)

Preface

According to legend, Hippocrates advocated the use of willow bark, *Salix alba*, to ameliorate the pain of childbirth and fever. Salicylates were later isolated as the active component in willow bark and conjugated with an acetyl group to generate aspirin. Today, aspirin is the most well-known nonsteroidal antiinflammatory drug and has numerous recognized clinical benefits, ranging from reducing the risk of heart attack and colon polyposis to its frequently exploited antiinflammatory and pain-relief actions, by decreasing the production of prostaglandins and thromboxane from arachidonic acid. The other arachidonate-derived lipid mediators leukotriene B_4 and cysteinyl leukotrienes have also attracted research interest and are now well known, particularly in the immunology field, as a potent leukocyte attractant and a bronchoconstrictor, respectively.

Recent advances in biochemistry and molecular biology have enabled us to purify and cDNA-clone multiple biosynthetic enzymes and receptors responsible for the production and recognition of individual lipid mediators. Numerous lines of transgenic and knockout mice in relation to lipid mediators have been established and analyzed to unveil their *in vivo* roles. Given that some lipid mediators were shown to cause inflammatory, immune, or oncogenic disorders, some of these enzymes and receptors are now recognized as targets for new drugs. Furthermore, whole sequencing of the human genome led us to discover many orphan receptors, some of which were later identified as receptors for lysophospholipids or their derivatives, such as platelet-activating factor, lysophosphatidic acid, and sphingosine 1-phosphate, and their physiological and pathophysiological roles are currently being extensively analyzed. It has also become evident that omega-3 polyunsaturated fatty acids, which have long been believed to be beneficial for our health, are metabolized to another class of lipid mediators (the so-called pro-resolving lipid mediators), which include resolvins and protectins. The molecular nature of lysophospholipid acyltransferases, which are involved in the remodeling and asymmetry formation of membrane phospholipids, has finally been identified. In addition to progress in protein chemistry and molecular biology, the development of highly sensitive mass spectrometry has greatly contributed to the direct identification of novel lipid mediators that are lowly abundant in tissues. Recently, imaging mass spectrometry successfully elucidated the intra-tissue and intracellular distribution of a particular class of lipids.

We are proud to say that many Japanese scientists contributed to the recent progress in lipid mediator research. In 2010, we started to organize a group of Japanese lipid researchers with the support of the research grant "Lipid machineries" from the Ministry of Education, Culture, Sports, Science, and Technology in Japan (MEXT). At the end of this research term, we are pleased to publish this book entitled *Bioactive Lipid Mediators: Current Reviews and Protocols*, which contains 22 comprehensive reviews of various lipid mediators and seven up-to-date protocols to analyze lipid mediators. No such comprehensive book on lipid mediators has been previously published, and we therefore believe that this book will greatly contribute to the understanding of the roles of lipid mediators in physiology and pathology, the determination of research targets for the near future, and the commencement of drug screening to develop new drugs that target lipid mediators.

We sincerely thank all the authors for their invaluable contributions to this book. We would like to dedicate this book to two prominent lipid researchers, Professors Ichiro Kudo and Santosh Nigam, who opened up this field and trained numerous young researchers over their lifetimes.

Tokyo, Japan

Takehiko Yokomizo
Makoto Murakami

Contents

Part I
Enzymes and Receptors for Lipid Mediators

Chapter 1
Lysophospholipid Acyltransferases

Hideo Shindou, Takeshi Harayama, and Daisuke Hishikawa

Abstract Glycerophospholipids are the main components of cellular membranes. Saturated (also monounsaturated) fatty acids and polyunsaturated fatty acids are usually esterified at the *sn*-1 and *sn*-2 position, respectively, in an asymmetrical manner. Using acyl-CoAs as donors, fatty acids of glycerophospholipids are regulated by lysophospholipid acyltransferases in a *de novo* pathway (Kennedy pathway) and a remodeling pathway (Lands' cycle) to generate membrane diversity. Both pathways were reported in the 1950s. Fourteen lysophospholipid acyltransferases in the 1-acylglycerol-3-phosphate *O*-acyltransferase (AGPAT) and membrane-bound *O*-acyltransferases (MBOAT) families have been identified to date. In this section, recent studies reporting the cloning and characterization of mammalian lysophospholipid acyltransferases are summarized.

Keywords Phospholipid • Remodeling • Lands' cycle • Kennedy pathway • LPLAT

1.1 Introduction

Glycerophospholipids (phospholipids) are important not only as structural and functional components of cellular membranes, but also as the precursors of various lipid mediators, such as platelet-activating factor (PAF) and eicosanoids [1, 2]. Phospholipids are also the main components of pulmonary surfactant: they are composed of two fatty acids and a polar head on a glycerol backbone. Tissues and cells contain several phospholipids such as phosphatidic acid (PA), phosphatidylcholine

H. Shindou (✉)
Department of Lipid Signaling, Research Institute, National Center for Global Health and Medicine, 1-21-1 Toyama, Shinjuku-ku, Tokyo 162-8655, Japan

CREST, Japan Science and Technology Agency,
4-1-8 Honcho, Kawaguchi, Saitama 332-0012, Japan
e-mail: hshindou-tky@umin.net

T. Harayama • D. Hishikawa
Department of Lipid Signaling, Research Institute, National Center for Global Health and Medicine, 1-21-1 Toyama, Shinjuku-ku, Tokyo 162-8655, Japan

T. Yokomizo, M. Murakami (eds.), *Bioactive Lipid Mediators*,
DOI 10.1007/978-4-431-55669-5_1

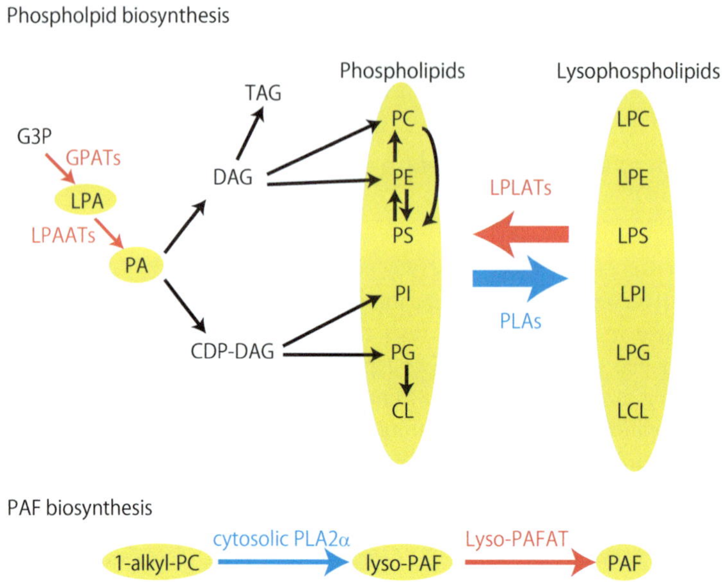

Fig. 1.1 Phospholipid biosynthetic pathways. Acyltransferase steps are indicated in *red*, phospholipase (PLA) steps to release fatty acids in *blue*, and phospholipids in *yellow*

(PC), phosphatidylethanolamine (PE), phosphatidylserine (PS), phosphatidylinositol (PI), phosphatidylglycerol (PG), and cardiolipin (CL) [3, 4]. Phospholipids are biosynthesized by two pathways that were proposed in the 1950s: the first is the de novo pathway (Kennedy pathway) [5], and the second is the remodeling pathway (Lands' cycle) [6]. Saturated and monounsaturated fatty acids are usually esterified at the *sn*-1 position, whereas polyunsaturated acyl groups are esterified at the *sn*-2 position. Using acyl-CoAs as donors, phospholipids are first produced from glycerol-3-phosphate (G3P) by the *de novo* pathway. Next, in the remodeling pathway, the fatty acids of phospholipids are maturated by the coordinated actions of phospholipase A_2s (PLA_2s) and lysophospholipid acyltransferases (LPLATs) (Fig. 1.1) [4, 7, 8].

Recently, several LPLATs were identified in the 1-acylglycerol-3-phosphate *O*-acyltransferase (AGPAT) and membrane-bound *O*-acyltransferase (MBOAT) families. Both these families have motifs that are essential for LPLAT activities. Site-directed mutagenesis revealed four AGPAT motifs ([1] HxxxxD, [2] GxxFxxR, [3] xxEGxx, and [4] xxxxPxx), and four MBOAT motifs ([A] WD, [B] WHGxxxGYxxxF, [C] YxxxxF, and [D] YxxxYFxxH) [7, 9, 10]. AGPAT motif 4 consists of a conserved proline surrounded by hydrophobic amino acids. Although the AGPAT and MBOAT motifs are known to differ completely, more detailed structural information will be provided by future protein crystallization analyses.

Many of the LPLATs have been assigned multiple names because several groups identified them independently or registered different names for the same enzyme based on activity or sequence homology [8]. To eliminate confusion regarding the nomenclature, the proposed names based on their substrate specificities and in order of their publication are summarized in Table 1.1.

Table 1.1 Summary of lysophospholipid acyltransferases (LPLATs) from the 1-acylglycerol-3-phosphate *O*-acyltransferase (AGPAT) and membrane-bound *O*-acyltransferases (MBOAT) families. Other names were registered in HomoloGene of the NCBI database

Name	Other name	Product	Mouse	Human	Family
GPAT1	GPAM; GPAT; GPAT1; RP11-426E5.2	LPA	NP_032175.2	NP_065969	AGPAT
GPAT2	xGPAT1; CT123	LPA	NP_001074558.1	NP_997211.2	AGPAT
GPAT3	AGPAT8; AGPAT9; LPAATθ; MAG1; HMFN0839	LPA	NP_766303.1	NP_116106.2	AGPAT
GPAT4	AGPAT4; LPAATδ; dJ473J16.2; RP3-473 J16.2	LPA	NP_061213.2	NP_848934	AGPAT
LPAAT1	AGPAT1; LPAATα; G15; LPAATA; 1-AGPAT1	PA	NP_001156851.1	NP_006402	AGPAT
LPAAT2	AGPAT2; LPAATβ; BSCL; BSCL1; LPAAB; 1-AGPAT2	PA	NP_080488.1	NP_006403.2	AGPAT
LPAAT3	AGPAT3; LPAATγ	PA	NP_443747.2	NP_001032642.1	AGPAT
LPAAT4	AGPAT4; LPAATδ; dJ473J16.2; RP3-473 J16.2	PA	NP_080920.2	NP_064518.1	AGPAT
AGPAT5	LPAATε	PA?	NP_081068.1	NP_060831.2	AGPAT
LPCAT1	AYTL2; PFAAP3	PC, PAF	NP_663351.3	NP_079106.3	AGPAT
LPCAT2	LysoPAFAT; AYTL1	PAF, PC	NP_766602.1	NP_060309.2	AGPAT
LPCAT3	MBOAT5; C3F; LPCAT; LPSAT; OACT5; nessy; LPLAT 5	PC, PS, PE	NP_660112.1	NP_005759.4	MBOAT
LPCAT4	MBOAT2; OACT2	PC, PE	NP_997089.1	NP_705841.2	MBOAT
LPEAT1	MBOAT1; LPSAT; OACT1; LPLAT 1; dJ434O11.1	PE, PS	NP_705774.1	NP_001073949.1	MBOAT
LPEAT2	LPCAT4; AYTL3; AGPAT7; LPAATη	PE	NP_997089.1	NP_705841.2	AGPAT
LPIAT1	MBOAT7; BB1; LRC4; LENG4; LPIAT; MBOA7; OACT7; hMBOA-7	PI	NP_084210.2	NP_077274.3	MBOAT
LPGAT1	NET8; NET8; FAM34A; FAM34A1	PG	NP_001128301.1	NP_055688.1	AGPAT
LCLAT1	AGPAT8; LYCAT; ALCAT1; UNQ1849; HSRG1849	CL, PI, PG	NP_001171439.1	NP_872357.2	AGPAT

1.2 Lysophosphatidic Acid Acyltransferase (LPAAT) Enzymes

Lysophosphatidic acid (LPA) is the substrate for LPAAT enzymes and is biosynthesized as part of the acyltransferase reaction. Four mammalian GPATs, which synthesize LPA from glycerophospholipids (GP), have been cloned from the AGPAT family [2, 4]. GPAT1 and GPAT2 function in the outer mitochondrial membrane, whereas GPAT3 and GPAT4 are localized to the endoplasmic reticulum (ER) [4]. GPAT4 is also found in lipid droplets [11]. The four GPATs preferentially use saturated and monounsaturated fatty acyl-CoAs to produce LPA, which is then converted to PA by LPAAT. To date, four LPAATs (LPAAT1, -2, -3, and -4) have been cloned and characterized. The representative LPAAT-catalyzed reaction is shown in Fig. 1.2.

1.2.1 LPAAT1 and LPAAT2

Human LPAAT1 and LPAAT2 were cloned based on their homology with yeast, *Escherichia coli*, and coconut AGPATs. LPAAT1 has a higher activity toward 14:0-, 16:0-, and 18:2-CoAs, whereas LPAAT2 exhibited higher activity toward 20:4-CoA compared with 16:0- or 18:0-CoA. Both mRNAs are found in a broad range of tissues [4]. LPAAT2 mutations have been linked to congenital generalized lipodystrophy (also known as Berardinelli–Seip syndrome) [12], suggesting that LPAAT2 might be involved in triacylglycerol (TAG) synthesis and storage in adipocytes.

1.2.2 LPAAT3

LPAAT3 is mainly expressed in the testis and is upregulated in an age-dependent manner [13, 14]. In a biochemical assay, LPAAT3 exhibited LPAAT activity with a preference for polyunsaturated fatty acyl-CoAs (PUFA-CoAs) such as 22:6-CoA [15]. Thus, docosahexaenoic acid (DHA, 22:6) could be inserted into phospholipid via the Kennedy pathway. The induction of LPAAT3 during germ cell development might contribute to the accumulation of PUFAs in testicular phospholipids, suggesting that it has a potential role in proper sperm cell differentiation and maturation. LPAAT3 is also reported to affect Golgi structure and function because Golgi membrane tubule formation and trafficking were inhibited by overexpression of LPAAT3 [16]. In addition, LPAAT3 was detected in lipid droplets [11].

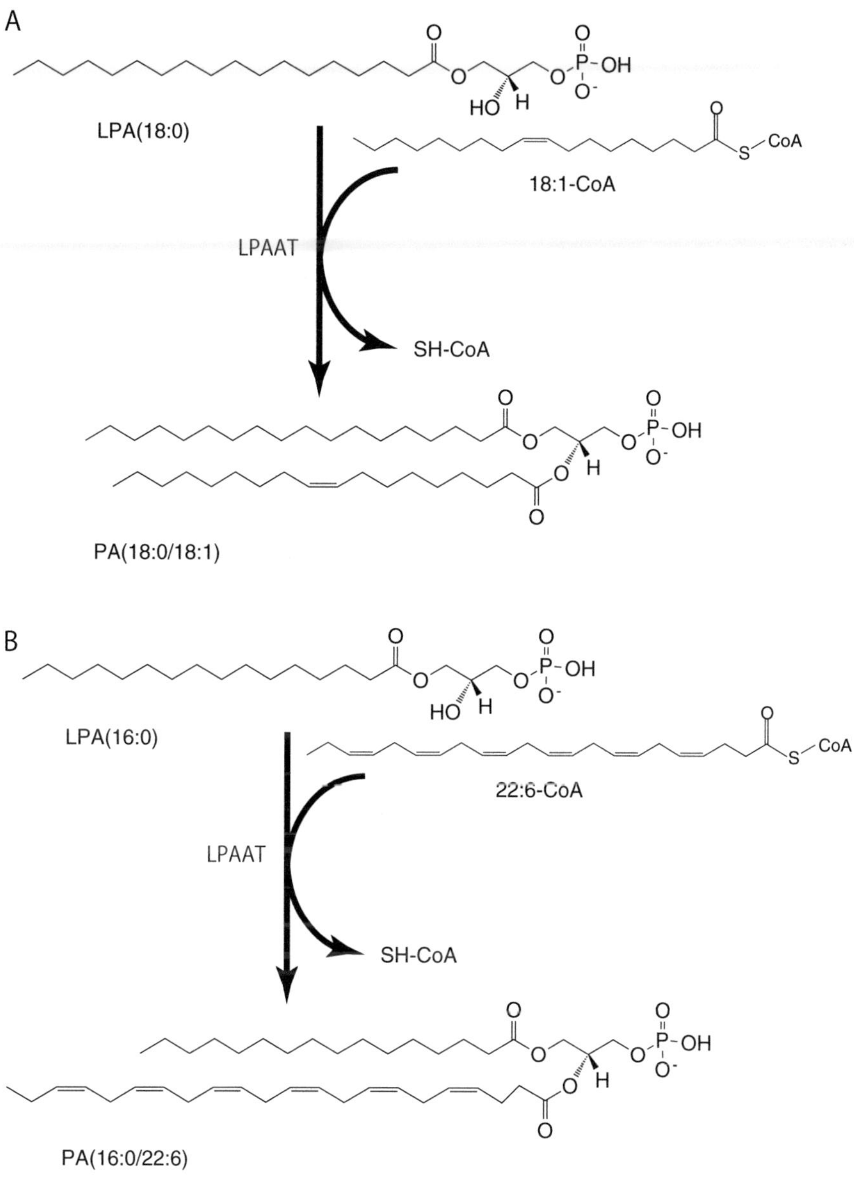

Fig. 1.2 Representative LPAAT-catalyzed reactions: PA (18:0/18:1) (**a**); PA (16:0/22:6) (**b**)

1.2.3 LPAAT4

Recently, LPAAT4 was also reported to possess LPAAT activity with 22:6-CoA [17]. LPAAT4 mRNA is expressed predominantly in the brain. The brain contains an abundant amount of DHA-containing phospholipids. Therefore, LPAAT4 might have important roles in brain function.

1.2.4 Putative LPAAT

AGPAT5 (also called LPAATε) was reported to be a LPAAT and lyso-PE (LPE) AT, but it has not yet been analyzed in detail [18, 19]. AGPAT5 also contains the AGPAT motifs. Further studies are needed to identify the AGPAT5 roles.

1.3 Lysophosphatidylcholine Acyltransferase (LPCAT) Enzymes

Phosphatidylcholine (PC) is biosynthesized from LPC by LPCATs in the remodeling pathway. Four LPCATs have been reported to date: LPCAT1 and LPCAT2 from the AGPAT family, and LPCAT3 and LPCAT4 from the MBOAT family. The representative LPCAT-catalyzed reaction is shown in Fig. 1.3.

1.3.1 LPCAT1

LPCAT1 was the first enzyme identified as having LPCAT activity [20, 21]. LPCAT1 preferentially uses 16:0-CoA to produce dipalmitoyl-PC (DPPC), which is the main component of the pulmonary surfactant that prevents alveolar collapse, small airway closure, and alveolar flooding by decreasing surface tension. Alveolar type II cells produce the pulmonary surfactant that is essential for respiration. Consistent with this, LPCAT1 is expressed mainly in the lung, particularly in alveolar type II cells, and its mRNA expression is upregulated during the perinatal period [21].

Pulmonary surfactant deficiency is an important contributing factor during the pathogenesis of infant respiratory distress syndrome (IRDS), acute respiratory distress syndrome (ARDS), bronchial asthma, and bronchiolitis because pulmonary surfactant plays a critical role in respiratory physiology [22]. Recently, several groups reported the function of LPCAT1 in the lung. LPCAT1 gene trap mice exhibited a decreased level of saturated PC and increased perinatal mortality because of respiratory failure [23]. In addition, LPCAT1-knockout mice had low levels of DPPC, as well as higher sensitivity for acute lung injury than control mice [24]. These reports suggest that the saturated PC generated by LPCAT1 is important for lung surfactant production and function.

Retinal degeneration and visual dysfunction were also found in a mouse strain with a LPCAT1 mutation (*rd11*) [25]. LPCAT1 mRNA levels decreased in the retina and the brain in response to the onset of diabetes in Ins2(Akita) and *db/db* mice, mouse models of type 1 and type 2 diabetes, respectively [26]. LPCAT1 expression was reported in colorectal cancer and was also correlated with the progression of prostate cancer [27, 28].

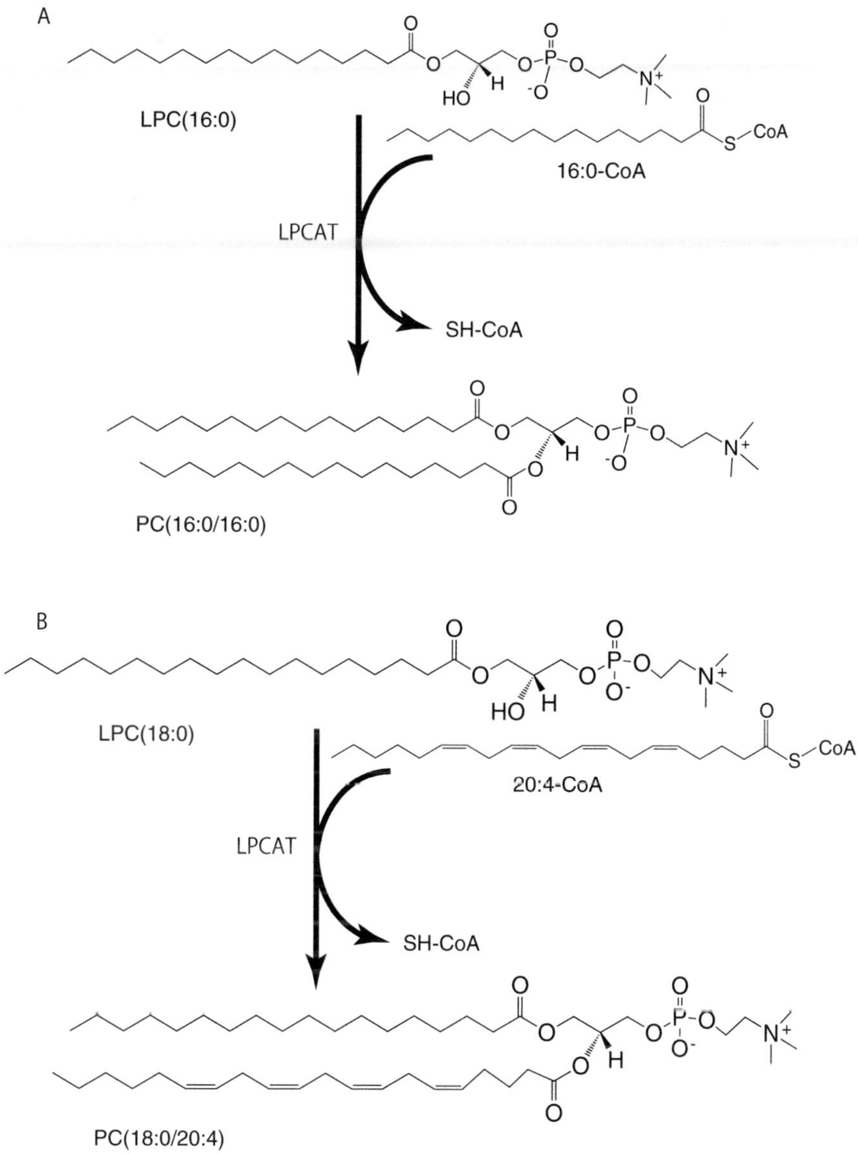

Fig. 1.3 Representative LPCAT-catalyzed reaction: PC (16:0/16:0, DPPC) (**a**); PC (18:0/20:4) (**b**)

LPCAT1 also has lyso-PAF acetyltransferase activity to produce PAF. PAF production by LPCAT1 is described as "lyso-PAF acetyltransferase activity." However, additional studies are required to discover the biological roles of LPCAT1-generated PAF.

1.3.2 LPCAT2

The main product of the LPCAT2-mediated reaction is thought to be PAF, although LPCAT2 also possesses LPCAT (lyso-PAF acyltransferase) activity with a preference for 20:4-CoA [29]. Therefore, LPCAT2 is not only important during the biosynthesis of PAF, but also for membrane homeostasis in inflammatory cells. However, it is important to identify and characterize the binding sites for each substrate (acetyl-CoA and arachidonoyl-CoA). Both LPCAT1 and LPCAT2 are localized in the endoplasmic reticulum (ER) and are also found on the surface of lipid droplets [20, 29, 30]. PAF biosynthetic activity of LPCAT2 is described as "lyso-PAF acetyltransferase activity."

1.3.3 LPCAT3

LPCAT3 mRNA is expressed ubiquitously, and the protein exhibits LPCAT, LPEAT, and LPSAT activities with PUFA-CoAs such as 20:4-CoA and 18:2-CoA [31, 32]. LPCAT3 knockdown in HEK293 cells induced apoptosis and altered cellular morphology [33]. Liver-specific LPCAT3 knockdown in mice showed increased levels of LPC, which led to decreased hepatic triglyceride levels and increased triglyceride-rich lipoprotein production and apolipoprotein-B secretion [34]. LPCAT3 knockdown in mammalian cells also enhanced the palmitic acid-induced unfolded protein response [35]. LPCAT3 is induced by agonists for peroxisome proliferator-activator receptor-α and liver X receptor, as well as during the differentiation of C3H10T1/2 cells into adipocyte-like cells [32, 36, 37]. LPCAT1, LPCAT2, LPCAT3, and LPCAT4 were induced in a model of nonalcoholic steatohepatitis [38]. In addition, hepatic LPCAT3 and LPCAT4 were induced by treatment with fibrates [39]. Interestingly, the *Drosophila* orthologue of LPCAT3, *nessy*, is controlled by ultrabithorax (Ubx), homeobox (Hox), and other Hox proteins during *Drosophila* embryogenesis [40].

1.3.4 LPCAT4

LPCAT4 has both LPCAT and LPEAT activities [31, 41]. LPCAT4 preferentially uses 18:1-CoA as a donor. Mouse LPCAT4 mRNA is highly expressed in the epididymis, brain, testis, and ovary. However, biological roles of LPCAT4 remain unclear.

1.4 Lyso-PAF Acetyltransferase Enzymes

The lyso-PAF acetyltransferase-catalyzed reaction is shown in Fig. 1.4.

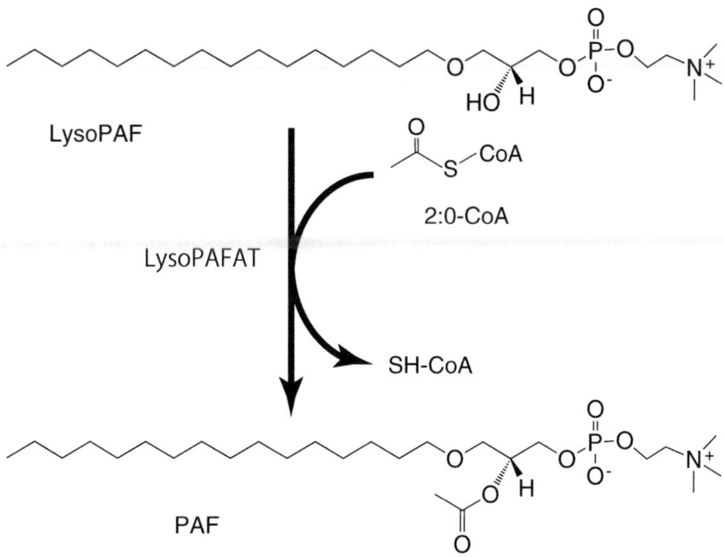

Fig. 1.4 Representative lyso-PAF acetyltransferase-catalyzed reaction: PAF

1.4.1 LPCAT1 and LPCAT2

In response to extracellular stimuli, PAF is synthesized rapidly and has important roles as a potent pro-inflammatory lipid mediator [42]. Specifically, it triggers various cellular functions via its G protein-coupled receptor, PAF receptor [43]. PAF is biosynthesized via the activation of acetyl-CoA:lyso-PAF acetyltransferase (lyso-PAFAT). To date, two lyso-PAFAT enzymes have been reported: LPCAT1 and LPCAT2 [9, 29]. LPCAT1 is mainly expressed in the lung, whereas LPCAT2 is mainly observed in inflammatory cells such as peritoneal macrophages. Before the identification of LPCAT1 and LPCAT2, it was reported that endogenous lyso-PAFAT activity is enhanced in three distinct ways in mouse macrophages: (i) rapid activation (30 s) after PAF stimulation and (ii, iii) minutes to hours after lipopolysaccharide stimulation [44]. However, three distinct mechanisms have been clarified. (1) After G protein-coupled receptor (PAF receptor and ATP receptor) stimulation for 30 s, LPCAT2 phosphorylation at Ser34 in the conventional protein kinase C (cPKC) pathway enhances lyso-PAF acyltransferase activity [45]. (2) After lipopolysaccharide stimulation for 30 min, LPCAT2 is phosphorylated at Ser34 and activated via the p38 MAP kinase (MAPK)/MAPK-activated protein kinase 2 (MK2) pathway [46]. (3) LPCAT2 mRNA is upregulated within 16–24 h of lipopolysaccharide stimulation [29].

In contrast, LPCAT1 is unaffected by lipopolysaccharide, PAF, or ATP stimulation. Therefore, LPCAT1 is a constitutively expressed lyso-PAF acetyltransferase, whereas LPCAT2 is an inducible lyso-PAF acetyltransferase. This relationship is similar to cyclooxygenase 1 and 2, which are constitutively expressed and inducible enzymes, respectively [47]. Recently, the LPCAT2-specific inhibitor TSI-01 was identified from 174,131 compounds. TSI-01 might form a better basis for antiinflammatory drugs in PAF-related diseases [48].

LPCAT2 expression is elevated in several conditions. LPCAT2 mRNA was also reported to be upregulated in the ipsilateral spinal cord in a rat model of peripheral nerve injury [49]. In addition, LPCAT2 was increased in the spinal cords of mice with experimental allergic encephalomyelitis (EAE), which is a model of multiple sclerosis [50]. During this condition, inflammatory cells infiltrate into the central nervous system (CNS).

1.5 Lyso-PE Acyltransferase (LPEAT) Enzymes

A representative LPEAT-catalyzed reaction is shown in Fig. 1.5.

1.5.1 LPEAT1

LPEAT1 exhibits both LPEAT and LPSAT activities and has a preference for 18:1-CoA as a donor [31, 41]. LPEAT1 mRNA is highly expressed in the stomach, epididymis, and colon. Furthermore, the human LPEAT1 gene, which is located on chromosome 6, was disrupted in a brachydactyly-syndactyly syndrome patient [51], suggesting that LPEAT1 contributes to the turnover of phospholipids during normal development and organogenesis.

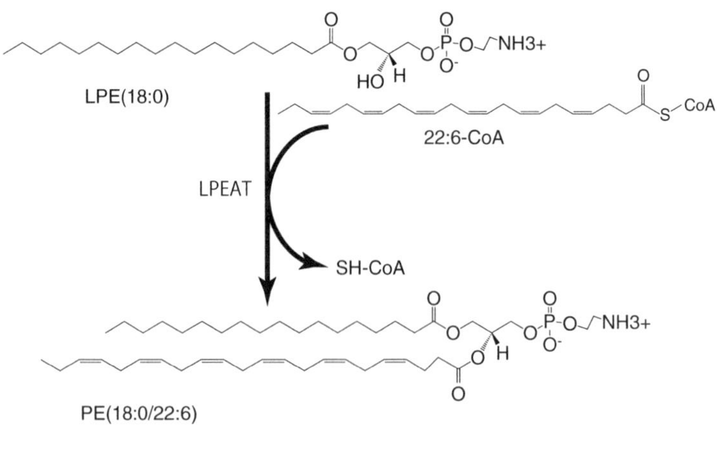

Fig. 1.5 Representative LPEAT-catalyzed reaction: PE (18:0/22:6)

1.5.2 LPEAT2

LPEAT2 was identified from the AGPAT family. Although LPEAT2 exhibits LPEAT, LPGAT, LPSAT, and LPCAT activities using 18:1-CoA or 20:4-CoA as the acyl donor *in vitro*, only LPEAT activity was decreased by its siRNA transfection in HEK293T cells [52]. LPEAT2 is expressed at high levels in the brain, suggesting that it might be important for the biogenesis of brain PE. However, the reported biochemical activities of LPEAT2 are inconsistent with the brain PE composition [53]. It has been reported that hepatic LPEAT2 expression increases on exposure to lithocholic acid exposure [54]. Confusingly, LPEAT2 is also called LPCAT4 (see Table 1.1).

1.5.3 LPCAT3 and LPCAT4

LPCAT3 and LPCAT4 both exhibit LPEAT activity [31, 41]; however, their biological roles as LPEAT are yet to be clarified. The regulation of these enzymes is described earlier in the "LPCAT enzymes" section.

1.6 Lyso-PS Acyltransferase (LPSAT) Enzymes

The representative LPSAT-catalyzed reaction is shown in Fig. 1.6.

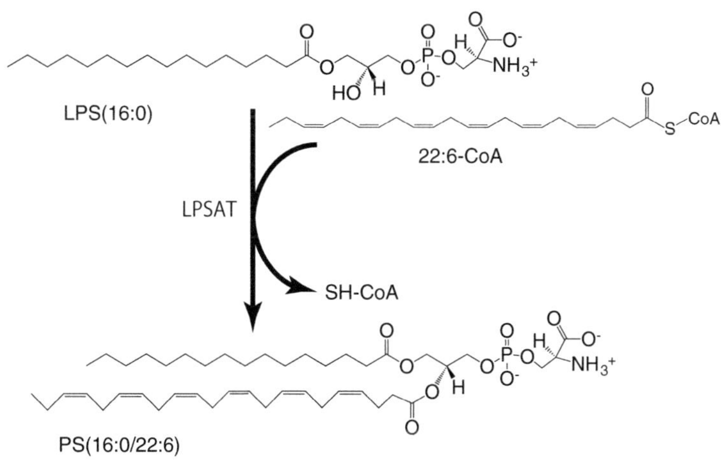

Fig. 1.6 Representative LPSAT-catalyzed reaction: PS (16:0/22:6)

1.6.1 *LPCAT3 and LPCAT4*

LPCAT3 and LPCAT4 both exhibit LPSAT activity [31, 41]; however, their biological roles as LPSAT are yet to be clarified. The regulation of these enzymes is described earlier in the "LPCAT enzymes" section.

1.7 Lyso-PI Acyltransferase (LPIAT) Enzymes

The representative LPIAT-catalyzed reaction is shown in Fig. 1.7.

1.7.1 *LPIAT1*

LPIAT1 is the first reported LPIAT and was identified from the MBOAT family [55]. LPIAT1 prefers 20:4-CoA and 20:5-CoA as donors. In LPIAT1-KO mice, PI and PI phosphates containing arachidonic acid decreased, and the mice exhibited abnormal brain morphology, delayed neural migration, and reduced neurite outgrowth. LPIAT1-KO mice were significantly smaller than their littermates and were born at a frequency less than would be expected from the Mendelian ratio [56, 57]. Arachidonic acid containing PI produced by LPIAT1 is converted to phosphoinositides. They play an important role in brain development [56]. A *Caenorhabditis elegans* LPIAT1 (mboa-7) mutant showed a "bag of worms" phenotype, whereby the embryos hatched within the mother, leaving a cuticle sack containing multiple wriggling larvae [55].

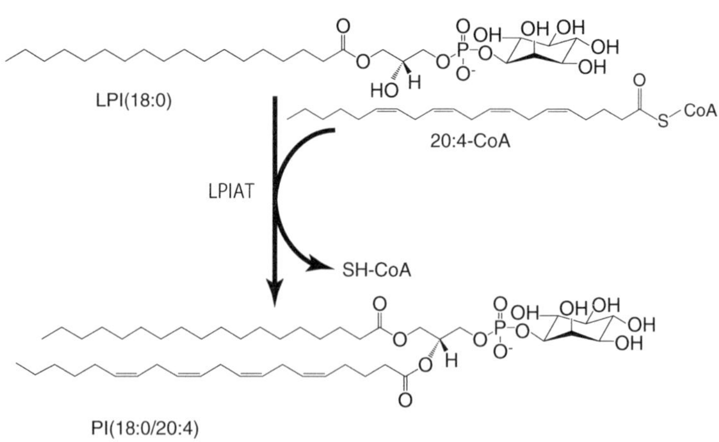

Fig. 1.7 Representative LPIAT-catalyzed reaction: PI (18:0/20:4)

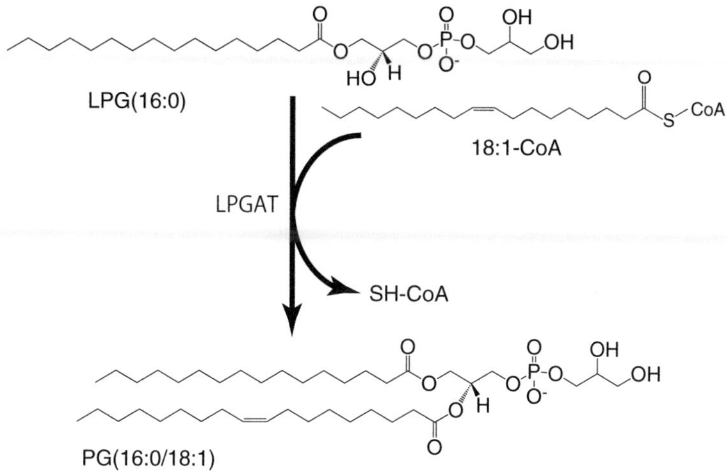

Fig. 1.8 Representative LPGAT-catalyzed reaction: PG (16:0/18:1)

1.7.2 Lyso-CL Acyltransferase (LCLAT)1

LCLAT1, a member of the AGPAT family, has LPIAT activity and uses 18:1-CoA as the donor [58]. LCLAT1 can also incorporate 18:0-CoA into the *sn*-1 position of LPI [59], as is described in detail in the "LCLAT enzyme" section.

1.8 Lyso-PG Acyltransferase (LPGAT) Enzyme

The representative LPGAT-catalyzed reaction is shown in Fig. 1.8.

1.8.1 LPGAT1

PG is synthesized from LPG by LPGAT enzyme during the Lands' cycle. LPGAT1 was cloned as an LPGAT enzyme from the AGPAT family [60]. Human LPGAT1 has a preference for 16:0-, 18:0-, and 18:1-CoAs as donors and is widely expressed in several tissues. PG is a precursor for the synthesis of CL.

1.9 LCLAT Enzyme

Representative LCLAT-catalyzed reaction is shown in Fig. 1.9.

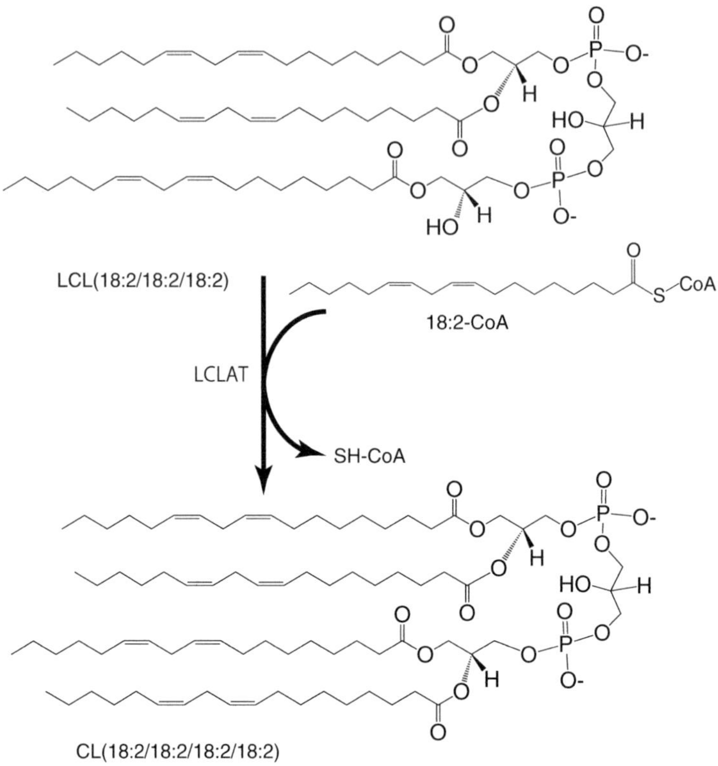

Fig. 1.9 Representative LCLAT-catalyzed reaction: CL (18:2/18:2/18:2/18:2)

1.9.1 LCLAT1

CL, a dimeric glycerophospholipid, is remodeled from dilyso-CL and monolyso-CL by LCLAT. Normal CL contains four linoleoyl group (C18:2) fatty acyl chains. Mouse LCLAT1 was first identified from the AGPAT family as a LCLAT enzyme [58]. Subsequently, it was also reported to possess LPIAT and LPGAT activities [61]. A previous study used site-directed mutagenesis to identify a putative LPI recognition site. Transfection with LCLAT1 siRNA decreased LPIAT and LPGAT activity but not LCLAT activity. It is possible that LCLAT1 functions as both LPIAT and LPGAT enzymes *in vivo*. An additional report showed that LCLAT1 was localized in the mitochondria-associated membrane (MAM) [62]. LCLAT1 overexpression in cells increased the amount of polyunsaturated fatty acid that contained CL and also induced oxidative stress and mitochondrial dysfunction. LCLAT1-KO mice are protected from diet-induced obesity, insulin resistance, and hypertrophic cardiomyopathy [34, 62, 63].

1.10 Conclusion

The Kennedy pathway and the Lands' cycle were first proposed in the 1950s. More than ten LPLATs have been identified during the past decade, resulting in significant advancement of the LPLAT field. However, the nomenclature should be standardized in the international conferences to bring about progress in phospholipid research because most enzymes have several confusing names. It is possible that additional LPLATs with preferences for different substrates might contribute to the generation of membrane diversity and will be identified in future studies. The redundant and pleiotropic substrate preferences of LPLATs might help regulate membrane diversity in tissues, which could be changed in response to external stimuli (Fig. 1.10). Further *in vivo* studies are needed to elucidate the biological roles of LPLATs and to understand the biological significance of membrane diversity and asymmetry.

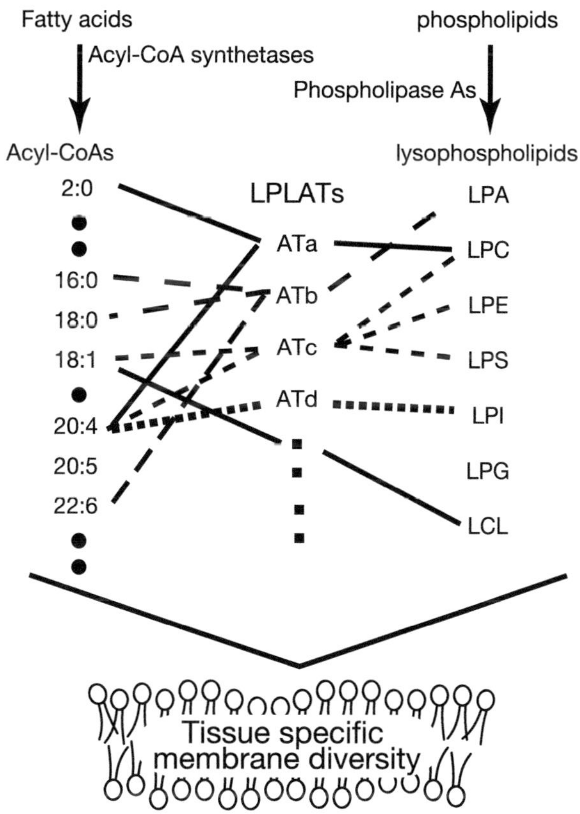

Fig. 1.10 Generation of tissue-specific membrane diversity. Several LPLATs biosynthesize phospholipids using acyl-CoAs and lysophospholipids. Expression pattern of LPLATs, acyl-CoA synthetases, and phospholipase As determines phospholipid composition in individual cells

Acknowledgments We are grateful to Prof. Takao Shimizu and all members of Shimizu's laboratory (National Center for Global Health and Medicine, and The University of Tokyo) for their valuable suggestions.

Note This work is supported by CREST, the Japan Science and Technology Agency (H.S.), a grant-in-aid for Scientific Research (C) (H.S.), and a Grant-in-Aid for Young Scientists (B) (D.H.) from the Ministry of Education, Culture, Sports, Science and Technology (MEXT) of Japan.

References

1. Shimizu T (2009) Lipid mediators in health and disease: enzymes and receptors as therapeutic targets for the regulation of immunity and inflammation. Annu Rev Pharmacol 49:123–150
2. Shindou H, Shimizu T (2009) Acyl-CoA: lysophospholipid acyltransferases. J Biol Chem 284:1–5
3. van Meer G, Voelker DR, Feigenson GW (2008) Membrane lipids: where they are and how they behave. Nat Rev Mol Cell Biol 9:112–124
4. Yamashita A, Hayashi Y, Nemoto-Sasaki Y, Ito M, Oka S, Tanikawa T, Waku K, Sugiura T (2013) Acyltransferases and transacylases that determine the fatty acid composition of glycerolipids and the metabolism of bioactive lipid mediators in mammalian cells and model organisms. Prog Lipid Res 53C:18–81
5. Kennedy EP, Weiss SB (1956) The function of cytidine coenzymes in the biosynthesis of phospholipids. J Biol Chem 222:193–214
6. Lands WE (1958) Metabolism of glycerolipides; a comparison of lecithin and triglyceride synthesis. J Biol Chem 231:883–888
7. Shindou H, Hishikawa D, Harayama T, Eto M, Shimizu T (2013) Generation of membrane diversity by lysophospholipid acyltransferases. J Biochem 154:21–28
8. Hishikawa D, Hashidate T, Shimizu T, Shindou H (2014) Diversity and function of membrane glycerophospholipids generated by the remodeling pathway in mammalian cells. J Lipid Res 55:799–807
9. Harayama T, Shindou H, Ogasawara R, Suwabe A, Shimizu T (2008) Identification of a novel noninflammatory biosynthetic pathway of platelet-activating factor. J Biol Chem 283:11097–11106
10. Shindou H, Eto M, Morimoto R, Shimizu T (2009) Identification of membrane O-acyltransferase family motifs. Biochem Biophys Res Commun 383:320–325
11. Wilfling F, Wang H, Haas JT, Krahmer N, Gould TJ, Uchida A, Cheng JX, Graham M, Christiano R, Frohlich F, Liu X, Buhman KK, Coleman RA, Bewersdorf J, Farese RV Jr, Walther TC (2013) Triacylglycerol synthesis enzymes mediate lipid droplet growth by relocalizing from the ER to lipid droplets. Dev Cell 24:384–399
12. Agarwal AK, Arioglu E, De Almeida S, Akkoc N, Taylor SI, Bowcock AM, Barnes RI, Garg A (2002) AGPAT2 is mutated in congenital generalized lipodystrophy linked to chromosome 9q34. Nat Genet 31:21–23
13. Yuki K, Shindou H, Hishikawa D, Shimizu T (2009) Characterization of mouse lysophosphatidic acid acyltransferase 3: an enzyme with dual functions in the testis. J Lipid Res 50:860–869
14. Koeberle A, Shindou H, Harayama T, Yuki K, Shimizu T (2012) Polyunsaturated fatty acids are incorporated into maturating male mouse germ cells by lysophosphatidic acid acyltransferase 3. FASEB J 26:169–180
15. Koeberle A, Shindou H, Harayama T, Shimizu T (2010) Role of lysophosphatidic acid acyltransferase 3 for the supply of highly polyunsaturated fatty acids in TM4 Sertoli cells. FASEB J 24:4929–4938

16. Schmidt J, Brown W (2009) Lysophosphatidic acid acyltransferase 3 regulates Golgi complex structure and function. J Cell Biol 186:211–218
17. Eto M, Shindou H, Shimizu T (2014) A novel lysophosphatidic acid acyltransferase enzyme (LPAAT4) with a possible role for incorporating docosahexaenoic acid into brain glycerophospholipids. Biochem Biophys Res Commun 443:718–724
18. Lu B, Jiang YJ, Zhou Y, Xu FY, Hatch GM, Choy PC (2005) Cloning and characterization of murine 1-acyl-sn-glycerol 3-phosphate acyltransferases and their regulation by PPARalpha in murine heart. Biochem J 385:469–477
19. Prasad SS, Garg A, Agarwal AK (2011) Enzymatic activities of the human AGPAT isoform 3 and isoform 5: localization of AGPAT5 to mitochondria. J Lipid Res 52:451–462
20. Nakanishi H, Shindou H, Hishikawa D, Harayama T, Ogasawara R, Suwabe A, Taguchi R, Shimizu T (2006) Cloning and characterization of mouse lung-type acyl-CoA:lysophosphatidylcholine acyltransferase 1 (LPCAT1): expression in alveolar type II cells and possible involvement in surfactant production. J Biol Chem 281:20140–20147
21. Chen X, Hyatt BA, Mucenski ML, Mason RJ, Shannon JM (2006) Identification and characterization of a lysophosphatidylcholine acyltransferase in alveolar type II cells. Proc Natl Acad Sci U S A 103:11724–11729
22. Stevens TP, Sinkin RA (2007) Surfactant replacement therapy. Chest 131:1577–1582
23. Bridges JP, Ikegami M, Brilli LL, Chen X, Mason RJ, Shannon JM (2010) LPCAT1 regulates surfactant phospholipid synthesis and is required for transitioning to air breathing in mice. J Clin Invest 120:1736–1748
24. Harayama T, Eto M, Shindou H, Kita Y, Otsubo E, Hishikawa D, Ishii S, Sakimura K, Mishina M, Shimizu T (2014) Lysophospholipid acyltransferases mediate phosphatidylcholine diversification to achieve the physical properties required in vivo. Cell Metab 20:295–305
25. Friedman JS, Chang B, Krauth DS, Lopez I, Waseem NH, Hurd RE, Feathers KL, Branham KE, Shaw M, Thomas GE, Brooks MJ, Liu C, Bakeri HA, Campos MM, Maubaret C, Webster AR, Rodriguez IR, Thompson DA, Bhattacharya SS, Koenekoop RK, Heckenlively JR, Swaroop A (2010) Loss of lysophosphatidylcholine acyltransferase 1 leads to photoreceptor degeneration in rd11 mice. Proc Natl Acad Sci U S A 107:15523–15528
26. Cheng L, Han X, Shi Y (2009) A regulatory role of LPCAT1 in the synthesis of inflammatory lipids, PAF and LPC, in the retina of diabetic mice. Am J Physiol Endocrinol Metab 297:E1276–E1282
27. Mansilla F, da Costa KA, Wang S, Kruhoffer M, Lewin TM, Orntoft TF, Coleman RA, Birkenkamp-Demtroder K (2009) Lysophosphatidylcholine acyltransferase 1 (LPCAT1) overexpression in human colorectal cancer. J Mol Med (Berl) 87:85–97
28. Zhou X, Lawrence TJ, He Z, Pound CR, Mao J, Bigler SA (2012) The expression level of lysophosphatidylcholine acyltransferase 1 (LPCAT1) correlates to the progression of prostate cancer. Exp Mol Pathol 92:105–110
29. Shindou H, Hishikawa D, Nakanishi H, Harayama T, Ishii S, Taguchi R, Shimizu T (2007) A single enzyme catalyzes both platelet-activating factor production and membrane biogenesis of inflammatory cells. Cloning and characterization of acetyl-CoA:LYSO-PAF acetyltransferase. J Biol Chem 282:6532–6539
30. Moessinger C, Kuerschner L, Spandl J, Shevchenko A, Thiele C (2011) Human lysophosphatidylcholine acyltransferases 1 and 2 are located in lipid droplets where they catalyze the formation of phosphatidylcholine. J Biol Chem 286:21330–21339
31. Hishikawa D, Shindou H, Kobayashi S, Nakanishi H, Taguchi R, Shimizu T (2008) Discovery of a lysophospholipid acyltransferase family essential for membrane asymmetry and diversity. Proc Natl Acad Sci U S A 105:2830–2835
32. Zhao Y, Chen YQ, Bonacci TM, Bredt DS, Li S, Bensch WR, Moller DE, Kowala M, Konrad RJ, Cao G (2008) Identification and characterization of a major liver lysophosphatidylcholine acyltransferase. J Biol Chem 283:8258–8265
33. Jain S, Zhang X, Khandelwal PJ, Saunders AJ, Cummings BS, Oelkers P (2009) Characterization of human lysophospholipid acyltransferase 3. J Lipid Res 50:1563–1570

34. Li Z, Ding T, Pan X, Li Y, Li R, Sanders PE, Kuo MS, Hussain MM, Cao G, Jiang XC (2012) Lysophosphatidylcholine acyltransferase 3 knockdown-mediated liver lysophosphatidylcholine accumulation promotes very low density lipoprotein production by enhancing microsomal triglyceride transfer protein expression. J Biol Chem 287:20122–20131
35. Ariyama H, Kono N, Matsuda S, Inoue T, Arai H (2010) Decrease in membrane phospholipid unsaturation induces unfolded protein response. J Biol Chem 285:22027–22035
36. Demeure O, Lecerf F, Duby C, Desert C, Ducheix S, Guillou H, Lagarrigue S (2011) Regulation of LPCAT3 by LXR. Gene (Amst) 470:7–11
37. Eto M, Shindou H, Koeberle A, Harayama T, Yanagida K, Shimizu T (2012) Lysophosphatidylcholine acyltransferase 3 is the key enzyme for incorporating arachidonic acid into glycerophospholipids during adipocyte differentiation. Int J Mol Sci 13:16267–16280
38. Tanaka N, Matsubara T, Krausz KW, Patterson AD, Gonzalez FJ (2012) Disruption of phospholipid and bile acid homeostasis in mice with nonalcoholic steatohepatitis. Hepatology 56:118–129
39. Yamazaki T, Wakabayashi M, Ikeda E, Tanaka S, Sakamoto T, Mitsumoto A, Kudo N, Kawashima Y (2012) Induction of 1-acylglycerophosphocholine acyltransferase genes by fibrates in the liver of rats. Biol Pharm Bull 35:1509–1515
40. Maurel-Zaffran C, Chauvet S, Jullien N, Miassod R, Pradel J, Aragnol D (1999) nessy, an evolutionary conserved gene controlled by Hox proteins during *Drosophila* embryogenesis. Mech Dev 86:159–163
41. Gijón M, Riekhof W, Zarini S, Murphy R, Voelker D (2008) Lysophospholipid acyltransferases and arachidonate recycling in human neutrophils. J Biol Chem 283:30235–30245
42. Prescott SM, Zimmerman GA, McIntyre TM (1990) Platelet-activating factor. J Biol Chem 265:17381–17384
43. Ishii S, Shimizu T (2000) Platelet-activating factor (PAF) receptor and genetically engineered PAF receptor mutant mice. Prog Lipid Res 39:41–82
44. Shindou H, Ishii S, Yamamoto M, Takeda K, Akira S, Shimizu T (2005) Priming effect of lipopolysaccharide on acetyl-coenzyme A: lyso-platelet-activating factor acetyltransferase is MyD88 and TRIF independent. J Immunol 175:1177–1183
45. Morimoto R, Shindou H, Tarui M, Shimizu T (2014) Rapid production of platelet-activating factor is induced by protein kinase C alpha-mediated phosphorylation of lysophosphatidylcholine acyltransferase 2 protein. J Biol Chem 289:15566–15576
46. Morimoto R, Shindou H, Oda Y, Shimizu T (2010) Phosphorylation of lysophosphatidylcholine acyltransferase 2 at Ser34 enhances platelet-activating factor production in endotoxin-stimulated macrophages. J Biol Chem 285:29857–29862
47. Smith WL, Langenbach R (2001) Why there are two cyclooxygenase isozymes. J Clin Invest 107:1491–1495
48. Tarui M, Shindou H, Kumagai K, Morimoto R, Harayama T, Hashidate T, Kojima H, Okabe T, Nagano T, Nagase T, Shimizu T (2014) Selective inhibitors of a PAF biosynthetic enzyme lysophosphatidylcholine acyltransferase 2. J Lipid Res 55:1386–1396
49. Okubo M, Yamanaka H, Kobayashi K, Kanda H, Dai Y, Noguchi K (2012) Up-regulation of platelet-activating factor synthases and its receptor in spinal cord contribute to development of neuropathic pain following peripheral nerve injury. Mol Pain 8:8
50. Kihara Y, Yanagida K, Masago K, Kita Y, Hishikawa D, Shindou H, Ishii S, Shimizu T (2008) Platelet-activating factor production in the spinal cord of experimental allergic encephalomyelitis mice via the group IVA cytosolic phospholipase A2-lyso-PAFAT axis. J Immunol 181:5008–5014
51. Dauwerse JG, de Vries BB, Wouters CH, Bakker E, Rappold G, Mortier GR, Breuning MH, Peters DJ (2007) A t(4;6)(q12;p23) translocation disrupts a membrane-associated O-acetyl transferase gene (MBOAT1) in a patient with a novel brachydactyly-syndactyly syndrome. Eur J Hum Genet 15:743–751

52. Cao J, Shan D, Revett T, Li D, Wu L, Liu W, Tobin JF, Gimeno RE (2008) Molecular identification of a novel mammalian brain isoform of acyl-CoA:lysophospholipid acyltransferase with prominent ethanolamine lysophospholipid acylating activity, LPEAT2. J Biol Chem 283:19049–19057
53. Yabuuchi H, O'Brien JS (1968) Positional distribution of fatty acids in glycerophosphatides of bovine gray matter. J Lipid Res 9:65–67
54. Matsubara T, Tanaka N, Sato M, Kang DW, Krausz KW, Flanders KC, Ikeda K, Luecke H, Wakefield LM, Gonzalez FJ (2012) TGF-beta-SMAD3 signaling mediates hepatic bile acid and phospholipid metabolism following lithocholic acid-induced liver injury. J Lipid Res 53:2698–2707
55. Lee HC, Inoue T, Imae R, Kono N, Shirae S, Matsuda S, Gengyo-Ando K, Mitani S, Arai H (2008) *Caenorhabditis elegans* mboa-7, a member of the MBOAT family, is required for selective incorporation of polyunsaturated fatty acids into phosphatidylinositol. Mol Biol Cell 19:1174–1184
56. Lee HC, Inoue T, Sasaki J, Kubo T, Matsuda S, Nakasaki Y, Hattori M, Tanaka F, Udagawa O, Kono N, Itoh T, Ogiso H, Taguchi R, Arita M, Sasaki T, Arai H (2012) LPIAT1 regulates arachidonic acid content in phosphatidylinositol and is required for cortical lamination in mice. Mol Biol Cell 23:4689–4700
57. Anderson KE, Kielkowska A, Durrant TN, Juvin V, Clark J, Stephens LR, Hawkins PT (2013) Lysophosphatidylinositol-acyltransferase-1 (LPIAT1) is required to maintain physiological levels of PtdIns and PtdInsP2 in the mouse. PLoS One 8:e58425
58. Cao J, Liu Y, Lockwood J, Burn P, Shi Y (2004) A novel cardiolipin-remodeling pathway revealed by a gene encoding an endoplasmic reticulum-associated acyl-CoA:lysocardiolipin acyltransferase (ALCAT1) in mouse. J Biol Chem 279:31727–31734
59. Imae R, Inoue T, Nakasaki Y, Uchida Y, Ohba Y, Kono N, Nakanishi H, Sasaki T, Mitani S, Arai H (2012) LYCAT, a homologue of *C. elegans* acl-8, acl-9, and acl-10, determines the fatty acid composition of phosphatidylinositol in mice. J Lipid Res 53:335–347
60. Yang Y, Cao J, Shi Y (2004) Identification and characterization of a gene encoding human LPGAT1, an endoplasmic reticulum-associated lysophosphatidylglycerol acyltransferase. J Biol Chem 279:55866–55874
61. Zhao Y, Chen YQ, Li S, Konrad RJ, Cao G (2009) The microsomal cardiolipin remodeling enzyme acyl-CoA lysocardiolipin acyltransferase is an acyltransferase of multiple anionic lysophospholipids. J Lipid Res 50:945–956
62. Li J, Romestaing C, Han X, Li Y, Hao X, Wu Y, Sun C, Liu X, Jefferson LS, Xiong J, Lanoue KF, Chang Z, Lynch CJ, Wang H, Shi Y (2010) Cardiolipin remodeling by ALCAT1 links oxidative stress and mitochondrial dysfunction to obesity. Cell Metab 12:154–165
63. Liu X, Ye B, Miller S, Yuan H, Zhang H, Tian L, Nie J, Imae R, Arai H, Li Y, Cheng Z, Shi Y (2012) Ablation of ALCAT1 mitigates hypertrophic cardiomyopathy through effects on oxidative stress and mitophagy. Mol Cell Biol 32:4493–4504

Chapter 2
Phospholipase A_2

Makoto Murakami and Yoshitaka Taketomi

Abstract Phospholipase A_2s (PLA_2s) are a group of enzymes that hydrolyze the *sn*-2 position of phospholipids to generate fatty acids and lysophospholipids, which serve as lipid mediators or their precursors. Mammalian genomes encode genes for more than 30 PLA_2s or related enzymes, which are subdivided into several groups on the basis of their structures, enzymatic properties, and evolutional relationships. Among them, the Ca^{2+}-dependent cytosolic PLA_2 ($cPLA_2$), Ca^{2+}-independent PLA_2 ($iPLA_2$), and secreted PLA_2 ($sPLA_2$) families are regarded as the "big three." From a general point of view, $cPLA_2\alpha$ (the prototypic $cPLA_2$) plays a major role in the initiation of arachidonic acid (AA) metabolism, the $iPLA_2$ family contributes to membrane homeostasis or energy metabolism, and the $sPLA_2$ family affects various biological events by modulating extracellular phospholipid milieus in response to given microenvironmental cues. In this chapter, we overview current understanding of the biological functions of PLA_2s as revealed by gene-manipulated mice and human diseases.

Keywords Arachidonic acid • Eicosanoid • Fatty acid • Glycerophospholipid • Immunity • Inflammation • Lipoprotein • Lysophospholipid • Metabolic disease • Phospholipase A_2

2.1 Introduction

Phospholipase A_2 (PLA_2) catalyzes the hydrolysis of the *sn*-2 position of membrane glycerophospholipids to liberate free fatty acids and lysophospholipids. To date, more than 30 enzymes that possess PLA_2 or related activities have been identified in

M. Murakami (✉) • Y. Taketomi
Lipid Metabolism Project, Tokyo Metropolitan Institute of Medical Science,
2-1-6 Kamikitazawa, Setagaya, Tokyo 156-8506, Japan
e-mail: murakami-mk@igakuken.or.jp

T. Yokomizo, M. Murakami (eds.), *Bioactive Lipid Mediators*,
DOI 10.1007/978-4-431-55669-5_2

mammals, and these have been subdivided into several groups based on their structures, catalytic mechanisms, localizations. and evolutionary relationships. The $cPLA_2$ family contains 6 enzymes ($cPLA_2\alpha$–ζ), which (except for $cPLA_2\gamma$) contain an N-terminal C2 domain for Ca^{2+}-dependent association with the membrane. The $iPLA_2$ or patatin-like phospholipase domain-containing lipase (PNPLA) family includes 9 enzymes, some of which act principally on phospholipids and others on neutral lipids such as triglyceride (TG). The $sPLA_2$ family, in which 10 catalytically active enzymes have been identified, are low molecular weight, extracellular enzymes that require Ca^{2+} of the mM order for optimal enzymatic activity. Because of this diversity, PLA_2 enzymes have been implicated in various biological processes such as lipid mediator production, membrane remodeling, and energy metabolism.

During the past few decades, studies of various PLA_2 transgenic and/or knockout mice as well as human diseases with PLA_2 gene mutations have provided new insights into the emerging biological roles of individual PLA_2s. The functions of individual PLA_2s may not simply reflect changes in lipid mediator signaling, or more particularly eicosanoid signaling, but may also be attributable to hydrolysis of one or a combination of various target membrane lipids. Herein, we focus on the pathophysiology of various PLA_2s as revealed by information from transgenic or knockout mice, as well as human diseases.

2.2 The $cPLA_2$ Family

2.2.1 General Aspects of $cPLA_2$s

Enzymes belonging to the $cPLA_2$ family are characterized by the presence of a C2 domain in their N-terminal region, with the exception of $cPLA_2\gamma$, which lacks this domain. Evolutionarily, the $cPLA_2$ family emerged from the ancestral $iPLA_2$ family at the branching point of vertebrates, correlating with the development of eicosanoid signaling cascades. $cPLA_2\alpha$ is no doubt the best-known PLA_2, with a major role in releasing arachidonic acid (AA), a precursor of eicosanoids (prostaglandins, PGs, and leukotrienes, LTs), from cellular membrane phospholipids.

2.2.2 $cPLA_2\alpha$

$cPLA_2\alpha$, also known as group IVA PLA_2, is localized in the cytosol of resting cells, and in response to an increase in cytosolic Ca^{2+} levels after cell activation, it translocates to the perinuclear or, more specifically, the Golgi membranes to encounter its preferred substrate, AA-containing phosphatidylcholine (PC). Ceramide-1-phosphate or phosphoinositide-4,5-bisphosphate (PIP_2) enhances the membrane

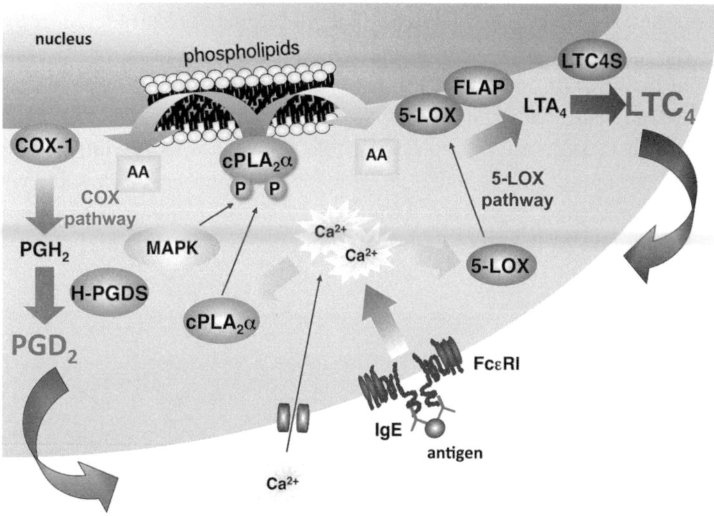

Fig. 2.1 Activation of cytosolic PLA_2 ($cPLA_2$)α in mast cells (MCs). In response to Ca^{2+} influx following FcεRI activation with IgE and cognate antigen, $cPLA_2\alpha$ translocates from the cytosol to the perinuclear membrane and is phosphorylated by mitogen-activated protein kinase (MAPK) for optimal activation. The arachidonic acid (AA) released from membrane phospholipids by $cPLA_2\alpha$ is then converted to PGD_2 by the sequential action of cyclooxygenase (COX)-1 (or COX-2 when the cells are primed by particular stimuli) and hematopoietic PGD_2 synthase (H-PGDS) to PGD_2 or by the sequential action of 5-lipoxygenase (5-LOX) incorporation with 5-LOX-activating protein (FLAP) and LTC_4 synthase (LTC4S) to LTC_4

interaction of $cPLA_2\alpha$. Mitogen-activated protein kinases phosphorylate Ser^{505} on $cPLA_2\alpha$, leading to its activation. The AA released by $cPLA_2\alpha$ is then converted to PGs and LTs by cyclooxygenases and 5-lipoxygenase, respectively. As an example, the $cPLA_2\alpha$ activation mechanism in mast cells (MCs) is shown in Fig. 2.1.

Mice lacking $cPLA_2\alpha$ (*Pla2g4a*$^{-/-}$) exhibit a number of striking phenotypes that can be explained by defects in pathways involving PGs, LTs, or platelet-activating factor (PAF). For instance, *Pla2g4a*$^{-/-}$ mice are protected from asthma, acute respiratory distress syndrome, and pulmonary fibrosis, which can be explained by marked reductions of detrimental lipid mediators such as LTs and PAF [1–3]. *Pla2g4a*$^{-/-}$ mice or wild-type mice treated with a $cPLA_2\alpha$ inhibitor are less susceptible to experimental autoimmune encephalomyelitis or collagen-induced arthritis [4, 5], are protected from brain injuries caused by ischemia or Aβ amyloid [6, 7], and have reduced incidences of intestinal and lung cancer [8, 9], all of which can be attributed to reduced PGE_2 signaling. Consistent with the protective role of PGE_2 in the gastrointestinal mucosa, the intestinal epithelium of *Pla2g4a*$^{-/-}$ mice has numerous small ulcerative lesions [9]. The impairment of female fertility observed in *Pla2g4a*$^{-/-}$ mice suggests that $cPLA_2\alpha$ has an important role in parturition and implantation by providing $PGF_{2\alpha}$ and PGE_2 [10, 7]. Because of reduced production of thromboxane A_2 (TXA_2) by platelets, *Pla2g4a*$^{-/-}$ mice are protected from thromboembolism and have prolonged bleeding times [11]. Furthermore, ablation or knockdown of $cPLA_2\alpha$ ameliorates metabolic disorders including atherosclerosis, hepatic fibrosis, insulin resistance, and adipose tissue inflammation [12–14]. In all cases, the levels of lipid mediators responsible for the corresponding pathophysiological events are markedly reduced in *Pla2g4a*$^{-/-}$ mice relative to wild-type mice. In humans, an inherited *PLA2G4A* mutation is linked to impaired eicosanoid biosynthesis, ulceration of the small intestine, and platelet dysfunction [15].

2.2.3 Other $cPLA_2$s

$cPLA_2\beta$, δ, ε, and ζ (group IVB, IVD, IVE, and IVF PLA_2s) map to the same chromosomal locus and are therefore evolutionally more related [16]. $cPLA_2\beta$ is a dual PLA_1/PLA_2 enzyme, although $cPLA_2\delta$ has a robust PLA_1 activity in preference to PLA_2 activity. $cPLA_2\gamma$ (group IVC PLA_2) is unique in that it lacks the C2 domain and displays lysophospholipase and transacylase activities in addition to PLA_2 activity [17]. The *in vivo* functions of these $cPLA_2$ isoforms are entirely unknown because knockout studies have yet to be performed. A recent study has shown that $cPLA_2\varepsilon$ may drive recycling through clathrin-independent endocytosis [18].

2.3 The $iPLA_2$/PNPLA Family

2.3.1 General Aspects of $iPLA_2$s

The human genome encodes nine $iPLA_2$/PNPLA enzymes, which share a protein motif known as the "patatin domain" with an unusual folding topology that differs from classical lipases (Fig. 2.2). The $cPLA_2$ and $iPLA_2$ families seem to have evolved from a common ancestral gene, as their catalytic domains are commonly characterized by a three-layer $\alpha/\beta/\alpha$ architecture employing a conserved Ser/Asp catalytic dyad instead of the classical catalytic triad [19]. $iPLA_2$/PNPLA enzymes are found in virtually all eukaryotes including yeast, plants, invertebrates, and vertebrates, suggesting that they possess fundamental roles in cellular lipid metabolism conserved in the eukaryote kingdom. The designation "PNPLA" appears to be more appropriate than "$iPLA_2$," as some of the isoforms have enzymatic activities apparently distinct from *bona fide* PLA_2 activity. For instance, $iPLA_2\zeta$/PNPLA2 functions as a major TG lipase in adipose and many other tissues, whereas $iPLA_2\varepsilon$/PNPLA3 may act mainly as an acyltransferase or transacylase for accumulation of TG, particularly in the liver [20]. Here, we focus on two particular $iPLA_2$s, $iPLA_2\beta$/PNPLA9 and $iPLA_2\gamma$/PNPLA8, which have robust PLA_2 activity.

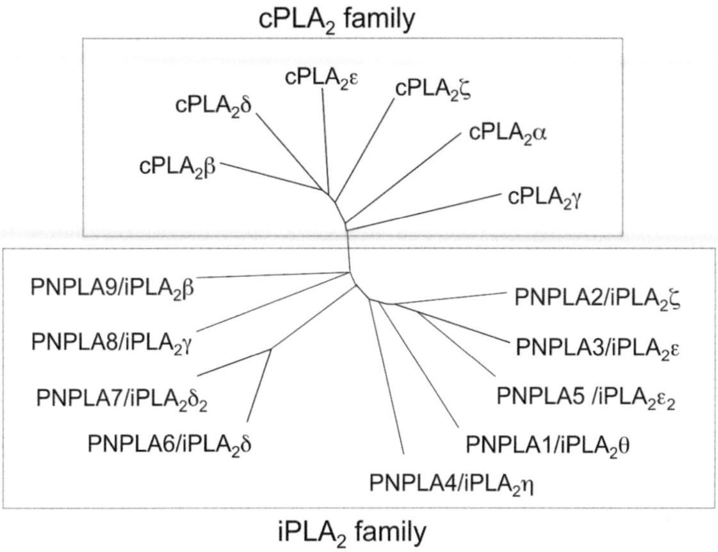

Fig. 2.2 Evolutional relationship between the $cPLA_2$ and $iPLA_2$ families. The $iPLA_2$ family is present in all eukaryotes, whereas the $cPLA_2$ family emerged from the $iPLA_2$ family at the stage of divergence of vertebrates

2.3.2 *iPLA₂β*

$iPLA_2\beta$ (PNPLA9 or group VIA PLA_2), the best characterized $iPLA_2$, has long been thought to be involved in homeostatic phospholipid remodeling through deacylation of phospholipids in the Lands' cycle. Indeed, the composition of phospholipids, particularly those containing docosahexaenoic acid (DHA), is noticeably altered in the brain (but not other tissues) of mice lacking $iPLA_2\beta$ (*Pla2g6*$^{-/-}$) [21]. Notably, human *PLA2G6* mutations are associated with neurodegenerative diseases such as infantile neuroaxonal dystrophy (INAD), neurodegeneration with brain iron accumulation (NBIA), and Schindler's disease, which share the distinctive pathological feature of axonal degeneration with spheroid bodies in the nervous system [22]. Similar neurodegenerative phenotypes are also evident in *Pla2g6*$^{-/-}$ mice or *Pla2g6* mutant mice (*Pla2g6-inad*, in which the *Pla2g6* gene harbors a point mutation), which show motor dysfunction caused by widespread degeneration of axons and synapses, accompanied by the appearance of spheroids and vacuoles [23, 24]. $iPLA_2\beta$ has also been proposed to have more diverse signaling roles. These *Pla2g6*$^{-/-}$ phenotypes include male infertility [25], defective opening of store-operated Ca^{2+} entry, probably caused by reduced production of lysophosphatidylcholine (LPC) [26], impaired insulin secretion by pancreatic β-cells [27], reduced apoptosis [28], decreased eicosanoid generation in vascular cells [29], and protection from ovarian cancer, possibly through reduction of lysophosphatidic acid (LPA) [30]. In most cases, however, the $iPLA_2\beta$-driven lipid metabolic processes underlying these events are poorly characterized.

2.3.3 iPLA$_2\gamma$

iPLA$_2\gamma$, also known as PNPLA8 or group VIB PLA$_2$, is localized to mitochondria or peroxisomes and displays PLA$_2$ or PLA$_1$ activity depending on the substrates involved [31]. Mice null for iPLA$_2\gamma$ (*Pnpla8*$^{-/-}$) exhibit bioenergetic dysfunctional phenotypes, including growth retardation, cold intolerance, reduced exercise endurance, increased mortality from cardiac stress, and abnormal mitochondrial function with an altered cardiolipin composition [32]. Furthermore, *Pnpla8*$^{-/-}$ mice are resistant to diet-induced obesity, fatty liver, and hyperlipidemia [33, 34]. These mice also display lipodystrophy, impaired glucose-stimulated insulin secretion, and decreased mitochondrial β-oxidation. Myocardium-specific *Pnpla8*-transgenic mice show a dramatic reduction of myocardial phospholipid mass, marked accumulation of TG, impaired mitochondrial function, and hemodynamic dysfunction [35]. Thus, iPLA$_2\gamma$ appears to be crucial for maintaining efficient bioenergetic mitochondrial function by tailoring mitochondrial lipid metabolism. However, considering that defective β-oxidation usually leads to increased fat accumulation in peripheral tissues, the protective effect of iPLA$_2\gamma$ ablation against diet-induced metabolic disorders might involve an as yet unknown mechanism. *Pnpla8*$^{-/-}$ mice also display a profound alteration in hippocampal mitochondrial homeostasis, leading to cognitive dysfunction [36]. The *Pnpla8*$^{-/-}$ hippocampus has an increased level of cardiolipin and a decrease of plasmalogen, implying a function of iPLA$_2\gamma$ in remodeling of these phospholipids. Overall, the neurological abnormalities in *Pnpla8*$^{-/-}$ mice are reminiscent of features in patients with Barth syndrome, a disease caused by disturbed cardiolipin metabolism [37].

2.4 The sPLA$_2$ Family

2.4.1 General Aspects of sPLA$_2$s

More than one third of the PLA$_2$ enzymes belong to the sPLA$_2$ family, which contains ten catalytically active isoforms (IB, IIA, IIC, IID, IIE, IIF, III, V, X, XIIA). Individual sPLA$_2$s exhibit unique tissue and cellular localizations and enzymatic properties, suggesting their distinct pathophysiological roles. Classical group I/II/V/X sPLA$_2$s are closely related, 14- to 19-kDa secreted enzymes with a highly conserved Ca^{2+}-binding loop and a His/Asp catalytic dyad as well as conserved disulfide bonds, whereas group III and XII sPLA$_2$s are atypical and classified into distinct classes. As sPLA$_2$s are secreted, their target membranes should reside in the extracellular spaces. Individual sPLA$_2$s contribute to various biological events through production of lipid mediators, promotion of membrane remodeling, modification of extracellular noncellular lipid components such as surfactant, microparticles, and lipoproteins, or degradation of foreign phospholipids such as those originating from microbes and dietary components. Here we overview the

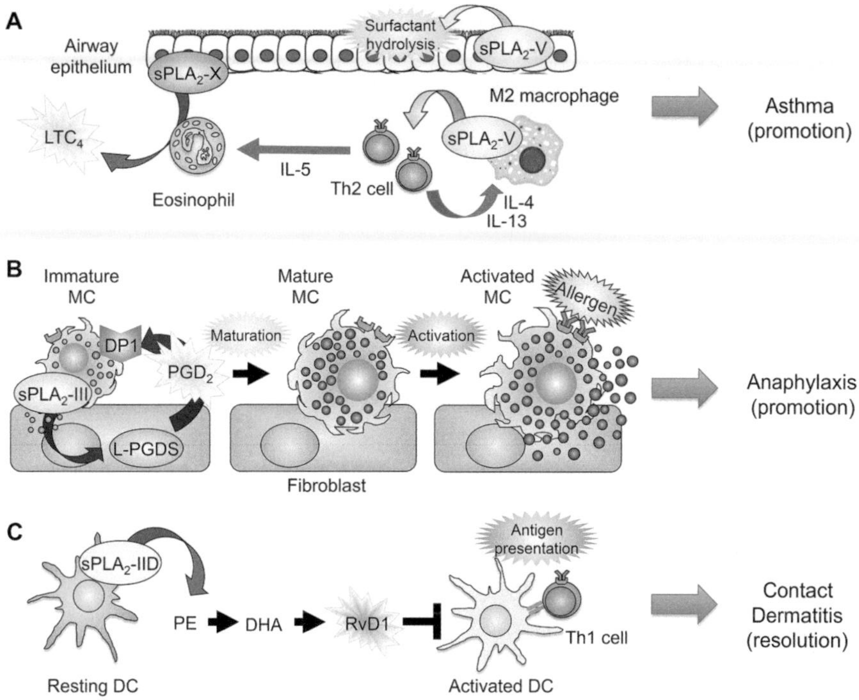

Fig. 2.3 Several examples of $sPLA_2$-driven lipid networks. (**a**) $sPLA_2$-V in M2 macrophages facilitates the Th2 response and that in airway epithelial cells degrades lung surfactant. $sPLA_2$-X from the airway epithelium acts on eosinophils to augment LTC_4 generation. Accordingly, the two $sPLA_2$s independently promote asthma. (**b**) $sPLA_2$-III released from immature MCs acts on fibroblasts to promote L-PGDS-dependent generation of PGD_2, which in turn acts on the PGD_2 receptor DP1 on MCs to promote MC maturation. Accordingly, PLA2G3 facilitates MC-dependent anaphylaxis. Activation of $cPLA_2\alpha$ in mature MCs is highlighted in Fig. 2.1. (**c**) In lymph nodes, $sPLA_2$-IID in DCs hydrolyzes PE to release DHA, which is then converted to the pro-resolving lipid mediator resolving D1 (RvD1) to sequester Th1 immunity. Accordingly, $sPLA_2$-IID ameliorates Th1-dependent contact dermatitis

pathophysiological functions of several classical $sPLA_2$s (IB, IIA, IID, IIE, V, X) and an atypical $sPLA_2$ (group III), as revealed by their transgenic overexpression or gene targeting in mice. Several examples of these $sPLA_2$-mediated lipid networks are illustrated in Fig. 2.3.

2.4.2 $sPLA_2$-IB

$sPLA_2$-IB, often called "pancreatic $sPLA_2$," is abundantly expressed in the pancreas. After secretion into the duodenal lumen, an N-terminal heptapeptide of the inactive zymogen is cleaved by trypsin to yield an active enzyme. The main role of this enzyme is digestion of dietary and biliary phospholipids. Gene disruption of $sPLA_2$-IB (*Pla2g1b*$^{-/-}$) results in decreased phospholipid digestion and absorption in the gut [38]. The reduced intestinal production of LPC, which is a causal factor for hepatic insulin resistance, results in protection from diet-induced obesity, glucose intolerance, hyperlipidemia, and atherosclerosis in *Pla2g1b*$^{-/-}$ mice [39–41].

2.4.3 $sPLA_2$-IIA

$sPLA_2$-IIA is the only isozyme detectable in the circulation, particularly under pathological conditions. It is often referred to as "inflammatory $sPLA_2$" as its levels in sera or inflammatory exudates correlate with the severity of inflammatory diseases, and it is robustly induced by pro-inflammatory stimuli in various cells [42]. However, the precise role of $sPLA_2$-IIA in inflammation remains debatable, as a natural mutation in its gene (*Pla2g2a*) in C57BL/6 and 129Sv mice [43] prevents adequate evaluation of its functions by gene targeting. So far, therefore, most of the *in vivo* functions of $sPLA_2$-IIA have currently been addressed mainly using *Pla2g2a*-transgenic mice.

Pla2g2a-transgenic mice have skin abnormalities manifested by hair loss and epidermal hyperplasia [44] and by increased carcinogen-induced skin cancer [45]. In line with clinical evidence that the serum level of $sPLA_2$-IIA correlates with cardiovascular diseases [46], *Pla2g2a*-transgenic mice develop advanced atherosclerotic lesions [47]. Given that atherosclerosis represents chronic inflammation in the aorta, $sPLA_2$-IIA can be regarded as a pro-inflammatory enzyme in atherosclerosis. The most probable physiological role of $sPLA_2$-IIA is degradation of bacterial membranes, thereby providing a first line of antimicrobial defense [48]. $sPLA_2$-IIA is capable of hydrolyzing phosphatidylethanolamine (PE) and phosphatidylglycerol in marked preference to PC, which can account for the preferential action of this enzyme on bacteria rather than on mammalian cells. Accordingly, *Pla2g2a*-transgenic mice or wild-type mice treated with $sPLA_2$-IIA are resistant to pneumonia and sepsis following bacterial infection [49]. For this reason, $sPLA_2$-IIA is often referred to as a "bactericidal $sPLA_2$."

Mouse strains with natural disruption of the *Pla2g2a* gene (see foregoing) are more sensitive to intestinal tumorigenesis [50]. Transgenic transfer of the *Pla2g2a* gene into these strains reduces the incidence of intestinal polyposis [51]. Thus, $sPLA_2$-IIA appears to have an antitumorigenic role in the gastrointestinal tract. Presumably, bactericidal $sPLA_2$-IIA may affect the gastrointestinal microflora, thereby influencing tumor development. On the other hand, $sPLA_2$-IIA expression is positively correlated with the malignancy of prostate cancer [52], revealing distinct impacts of $sPLA_2$-IIA on different types of cancer. Recently, the mutated *Pla2g2a* allele in the C57BL/6 mouse strain was delivered into the BALB/c mouse strain to produce *Pla2g2a*$^{-/-}$ BALB/c mice. Autoantibody-induced arthritis is attenuated in these *Pla2g2a*$^{-/-}$ mice relative to *Pla2g2a*-sufficent mice, whereas it is conversely aggravated in *Pla2g2a*-transgenic mice [53]. This study has provided the first compelling evidence for the pro-inflammatory role of sPLA$_2$-IIA. Recently, it has been shown that sPLA$_2$-IIA targets phospholipids in extracellular mitochondria, and thereby amplifies inflammation by producing eicosanoids as well as mitochondrial DNA, a kind of danger-associated molecular pattern (DAMP) [54].

2.4.4 sPLA$_2$-IID

sPLA$_2$-IID is structurally most related to sPLA$_2$-IIA and is expressed preferentially in dendritic cells (DCs) in secondary lymphoid organs [55], suggesting its immunoregulatory role. In a model of Th1-dependent contact dermatitis, resolution of inflammation is compromised in skin and lymph nodes of mice lacking sPLA$_2$-IID (*Pla2g2d*$^{-/-}$) [55]. sPLA$_2$-IID in regional lymph nodes mobilizes a pool of polyunsaturated fatty acids that can be metabolized to pro-resolving lipid mediators such as DHA-derived resolvin D1, which reduces Th1 cytokine production and DC activation. sPLA$_2$-IID preferentially hydrolyzes DHA-containing PE in lymph node membranes. In accordance with its antiinflammatory role, sPLA$_2$-IID expression in DCs is downregulated after cell activation. Furthermore, administration of sPLA$_2$-IID-Fc protein attenuates autoimmune diseases in mice [56]. Together, the existing data suggest that sPLA$_2$-IID is a "resolving sPLA$_2$" that ameliorates inflammation by mobilizing DHA-derived pro-resolving lipid mediators.

2.4.5 sPLA$_2$-IIE

Similar to sPLA$_2$-IID, sPLA$_2$-IIE is structurally most homologous to sPLA$_2$-IIA. Expression of sPLA$_2$-IIE is markedly induced in adipocytes during adipogenesis *in vitro* and after high-fat feeding *in vivo*. Mice deficient in sPLA$_2$-IIE (*Pla2g2e*$^{-/-}$) are modestly protected from diet-induced obesity, fatty liver, and hyperlipidemia [57]. sPLA$_2$-IIE preferentially hydrolyzes minor lipoprotein phospholipids, phosphatidylserine (PS), and PE, with no apparent fatty acid selectivity. As such, sPLA$_2$-IIE alters lipid composition in lipoproteins, thereby affecting fat deposition in adipose tissue and liver. Thus, sPLA$_2$-IIE is a "metabolic sPLA$_2$" that controls systemic metabolic states by modulating lipoprotein phospholipids. These findings shed light on the importance of the minor lipoprotein phospholipids (PS and PE) in metabolic regulation.

2.4.6 sPLA$_2$-V

Because sPLA$_2$-V is able to hydrolyze PC more efficiently than sPLA$_2$-IIA, most investigators in this research field have been interested in the potential roles of this enzyme in inflammation in the context of AA metabolism. Indeed, zymosan-induced peritonitis or lipopolysaccharide (LPS)-induced air pouch inflammation is partially attenuated in mice lacking sPLA$_2$-V (*Pla2g5*$^{-/-}$) [58, 59]. sPLA$_2$-V is highly expressed in the myocardium, and *Pla2g5*$^{-/-}$ mice exhibit a markedly decreased infarct size in a myocardial ischemia and reperfusion model [60]. sPLA$_2$-V is expressed in bronchial epithelial cells and alveolar macrophages, and *Pla2g5*$^{-/-}$ mice are protected from airway disorders such as antigen-induced asthma and LPS-induced respiratory distress syndrome [61, 62]. Moreover, in keeping with the view that hydrolysis of phospholipids in low density lipoprotein (LDL) by sPLA$_2$-V can promote foam cell formation by macrophages *in vitro* [63], *Ldlr*$^{-/-}$ mice transplanted with *Pla2g5*$^{-/-}$ bone marrow cells are partially protected from atherosclerosis [64]. Although most of these studies support the offensive roles of sPLA$_2$-V, the underlying mechanisms by which sPLA$_2$-V regulates each of these pathologies have remained controversial. Of note, sPLA$_2$-V prefers phospholipids bearing fatty acids with a lower degree of unsaturation (e.g., oleate and linoleate) to those containing highly polyunsaturated fatty acids (e.g., AA and DHA), making it unclear whether sPLA$_2$-V indeed mobilizes AA-derived eicosanoids *in vivo*. Because increased inflammation is generally accompanied by cPLA$_2\alpha$ activation, the observed changes in eicosanoid levels in *Pla2g5*$^{-/-}$ mice might simply reflect disease-associated changes in cPLA$_2\alpha$ activation, rather than hydrolytic liberation of AA by sPLA$_2$-V. Indeed, transgenic overexpression of sPLA$_2$-V leads to respiratory distress and neonatal death without alterations in pulmonary eicosanoid levels [65]. This phenotype has been ascribed to aberrant hydrolysis of surfactant phospholipids (dipalmitoyl-PC) and is apparently eicosanoid independent.

The roles of sPLA$_2$-V in inflammation have been proven to be more complex. Although sPLA$_2$-V was thought to be induced by pro-inflammatory stimuli (as in the case of sPLA$_2$-IIA), it has recently become obvious that its expression is induced by the Th2 cytokines IL-4 and IL-13, rather than proinflammatory stimuli including LPS, zymosan, and Th1 cytokines, which decrease sPLA$_2$-V expression [57, 66]. sPLA$_2$-V is expressed in IL-4-driven M2 macrophages and Th2 cells, which facilitate Th2-type inflammation while attenuating Th1 or Th17 immunity. Importantly, Th2 responses, as monitored by IL-4 expression and IgE production, are greatly reduced in *Pla2g5*$^{-/-}$ mice, thus underscoring the reduced asthmatic phenotype from the lack of sPLA$_2$-V. Thus, sPLA$_2$-V appears to function in at least two regulatory steps in asthma: (1) in antigen-presenting cells to regulate antigen processing and thereby the Th2 response, and (2) in airway-resident cells to promote airway inflammation that may involve surfactant degradation. *Pla2g5*$^{-/-}$ mice are more susceptible to *Candida* infection (Th1 immunity) and arthritis (Th17 immunity) [53, 67], which could also be partly explained by the ability of sPLA$_2$-V to promote Th2 immunity (and therefore to suppress Th1/Th17 immunity).

The function of sPLA$_2$-V as a "Th2-prone sPLA$_2$" also influences obesity, as Th2 or M2 response dampens adipose tissue inflammation. In obesity, sPLA$_2$-V is induced in hypertrophic adipocytes [57]. When fed a high-fat diet, *Pla2g5*$^{-/-}$ mice display hyperlipidemia with higher plasma levels of lipid-rich LDL and increased obesity, fatty liver, and insulin resistance. sPLA$_2$-V has a protective function against metabolic disorders by hydrolyzing and thereby normalizing PC in LDL and by tipping the immune balance toward an Th2/M2 state that counteracts adipose tissue inflammation. Mechanistically, $sPLA_2$-V-driven oleate and linoleate from PC in LDL dampen M1 macrophage polarization by saturated fatty acids (e.g., palmitate), probably through attenuation of endoplasmic reticulum stress. Clinically also, $sPLA_2$-V expression in human visceral adipose tissue inversely correlates with plasma LDL levels. These studies underscore the physiological relevance of lipoprotein hydrolysis by $sPLA_2$s, highlight two adipocyte-driven "metabolic $sPLA_2$s" ($sPLA_2$-IIE and $sPLA_2$-V) as integrated regulators of immune and metabolic responses, and bring about a paradigm shift toward a better understanding of the roles of the $sPLA_2$ family as a metabolic coordinator.

2.4.7 *$sPLA_2$-X*

As in the case of $sPLA_2$-IB, $sPLA_2$-X is synthesized as a zymogen, and removal of an N-terminal pro-peptide produces an active mature enzyme [68]. Among the $sPLA_2$s, $sPLA_2$-X has the highest binding affinity for PC and thus exhibits the most potent ability to hydrolyze plasma membrane phospholipids in intact cells [69]. Because of this property, many investigators have speculated that $sPLA_2$-X has a major role in inflammation. In line with this scenario, mice lacking $sPLA_2$-X (*Pla2g10*$^{-/-}$) are refractory to antigen-induced asthma, with markedly reduced infiltration of eosinophils and lymphocytes, attenuated goblet cell hyperplasia and smooth muscle layer thickening, and decreased levels of Th2 cytokines and pro-asthmatic eicosanoids [70]. The attenuated asthmatic responses in *Pla2g10*$^{-/-}$ mice are fully restored by knock-in of human $sPLA_2$-X, and treatment of the knock-in mice with an inhibitor specific for human $sPLA_2$-X suppresses airway inflammation [71]. Mechanistically, $sPLA_2$-X secreted from the airway epithelium may act on infiltrating eosinophils to augment LT production in a process involving LPC-dependent activation of $cPLA_2\alpha$ [72]. *Pla2g10*$^{-/-}$ mice are also protected from the early phase of influenza infection [73], further highlighting the role of this enzyme in the airway. Moreover, $sPLA_2$-X is one of the major $sPLA_2$ isoforms detected in the airway of patients with asthma [74], thus directing attention to $sPLA_2$-X as a novel therapeutic target for asthma. In contrast to $sPLA_2$-V, however, $sPLA_2$-X does not influence the Th2 response itself, as antigen-sensitized *Pla2g10*$^{-/-}$ mice have normal IgE and IL-4 levels.

Several phenotypes have been reported for *Pla2g10*$^{-/-}$ mice, but the data are controversial. These phenotypes include protection from myocardial infarction or aneurysm [75, 76], exacerbation or attenuation of atherosclerosis [77, 78], increased or decreased adiposity [79, 80], altered macrophage responses [81], and lower response to peripheral pain [79]. In some of these studies, experiments were performed under the assumption that $sPLA_2$-X is expressed in immune cells such as neutrophils and macrophages. However, the expression of $sPLA_2$-X in such immune cells is very low or almost undetectable [75, 79], raising a question as to the physiological relevance of studies involving adoptive transfer of *Pla2g10*$^{-/-}$ bone marrow-derived cells. Rather, the possibility that paracrine $sPLA_2$-X may alter the properties of inflammatory cells should be taken into account. Because $sPLA_2$-X is abundantly expressed in the gut epithelium, it is likely that the decreased digestion and absorption of dietary and biliary phospholipids are eventually linked to reduced fat accumulation in adipose tissue of *Pla2g10*$^{-/-}$ mice [79], a situation similar to *Pla2g1b*$^{-/-}$ mice (see foregoing).

$sPLA_2$-X is most abundantly expressed in the testis, where it is stored in acrosomes (secretory granules) in the head of sperm cells [82]. *Pla2g10*$^{-/-}$ spermatozoa display an impaired acrosome reaction and low fertility despite showing a normal number and motility [83, 82]. Thus, $sPLA_2$-X plays a specific role in sperm activation, boosting the acrosome reaction by producing LPC from sperm membranes in a paracrine or autocrine manner. Last, a striking skin phenotype characterized by alopecia in *Pla2g10*-transgenic mice points to a unique role of $sPLA_2$-X in hair homeostasis [84]. Although grossly the coat hairs of *Pla2g10*$^{-/-}$ mice appear normal, they have ultrastructural abnormalities including a hypoplasic outer root sheath and reduced melanin granules in their hair follicles.

2.4.8 $sPLA_2$-III

$sPLA_2$-III, an atypical $sPLA_2$, more closely resembles bee venom $sPLA_2$ rather than other mammalian $sPLA_2$s [85]. Transgenic overexpression of $sPLA_2$-III in mice with an *ApoE*$^{-/-}$ background results in increased atherosclerosis from accelerated LDL hydrolysis and increased TXA_2 synthesis [86]. These mice also develop systemic inflammation as they age because of elevated eicosanoid formation [87]. Thus, beyond the overexpression strategy, $sPLA_2$-III has a pro-inflammatory potential.

$sPLA_2$-III is highly expressed in the epididymal epithelium. Studies using mice lacking $sPLA_2$-III (*Pla2g3*$^{-/-}$) have revealed that epididymal $sPLA_2$-III acts on immature sperm cells passing through the duct in a paracrine manner to regulate phospholipid remodeling. During epididymal transit of spermatozoa, PC in the sperm membrane undergoes a dramatic shift in its acyl groups from oleate, linoleate, and AA to docosapentaenoic acid (DPA) and DHA, and the increased proportion of DPA/DHA consequently contributes to increased sperm membrane fluidity and thereby sperm motility. In *Pla2g3*$^{-/-}$ mice, this sperm membrane remodeling is severely compromised. Accordingly, spermatozoa from *Pla2g3*$^{-/-}$ mice have a low DPA/DHA content, aberrant acrosomes and flagella with an abnormal axoneme configuration, and display hypomotility and reduced fertility [88]. Thus, the two "reproductive $sPLA_2$s" ($sPLA_2$-III and $sPLA_2$-X), which are expressed in different locations within the male genital organs, exert nonredundant but interrelated functions in two major steps of male fertility, the former during sperm maturation in the epididymis and the latter during capacitation and the acrosome reaction, likely after ejaculation in the uterus and oviduct.

Microenvironmental alterations in MC phenotypes affect susceptibility to allergy, yet the mechanisms underlying the proper maturation of MCs toward an allergy-sensitive phenotype were poorly understood. $sPLA_2$-III is stored in and released from MC granules, and MC-associated passive and active anaphylactic responses are markedly attenuated in *Pla2g3*$^{-/-}$ mice, whereas they are augmented in *Pla2g3*-transgenic mice [89]. Tissue MCs in *Pla2g3*$^{-/-}$ mice are immature and are therefore resistant to IgE-dependent and -independent activation. Similar MC abnormalities are also seen in mice lacking lipocalin-type prostaglandin D_2 (PGD_2) synthase (L-PGDS) or those lacking the PGD_2 receptor DP1, suggesting their functional relationship. Indeed, genetic or pharmacological ablation of DP1 in MCs or L-PGDS in fibroblasts phenocopies that of $sPLA_2$-III in MCs in terms of defective MC maturation and anaphylaxis. Taken together, the data suggest that $sPLA_2$-III secreted from immature MCs is coupled with fibroblastic L-PGDS to provide microenvironmental PGD_2, which in turn promotes MC maturation via DP1. The $sPLA_2$-III/L-PGDS/DP1 paracrine loop is a novel lipid-orchestrated mechanism, providing a missing microenvironmental cue that underlies the proper maturation of MCs.

2.5 Concluding Remarks

With the growing list of knockout and transgenic mouse strains for PLA_2s, much progress has been made in delineating the physiological functions of each PLA_2. It is now becoming obvious that $cPLA_2\alpha$ is a central regulator of AA metabolism, supported by the view that the molecular evolution of $cPLA_2\alpha$ coincided with that of eicosanoid receptors when vertebrates evolved, that the $iPLA_2$ family is a fundamental regulator of membrane homeostasis and energy metabolism, and that individual $sPLA_2$s exert unique and tissue-specific biological functions by acting on extracellular phospholipids, which include adjacent cell membranes, noncellular lipid components, and foreign phospholipids such as those in microbes and the diet. The diversity of target phospholipids and products may explain why each PLA_2 family contains many isoforms. Further advances in this research field and their integration for therapeutic applications are likely to benefit from improved, time- and space-resolved lipidomics technology that will allow monitoring of individual PLA_2s and their associated forms of lipid metabolism within specific tissue niches. Hopefully, the next decade will yield a comprehensive map of the PLA_2-driven lipid networks, which will allow the therapeutic application of inhibitors for some PLA_2s central to human diseases.

Acknowledgments This work was supported by grants-in aid for Scientific Research from the Ministry of Education, Culture, Sports, Science and Technology of Japan and CREST from the Japan Science and Technology Agency (JST).

References

1. Nagase T, Uozumi N, Ishii S, Kita Y, Yamamoto H, Ohga E, Ouchi Y, Shimizu T (2002) A pivotal role of cytosolic phospholipase A_2 in bleomycin-induced pulmonary fibrosis. Nat Med 8(5):480–484. doi:10.1038/nm0502-480
2. Nagase T, Uozumi N, Ishii S, Kume K, Izumi T, Ouchi Y, Shimizu T (2000) Acute lung injury by sepsis and acid aspiration: a key role for cytosolic phospholipase A_2. Nat Immunol 1(1): 42–46. doi:10.1038/76897
3. Uozumi N, Kume K, Nagase T, Nakatani N, Ishii S, Tashiro F, Komagata Y, Maki K, Ikuta K, Ouchi Y, Miyazaki J, Shimizu T (1997) Role of cytosolic phospholipase A_2 in allergic response and parturition. Nature (Lond) 390(6660):618–622. doi:10.1038/37622
4. Marusic S, Leach MW, Pelker JW, Azoitei ML, Uozumi N, Cui J, Shen MW, DeClercq CM, Miyashiro JS, Carito BA, Thakker P, Simmons DL, Leonard JP, Shimizu T, Clark JD (2005) Cytosolic phospholipase $A_2\alpha$-deficient mice are resistant to experimental autoimmune encephalomyelitis. J Exp Med 202(6):841–851. doi:10.1084/jem.20050665
5. Hegen M, Sun L, Uozumi N, Kume K, Goad ME, Nickerson-Nutter CL, Shimizu T, Clark JD (2003) Cytosolic phospholipase $A_2\alpha$-deficient mice are resistant to collagen-induced arthritis. J Exp Med 197(10):1297–1302. doi:10.1084/jem.20030016
6. Kriem B, Sponne I, Fifre A, Malaplate-Armand C, Lozaćh-Pillot K, Koziel V, Yen-Potin FT, Bihain B, Oster T, Olivier JL, Pillot T (2005) Cytosolic phospholipase A_2 mediates neuronal apoptosis induced by soluble oligomers of the amyloid β peptide. FASEB J 19(1):85–87. doi:10.1096/fj.04-1807fje
7. Bonventre JV, Huang Z, Taheri MR, O'Leary E, Li E, Moskowitz MA, Sapirstein A (1997) Reduced fertility and postischaemic brain injury in mice deficient in cytosolic phospholipase A_2. Nature (Lond) 390(6660):622–625. doi:10.1038/37635
8. Weiser-Evans MC, Wang XQ, Amin J, Van Putten V, Choudhary R, Winn RA, Scheinman R, Simpson P, Geraci MW, Nemenoff RA (2009) Depletion of cytosolic phospholipase A_2 in bone marrow-derived macrophages protects against lung cancer progression and metastasis. Cancer Res 69(5):1733–1738. doi:10.1158/0008-5472.CAN-08-3766
9. Takaku K, Sonoshita M, Sasaki N, Uozumi N, Doi Y, Shimizu T, Taketo MM (2000) Suppression of intestinal polyposis in $Apc^{\Delta 716}$ knockout mice by an additional mutation in the cytosolic phospholipase A_2 gene. J Biol Chem 275(44):34013–34016. doi:10.1074/jbc.C000585200
10. Song H, Lim H, Paria BC, Matsumoto H, Swift LL, Morrow J, Bonventre JV, Dey SK (2002) Cytosolic phospholipase $A_2\alpha$ deficiency is crucial for 'on-time' embryo implantation that directs subsequent development. Development (Camb) 129(12):2879–2889
11. Wong DA, Kita Y, Uozumi N, Shimizu T (2002) Discrete role for cytosolic phospholipase $A_2\alpha$ in platelets: studies using single and double mutant mice of cytosolic and group IIA secretory phospholipase A_2. J Exp Med 196(3):349–357. doi:10.1084/jem.20011443
12. Hadad N, Burgazliev O, Elgazar-Carmon V, Solomonov Y, Wueest S, Item F, Konrad D, Rudich A, Levy R (2013) Induction of cytosolic phospholipase $a_2\alpha$ is required for adipose neutrophil infiltration and hepatic insulin resistance early in the course of high-fat feeding. Diabetes 62(9):3053–3063. doi:10.2337/db12-1300
13. Ishihara K, Miyazaki A, Nabe T, Fushimi H, Iriyama N, Kanai S, Sato T, Uozumi N, Shimizu T, Akiba S (2012) Group IVA phospholipase A_2 participates in the progression of hepatic fibrosis. FASEB J 26(10):4111–4121. doi:10.1096/fj.12-205625
14. Ii H, Yokoyama N, Yoshida S, Tsutsumi K, Hatakeyama S, Sato T, Ishihara K, Akiba S (2009) Alleviation of high-fat diet-induced fatty liver damage in group IVA phospholipase A2-knockout mice. PLoS One 4(12):e8089. doi:10.1371/journal.pone.0008089
15. Adler DH, Cogan JD, Phillips JA 3rd, Schnetz-Boutaud N, Milne GL, Iverson T, Stein JA, Brenner DA, Morrow JD, Boutaud O, Oates JA (2008) Inherited human $cPLA_2\alpha$ deficiency is associated with impaired eicosanoid biosynthesis, small intestinal ulceration, and platelet dysfunction. J Clin Invest 118(6):2121–2131. doi:10.1172/JCI30473

16. Ohto T, Uozumi N, Hirabayashi T, Shimizu T (2005) Identification of novel cytosolic phospholipase A_2s, murine cPLA$_2\delta$, ε, and ζ, which form a gene cluster with cPLA$_2\beta$. J Biol Chem 280(26):24576–24583. doi:10.1074/jbc.M413711200
17. Underwood KW, Song C, Kriz RW, Chang XJ, Knopf JL, Lin LL (1998) A novel calcium-independent phospholipase A_2, cPLA$_2\gamma$, that is prenylated and contains homology to cPLA$_2$. J Biol Chem 273(34):21926–21932. doi:10.1074/jbc.273.34.21926
18. Capestrano M, Mariggio S, Perinetti G, Egorova AV, Iacobacci S, Santoro M, Di Pentima A, Iurisci C, Egorov MV, Di Tullio G, Buccione R, Luini A, Polishchuk RS (2014) Cytosolic phospholipase $A_2\varepsilon$ drives recycling through the clathrin-independent endocytic route. J Cell Sci 127(pt 5):977–993. doi:10.1242/jcs.136598
19. Ghosh M, Loper R, Ghomashchi F, Tucker DE, Bonventre JV, Gelb MH, Leslie CC (2007) Function, activity, and membrane targeting of cytosolic phospholipase $A_2\zeta$ in mouse lung fibroblasts. J Biol Chem 282(16):11676–11686. doi:10.1074/jbc.M608458200
20. Kienesberger PC, Oberer M, Lass A, Zechner R (2009) Mammalian patatin domain containing proteins: a family with diverse lipolytic activities involved in multiple biological functions. J Lipid Res 50(suppl):S63–S68. doi:10.1194/jlr.R800082-JLR200
21. Basselin M, Rosa AO, Ramadan E, Cheon Y, Chang L, Chen M, Greenstein D, Wohltmann M, Turk J, Rapoport SI (2010) Imaging decreased brain docosahexaenoic acid metabolism and signaling in iPLA$_2\beta$ (VIA)-deficient mice. J Lipid Res 51(11):3166–3173. doi:10.1194/jlr.M008334
22. Morgan NV, Westaway SK, Morton JE, Gregory A, Gissen P, Sonek S, Cangul H, Coryell J, Canham N, Nardocci N, Zorzi G, Pasha S, Rodriguez D, Desguerre I, Mubaidin A, Bertini E, Trembath RC, Simonati A, Schanen C, Johnson CA, Levinson B, Woods CG, Wilmot B, Kramer P, Gitschier J, Maher ER, Hayflick SJ (2006) PLA2G6, encoding a phospholipase A_2, is mutated in neurodegenerative disorders with high brain iron. Nat Genet 38(7):752–754. doi:10.1038/ng1826
23. Shinzawa K, Sumi H, Ikawa M, Matsuoka Y, Okabe M, Sakoda S, Tsujimoto Y (2008) Neuroaxonal dystrophy caused by group VIA phospholipase A_2 deficiency in mice: a model of human neurodegenerative disease. J Neurosci 28(9):2212–2220. doi:10.1523/JNEUROSCI.4354-07.2008
24. Malik I, Turk J, Mancuso DJ, Montier L, Wohltmann M, Wozniak DF, Schmidt RE, Gross RW, Kotzbauer PT (2008) Disrupted membrane homeostasis and accumulation of ubiquitinated proteins in a mouse model of infantile neuroaxonal dystrophy caused by PLA2G6 mutations. Am J Pathol 172(2):406–416. doi:10.2353/ajpath.2008.070823
25. Bao S, Miller DJ, Ma Z, Wohltmann M, Eng G, Ramanadham S, Moley K, Turk J (2004) Male mice that do not express group VIA phospholipase A_2 produce spermatozoa with impaired motility and have greatly reduced fertility. J Biol Chem 279(37):38194–38200. doi:10.1074/jbc.M406489200
26. Bolotina VM (2008) Orai, STIM1 and iPLA$_2\beta$: a view from a different perspective. J Physiol 586(13):3035–3042. doi:10.1113/jphysiol.2008.154997
27. Bao S, Jacobson DA, Wohltmann M, Bohrer A, Jin W, Philipson LH, Turk J (2008) Glucose homeostasis, insulin secretion, and islet phospholipids in mice that overexpress iPLA$_2\beta$ in pancreatic beta-cells and in iPLA$_2\beta$-null mice. Am J Physiol Endocrinol Metab 294(2):E217–E229. doi:10.1152/ajpendo.00474.2007
28. Bao S, Li Y, Lei X, Wohltmann M, Jin W, Bohrer A, Semenkovich CF, Ramanadham S, Tabas I, Turk J (2007) Attenuated free cholesterol loading-induced apoptosis but preserved phospholipid composition of peritoneal macrophages from mice that do not express group VIA phospholipase A_2. J Biol Chem 282(37):27100–27114. doi:10.1074/jbc.M701316200
29. Moon SH, Jenkins CM, Mancuso DJ, Turk J, Gross RW (2008) Smooth muscle cell arachidonic acid release, migration, and proliferation are markedly attenuated in mice null for calcium-independent phospholipase $A_2\beta$. J Biol Chem 283(49):33975–33987. doi:10.1074/jbc.M805817200

30. Li H, Zhao Z, Wei G, Yan L, Wang D, Zhang H, Sandusky GE, Turk J, Xu Y (2010) Group VIA phospholipase A_2 in both host and tumor cells is involved in ovarian cancer development. FASEB J 24(10):4103–4116. doi:10.1096/fj.10-161356
31. Yan W, Jenkins CM, Han X, Mancuso DJ, Sims HF, Yang K, Gross RW (2005) The highly selective production of 2-arachidonoyl lysophosphatidylcholine catalyzed by purified calcium-independent phospholipase $A_2\gamma$: identification of a novel enzymatic mediator for the generation of a key branch point intermediate in eicosanoid signaling. J Biol Chem 280(29):26669–26679. doi:10.1074/jbc.M502358200
32. Mancuso DJ, Sims HF, Han X, Jenkins CM, Guan SP, Yang K, Moon SH, Pietka T, Abumrad NA, Schlesinger PH, Gross RW (2007) Genetic ablation of calcium-independent phospholipase $A_2\gamma$ leads to alterations in mitochondrial lipid metabolism and function resulting in a deficient mitochondrial bioenergetic phenotype. J Biol Chem 282(48):34611–34622. doi:10.1074/jbc.M707795200
33. Song H, Wohltmann M, Bao S, Ladenson JH, Semenkovich CF, Turk J (2010) Mice deficient in group VIB phospholipase A_2 (iPLA$_2\gamma$) exhibit relative resistance to obesity and metabolic abnormalities induced by a Western diet. Am J Physiol Endocrinol Metab 298(6):E1097–E1114. doi:10.1152/ajpendo.00780.2009
34. Mancuso DJ, Sims HF, Yang K, Kiebish MA, Su X, Jenkins CM, Guan S, Moon SH, Pietka T, Nassir F, Schappe T, Moore K, Han X, Abumrad NA, Gross RW (2010) Genetic ablation of calcium-independent phospholipase $A_2\gamma$ prevents obesity and insulin resistance during high fat feeding by mitochondrial uncoupling and increased adipocyte fatty acid oxidation. J Biol Chem 285(47):36495–36510. doi:10.1074/jbc.M110.115766
35. Mancuso DJ, Han X, Jenkins CM, Lehman JJ, Sambandam N, Sims HF, Yang J, Yan W, Yang K, Green K, Abendschein DR, Saffitz JE, Gross RW (2007) Dramatic accumulation of triglycerides and precipitation of cardiac hemodynamic dysfunction during brief caloric restriction in transgenic myocardium expressing human calcium-independent phospholipase $A_2\gamma$. J Biol Chem 282(12):9216–9227. doi:10.1074/jbc.M607307200
36. Mancuso DJ, Kotzbauer P, Wozniak DF, Sims HF, Jenkins CM, Guan S, Han X, Yang K, Sun G, Malik I, Conyers S, Green KG, Schmidt RE, Gross RW (2009) Genetic ablation of calcium-independent phospholipase $A_2\gamma$ leads to alterations in hippocampal cardiolipin content and molecular species distribution, mitochondrial degeneration, autophagy, and cognitive dysfunction. J Biol Chem 284(51):35632–35644. doi:10.1074/jbc.M109.055194
37. Bione S, D'Adamo P, Maestrini E, Gedeon AK, Bolhuis PA, Toniolo D (1996) A novel X-linked gene, G4.5. is responsible for Barth syndrome. Nat Genet 12(4):385–389. doi:10.1038/ng0496-385
38. Huggins KW, Boileau AC, Hui DY (2002) Protection against diet-induced obesity and obesity-related insulin resistance in group 1B PLA$_2$-deficient mice. Am J Physiol Endocrinol Metab 283(5):E994–E1001. doi:10.1152/ajpendo.00110.2002
39. Hollie NI, Konaniah ES, Goodin C, Hui DY (2014) Group 1B phospholipase A_2 inactivation suppresses atherosclerosis and metabolic diseases in LDL receptor-deficient mice. Atherosclerosis 234(2):377–380. doi:10.1016/j.atherosclerosis.2014.03.027
40. Hollie NI, Hui DY (2011) Group 1B phospholipase A_2 deficiency protects against diet-induced hyperlipidemia in mice. J Lipid Res 52(11):2005–2011. doi:10.1194/jlr.M019463
41. Labonte ED, Kirby RJ, Schildmeyer NM, Cannon AM, Huggins KW, Hui DY (2006) Group 1B phospholipase A_2-mediated lysophospholipid absorption directly contributes to postprandial hyperglycemia. Diabetes 55(4):935–941. doi:10.2337/diabetes.55.04.06.db05-1286
42. Scott DL, White SP, Browning JL, Rosa JJ, Gelb MH, Sigler PB (1991) Structures of free and inhibited human secretory phospholipase A_2 from inflammatory exudate. Science 254(5034):1007–1010
43. MacPhee M, Chepenik KP, Liddell RA, Nelson KK, Siracusa LD, Buchberg AM (1995) The secretory phospholipase A_2 gene is a candidate for the *Mom1* locus, a major modifier of *Apc*^{Min}-induced intestinal neoplasia. Cell 81(6):957–966. doi:10.1016/0092-8674(95)90015-2

44. Grass DS, Felkner RH, Chiang MY, Wallace RE, Nevalainen TJ, Bennett CF, Swanson ME (1996) Expression of human group II PLA_2 in transgenic mice results in epidermal hyperplasia in the absence of inflammatory infiltrate. J Clin Invest 97(10):2233–2241. doi:10.1172/JCI118664
45. Mulherkar R, Kirtane BM, Ramchandani A, Mansukhani NP, Kannan S, Naresh KN (2003) Expression of enhancing factor/phospholipase A_2 in skin results in abnormal epidermis and increased sensitivity to chemical carcinogenesis. Oncogene 22(13):1936–1944. doi:10.1038/sj.onc.1206229
46. Kugiyama K, Ota Y, Takazoe K, Moriyama Y, Kawano H, Miyao Y, Sakamoto T, Soejima H, Ogawa H, Doi H, Sugiyama S, Yasue H (1999) Circulating levels of secretory type II phospholipase A_2 predict coronary events in patients with coronary artery disease. Circulation 100(12):1280–1284. doi:10.1161/01.CIR.100.12.1280
47. Ivandic B, Castellani LW, Wang XP, Qiao JH, Mehrabian M, Navab M, Fogelman AM, Grass DS, Swanson ME, de Beer MC, de Beer F, Lusis AJ (1999) Role of group II secretory phospholipase A_2 in atherosclerosis: 1. Increased atherogenesis and altered lipoproteins in transgenic mice expressing group IIa phospholipase A_2. Arterioscler Thromb Vasc Biol 19(5):1284–1290. doi:10.1161/01.ATV.19.5.1284
48. Weinrauch Y, Elsbach P, Madsen LM, Foreman A, Weiss J (1996) The potent anti-Staphylococcus *aureus* activity of a sterile rabbit inflammatory fluid is due to a 14-kD phospholipase A_2. J Clin Invest 97(1):250–257. doi:10.1172/JCI118399
49. Laine VJ, Grass DS, Nevalainen TJ (1999) Protection by group II phospholipase A_2 against *Staphylococcus aureus*. J Immunol 162(12):7402–7408
50. Cormier RT, Hong KH, Halberg RB, Hawkins TL, Richardson P, Mulherkar R, Dove WF, Lander ES (1997) Secretory phospholipase *Pla2g2a* confers resistance to intestinal tumorigenesis. Nat Genet 17(1):88–91. doi:10.1038/ng0997-88
51. Kennedy BP, Soravia C, Moffat J, Xia L, Hiruki T, Collins S, Gallinger S, Bapat B (1998) Overexpression of the nonpancreatic secretory group II PLA_2 messenger RNA and protein in colorectal adenomas from familial adenomatous polyposis patients. Cancer Res 58(3):500–503
52. Mirtti T, Laine VJ, Hiekkanen H, Hurme S, Rowe O, Nevalainen TJ, Kallajoki M, Alanen K (2009) Group IIA phospholipase A as a prognostic marker in prostate cancer: relevance to clinicopathological variables and disease-specific mortality. APMIS 117(3):151–161. doi:10.1111/j.1600-0463.2008.00002.x
53. Boilard E, Lai Y, Larabee K, Balestrieri B, Ghomashchi F, Fujioka D, Gobezie R, Coblyn JS, Weinblatt ME, Massarotti EM, Thornhill TS, Divangahi M, Remold H, Lambeau G, Gelb MH, Arm JP, Lee DM (2010) A novel anti-inflammatory role for secretory phospholipase A_2 in immune complex-mediated arthritis. EMBO Mol Med 2(5):172–187. doi:10.1002/emmm.201000072
54. Boudreau LH, Duchez AC, Cloutier N, Soulet D, Martin N, Bollinger J, Pare A, Rousseau M, Naika GS, Levesque T, Laflamme C, Marcoux G, Lambeau G, Farndale RW, Pouliot M, Hamzeh-Cognasse H, Cognasse F, Garraud O, Nigrovic PA, Guderley H, Lacroix S, Thibault L, Semple JW, Gelb MH, Boilard E (2014) Platelets release mitochondria serving as substrate for bactericidal group IIA-secreted phospholipase A_2 to promote inflammation. Blood 124(14):2173–2183. doi:10.1182/blood-2014-05-573543
55. Miki Y, Yamamoto K, Taketomi Y, Sato H, Shimo K, Kobayashi T, Ishikawa Y, Ishii T, Nakanishi H, Ikeda K, Taguchi R, Kabashima K, Arita M, Arai H, Lambeau G, Bollinger JM, Hara S, Gelb MH, Murakami M (2013) Lymphoid tissue phospholipase A_2 group IID resolves contact hypersensitivity by driving antiinflammatory lipid mediators. J Exp Med 210(6):1217–1234. doi:10.1084/jem.20121887
56. von Allmen CE, Schmitz N, Bauer M, Hinton HJ, Kurrer MO, Buser RB, Gwerder M, Muntwiler S, Sparwasser T, Beerli RR, Bachmann MF (2009) Secretory phospholipase A_2-IID is an effector molecule of $CD4^+CD25^+$ regulatory T cells. Proc Natl Acad Sci U S A 106(28):11673–11678. doi:10.1073/pnas.0812569106

57. Sato H, Taketomi Y, Ushida A, Isogai Y, Kojima T, Hirabayashi T, Miki Y, Yamamoto K, Nishito Y, Kobayashi T, Ikeda K, Taguchi R, Hara S, Ida S, Miyamoto Y, Watanabe M, Baba H, Miyata K, Oike Y, Gelb MH, Murakami M (2014) The adipocyte-inducible secreted phospholipases PLA2G5 and PLA2G2E play distinct roles in obesity. Cell Metab 20(1):119–132. doi:10.1016/j.cmet.2014.05.002
58. Lapointe S, Brkovic A, Cloutier I, Tanguay JF, Arm JP, Sirois MG (2010) Group V secreted phospholipase A_2 contributes to LPS-induced leukocyte recruitment. J Cell Physiol 224(1):127–134. doi:10.1002/jcp.22106
59. Satake Y, Diaz BL, Balestrieri B, Lam BK, Kanaoka Y, Grusby MJ, Arm JP (2004) Role of group V phospholipase A_2 in zymosan-induced eicosanoid generation and vascular permeability revealed by targeted gene disruption. J Biol Chem 279(16):16488–16494. doi:10.1074/jbc.M313748200
60. Yano T, Fujioka D, Saito Y, Kobayashi T, Nakamura T, Obata JE, Kawabata K, Watanabe K, Watanabe Y, Mishina H, Tamaru S, Kugiyama K (2011) Group V secretory phospholipase A_2 plays a pathogenic role in myocardial ischaemia-reperfusion injury. Cardiovasc Res 90(2):335–343. doi:10.1093/cvr/cvq399
61. Giannattasio G, Fujioka D, Xing W, Katz HR, Boyce JA, Balestrieri B (2010) Group V secretory phospholipase A_2 reveals its role in house dust mite-induced allergic pulmonary inflammation by regulation of dendritic cell function. J Immunol 185(7):4430–4438. doi:10.4049/jimmunol.1001384
62. Munoz NM, Meliton AY, Meliton LN, Dudek SM, Leff AR (2009) Secretory group V phospholipase A_2 regulates acute lung injury and neutrophilic inflammation caused by LPS in mice. Am J Physiol Lung Cell Mol Physiol 296(6):L879–L887. doi:10.1152/ajplung.90580.2008
63. Gesquiere L, Cho W, Subbaiah PV (2002) Role of group IIa and group V secretory phospholipases A_2 in the metabolism of lipoproteins. Substrate specificities of the enzymes and the regulation of their activities by sphingomyelin. Biochemistry 41(15):4911–4920
64. Bostrom MA, Boyanovsky BB, Jordan CT, Wadsworth MP, Taatjes DJ, de Beer FC, Webb NR (2007) Group v secretory phospholipase A_2 promotes atherosclerosis: evidence from genetically altered mice. Arterioscler Thromb Vasc Biol 27(3):600–606. doi:10.1161/01.ATV.0000257133.60884.44
65. Ohtsuki M, Taketomi Y, Arata S, Masuda S, Ishikawa Y, Ishii T, Takanezawa Y, Aoki J, Arai H, Yamamoto K, Kudo I, Murakami M (2006) Transgenic expression of group V, but not group X, secreted phospholipase A_2 in mice leads to neonatal lethality because of lung dysfunction. J Biol Chem 281(47):36420–36433. doi:10.1074/jbc.M607975200
66. Ohta S, Imamura M, Xing W, Boyce JA, Balestrieri B (2013) Group V secretory phospholipase A_2 is involved in macrophage activation and is sufficient for macrophage effector functions in allergic pulmonary inflammation. J Immunol 190(12):5927–5938. doi:10.4049/jimmunol.1203202
67. Balestrieri B, Maekawa A, Xing W, Gelb MH, Katz HR, Arm JP (2009) Group V secretory phospholipase A_2 modulates phagosome maturation and regulates the innate immune response against *Candida albicans*. J Immunol 182(8):4891–4898. doi:10.4049/jimmunol.0803776
68. Cupillard L, Koumanov K, Mattei MG, Lazdunski M, Lambeau G (1997) Cloning, chromosomal mapping, and expression of a novel human secretory phospholipase A_2. J Biol Chem 272(25):15745–15752. doi:10.1074/jbc.272.25.15745
69. Murakami M, Koduri RS, Enomoto A, Shimbara S, Seki M, Yoshihara K, Singer A, Valentin E, Ghomashchi F, Lambeau G, Gelb MH, Kudo I (2001) Distinct arachidonate-releasing functions of mammalian secreted phospholipase A_2s in human embryonic kidney 293 and rat mastocytoma RBL-2H3 cells through heparan sulfate shuttling and external plasma membrane mechanisms. J Biol Chem 276(13):10083–10096. doi:10.1074/jbc.M007877200
70. Henderson WR Jr, Chi EY, Bollinger JG, Tien YT, Ye X, Castelli L, Rubtsov YP, Singer AG, Chiang GK, Nevalainen T, Rudensky AY, Gelb MH (2007) Importance of group X-secreted phospholipase A_2 in allergen-induced airway inflammation and remodeling in a mouse asthma model. J Exp Med 204(4):865–877. doi:10.1084/jem.20070029

71. Henderson WR Jr, Oslund RC, Bollinger JG, Ye X, Tien YT, Xue J, Gelb MH (2011) Blockade of human group X secreted phospholipase A_2 (GX-sPLA_2)-induced airway inflammation and hyperresponsiveness in a mouse asthma model by a selective GX-sPLA_2 inhibitor. J Biol Chem 286(32):28049–28055. doi:10.1074/jbc.M111.235812
72. Lai Y, Oslund RC, Bollinger JG, Henderson WR Jr, Santana LF, Altemeier WA, Gelb MH, Hallstrand TS (2010) Eosinophil cysteinyl leukotriene synthesis mediated by exogenous secreted phospholipase A_2 group X. J Biol Chem 285(53):41491–41500. doi:10.1074/jbc.M110.153338
73. Kelvin AA, Degousee N, Banner D, Stefanski E, Leomicronn AJ, Angoulvant D, Paquette SG, Huang SS, Danesh A, Robbins CS, Noyan H, Husain M, Lambeau G, Gelb M, Kelvin DJ, Rubin BB (2014) Lack of group X secreted phospholipase A_2 increases survival following pandemic H1N1 influenza infection. Virology 454-455:78–92. doi:10.1016/j.virol.2014.01.030
74. Hallstrand TS, Lai Y, Ni Z, Oslund RC, Henderson WR Jr, Gelb MH, Wenzel SE (2011) Relationship between levels of secreted phospholipase A_2 groups IIA and X in the airways and asthma severity. Clin Exp Allergy 41(6):801–810. doi:10.1111/j.1365-2222.2010.03676.x
75. Watanabe K, Fujioka D, Saito Y, Nakamura T, Obata JE, Kawabata K, Watanabe Y, Mishina H, Tamaru S, Hanasaki K, Kugiyama K (2012) Group X secretory PLA_2 in neutrophils plays a pathogenic role in abdominal aortic aneurysms in mice. Am J Physiol Heart Circ Physiol 302(1):H95–H104. doi:10.1152/ajpheart.00695.2011
76. Fujioka D, Saito Y, Kobayashi T, Yano T, Tezuka H, Ishimoto Y, Suzuki N, Yokota Y, Nakamura T, Obata JE, Kanazawa M, Kawabata K, Hanasaki K, Kugiyama K (2008) Reduction in myocardial ischemia/reperfusion injury in group X secretory phospholipase A_2-deficient mice. Circulation 117(23):2977–2985. doi:10.1161/CIRCULATIONAHA.107.743997
77. Ait-Oufella H, Herbin O, Lahoute C, Coatrieux C, Loyer X, Joffre J, Laurans L, Ramkhelawon B, Blanc-Brude O, Karabina S, Girard CA, Payre C, Yamamoto K, Binder CJ, Murakami M, Tedgui A, Lambeau G, Mallat Z (2013) Group X secreted phospholipase A_2 limits the development of atherosclerosis in LDL receptor-null mice. Arterioscler Thromb Vasc Biol 33(3):466–473. doi:10.1161/ATVBAHA.112.300309
78. Zack M, Boyanovsky BB, Shridas P, Bailey W, Forrest K, Howatt DA, Gelb MH, de Beer FC, Daugherty A, Webb NR (2011) Group X secretory phospholipase A_2 augments angiotensin II-induced inflammatory responses and abdominal aortic aneurysm formation in *apoE-deficient* mice. Atherosclerosis 214(1):58–64. doi:10.1016/j.atherosclerosis.2010.08.054
79. Sato H, Isogai Y, Masuda S, Taketomi Y, Miki Y, Kamei D, Hara S, Kobayashi T, Ishikawa Y, Ishii T, Ikeda K, Taguchi R, Ishimoto Y, Suzuki N, Yokota Y, Hanasaki K, Suzuki-Yamamoto T, Yamamoto K, Murakami M (2011) Physiological roles of group X-secreted phospholipase A_2 in reproduction, gastrointestinal phospholipid digestion, and neuronal function. J Biol Chem 286(13):11632–11648. doi:10.1074/jbc.M110.206755
80. Li X, Shridas P, Forrest K, Bailey W, Webb NR (2010) Group X secretory phospholipase A_2 negatively regulates adipogenesis in murine models. FASEB J 24(11):4313–4324. doi:10.1096/fj.10-154716
81. Shridas P, Bailey WM, Talbott KR, Oslund RC, Gelb MH, Webb NR (2011) Group X secretory phospholipase A_2 enhances TLR4 signaling in macrophages. J Immunol 187(1):482–489. doi:10.4049/jimmunol.1003552
82. Escoffier J, Jemel I, Tanemoto A, Taketomi Y, Payre C, Coatrieux C, Sato H, Yamamoto K, Masuda S, Pernet-Gallay K, Pierre V, Hara S, Murakami M, De Waard M, Lambeau G, Arnoult C (2010) Group X phospholipase A_2 is released during sperm acrosome reaction and controls fertility outcome in mice. J Clin Invest 120(5):1415–1428. doi:10.1172/JCI40494
83. Escoffier J, Pierre VJ, Jemel I, Munch L, Boudhraa Z, Ray PF, De Waard M, Lambeau G, Arnoult C (2011) Group X secreted phospholipase A_2 specifically decreases sperm motility in mice. J Cell Physiol 226(10):2601–2609. doi:10.1002/jcp.22606
84. Yamamoto K, Taketomi Y, Isogai Y, Miki Y, Sato H, Masuda S, Nishito Y, Morioka K, Ishimoto Y, Suzuki N, Yokota Y, Hanasaki K, Ishikawa Y, Ishii T, Kobayashi T, Fukami K, Ikeda K, Nakanishi H, Taguchi R, Murakami M (2011) Hair follicular expression and function of group

X secreted phospholipase A_2 in mouse skin. J Biol Chem 286(13):11616–11631. doi:10.1074/jbc.M110.206714
85. Valentin E, Ghomashchi F, Gelb MH, Lazdunski M, Lambeau G (2000) Novel human secreted phospholipase A_2 with homology to the group III bee venom enzyme. J Biol Chem 275(11):7492–7496
86. Sato H, Kato R, Isogai Y, Saka G, Ohtsuki M, Taketomi Y, Yamamoto K, Tsutsumi K, Yamada J, Masuda S, Ishikawa Y, Ishii T, Kobayashi T, Ikeda K, Taguchi R, Hatakeyama S, Hara S, Kudo I, Itabe H, Murakami M (2008) Analyses of group III secreted phospholipase A_2 transgenic mice reveal potential participation of this enzyme in plasma lipoprotein modification, macrophage foam cell formation, and atherosclerosis. J Biol Chem 283(48):33483–33497. doi:10.1074/jbc.M804628200
87. Sato H, Taketomi Y, Isogai Y, Masuda S, Kobayashi T, Yamamoto K, Murakami M (2009) Group III secreted phospholipase A_2 transgenic mice spontaneously develop inflammation. Biochem J 421(1):17–27. doi:10.1042/BJ20082429
88. Sato H, Taketomi Y, Isogai Y, Miki Y, Yamamoto K, Masuda S, Hosono T, Arata S, Ishikawa Y, Ishii T, Kobayashi T, Nakanishi H, Ikeda K, Taguchi R, Hara S, Kudo I, Murakami M (2010) Group III secreted phospholipase A_2 regulates epididymal sperm maturation and fertility in mice. J Clin Invest 120(5):1400–1414. doi:10.1172/JCI40493
89. Taketomi Y, Ueno N, Kojima T, Sato H, Murase R, Yamamoto K, Tanaka S, Sakanaka M, Nakamura M, Nishito Y, Kawana M, Kambe N, Ikeda K, Taguchi R, Nakamizo S, Kabashima K, Gelb MH, Arita M, Yokomizo T, Nakamura M, Watanabe K, Hirai H, Nakamura M, Okayama Y, Ra C, Aritake K, Urade Y, Morimoto K, Sugimoto Y, Shimizu T, Narumiya S, Hara S, Murakami M (2013) Mast cell maturation is driven via a group III phospholipase A2-prostaglandin D_2-DP1 receptor paracrine axis. Nat Immunol 14(6):554–563. doi:10.1038/ni.2586

Chapter 3
Prostaglandin Terminal Synthases as Novel Drug Targets

Shuntaro Hara

Abstract Prostanoids are lipid metabolites of ω3 and ω6 20-carbon essential fatty acids such as arachidonic acid and have a broad range of biological activities. Three kinds of enzymes—phospholipase A_2 (PLA_2), cyclooxygenase (COX), and prostaglandin (PG) terminal synthase—are involved in the biosynthesis of prostanoids. Arachidonic acid released from membrane glycerophospholipids by PLA_2 enzymes is then supplied to either of the two COX isozymes, COX-1 and COX-2. The COX metabolite PGH_2 is then converted to each prostanoid by specific PG terminal synthases. Nonsteroidal antiinflammatory drugs (NSAIDs) exert their antiinflammatory and antitumor effects by inhibiting COX and thereby reducing prostanoid production. However, gastrointestinal, renal, and the recently reported cardiovascular side effects associated with the pharmacological inhibition of the COX enzymes have led to renewed attention to other potential targets for NSAIDs. As new methods appear for the selective modulation of prostanoid production, PG terminal synthases have gained attention as a novel target for NSAIDs. To date, multiple PG terminal synthases have been identified, and mice with specific deletions in each of these PG terminal synthases have been engineered. In this review, we summarize the current understanding of the *in vivo* roles of PG terminal synthases by knockout mouse studies.

Keywords Prostaglandin • Prostacyclin • Thromboxane • Prostaglandin terminal synthase • Prostaglandin E synthase • Prostaglandin D synthase • Prostacyclin synthase • Thromboxane synthase • Nonsteroidal antiinflammatory drugs • Knockout mice

S. Hara, Ph.D. (✉)
Division of Health Chemistry, Department of Healthcare and Regulatory Sciences, School of Pharmacy, Showa University, 1-5-8 Hatanodai, Shinagawa-ku, Tokyo 142-8555, Japan
e-mail: haras@pharm.showa-u.ac.jp

T. Yokomizo, M. Murakami (eds.), *Bioactive Lipid Mediators*,
DOI 10.1007/978-4-431-55669-5_3

3.1 Introduction

Prostanoids are cyclic and oxygenated metabolites of ω3 and ω6 20-carbon essential fatty acids and have a broad range of biological activities. Prostanoids include what are sometimes referred to as the "classical" prostaglandins (PGs), such as PGD, PGE, and PGF, as well as prostacyclin (PGI) and thromboxane (TX). The most abundant prostanoids are the "2-series" compounds such as PGE_2 (the "2" denotes the number of carbon–carbon double bonds in the prostanoid) that are formed from arachidonic acid.

Three kinds of enzymes—phospholipase A_2 (PLA_2), cyclooxygenase (COX), and PG terminal synthase—are involved in the biosynthesis of prostanoids (Fig. 3.1), and each of the three enzymatic steps involves multiple enzymes that can act in different phases of cell activation [1–3]. Arachidonic acid released from membrane glycerophospholipids by PLA_2 enzymes is then supplied to either of the two COX isozymes, COX-1 or COX-2. The constitutive COX-1 contributes mainly to immediate prostanoid generation, whereas the inducible COX-2 mediates delayed prostanoid generation. The COX metabolite PGH_2 is then converted to each prostanoid by specific PG terminal synthases.

Nonsteroidal antiinflammatory drugs (NSAIDs) exert their antiinflammatory and antitumor effects by inhibiting COX and thereby reducing prostanoid production. Long-term application of NSAIDs is associated with severe side effects, mainly gastrointestinal injury and renal irritations, apparently because of impaired COX-1-dependent prostanoid biosynthesis [1]. Although COX-2 selective inhibitors show reduced gastrointestinal complications, recent clinical trials have indicated that these inhibitors pose significantly increased cardiovascular risk. Specific inhibition of COX-2 alters the balance between platelet-derived TXA_2 and endothelium-derived PGI_2, leading to increases in the risk of thrombosis caused by altered vascular tone [4]. Thus, to develop novel NSAIDs without adverse side effects, more selective modulation of prostanoid production appears to be desirable. For this reason, PG terminal synthases are currently gaining attention as a novel target for NSAIDs.

To date, multiple PG terminal synthases have been identified (Table 3.1). Recently, mice with specific deletions in each of these synthases have been engineered. In this review, we summarize the current understanding of the *in vivo* roles of PG terminal synthases by knockout (KO) mouse studies. For details of the biochemical properties of PG terminal synthases, please refer to the review by Smith et al. [3].

3.2 PGE Synthase

PGE synthase (PGES), which converts PGH_2 to PGE_2, is known to constitute a group of at least three structurally and biologically distinct enzymes [5]. Two of them are membrane bound and have been designated as microsomal PGES-1

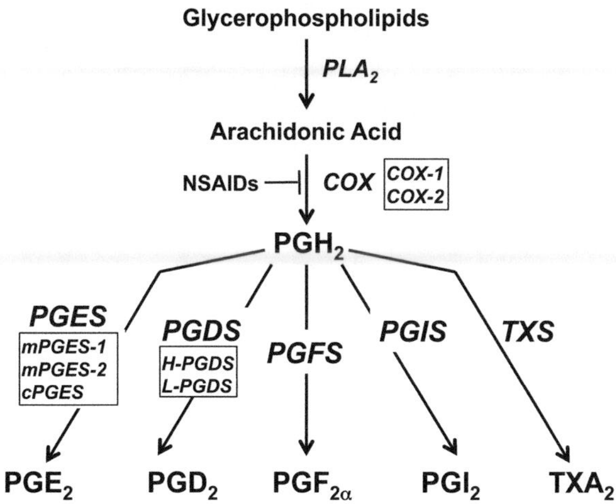

Fig. 3.1 Prostanoid biosynthetic pathway. Arachidonic acid released from membrane glycerophospholipids by phospholipase A_2 (PLA_2) enzymes is then supplied to either of the two cyclooxygenase (COX) isozymes, COX-1 or COX-2. The COX metabolite prostaglandin (PG) H_2 is then converted to each prostanoid [PGE_2, PGD_2, $PGF_{2\alpha}$, PGI_2, and thromboxane (TX) A_2] by specific PG terminal synthases

(mPGES-1) and mPGES-2. The third one is a cytosolic enzyme referred to as cytosolic PGES (cPGES).

3.2.1 mPGES-1

mPGES-1 was identified as the first PGES by Jakobsson et al. [6]. Murakami et al. also cloned rat and mouse orthologues of this protein and showed that this enzyme is identical to a membrane-associated PGES, which our group had originally detected in lipopolysaccharide (LPS)-stimulated macrophages [7]. mPGES-1 consists of 152 or 153 amino acids and belongs to the MAPEG (membrane-associated proteins involved in eicosanoid and glutathione metabolism) family. Among PGES enzymes, only mPGES-1 is markedly induced by pro-inflammatory stimuli, is downregulated by antiinflammatory glucocorticoids, and is functionally coupled with COX-2 in marked preference to COX-1. Because both COX-2 and mPGES-1 are present in the perinuclear membrane, colocalization of these two enzymes in the same subcellular compartment may allow efficient transfer of the unstable substrate PGH_2 between them. Steady-state expression of mPGES-1 in normal tissues is very low. Induction of mPGES-1 expression has been observed in various processes in which COX-2-derived PGE_2 has been shown to play a critical role, such as inflammation, fever, pain, tissue repair, and cancer [5, 8].

Table 3.1 Characteristics of multiple prostaglandin (PG) terminal synthases

		Tissue distribution	Subcellular localization	Molecular weight (kDa)	Structural characteristics	Phenotypes of knockout (KO) mice
PGES	mPGES-1	Inflamed tissues, cancerous tissues, kidney	Perinuclear membrane	16	MAPEG family	Impaired inflammatory reactions, suppression of carcinogenesis, etc.
	mPGES-2	Almost all tissues	Cytosol, Golgi	33	Glutaredoxin/ thioredoxin-like	–
	cPGES	Almost all tissues	Cytosol	23	–	Perinatal-lethal
PGDS	H-PGDS	Mast cells, microglia, Th2 cells, etc.	Cytosol	23	Glutathione-*S*-transferase	Suppression of neuroinflammation and demyelination, delayed resolution of inflammation, suppression of carcinogenesis, etc.
	L-PGDS	Brain, heart, male genital organs, etc.	Secreted	26	Lipocalin	Impaired sleep regulation and pain sensation, atherosclerosis and obesity, etc.
PGIS	PGIS	Vascular endothelial and smooth muscle cells, etc.	Endoplasmic reticulum	52	Cytochrome P450	High blood pressure, ischemic renal disorders
TXS	TXS	Platelets, bronchial epithelium, etc.	Endoplasmic reticulum	60	Cytochrome P450	Mild hemostatic defect, impaired wound healing

mPGES-1 KO mice were established in 2002 and have been used to demonstrate the involvement of mPGES-1 in various kinds of diseases [9, 10]. mPGES-1 KO mice are generally protected against a variety of inflammatory disease phenotypes, including collagen- or anti-collagen antibody-induced arthritis, LPS-induced bone loss, and antigen-induced edema [10–12]. In a collagen-induced arthritis model, reduced inflammation in the mPGES-1 KO mice was associated with a failure to produce antibody against type II collagen, suggesting that mPGES-1 has a role in the development of a humoral immune response [13]. We further found that mPGES-1 KO mice displayed significantly reduced accumulation of exudate and impaired leukocyte migration into the pleural cavity during carrageenan-induced paw edema formation [5]. The formation of inflammatory granulation tissue and attendant angiogenesis in the dorsum induced by subcutaneous implantation of a cotton thread were also significantly reduced in mPGES-1 KO mice [11]. Furthermore, mPGES-1 KO mice exhibited reductions in pain, fever, and other symptoms associated with inflammatory diseases [14, 15]. It is noteworthy that genetic deletion of mPGES-1 in mice does not adversely affect cardiovascular functions. These studies suggested the possibility that pharmacological targeting of mPGES-1 may ultimately prove to less toxic and perhaps more effective than the traditional NSAIDs for controlling acute inflammatory diseases. New drug candidates have recently been developed for targeting mPGES-1. Some of them have been shown to suppress inflammatory reactions in animal models. Xu et al. reported that when tested in the guinea pig and a knock-in mouse expressing human mPGES-1, MF63 inhibited LPS-induced pyresis, hyperalgesia, and iodoacetate-induced osteoarthritic pain, although it did not cause gastrointestinal toxic effects [16]. Leclerc et al. reported that their compound II attenuated both the acute and delayed inflammatory responses in rat adjuvant-induced arthritis [17].

The role of mPGES-1-derived PGE_2 in brain diseases, including ischemic injury and several neurodegenerative diseases, has also been established in models using mPGES-1 KO mice. In mPGES-1 KO mice, infarction, edema, and apoptotic cell death in the cortex after ischemia were all reduced compared with those in wild-type (WT) mice [18]. The behavioral neurological dysfunctions observed after ischemia in WT mice were also significantly ameliorated in mPGES-1 KO mice. Furthermore, mPGES-1 KO mice had less severe symptoms of experimental autoimmune encephalomyelitis [19]. We further found that mPGES-1 deletion reduced the accumulation of microglia around senile plaques and attenuated learning impairments in Tg2576 mice, a transgenic Alzheimer's disease mouse model [20].

It has also been shown that whereas the selective inhibition or KO of COX-2 accelerated thrombogenesis and elevated blood pressure in mice, the deletion of mPGES-1 had no such effect and restrained atherogenesis, the proliferative response to vascular injury, and angiotensin-induced aortic aneurysm formation in mice [21–23]. mPGES-1 inhibitors are thus expected to be applicable as therapeutic agents for inflammatory neurological or cardiovascular diseases.

In addition to COX-2, mPGES-1 levels are increased within a number of human cancers, and the tumorigenic potential of mPGES-1 has been suggested by several studies using cell culture systems. Transfection of mPGES-1 in combination with COX-2, but not with COX-1, into HEK293 cells led to cellular transformation with a concomitant increase in PGE_2 [24]. Furthermore, the COX-2/mPGES-1-cotransfected cells formed a number of large colonies in soft agar culture and were tumorigenic when implanted into nude mice. It was also shown that transgenic mice overexpressing both COX-2 and mPGES-1 developed metaplasia, hyperplasia, and tumorous growth in the glandular stomach with heavy macrophage infiltration [25]. We further showed that PGE_2 synthesis, cell proliferation, and invasive activity *in vitro* and xenograft formation *in vivo* were reduced by mPGES-1 knockdown and conversely enhanced by mPGES-1 overexpression in Lewis lung carcinoma (LLC) cells [26]. In addition to cancer cell-associated mPGES-1, host-associated mPGES-1 also contributes to tumor growth, invasion, and metastasis. LLC tumors grafted subcutaneously into mPGES-1 KO mice grew more slowly than did those grafted into WT mice, with concomitant decreases in the density of microvascular networks. Lung metastasis of intravenously injected LLC cells was also significantly less apparent in mPGES-1 KO mice than in WT mice. mPGES-1-driven PGE_2 signaling on host stromal cells may be functionally linked to the induction of potent pro-angiogenic and matrix-degrading factors, which in turn would facilitate tumor development.

An effect of mPGES-1 deficiency on intestinal tumorigenesis has also been reported. Nakanishi et al. showed that the genetic deletion of mPGES-1 ameliorated the development of intestinal tumors in both $Apc^{\Delta 14}$-dependent and azoxymethane-induced models [27, 28]. We also found that mPGES-1 deletion suppressed azoxymethane-induced colon carcinogenesis with reduced nuclear translocation of β-catenin and altered expression profiles of chemokines/cytokines [29]. It was noteworthy that genetic ablation of mPGES-1 resulted not only in the suppression of carcinogenic PGE_2 production, but also in the reciprocal upregulation of anticarcinogenic PGD_2 and PGI_2 production. In light of the fact that inhibition of COX-2 suppresses both pro- and antitumorigenic PGs, an mPGES-1-specific inhibitor, which blunts pro-tumorigenic PGE_2 while increasing antitumorigenic PGD_2 and PGI_2, is expected to be a more effective anticarcinogenic agent than a COX-2-specific inhibitor.

It is also important to remember that mPGES-1-derived PGE_2 is crucial for maintaining gastrointestinal mucosal homeostasis. We found, in a mouse model of dextran sodium sulfate-induced colitis, that mPGES-1 deficiency induced severe epithelial loss and crypt abscesses in the colon [5]. mPGES-1-specific inhibitors may worsen inflammatory bowel disease, including Crohn's disease and ulcerative colitis. It has also been reported that healing of acetic acid-induced ulcers was significantly delayed in mPGES-1 KO mice compared with WT mice, accompanied with reduced angiogenesis in ulcer granulation tissues [30]. These results indicated that mPGES-1 is involved in the wound-healing process in gastric ulcers. Therefore, any mPGES-1 inhibitors that may be developed would need to be used carefully in patients with gastrointestinal ulcers.

3.2.2 mPGES-2

mPGES-2 was initially purified from a microsomal fraction of bovine heart, and cDNAs encoding human and monkey homologues were subsequently identified [31]. mPGES-2 is a 41-kDa protein consisting of 378 to 385 amino acids that is structurally distinct from mPGES-1. mPGES-2 has an N-terminal hydrophobic domain, followed by a glutaredoxin/thioredoxin homology region, in which the consensus thioredoxin homology sequence of Cys^{110}-X-X-Cys^{113} is present. mPGES-2 is synthesized as a Golgi membrane-associated protein, and the proteolytic removal of the N-terminal hydrophobic domain leads to the formation of a mature cytosolic enzyme. When transfected in several cell lines, mPGES-2 is coupled with both COX-1 and COX-2, leading to PGE_2 production [32]. The transcript for mPGES-2 is more abundantly distributed in the brain, heart, skeletal muscle, kidney, and liver than in other tissues; this differs from the expression profile of mPGES-1. mPGES-2 expression is rather constitutive in various cells and tissues and is not elevated appreciably during inflammation or tissue damage. This finding suggests that the operation of these two enzymes is not always redundant, but rather that both enzymes exhibit tissue-specific functions.

mPGES-2 KO mice showed no specific phenotype and no alteration in PGE_2 levels in several tissues (including liver, kidney, heart, and brain) or in LPS-stimulated macrophages [33]. These results suggest that mPGES-2 is not involved in PGE_2 synthesis under the physiological and pathological conditions tested thus far. However, the possibility of tissue-specific or particular pathological roles of mPGES-2 has not yet been ruled out.

3.2.3 cPGES

Our group purified cPGES as a cytosolic form of PGES from LPS-treated rat brains, and sequence analysis of the 23-kDa purified protein revealed that it is identical to the heat shock protein 90 (Hsp90)-associated protein p23, which had been originally implicated as a cofactor for the molecular chaperone function of Hsp90 [34]. cPGES is directly associated with and phosphorylated by protein kinase CK2. In activated cells, CK2-directed phosphorylation of cPGES occurs in parallel with increased cPGES enzymatic activity and PGE_2 production, and these processes are facilitated by interaction with Hsp90 [35].

cPGES is expressed ubiquitously and in abundance in the cytosol of various tissues and cells. Cotransfection and antisense experiments have indicated that cPGES is capable of converting COX-1-, but not COX-2-, derived PGH_2 to PGE_2 in cells, particularly during the immediate PGE_2 biosynthetic response elicited by Ca^{2+}-evoked stimuli [34]. Localization of cPGES in the cytosol may allow coupling with proximal COX-1 in the ER in preference to distal COX-2 in the perinuclear envelope, although other regulatory mechanisms could also be involved.

To elucidate the *in vivo* function of cPGES, cPGES KO mice were developed, but cPGES KO mice were perinatal lethal with poor lung development, delayed skin maturation, and growth retardation [36–38]. The lung phenotype was consistent with a defective glucocorticoid response and with p23 acting as a co-chaperone for the glucocorticoid receptor/Hsp90 complex. In fibroblasts and tissues from cPGES KO mice, the expression of glucocorticoid-responsive genes, as well as the glucocorticoid transcriptional activation of reporter plasmids, was reduced. Defective nuclear translocation of the glucocorticoid receptor in cPGES-deficient fibroblasts was also observed. Although PGE_2 levels were reduced in lung and other tissues in cPGES KO mice, primary fibroblasts from these mice showed increased, rather than decreased, PGE_2 production. We found that cPGES-deficient fibroblasts decreased the expression of the PGE_2-degrading enzyme, 15-hydroxyprostaglandin dehydrogenase, which catalyzes the inactivating conversion of the PGE_2 15-OH to a 15-keto group, compared with that of WT fibroblasts [39]. These results suggested that the PGE_2-inactivating pathway may be controlled by the PGE_2 biosynthetic enzyme, cPGES.

3.3 PGD Synthase

PGD synthase (PGDS) catalyzes the isomerization of PGH_2 to PGD_2 and occurs in two distinct types [3]. One is hematopoietic PGDS (H-PGDS), which is found in mast cells, Th2 cells, and microglia, and the other is lipocalin-type PGDS (L-PGDS), which is localized in the brain, male genital organs, and cardiovascular tissues, including the human heart.

3.3.1 H-PGDS

H-PGDS was originally purified from rat spleen by Christ-Hazelhof and Nugteren as a 26-kDa, cytosolic, monomeric glutathione-requiring enzyme [40]. Sequences of full-length cDNAs for the human and mouse H-PGDS were obtained by Kanaoka et al. [41]. The cDNA encodes a protein composed of 199 amino acid residues, which is identified as a vertebrate homologue of class σ glutathione-*S*-transferase. The N-terminal methionine is cleaved from the mature protein. The X-ray crystal structure analysis of the human recombinant H-PGDS revealed that this enzyme is a 45- to 49-kDa homodimeric protein that binds one molecule of reduced glutathione per monomer and one Mg^{2+} ion per dimer [41].

H-PGDS-derived PGD_2 is involved in a variety of allergic and nonallergic disorders. H-PGDS is expressed in infiltrated leukocytes in the nasal mucosa of patients with polyposis or allergic rhinitis, and in necrotic muscle fibers of patients with Duchenne's muscular dystrophy and polymyositis. In addition to gene deletion of

H-PGDS, oral administration of an oral active H-PGDS inhibitor (HQL-79) suppressed the astrogliosis, neuroinflammation, and demyelination seen in the genetic demyelinating *twitcher* mice, an animal model of human Krabbe's disease, and the expansion of muscular necrosis in mdx mice, an animal model of Duchenne's muscular dystrophy [42, 43]. These results suggested that inhibition of PGDS would be an effective therapy for neuroinflammation and muscular dystrophy. HQL-79 administration also ameliorated airway inflammation in WT and human H-PGDS-transgenic mice.

Although PGD_2 promotes allergic inflammation via the PGD_2 receptor DP2, PGD_2 is initially converted to 15-deoxy-$\Delta^{12,14}$-PGJ_2, a putative ligand for peroxisome proliferator-activated receptor-γ (PPARγ). Studies using H-PGDS KO mice indicated that H-PGDS plays a central role in controlling the onset of acute inflammation and its resolution. H-PGDS synthesizes 15-deoxy-$\Delta^{12,14}$-PGJ_2 and together with PGD_2 working on the DP1 receptor, the 15-deoxy-$\Delta^{12,14}$-PGJ_2 controls the balance of pro- versus antiinflammatory cytokines as well as macrophage clearance through draining lymphatics. In H-PGDS KO mice, zymosan-induced peritonitis was more severe during the onset phase and resolution was impaired [44]. 15-Deoxy-$\Delta^{12,14}$-PGJ_2 not only can activate PPARγ but also can inhibit a range of pro-inflammatory signaling pathways, including NF-κB. Trivedi et al. investigated the severity and duration of a delayed-type hypersensitivity reaction in H-PGDS and transgenic mice and found that H-PGDS KO mice displayed a more severe inflammatory response that failed to resolve, characterized histologically as persistent acute inflammation, whereas transgenic mice had little detectable inflammation [45]. Thus, H-PGDS-derived PGD_2 is likely to have a complex role by acting as both a pro- and antiinflammatory mediator, depending on the target cell and stimulus.

H-PGDS-derived PGD_2 also functions as an antitumorigenic factor. Gene deletion of H-PGDS in *Apc*$^{Min/+}$ mice induced an increase in the number of intestinal adenomas, whereas *Apc*$^{Min/+}$ mice with transgenic human H-PGDS tended to have fewer intestinal adenomas [46]. Furthermore, host H-PGDS deficiency enhanced the progression of subcutaneously grafted LLC, accompanied by increased vascular leakage, angiogenesis, and monocyte/mast cell infiltration. Mast cell H-PGDS-derived PGD_2 suppresses vascular leakage and modulates TNF-α production, thereby shaping the tumor microenvironment [47].

3.3.2 L-PGDS

L-PGDS was isolated from the rat brain as a 26-kDa glutathione-independent PGDS [48]. The cDNA for L-PGDS was isolated from the rat brain by Urade et al. [49] and subsequently from humans and many other mammalian species, as well as from nonmammals such as chickens, frogs, and fish. The L-PGDS cDNA encodes a protein composed of 189 to 190 amino acid residues. L-PGDS is posttranslationally modified by cleavage of an N-terminal signal peptide of 24 and 22 amino acid residues from mouse and human enzymes, respectively. Two N-glycosylation sites are well conserved in all mammalian enzymes, and L-PGDS is highly glycosylated. A homology search revealed that L-PGDS is a member of the lipocalin gene family, which consists of small, secretory proteins that serve as transporters of various lipophilic ligands.

L-PGDS was originally identified in the brain and has been shown to be involved in sleep regulation. Human L-PGDS-overexpressing transgenic mice slept excessively after noxious stimulation such as tail clipping. L-PGDS KO mice failed to exhibit a rebound from excessive sleep after sleep deprivation [50]. In the CNS, L-PGDS is also considered to be involved in pain sensation. Intrathecal administration of PGE_2, an endogenous pain-producing substance, failed to elicit allodynia (touch-evoked pain), which is one typical phenomenon of neuropathic pain, although it did evoke thermal hyperalgesia in L-PGDS KO mice [51]. It was also found that the allodynic response induced by the γ-aminobutyric acid (GABA) A receptor antagonist bicuculline was selectively abolished in the L-PGDS KO mice.

L-PGDS is expressed in cardiovascular systems as well as nervous systems and has important vascular functions. A genetic polymorphism of L-PGDS has been identified in Japanese hypertensive patients with carotid atherosclerosis [52]. L-PGDS KO mice have been shown to exhibit nephropathy, atherosclerosis, and obesity [53].

The role of L-PGDS in various kinds of immune responses has also been established in models using L-PGDS transgenic and KO mice. When mice were infected with *Pseudomonas*, the clearance of *Pseudomonas* from the lung was improved in human L-PGDS-overexpressing transgenic mice and impaired in L-PGDS KO mice [54]. In a dextran sodium sulfate-induced colitis model, L-PGDS KO mice showed lower disease activity than WT mice [55]. Furthermore, we recently found that L-PGDS is coupled with group III secretory PLA_2 as an upstream enzyme to synthesize PGD_2 and drive mast cell maturation. In L-PGDS KO mice, the maturation of tissue mast cells was impaired and passive cutaneous anaphylaxis was suppressed [56].

3.4 PGF Synthase

PGE_2 and $PGF_{2\alpha}$ were the first prostanoids to be isolated from human semen. However, despite the long history of research on the physiological and pathological functions of $PGF_{2\alpha}$, the identity of PGF synthase (PGFS), which catalyzes PGH_2 to $PGF_{2\alpha}$*in vivo*, is unclear. Some enzymes belonging to the aldo-keto reductase (AKR) superfamily have been shown to exhibit PGFS activity [3, 57].

3.5 PGI Synthase

PGI synthase (PGIS) catalyzes the isomerization of PGH_2 to PGI_2. Ullrich and coworkers were the first to provide spectral evidence that PGIS is a cytochrome P450, but unlike most P450s, PGIS does not require NADPH and O_2 as cosubstrates [58]. We collaborated with Ullrich's research group to purify and characterize PGIS from bovine aorta, and then isolated cDNA for PGIS and designated this enzyme as CYP8A1 in the P450 family [59]. The PGIS cDNA encodes a 56-kDa protein consisting of 500 to 501 amino acid residues. In addition to mPGES-1, PGIS is functionally coupled with COX-2 in marked preference to COX-1, although this enzyme is constitutively expressed in vascular cells [60].

PGIS-derived PGI_2 is a strong vasodilator that inhibits the growth of vascular smooth muscle cells and is also the most potent endogenous inhibitor of platelet aggregation. Therefore, it has been considered to play an important role in cardiovascular diseases. We showed that overexpression of PGIS prevents neointimal formation after carotid balloon injury in rats [61]. Iwai et al. identified a repeat polymorphism in the promoter region of the human PGIS gene that is associated with promoter activity [62]. They further showed that this repeat polymorphism might be a risk factor for higher pulse pressure and consequently a risk factor for systolic hypertension in the Japanese population. The blood pressure of PGIS KO mice was significantly higher than that of WT mice [63]. Furthermore, PGIS KO mice developed ischemic renal disorders, including nephrosclerosis and renal infarction.

A role of PGIS in inflammatory disease and carcinogenesis has also been suggested. Pulmonary-specific PGIS-overexpressing mice were chemoprotected from developing lung tumors in a smoke-exposure model [64].

3.6 TX Synthase

TX synthase (TXS) catalyzes the isomerization of PGH_2 to TXA_2 with parallel production of malondialdehyde and 12-hydroxyheptadecatrienoic acid (12-HHT). In addition to PGIS, Ullrich and coworkers purified and characterized TXS from human platelets as a cytochrome P450 enzyme [65]. The cDNA for TXS was isolated from human platelets [66]. The TXS cDNA encodes a 60-kDa protein composed of 533 amino acid residues. Although both TXS and PGIS belong to the cytochrome P450 family, they share only 15 % sequence identity. TXS was designated as CYP5 in the P450 family.

TXA_2 is a potent stimulator of platelet activation and aggregation and vascular constriction. Gene deletion of TXS has been shown to cause a mild hemostatic defect and to protect mice against arachidonate-induced shock and death [67]. TXS inhibitors were originally considered to be promising antiplatelet agents, but clinical trials of various inhibitors yielded unsatisfactory results when compared with low-dose aspirin. Nonetheless, TXS inhibitors have been evaluated for other diseases involving TXA_2, such as bronchial asthma and pulmonary hypertension [68].

Very recently, 12-HHT, the other TXS-derived metabolite, was found to act on BLT2 receptors [69]. Specifically, Liu et al. reported that 12-HHT promoted epidermal wound healing by accelerating keratinocyte migration via the BLT2 receptors [70].

3.7 Conclusion

It has become apparent there are multiple PG terminal enzymes in mammalian cells and that distinct PG terminal enzymes may control the spatial and temporal production of prostanoids in different pathophysiological events in particular tissues and cells. Further investigation into the biochemical properties, transcriptional regulation, and *in vitro* and *in vivo* functions of each PG terminal enzyme may illuminate the potential utility of clinically targeting these enzymes.

References

1. Smith WL, DeWitt DL, Garavito RM (2000) Cyclooxygenases: structural, cellular, and molecular biology. Annu Rev Biochem 69:145–182
2. Kudo I, Murakami M (2002) Phospholipase A_2 enzymes. Prostaglandins Other Lipid Mediat 69:3–58
3. Smith WL, Urade Y, Jakobsson PJ (2011) Enzymes of the cyclooxygenase pathways of prostanoid biosynthesis. Chem Rev 111:5821–5865
4. Hara S, Kudo I (2006) COX-2 inhibitors and the risk of cardiovascular events. Jpn Med Assoc J 49:276–278
5. Hara S, Kamei D, Sasaki Y et al (2010) Prostaglandin E synthases: understanding their pathophysiological roles through mouse genetic models. Biochimie 92:651–659
6. Jakobsson PJ, Thorén S, Morgenstern R et al (1999) Identification of human prostaglandin E synthase: a microsomal, glutathione-dependent, inducible enzyme, constituting a potential novel drug target. Proc Natl Acad Sci U S A 96:7220–7225
7. Murakami M, Naraba H, Tanioka T et al (2000) Regulation of prostaglandin E_2 biosynthesis by inducible membrane-associated prostaglandin E_2 synthase that acts in concert with cyclooxygenase-2. J Biol Chem 275:32783–32792
8. Samuelsson B, Mogenstern R, Jakobsson PJ (2007) Membrane prostaglandin E synthase-1: a novel therapeutic target. Pharmacol Rev 59:207–224
9. Uematsu S, Matsumoto M, Takeda K et al (2002) Lipopolysaccharide-dependent prostaglandin E_2 production is regulated by the glutathione-dependent prostaglandin E_2 synthase gene induced by the Toll-like receptor 4/MyD88/NF-IL6 pathway. J Immunol 168:5811–5816
10. Trebino CE, Stock JL, Gibbons CP et al (2003) Impaired inflammatory and pain responses in mice lacking an inducible prostaglandin E synthase. Proc Natl Acad Sci U S A 100:9044–9049
11. Kamei D, Yamakawa K, Takegoshi Y et al (2004) Reduced pain hypersensitivity and inflammation in mice lacking microsomal prostaglandin E synthase-1. J Biol Chem 279:33684–33695

12. Inada M, Matsumoto C, Uematsu S et al (2006) Membrane-bound prostaglandin E synthase-1-mediated prostaglandin E_2 production by osteoblast plays a critical role in lipopolysaccharide-induced bone loss associated with inflammation. J Immunol 177:1879–1885
13. Kojima F, Kapoor M, Yang L et al (2008) Defective generation of a humoral immune response is associated with a reduced incidence and severity of collagen-induced arthritis in microsomal prostaglandin E synthase-1 null mice. J Immunol 180:8361–8368
14. Engblom D, Saha S, Engström L et al (2003) Microsomal prostaglandin E synthase-1 is the central switch during immune-induced pyresis. Nat Neurosci 6:1137–1138
15. Saha S, Engström L, Mackerlova L et al (2005) Impaired febrile responses to immune challenge in mice deficient in microsomal prostaglandin E synthase-1. Am J Physiol 288:R1100–R1107
16. Xu D, Rowland SE, Clark P et al (2008) MF63 [2-(6-chloro-1*H*-phenanthro[9,10-*d*]imidazole-2-yl)-isophthalonitrile], a selective microsomal prostaglandin E synthase-1 inhibitor, relieves ryresis and pain in preclinical models of inflammation. J Pharmacol Exp Ther 326:754–763
17. Leclerc P, Paweizik SC, Idborg H et al (2013) Characterization of a new mPGES-1 inhibitor in rat models of inflammation. Prostaglandins Other Lipid Mediat 102-103:1–12
18. Ikeda-Matsuo Y, Ota A, Fukada T et al (2006) Microsomal prostaglandin E synthase-1 is a critical factor of stroke-reperfusion injury. Proc Natl Acad Sci U S A 103:11790–11795
19. Kihara Y, Matsushita T, Kita Y et al (2009) Targeted lipidomics revealed mPGES-1-PGE2 as a therapeutic target for multiple sclerosis. Proc Natl Acad Sci U S A 106:21807–21812
20. Akitake Y, Nakatani Y, Kamei D et al (2013) Microsomal prostaglandin E synthase-1 is induced in Alzheimer's disease and its deletion mitigates Alzheimer's disease-like pathology in a mouse model. J Neurosci Res 91:909–919
21. Wang M, Zukas AM, Hui Y et al (2006) Deletion of microsomal prostaglandin E synthase-1 augments prostacyclin and retards atherogenesis. Proc Natl Acad Sci U S A 103:14507–14512
22. Wang M, Lee E, Song W et al (2008) Microsomal prostaglandin E synthase-1 deletion suppresses oxidative stress and angiotensin II-induced abdominal aortic aneurysm formation. Circulation 117:1302–1309
23. Wang M, Ihida-Stansbury K, Kothapalli D et al (2011) Microsomal prostaglandin E2 synthase-1 modulates the response to vascular injury. Circulation 123:631–639
24. Kamei D, Murakami M, Nakatani Y et al (2003) Potential role of microsomal prostaglandin E synthase-1 in tumorigenesis. J Biol Chem 278:19396–19405
25. Oshima H, Oshima M, Inaba K et al (2004) Hyperplastic gastric tumors induced by activated macrophages in COX-2/mPGES-1 transgenic mice. EMBO J 23:1669–1678
26. Kamei D, Murakami M, Sasaki Y et al (2010) Microsomal prostaglandin E synthase-1 in both cancer cells and hosts contributes to tumor growth, invasion and metastasis. Biochem J 425:361–371
27. Nakanishi M, Montrose DC, Clark P et al (2008) Genetic deletion of mPGES-1 suppresses intestinal tumorigenesis. Cancer Res 68:3251–3259
28. Nakanishi M, Menoret A, Tanaka T et al (2011) Selective PGE_2 suppression impairs colon carcinogenesis and modifies local mucosal immunity. Cancer Prev Res 4:1198–1208
29. Sasaki Y, Kamei D, Ishikawa Y et al (2012) Microsomal prostaglandin E synthase-1 is involved in multiple steps of colon carcinogenesis. Oncogene 31:2943–2952
30. Ae T, Ohno T, Hattori Y et al (2010) Role of microsomal prostaglandin E synthase-1 in the facilitation of angiogenesis and the healing of gastric ulcers. Am J Physiol 299:G1139–G1146
31. Tanikawa N, Ohmiya Y, Ohkubo H et al (2002) Identification and characterization of a novel type of membrane-associated prostaglandin E synthase. Biochem Biophys Res Commun 291:884–889
32. Murakami M, Nakashima K, Kamei D et al (2003) Cellular prostaglandin E_2 production by membrane-bound prostaglandin E synthase-2 via both cyclooxygenases-1 and -2. J Biol Chem 278:37937–37947

33. Jania LA, Chandrasekharan S, Backlund MG et al (2009) Microsomal prostaglandin E synthase-2 is not essential for *in vivo* prostaglandin E_2 biosynthesis. Prostaglandins Other Lipid Mediat 88:73–81
34. Tanioka T, Nakatani Y, Semmyo N et al (2000) Molecular identification of cytosolic prostaglandin E_2 synthase that is functionally coupled with cyclooxygenase-1 in immediate prostaglandin E_2 biosynthesis. J Biol Chem 275:32775–32782
35. Kobayashi T, Nakatani Y, Tanioka T et al (2004) Regulation of cytosolic prostaglandin E synthase by phosphorylation. Biochem J 381:59–69
36. Grad I, McKee TA, Ludwig SM et al (2006) The Hsp90 cochaperone p23 is essential for perinatal survival. Mol Cell Biol 26:8976–8983
37. Lovgren AK, Kovarova M, Koller BH (2007) cPGES/p23 is required for glucocorticoid receptor function and embryonic growth but not prostaglandin E_2 synthesis. Mol Cell Biol 27:4416–4430
38. Nakatani Y, Hokonohara Y, Kakuta S et al (2007) Knockout mice lacking cPGES/p23, a constitutively expressed PGE_2 synthetic enzyme, are perinatally lethal. Biochem Biophys Res Commun 362:387–392
39. Nakatani Y, Hokonohara Y, Tajima Y et al (2011) Involvement of the constitutive prostaglandin E synthase cPGES/p23 in expression of an initial prostaglandin E_2 inactivating enzyme, 15-PGDH. Prostaglandins Other Lipid Mediat 94:112–117
40. Christ-Hazelhof E, Nugteren DH (1979) Purification and characterisation of prostaglandin endoperoxide D-isomerase, a cytoplasmic, glutathione-requiring enzyme. Biochim Biophys Acta 572:43–51
41. Kanaoka Y, Ago H, Inagaki E et al (2000) Cloning and crystal structure of hematopoietic prostaglandin D synthase. Cell 90:1085–1095
42. Mohri I, Taniike M, Taniguchi H et al (2006) Prostaglandin D_2-mediated microglia/astrocyte interaction enhances astrogliosis and demyelination in *twitcher*. J Neurosci 26:4383–4393
43. Mohri I, Aritake K, Taniguchi H et al (2009) Inhibition of prostaglandin D synthase suppresses muscular necrosis. Am J Pathol 174:1735–1744
44. Rajakariar R, Hilliard M, Lawrence T et al (2007) Hematopoietic prostaglandin D_2 synthase controls the onset and resolution of acute inflammation through PGD_2 and 15-deoxyΔ12-14 PGJ_2. Proc Natl Acad Sci U S A 104:20979–20984
45. Trivedi SG, Newson J, Rajakariar R et al (2006) Essential role for hematopoietic prostaglandin D_2 synthase in the control of delayed type hypersensitivity. Proc Natl Acad Sci U S A 103:5179–5184
46. Park JM, Kanaoka Y, Eguchi N et al (2007) Hematopoietic prostaglandin D synthase suppresses intestinal adenomas in $Apc^{Min/+}$ mice. Cancer Res 67:881–889
47. Murata T, Aritake K, Matsumoto S et al (2011) Prostagladin D_2 is a mast cell-derived antiangiogenic factor in lung carcinoma. Proc Natl Acad Sci U S A 108:19802–19807
48. Urade Y, Fujimoto N, Hayaishi O (1985) Purification and characterization of rat brain prostaglandin D synthetase. J Biol Chem 260:12410–12415
49. Urade Y, Nagata A, Suzuki Y et al (1989) Primary structure of rat brain prostaglandin D synthetase deduced from cDNA sequence. J Biol Chem 264:1041–1045
50. Pinzar E, Kanaoka Y, Inui T et al (2000) Prostaglandin D synthase gene is involved in the regulation of non-rapid eye movement sleep. Proc Natl Acad Sci U S A 97:4903–4907
51. Eguchi N, Minami T, Shirafuji N et al (1999) Lack of tactile pain (allodynia) in lipocalin-type prostaglandin D synthase-deficient mice. Proc Natl Acad Sci U S A 96:726–730
52. Miwa Y, Takiuchi S, Kamide K et al (2004) Identification of gene polymorphism in lipocalin-type prostaglandin D synthase and its association with carotid atherosclerosis in Japanese hypertensive patients. Biochem Biophys Res Commun 322:428–433
53. Ragolia L, Palaia T, Hall CE et al (2005) Accelerated glucose intolerance, nephropathy, and atherosclerosis in prostaglandin D_2 synthase knock-out mice. J Biol Chem 280:29946–29955
54. Joo M, Kwon M, Sadikot RT et al (2007) Induction and function of lipocalin prostaglandin D synthase in host immunity. J Immunol 179:2565–2575

55. Hokari R, Kurihara C, Nagata N et al (2011) Increased expression of lipocalin-type-prostaglandin D synthase in ulcerative colitis and exacerbating role in murine colitis. Am J Physiol 300:G401–G408
56. Taketomi Y, Ueno N, Kojima T et al (2013) Mast cell maturation is driven via a group III phospholipase A_2-prostaglandin D_2-DP1 receptor paracrine axis. Nat Immunol 14:554–563
57. Watanabe K (2011) Recent reports about enzymes related to the synthesis of prostaglandin (PG) F_2 ($PGF_{2\alpha}$ and 9α, 11β-PGF_2). J Biochem 150:593–596
58. Ullrich V, Castle L, Weber P (1981) Spectral evidence for the cytochrome P450 nature of prostacyclin synthetase. Biochem Pharmacol 30:2033–2036
59. Hara S, Miyata A, Yokoyama C et al (1994) Isolation and molecular cloning of prostacyclin synthase from bovine endothelial cells. J Biol Chem 269:19897–19903
60. Ueno N, Murakami M, Tanioka T et al (2001) Coupling between cyclooxygenase, terminal prostanoid synthase, and phospholipase A_2. J Biol Chem 276:34918–34927
61. Todaka T, Yokoyama C, Yanamoto H et al (1999) Gene transfer of human prostacyclin synthase prevents neointimal formation after carotid balloon injury in rats. Stroke 30:419–426
62. Iwai N, Katsuya T, Ishikawa K et al (1999) Human prostacyclin synthase gene and hypertension: the Suita Study. Circulation 100:2231–2236
63. Yokoyama C, Yabuki T, Shimonishi M et al (2002) Prostacyclin-deficient mice develop ischemic renal disorders, including nephrosclerosis and renal infarction. Circulation 106:2397–23403
64. Keith RL, Miller YE, Hudish TM et al (2004) Pulmonary prostacyclin synthase overexpression chemoprevents tobacco smoke lung carcinogenesis in mice. Cancer Res 64:5897–5904
65. Nüsing R, Schneider-Voss S, Ullrich V (1990) Immunoaffinity purification of human thromboxane synthase. Arch Biochem Biophys 280:325–330
66. Yokoyama C, Miyata A, Ihara H et al (1991) Molecular cloning of human platelet thromboxane A synthase. Biochem Biophys Res Commun 178:1479–1484
67. Yu IS, Lin SR, Huang CC et al (2004) *TXAS*-deleted mice exhibit normal thrombopoiesis, defective hemostasis, and resistance to arachidonate-induced death. Blood 104:135–142
68. Dogné JM, de Leval X, Benoit P et al (2002) Therapeutic potential of thromboxane inhibitors in asthma. Expert Opin Invest Drugs 11:275–281
69. Okuno T, Iizuka Y, Okazaki H et al (2008) 12(S)-Hydroxyheptadeca-5Z, 8E, 10E-trienoic acid is a natural ligand for leukotriene B_4 receptor 2. J Exp Med 205:759–766
70. Liu M, Saeki K, Matsunobu T et al (2014) 12-Hydroxyheptadecatrienoic acid promotes epidermal wound healing by accelerating keratinocyte migration via the BLT2 receptor. J Exp Med 211:1063–1078

Chapter 4
Pathophysiological Roles of Prostanoid Receptors in the Central Nervous System

Tomoaki Inazumi and Yukihiko Sugimoto

Abstract Prostanoids comprising prostaglandins (PGs) and thromboxanes exert diverse actions by acting on their specific receptors. Recently, physiological roles of these receptors have been clarified using knockout mice for each receptor as well as receptor-selective agonists and antagonists. In the central nervous system (CNS), prostanoids have been shown to regulate not only fever, but also neuroinflammation, and to play a role in the pathogenesis of many neurodegenerative diseases. In this report, we review the recent research on the roles and molecular mechanisms of prostanoids and their receptors in the CNS and discuss their possibilities as therapeutic targets.

Keywords Prostanoid receptors • Central nervous system • Cyclooxygenase • Lipopolysaccharide • Fever • Innate immunity • Arachidonic acid • Monoacylglycerol lipase

4.1 Introduction

Prostanoids comprising four types of prostaglandins (PGE_2, PGD_2, $PGF_{2\alpha}$, and PGI_2) and thromboxane (TXA_2) are arachidonate metabolites synthesized by cyclooxygenase (COX) as the rate-limiting enzyme (Fig. 4.1). Prostanoids exert various physiological roles in various tissues and organs. In contrast to hormones, which exert systemic action, prostanoids work only locally because of their inactivating enzymes and exert their functions via selective receptors on the surface of neighboring cells [1]. Recently, knockout mice of the prostanoid-synthesizing enzymes or prostanoid receptors, as well as selective agonists and antagonists for each receptor, have been developed, and these tools have helped us to understand the physiological

T. Inazumi • Y. Sugimoto (✉)
Department of Pharmaceutical Biochemistry, Graduate School of Pharmaceutical Sciences, Kumamoto University, Oe-honmachi, Kumamoto 862-0973, Japan

Core Research for Evolutional Science and Technology, Japan Science and Technology Agency, Saitama 332-0012, Japan
e-mail: ysugi@kumamoto-u.ac.jp

T. Yokomizo, M. Murakami (eds.), *Bioactive Lipid Mediators*,
DOI 10.1007/978-4-431-55669-5_4

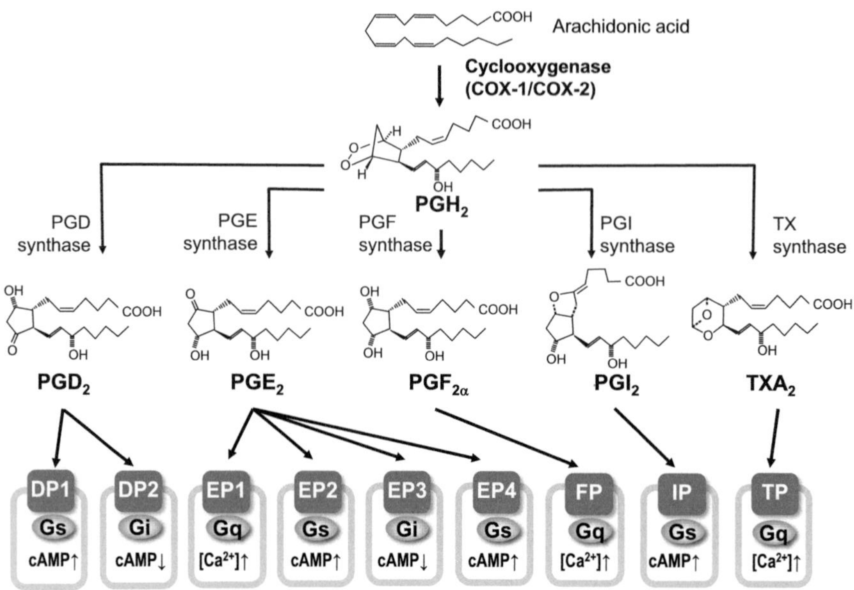

Fig. 4.1 Prostanoid synthesis and its receptors. PGH_2 is produced from arachidonic acid by COX-1/2, and is subsequently converted to different prostanoids by the action of their respective synthases. Prostanoids act on specific receptors on the plasma membrane and elicit changes in the levels of second messengers (cAMP or Ca^{2+})

roles and functional mechanisms of prostanoids at the molecular level [2]. It has been known that prostanoids regulate fever responses at the time of organism infection. Recently, functional mechanisms of the fever response by prostanoids in the CNS have been clarified, and it has also been shown that prostanoids are deeply involved in the development of many neurodegenerative diseases including Alzheimer's disease. In this review, we summarize these new physiological and pathophysiological roles and functional mechanisms of prostanoids in the CNS and discuss their possibilities as therapeutic targets.

4.2 Prostanoid Receptors

Pharmacological studies initially indicated the presence of selective receptors for each prostanoid [3, 4]. The first receptor was identified by Narumiya and his colleagues, who purified the TXA_2 receptor from human platelets [5], and subsequently, molecular identification of eight prostanoid receptors was achieved by cDNA cloning [6–12] (Fig. 4.1). These receptors are G protein-coupled receptors (GPCRs) with seven transmembrane domains and form a unique prostanoid receptor family. Recently, a second PGD_2-responding receptor was identified [13]. This receptor, abundantly expressed in Th2 cells, is called DP2 or chemoattractant receptor homologous molecule expressed on Th2 cell (CRTH2). The structure of this receptor is not similar to the other prostanoid receptors, and it belongs to the chemoattractant receptor family including the leukotriene B_4 receptors BLT1 and BLT2. Therefore, it was considered that this receptor originally responded to a substance different from PGD_2, and then came to recognize PGD_2 during the course of evolution. The signal transduction pathways of prostanoid receptors were revealed by various biochemical studies: the EP2, EP4, DP1, and IP receptors couple to Gs and stimulate adenylate cyclase [8, 10, 14], the EP1, FP, and TP receptors couple to Gq and elicit the elevation of intracellular [Ca^{2+}] [15–17], and the EP3 and DP2 receptors couple to Gi and contribute to the inhibition of adenylate cyclase and the stimulation of intracellular [Ca^{2+}] mobilization [13, 18]. This variety of receptor signaling contributes to the diverse physiological roles of prostanoids. In particular, the four PGE receptor subtypes EP1, EP2, EP3, and EP4 respond to PGE_2 and are used differently depending on the stimulation and cell type. In addition, recent studies revealed that the same receptor couples to different G proteins depending on cell type [19, 20], and EP2 and EP4 receptors stimulate the β-arrestin pathway independently of G proteins [21]. Additional detailed analyses are therefore required to clarify the functional mechanisms of prostanoids.

4.3 Molecular Mechanism of Fever Generation by PGE_2

A fever is elicited by cellular components of infectious organisms, such as lipopolysaccharide (LPS), or by non-infectious inflammatory substances. These exogenous stimulations first activate peripheral macrophages and induce the production of endogenous pyrogenic cytokines such as interleukin (IL)-1β and interleukin (IL)-6 [22]. Next, the produced cytokines act on the preoptic area (POA), which then stimulates the neural pathways that raise body temperature. Non-steroidal anti-inflammatory drugs (NSAIDs) including aspirin are selective COX inhibitors, and their powerful anti-pyrogenic effects suggest that prostanoids regulate fever generation [23, 24]. Indeed, the level of PGE_2 in the POA is elevated during the LPS-induced fever response, and NSAIDs treatment completely abolishes this increase in PGE_2 [25]. Moreover, intracerebral administration of PGE_2 induces a fever response [26]. Therefore, PGE_2 was suggested to work as a central mediator of fever, but the detailed molecular mechanisms involved had not been clarified. To identify the receptor involved in the generation of fever, Ushikubi et al. analyzed mice lacking each PGE receptor subtype. They found that only EP3 receptor-deficient mice fail to show a fever response induced by PGE_2, IL-1β (endogenous pyrogen), and LPS (exogenous pyrogen), although the EP3-deficient mice were normal in stress-induced thermogenesis [27]. Indeed, Sugimoto et al. revealed that the EP3 receptor transcript is abundantly expressed in the POA [28]. These studies demonstrated that PGE_2 mediates fever generation in response to both exogenous and endogenous pyrogens by acting on EP3 receptors in the POA.

Receptor-activator of nuclear factor κB ligand (RANKL) is an essential intercellular mediator that regulates bone remodeling, lymph node organogenesis, and the development of the mammary gland; it exerts its actions by acting on its receptor RANK [29]. Recently, Penninger and his colleagues revealed that RANKL/RANK signaling plays an essential role in both LPS-induced and cytokine-induced fever response [30]. Intracerebral administration of RANKL induces COX-2 expression and PGE_2 production and raises body temperature. Moreover, this fever response is blocked in NSAIDs-treated wild-type mice, EP3-deficient mice, and astrocyte-specific RANK-deficient mice. They also revealed that peripheral administration of LPS or IL-1β induces the expression of RANKL and RANK in the POA. These results demonstrated that RANKL signaling in astrocytes of the POA induces COX-2 expression and PGE_2 production, and PGE_2 elevates body temperature via the EP3 receptor (Fig. 4.2).

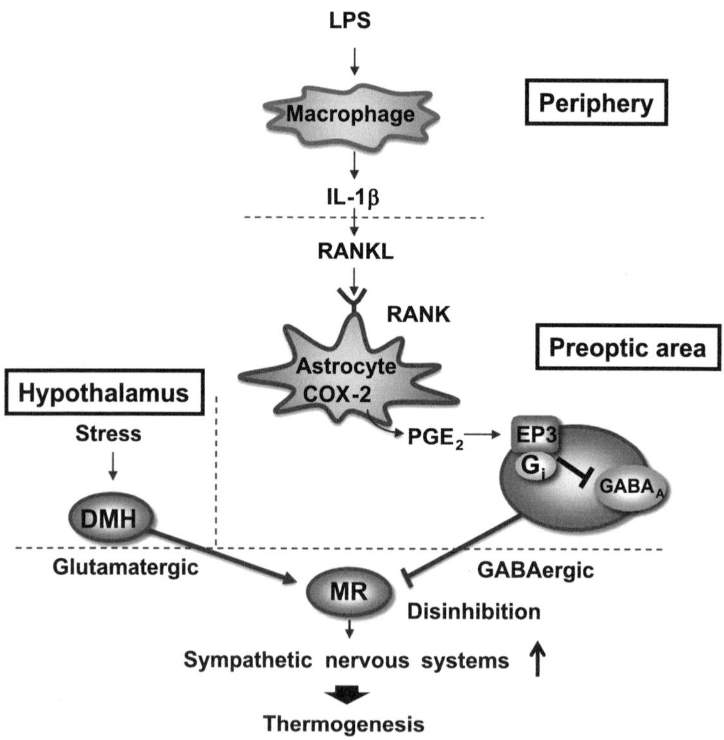

Fig. 4.2 Molecular mechanisms of inflammation-induced and stress-induced thermogenesis. Peripheral macrophages produce IL-1β (endogenous pyrogen) in response to lipopolysaccharide (LPS) (exogenous pyrogen) stimuli. The released IL-1β induces COX-2 expression in astrocytes in the preoptic area via the RANK-RANKL pathway. The produced PGE_2 decreases the expression of $GABA_A$ subunits via the EP3 receptor in a Gi-dependent manner, which provides direct GABAergic input to sympathetic neurons in the medullary raphe (*MR*) regions. As a result, MR neurons are disinhibited and activate brown adipose tissue-mediated thermogenesis. Stress signals activate dorsomedial hypothalamus (*DMH*) neurons, which also provide direct glutamatergic input to MR neurons to drive thermogenesis

Then, how does the EP3 receptor in the POA induce a fever response? Nakamura et al. showed that the EP3 receptor is expressed on the surface of neuronal cell bodies in the POA [31]. Moreover, they revealed that most EP3 receptor-positive neurons in the POA are also positive for glutamate decarboxylase 67, that is, they are inhibitory GABAergic neurons. They also revealed that administration of the $GABA_A$ agonist muscimol into the raphe nucleus abolishes the PGE_2-induced fever response [32]. Tsuchiya et al. isolated EP3-positive POA neurons from PGE2-administered mice and examined PGE_2-induced gene expression changes. They revealed that PGE_2 decreased the expression of the $GABA_A$ receptor in POA neurons [33]. PGE_2 is therefore thought to decrease $GABA_A$ receptor expression by acting on EP3 receptors on POA neurons and to attenuate the inhibitory control of GABAergic neurons to the raphe pallidus nucleus, leading to sympathetic nerve-mediated thermogenesis (Fig. 4.2).

Interestingly, a recent report demonstrated that stress signals activate dorsomedial hypothalamic neurons, which provide direct glutamatergic input to sympathetic premotor neurons in the medullary raphe region to drive thermogenesis [34]. Thus, inflammation-induced and stress-induced thermogenesis share a common output pathway but have very different input pathways (Fig. 4.2).

4.4 Regulatory Mechanism of Innate Immunity in the CNS by PGE_2

LPS activates innate immunity not only in peripheral tissues but also in the CNS during bacterial infection [35]. Microglia are mainly responsible for the host defense response in the CNS, but excessive activation of microglia induces a severe inflammatory state, leading to neurotoxicity [36]. Breyer and his colleagues revealed that neuronal damage induced by intracerebral LPS injection is blocked by NSAIDs treatment or EP2 receptor gene deficiency [37]. Moreover, they showed that EP2 receptor deficiency in microglia inhibits LPS-induced neuronal apoptosis in a co-culture system of neurons and microglia [38]. Recently, Andreasson et al. revealed that LPS induces the expression of the EP4 receptor in microglia. Moreover, they showed that an EP4 selective agonist inhibits the LPS-induced inflammatory response, and this inflammatory response is delayed in microglia-specific EP4-deficient mice [39]. These results demonstrated that PGE_2 produced by LPS stimuli activates innate immunity via the EP2 receptor of microglia in the early phase, and then inhibits the excessive inflammatory response by acting on the EP4 receptor in the late phase (Fig. 4.3). Both EP2 and EP4 receptors couple to the Gs protein and have similar signaling pathways: stimulation of adenylyl cyclase or the Gs-independent β-arrestin pathway [21, 40]. In this case, EP2 receptor signaling induces neuroinflammation, whereas EP4 receptor signaling mediates antiinflammatory effects. It remains to be elucidated as to how such different outcomes are elicited downstream of EP2 and EP4 receptors.

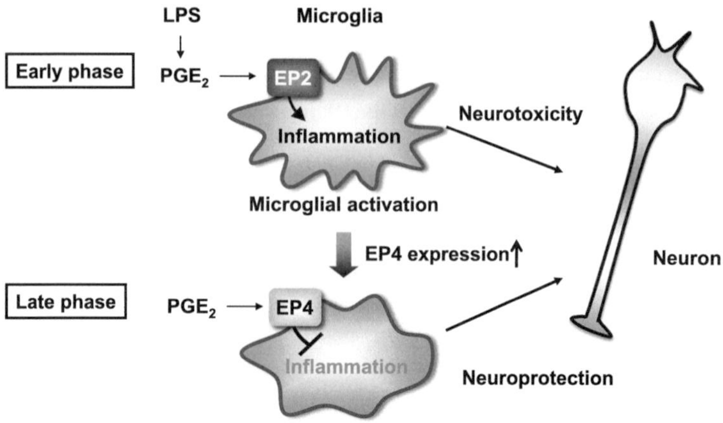

Fig. 4.3 Regulation of microglial activation by PGE_2 in the inflammatory response. In the early phase, LPS-induced PGE_2 activates innate immunity via microglial EP2. In contrast, in the late phase, EP4 expression is upregulated and EP4 inhibits inflammation

4.5 Novel Production Pathway of Arachidonic Acid in Brain

As already mentioned, prostanoids are arachidonate metabolites, and arachidonic acid is thought to be generated from membrane phospholipids by phospholipase A_2 (PLA_2) [41]. However, genetic deletion of the principal PLA_2 subtype, cytosolic PLA_2 ($cPLA_2$), does not change the arachidonic acid levels in the brain [42]. Recently, Nomura and his colleagues revealed that monoacylglycerol lipase (MAGL) produces arachidonic acid by hydrolyzing endocannabinoid 2-arachidonoylglycerol in some tissues including the brain [43] (Fig. 4.4). In MAGL-deficient mice, brain arachidonic acid levels were decreased, and brain PGE_2 production and microglial activation by LPS were markedly inhibited. In contrast, this inhibition was rarely seen in $cPLA_2$-deficient mice. They also showed that MAGL is largely responsible for PGE_2 production not only in the brain but also in the liver and lung, whereas $cPLA_2$ is responsible for PG production in the gut and spleen. These results demonstrate that different production pathways of arachidonic acid are utilized in different tissues.

Parkinson's disease and Alzheimer's disease are typical neurodegenerative diseases involving chronic inflammation [44], and NSAIDs treatment studies have shown that prostanoids contribute to the development of these disorders [45, 46]. Nomura et al. examined whether MAGL contributes to the production of prostanoids in these neurodegenerative diseases using an MAGL inhibitor [43, 47]. This MAGL inhibitor prevented both dopaminergic neuronal loss in the substantia nigra in a model of parkinsonism and neuroinflammation in a model of Alzheimer's disease, similarly to NSAIDs (Fig. 4.4). These results suggested that MAGL contributes to the production of prostanoids in the brain during the development of these neurodegenerative diseases.

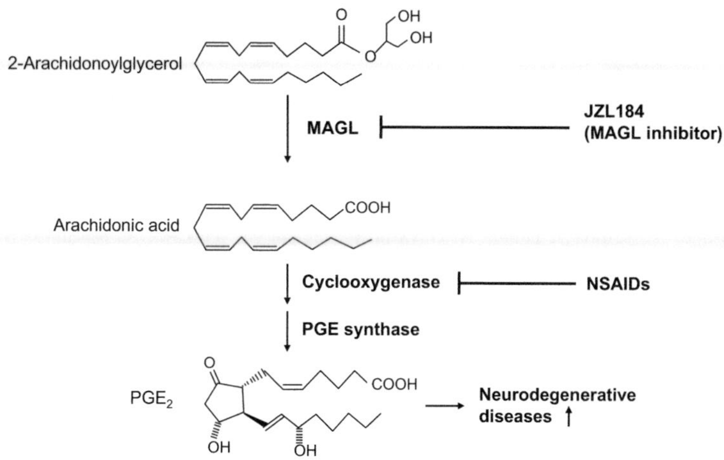

Fig. 4.4 Synthesis of arachidonic acid and PGE_2 in the brain. In the brain, MAGL converts 2-arachidonoylgycerol to arachidonic acid, which is further oxidized into PGE_2 by cyclooxygenase and PGE synthase. The MAGL inhibitor JZL184 was found to decrease brain PGE_2 levels and ameliorate the neuropathological phenotypes of neurodegenerative diseases

4.6 Perspectives

Recently, the physiological roles and functional mechanisms of prostanoids in the CNS have been clarified by studies on knockout mice of the enzymes involved in prostanoid synthesis and prostanoid receptors, or studies using receptor-selective agonists and antagonists. Widely used NSAIDs that inhibit prostanoid synthesis have severe problems of digestive or circulatory side effects. MAGL, which does not block PG synthesis in the gut, has the possibility of being a safer therapeutic target. However, prostanoids often act both positively and negatively via different receptors, and hence strategies inhibiting PG-synthesizing enzymes are not always appropriate. In this context, the development of selective agonists and antagonist of each PG receptor is expected to have great significance toward establishment of treatments against neurodegenerative diseases.

Acknowledgments We thank Professors Kazuhisa Nakayama, Atsushi Ichikawa, and Shuh Narumiya for precious advice. We also thank Dr. Soken Tsuchiya, Ms. Toshiko Sugimoto, and Ms. Yue Kanagawa for their continuous support. We are grateful to Dr. H. Akiko Popiel for careful reading of the manuscript.

References

1. Tai HH, Ensor CM, Tong M, Zhou H, Yan F (2002) Prostaglandin catabolizing enzymes. Prostaglandins Other Lipid Mediat 68-69:483–493
2. Sugimoto Y, Narumiya S (2007) Prostaglandin E receptors. J Biol Chem 282:11613–11617

3. Andersen NH, Eggerman TL, Harker LA, Wilson CH, De B (1980) On the multiplicity of platelet prostaglandin receptors. I. Evaluation of competitive antagonism by aggregometry. Prostaglandins 19:711–735
4. Gardiner PJ, Collier HO (1980) Specific receptors for prostaglandins in airways. Prostaglandins 19:819–841
5. Hirata M, Hayashi Y, Ushikubi F, Yokota Y, Kageyama R, Nakanishi S, Narumiya S (1991) Cloning and expression of cDNA for a human thromboxane A_2 receptor. Nature 349:617–620
6. Sugimoto Y, Namba T, Honda A, Hayashi Y, Negishi M, Ichikawa A, Narumiya S (1992) Cloning and expression of a cDNA for mouse prostaglandin E receptor EP3 subtype. J Biol Chem 267:6463–6466
7. Watabe A, Sugimoto Y, Honda A, Irie A, Namba T, Negishi M, Ito S, Narumiya S, Ichikawa A (1993) Cloning and expression of cDNA for a mouse EP1 subtype of prostaglandin E receptor. J Biol Chem 268:20175–20178
8. Honda A, Sugimoto Y, Namba T, Watabe A, Irie A, Negishi M, Narumiya S, Ichikawa A (1993) Cloning and expression of a cDNA for mouse prostaglandin E receptor EP2 subtype. J Biol Chem 268:7759–7762
9. Katsuyama M, Nishigaki N, Sugimoto Y, Morimoto K, Negishi M, Narumiya S, Ichikawa A (1995) The mouse prostaglandin E receptor EP2 subtype: cloning, expression, and northern blot analysis. FEBS Lett 372:151–156
10. Hirata M, Kakizuka A, Aizawa M, Ushikubi F, Narumiya S (1994) Molecular characterization of a mouse prostaglandin D receptor and functional expression of the cloned gene. Proc Natl Acad Sci U S A 91:11192–11196
11. Namba T, Oida H, Sugimoto Y, Kakizuka A, Negishi M, Ichikawa A, Narumiya S (1994) cDNA cloning of a mouse prostacyclin receptor: multiple signaling pathways and expression in thymic medulla. J Biol Chem 269:9986–9992
12. Sugimoto Y, Hasumoto K, Namba T, Irie A, Katsuyama M, Negishi M, Kakizuka A, Narumiya S, Ichikawa A (1994) Cloning and expression of a cDNA for mouse prostaglandin F receptor. J Biol Chem 269:1356–1360
13. Hirai H, Tanaka K, Yoshie O, Ogawa K, Kenmotsu K, Takamori Y, Ichimasa M, Sugamura K, Nakamura M, Takano S, Nagata K (2001) Prostaglandin D_2 selectively induces chemotaxis in T helper type 2 cells, eosinophils, and basophils via seven-transmembrane receptor CRTH2. J Exp Med 193:255–261
14. Tang DG, Grossi IM, Tang KQ, Diglio CA, Honn KV (1995) Inhibition of TPA and 12(S)-HETE-stimulated tumor cell adhesion by prostacyclin and its stable analogs: rationale for their antimetastatic effects. Int J Cancer 60:418–425
15. Tabata H, Tanaka S, Sugimoto Y, Kanki H, Kaneko S, Ichikawa A (2002) Possible coupling of prostaglandin E receptor EP1 to TRP5 expressed in *Xenopus laevis* oocytes. Biochem Biophys Res Commun 298:398–402
16. Ito S, Sakamoto K, Mochizuki-Oda N, Ezashi T, Miwa K, Okuda-Ashitaka E, Shevchenko VI, Kiso Y, Hayaishi O (1993) Prostaglandin $F_{2\alpha}$ receptor is coupled to Gq in cDNA-transfected Chinese hamster ovary cells. Biochem Biophys Res Commun 200:756–762
17. Dorn GW, Becker MW (1993) Thromboxane A_2 stimulated signal transduction in vascular smooth muscle. J Pharmacol Exp Ther 265:447–456
18. Namba T, Sugimoto Y, Negishi M, Irie A, Ushikubi F, Kakizuka A, Ito S, Ichikawa A, Narumiya S (1993) Alternative splicing of C-terminal tail of prostaglandin E receptor subtype EP3 determines G-protein specificity. Nature 365:166–170
19. Hatae N, Yamaoka K, Sugimoto Y, Negishi M, Ichikawa A (2002) Augmentation of receptor-mediated adenylyl cyclase activity by Gi-coupled prostaglandin receptor subtype EP3 in a $G\beta\gamma$ subunit-independent manner. Biochem Biophys Res Commun 290:162–168
20. Yamaoka K, Yano A, Kuroiwa K, Morimoto K, Inazumi T, Hatae N, Tabata H, Segi-Nishida E, Tanaka S, Ichikawa A, Sugimoto Y (2009) Prostaglandin EP3 receptor superactivates adenylyl cyclase via the Gq/PLC/Ca^{2+} pathway in a lipid raft-dependent manner. Biochem Biophys Res Commun 389:678–682

21. Buchanan FG, Gorden DL, Matta P, Shi Q, Matrisian LM, DuBois RN (2006) Role of β-arrestin 1 in the metastatic progression of colorectal cancer. Proc Natl Acad Sci U S A 103:1492–1497
22. Kluger MJ (1991) Fever: role of pyrogens and cryogens. Physiol Rev 71:93–127
23. Itami T, Ema M, Kanoh S (1986) Antipyretic mechanism of indomethacin in rabbits. J Pharmacobiodyn 9:271–275
24. Cao C, Matsumura K, Yamagata K, Watanabe Y (1997) Involvement of cyclooxygenase-2 in LPS-induced fever and regulation of its mRNA by LPS in the rat brain. Am J Physiol 272:R1712–R1725
25. Sehic E, Székely M, Ungar AL, Oladehin A, Blatteis CM (1996) Hypothalamic prostaglandin E_2 during lipopolysaccharide-induced fever in guinea pigs. Brain Res Bull 39:391–399
26. Saigusa T, Iriki M (1988) Regional differentiation of sympathetic nerve activity during fever caused by intracerebroventricular injection of PGE_2. Pflugers Arch 411:121–125
27. Ushikubi F, Segi E, Sugimoto Y, Murata T, Matsuoka T, Kobayashi T, Hizaki H, Tuboi K, Katsuyama M, Ichikawa A, Tanaka T, Yoshida N, Narumiya S (1998) Impaired febrile response in mice lacking the prostaglandin E receptor subtype EP3. Nature 395:281–284
28. Sugimoto Y, Shigemoto R, Namba T, Negishi M, Mizuno N, Narumiya S, Ichikawa A (1994) Distribution of the messenger RNA for the prostaglandin E receptor subtype EP3 in the mouse nervous system. Neuroscience 62:919–928
29. Leibbrandt A, Penninger JM (2008) RANK/RANKL: regulators of immune responses and bone physiology. Ann N Y Acad Sci 1143:123–150
30. Hanada R, Leibbrandt A, Hanada T, Kitaoka S, Furuyashiki T, Fujihara H, Trichereau J, Paolino M, Qadri F, Plehm R, Klaere S, Komnenovic V, Mimata H, Yoshimatsu H, Takahashi N, von Haeseler A, Bader M, Kilic SS, Ueta Y, Pifl C, Narumiya S, Penninger JM (2009) Central control of fever and female body temperature by RANKL/RANK. Nature 462:505–509
31. Nakamura K, Kaneko T, Yamashita Y, Hasegawa H, Katoh H, Ichikawa A, Negishi M (1999) Immunocytochemical localization of prostaglandin EP3 receptor in the rat hypothalamus. Neurosci Lett 260:117–120
32. Nakamura K, Matsumura K, Kaneko T, Kobayashi S, Katoh H, Negishi M (2002) The rostral raphe pallidus nucleus mediates pyrogenic transmission from the preoptic area. J Neurosci 22:4600–4610
33. Tsuchiya H, Oka T, Nakamura K, Ichikawa A, Saper CB, Sugimoto Y (2008) Prostaglandin E_2 attenuates preoptic expression of $GABA_A$ receptors via EP3 receptors. J Biol Chem 283:11064–11071
34. Kataoka N, Hioki H, Kaneko T, Nakamura K (2014) Psychological stress activates a dorsomedial hypothalamus-medullary raphe circuit driving brown adipose tissue thermogenesis and hyperthermia. Cell Metab 20:346–358
35. Qin L, Wu X, Block ML, Liu Y, Breese GR, Hong JS, Knapp DJ, Crews FT (2007) Systemic LPS causes chronic neuroinflammation and progressive neurodegeneration. Glia 55:453–462
36. Lehnardt S (2010) Innate immunity and neuroinflammation in the CNS: the role of microglia in Toll-like receptor-mediated neuronal injury. Glia 58:253–263
37. Montine TJ, Milatovic D, Gupta RC, Valyi-Nagy T, Morrow JD, Breyer RM (2002) Neuronal oxidative damage from activated innate immunity is EP2 receptor-dependent. J Neurochem 83:463–470
38. Shie FS, Montine KS, Breyer RM, Montine TJ (2005) Microglial EP2 is critical to neurotoxicity from activated cerebral innate immunity. Glia 52:70–77
39. Shi J, Johansson J, Woodling NS, Wang Q, Montine TJ, Andreasson K (2010) The prostaglandin E_2 E-prostanoid 4 receptor exerts anti-inflammatory effects in brain innate immunity. J Immunol 184:7207–7218
40. Chun KS, Lao HC, Trempus CS, Okada M, Langenbach R (2009) The prostaglandin receptor EP2 activates multiple signaling pathways and β-arrestin1 complex formation during mouse skin papilloma development. Carcinogenesis 30:1620–1627
41. Murakami M, Kudo I (2002) Phospholipase A_2. J Biochem 131:285–292

42. Rosenberger TA, Villacreses NE, Contreras MA, Bonventre JV, Rapoport SI (2003) Brain lipid metabolism in the $cPLA_2$ knockout mouse. J Lipid Res 44:109–117
43. Nomura DK, Morrison BE, Blankman JL, Long JZ, Kinsey SG, Marcondes MC, Ward AM, Hahn YK, Lichtman AH, Conti B, Cravatt BF (2011) Endocannabinoid hydrolysis generates brain prostaglandins that promote neuroinflammation. Science 334:809–813
44. Glass CK, Saijo K, Winner B, Marchetto MC, Gage FH (2010) Mechanisms underlying inflammation in neurodegeneration. Cell 140:918–934
45. Teismann P, Tieu K, Choi DK, Wu DC, Naini A, Hunot S, Vila M, Jackson-Lewis V, Przedborski S (2003) Cyclooxygenase-2 is instrumental in Parkinson's disease neurodegeneration. Proc Natl Acad Sci U S A 100:5473–5478
46. McKee AC, Carreras I, Hossain L, Ryu H, Klein WL, Oddo S, LaFerla FM, Jenkins BG, Kowall NW, Dedeoglu A (2008) Ibuprofen reduces Aβ, hyperphosphorylated tau and memory deficits in Alzheimer mice. Brain Res 1207:225–236
47. Piro JR, Benjamin DI, Duerr JM, Pi Y, Gonzales C, Wood KM, Schwartz JW, Nomura DK, Samad TA (2012) A dysregulated endocannabinoid-eicosanoid network supports pathogenesis in a mouse model of Alzheimer's disease. Cell Rep 1:617–623

Chapter 5
Lipoxygenases: A Chronological Perspective on the Synthesis of *S* and *R* Fatty Acid Hydroperoxides

Alan R. Brash

Abstract Lipoxygenase enzymes initiate, control, and terminate a free radical reaction of their polyunsaturated fatty acid substrate with molecular oxygen, typically forming a single chiral fatty acid hydroperoxide as product. The enzymes are best known for the synthesis of lipid mediators: for example, leukotrienes and resolvins in animals, jasmonates and short-chain aldehydes in plants. The first part of this review focuses on how and when the distinguishing features of these enzymes were discovered. Lipoxygenase activity was first detected around 1930, yet it took until the 1950s to recognize that the oxygen molecule is introduced with chirality (the products have optical activity), thus distinguishing the reactions from nonenzymatic lipid peroxidation. Specificity for particular polyunsaturated fatty acid substrates and the precise *S*-stereochemistry of the products were first established in the 1960s. In the 1980s the first lipoxygenases forming *R*-chirality hydroperoxides were recognized, ultimately leading to the discovery of human 12*R*-lipoxygenase in the late 1990s. Inactivating mutations in 12*R*-lipoxygenase (or epidermal lipoxygenase-3) are associated with congenital ichthyosis. These lipoxygenases are involved in mobilization of ceramides in the outer epidermis, and the final sections of this chapter give an overview of the current understanding of their role in contributing to formation of the epidermal water barrier.

Keywords Lipoxygenase • Dioxygenase • HPETE • HETE • Arachidonic acid • Linoleic acid • Stereo control • 12*R*-Lipoxygenase • Ichthyosis

This review takes a historical perspective in covering the actions of lipoxygenase (LOX) enzymes as catalysts of the *chiral* synthesis of fatty acid hydroperoxides. My earlier review covers many of the basics and remains valid for a background on LOX biochemistry [1], and a current review with Marcia E. Newcomer focuses on

---

A.R. Brash (✉)
Department of Pharmacology, and the Vanderbilt Institute of Chemical Biology, Vanderbilt University, Nashville, TN 37232, USA
e-mail: alan.brash@vanderbilt.edu

T. Yokomizo, M. Murakami (eds.), *Bioactive Lipid Mediators*,
DOI 10.1007/978-4-431-55669-5_5

the structure–function of LOX enzymes [2]. The perspective here is on the LOX-catalyzed synthesis of *R* and *S* configuration fatty acid hydroperoxides. In historical context, I describe discovery of *S*- and *R*-LOX enzymes and finally explain how we approached our current studies on the vital role of 12*R*-LOX in mammalian skin.

5.1 Characterization of Lipoxygenases

5.1.1 Early Days: Establishing the Existence of Lipoxygenases

There are numerous and occasionally very abundant LOX enzymes in plants, and all the early history of the LOX enzyme family—as well as many of the concepts of their mechanism of action—stemmed from studies with the enzymes in soybeans. Originally detected in the late 1920s and 1930s as an oxidizing activity in soybean flour [3], and named "lipoxidase" in 1932 [4], by 1940 it was recognized as an activity that reacted polyunsaturated fatty acids (PUFA) with molecular oxygen (O_2) to form fatty acid peroxides [5, 6]. By 1947 the work of Nobel laureate Hugo Theorell with Ralph Holman and Åke Åkeson resulted in purification and crystallization of the main enzyme in soybean flour [7], now known as the L-1 isozyme or soybean LOX-1. It is commercially available and is by far the most studied and most utilized LOX enzyme. It happens to use arachidonic acid, EPA, and DHA as excellent substrates, along with the fatty acids it would encounter in plants (linoleic, α-linolenic, and γ-linolenic acids), so this greatly enhances its utility as a model LOX enzyme. The Theorell preparation of soybean LOX-1 reacted with linoleic acid at 330 turnovers/s [7], around the top attainable rate for this enzyme, which is one of the most efficient known. In fact, for reasons that currently remain obscure, the best that many other LOX enzymes can manage is 10–100 times or even 1000 times slower than this. All the mammalian enzymes fall into the relatively slow category.

5.1.2 Oxygenation of Linoleic Acid

The reaction of soybean LOX with linoleic acid was outlined by Bergström and Holman in 1948 [8], and noting the conjugation of the double bonds in association with the substrate oxygenation and the likeness to the reactions of lipid peroxidation, which Bergström earlier had helped characterize [9]. This similarity to lipid peroxidation for a time raised a somewhat misleading concept. It was recognized that lipid peroxidation requires generation of a radical to start the process and then it is self-generating [10]: the radical is not lost as more and more molecules are peroxidized, hence the term "autoxidation." Around the 1940s and early 1950s it was considered possible that the soybean lipoxygenase generated a radical and then the process continued out of enzymatic control. A study of the enzymology by Al Tappel, Boyer, and Lundberg put paid to this misconception: they showed that soybean LOX could follow Michaelis–Menten kinetics and that the enzyme initiates each turnover of substrate to product [11]. Still, in modern days there remains something partly akin to the old concept: LOX enzymes require activation (generally an oxidation of the resting ferrous enzyme (Fe^{2+}) to ferric (Fe^{3+})) and only then do repetitive cycles of turnover occur (see [12, 13]). The difference in this modern understanding is that once activated, the LOX enzyme remains in control of each turnover of substrate to product.

5.1.3 Synthesis of a Chiral Fatty Acid Hydroperoxide

Of course, non-enzymatic lipid peroxidation proceeds with no stereo control and the fatty acid hydroperoxide products are racemic. In 1955 it was reported by Privett, Nickell, and Lundberg that the soybean lipoxygenase products of linoleic acid are optically active, the first evidence that a LOX enzyme forms a chiral product [14]. Knowing that the products exhibit optical rotation, however, does not signify the absolute configuration (*R* or *S*). By 1965 Hamberg and Samuelsson had defined the regio-specificity of soybean LOX-1 as oxygenation at the ω6 carbon of C18, C20, or C22 PUFA [15], and by 1967 they had determined the absolute configuration on the ω6-hydroperoxide of dihomo-γ-linolenic acid as 15*S*-hydroperoxy [16]. The equivalent ω6 oxygenation on other PUFAs gives the 13*S*-hydroperoxide of linoleic acid (13*S*-HPODE, 13*S*-hydroperoxy-octadecadienoic acid), 15*S*-HPETE from arachidonate, and 17*S*-HPDHA from docosahexaenoic acid.

By the early 1970s, several groups defined the characteristic stereochemistry of other plant LOX enzymes: Gardner and Weisleder identified the corn (maize) LOX product of linoleic acid as the 9*S*-hydroperoxide (9*S*-HPODE) [17], and in a classic methods paper on steric analysis of LOX products this was confirmed by Hamberg [18]. Galliard and Phillips identified 9*S*-HPODE as the main product of potato LOX [19], and Veldink, Vliegenthart and Boldingh showed that 9*S*-HPODE can be a minor by-product of soybean LOX preparations [20], as confirmed much later by Gardner [21].

5.1.4 Lipoxygenase, Not Lipoxidase

Lipoxygenases are fatty acid dioxygenases ($RH + O_2 \rightarrow RO_2H$): this is a covalently complete reaction and therefore in contrast to monooxygenases the transformation requires no reducing cofactor. Around the late 1960s, the nomenclature became standardized as lipoxygenase, indicating an enzyme that incorporates molecular oxygen (O_2) into its substrate. By contrast the term oxidase signifies an enzyme that reduces molecular oxygen, ultimately to water. For example, NADPH oxidases reduce O_2 to $HO_2\cdot$ (superoxide, in ionic form $O_2\cdot^-$) using electrons from NADPH, cytochrome *c* oxidase takes the electrons at the end of the respiratory chain and reduces O_2 to water, whereas the typical cytochrome P450 is both a monooxygenase and "mixed function oxidase:" the enzyme uses one atom of O_2 to oxygenate the substrate while the other O atom is reduced to water again using the reducing equivalents from NADPH ($RH + O_2 + NADPH\ (2H) \rightarrow ROH + H_2O + NADP^+$). Only Sigma still uses the term soybean lipoxidase (!).

5.1.5 Mid-1970s: Discovery of Lipoxygenases in Animals

The finding in 1971 that aspirin inhibits prostaglandin synthesis and the known effects of aspirin on blood platelet function [22] directed attention to the metabolism of arachidonic acid in platelets. A brilliant series of experiments by Mats Hamberg, Svensson and Samuelsson at the Karolinska Institutet not only revealed the existence of thromboxane A_2 of the cyclooxygenase pathway and its powerful pro-aggregatory effect on platelets [23], they also uncovered the first LOX enzyme in animal tissues, arachidonate 12*S*-lipoxygenase (product of the ALOX12 gene) [24]; platelet 12-LOX was discovered independently by Diederik Nugteren of Unilever [25]. We know from the work of Shozo Yamamoto's group and others that this platelet-type 12-LOX enzyme is widespread in many tissues [26, 27].

At much the same time Samuel Rapoport's group in Berlin was investigating a peroxidizing enzyme in peripheral blood reticulocytes induced by severe anemia, and reporting that it could oxygenate PUFA esterified in phospholipids [28]; this enzyme was identified as an arachidonate 15*S*-lipoxygenase [29, 30], and is now designated as the ALOX15 gene product, 15-LOX-1. Most mammals express the ALOX15 gene primarily as a 12-LOX, given the helpful designation leukocyte-type 12-LOX by Shozo Yamamoto [26, 27], and the enzyme is often referred to as a 12/15-LOX.

By 1976 the search for lipid mediators in mammalian leukocytes led to the discovery by Pierre Borgeat, Hamberg and Samuelsson at the Karolinska Institutet of 5*S*-lipoxygenase (from the ALOX5 gene) [31]. Soon the role of 5-LOX as the first committed step in leukotriene synthesis was revealed, along with the identity of the peptide leukotrienes as the slow-reacting substance of anaphylaxis, a major lipid mediator with bronchiolar contractile activity in asthma [32–34]. The development of leukotriene receptor antagonists and their clinical efficacy confirmed the important role of 5-LOX and the leukotrienes in asthma and other inflammatory diseases [35]. In more recent times the antitheses of these actions have also been appreciated with the recognition by Charles Serhan and colleagues of the involvement of various LOX products (lipoxins, resolvins, protectins) in promoting the resolution phase of inflammation [36].

For 20 years after the discovery in the mid-1970s of 12-LOX, 15-LOX-1, and 5-LOX, these three enzymes stood alone at the forefront of lipoxygenase pathophysiological studies in mammals [26, 37]. They each synthesize an *S*-configuration fatty acid hydroperoxide, the same as all the other known LOX enzymes of plants and animals at the time. It took until the late 1990s to recognize that three additional LOX genes occur in humans and other mammals: 15-LOX-2, 12*R*-LOX, and epidermal-LOX3 (eLOX3) (Fig. 5.1).

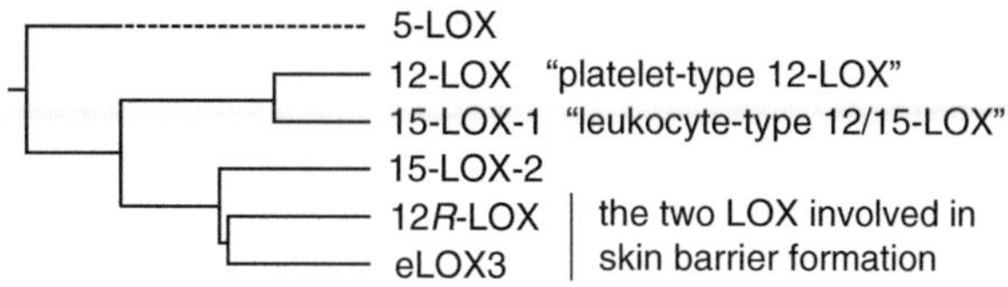

Fig. 5.1 The six human lipoxygenases

5.1.6 The Discovery of R-Lipoxygenases

In a curious finding at the time, in 1979 van Os, Rijke-Schilder, and Vliegenthart reported that the soybean LOX-2 enzyme formed not only 13*S*-HPODE from linoleic acid (same as the well-known soybean LOX-1 enzyme), but also 9*R*-HPODE, the first description of a specific LOX-catalyzed *R*-hydroperoxide formation, and later understood in terms of the fundamental basis of *S* and *R* stereocontrol in lipoxygenases (see following).

My interest in *R*-LOX biochemistry was sparked by a couple of unexpected occurrences. The first eye-opening event was a telephone conversation in 1985 with Gordon Bundy, a chemist at the Upjohn Company in Kalamazoo, Michigan. As an aside to the topic of our conversation on leukotrienes, Gordon told me he had uncovered *R*-HPETE synthesis "from an unusual source" and he asked me if I had ever heard of a lipoxygenase that synthesized a fatty acid *R*-hydroperoxide and did I know of any reason this was not possible? I replied that it did seem possible, noting that the cyclooxygenase enzyme initiates prostaglandin synthesis with an 11*R*-oxygenation of arachidonic acid [38]. Gordon Bundy's discovery, reported in the *Journal of Biological Chemistry* in 1986, described the synthesis of 8*R*-HPETE from arachidonic acid in the Caribbean coral *Pseudoplexaura porosa* [39].

My second introduction to *R*-LOX activity stemmed from the persistent letter writing of Laurent Meijer of the CNRS research laboratories in Roscoff, France, asking me to prepare 8*R*-HETE and 8*S*-HETE so he could test their activities in inducing the reinitiation of meiosis in starfish oocytes. Laurent had identified arachidonic acid and particularly 8-HETE as potent inducers of oocyte maturation (unpublished at the time, but related to the results of Meijer et al. [40]), and he had noticed that we resolved the 8-HETE enantiomers in a study of the mechanisms of autoxidation [41]. I prepared 8*R*- and 8*S*-HETE, and Laurent identified 8*R*-HETE as the active principle in inducing starfish oocyte maturation; I characterized the synthesis of specifically 8*R*-HETE in starfish oocyte homogenates, published in the *Journal of Biological Chemistry* in 1986 [42]. Later, Molly Hughes provided much circumstantial evidence for the existence of an 8*R*-HETE-specific G protein-coupled receptor (GPCR) in starfish oocytes [43].

Just before the Bundy and Meijer publications, E.J. Corey, Peter Lansbury, and Yasuji Yamada described the synthesis of a prostaglandin-related cyclopentenone fatty acid in the Japanese coral *Clavularia viridis* and postulated its synthesis via 8*R*-HPETE [44]. Bundy's and Corey's work had a background in the earlier discovery that the Caribbean coral *Plexaura homomalla* contained masses of prostaglandins

(~3 % of the dry weight) [45, 46]. During the 1970s this coral was used as a source of prostaglandins (hence the Upjohn interest in coral), and for biochemical interest the possibility had arisen that the coral prostaglandins were formed via a non-cyclo-oxygenase route [47, 48].

I decided to investigate the metabolism of arachidonic acid in the prostaglandin-containing coral *Plexaura homomalla*: this led to the discovery of 8*R*-LOX activity in the coral as well as transformation of 8*R*-HPETE to an extremely unstable but covalently complete epoxide known as an allene oxide [49]. Allene oxides can cyclize to give a prostaglandin-like 5-membered carbon ring, and this is a key step in the biosynthesis of fatty acid cyclopentenones in corals [50] and of jasmonic acid biosynthesis in plants [51, 52]. Despite the apparent similarity of the allene oxide-derived cyclopentenone to true prostaglandins, through many productive collaborations with Nigulas Samel and colleagues of Tallinn University of Technology in Estonia, the basis of coral prostaglandin synthesis was shown to be a cyclooxygenase-catalyzed pathway. This work led to the cloning and characterization of the coral cyclooxygenases, which are related quite closely (~50 % sequence identity) to mammalian COX-1 and COX-2 [53–55].

5.1.7 *Occurrence of* R-*HETE Biosynthesis*

By the late 1980s the synthesis of *R*-HETEs was identified from many sources, mostly invertebrate animals [49, 56–58]. One human connection was known, the occurrence of high concentrations of 12-HETE in the hyperproliferative and inflammatory skin of psoriasis, described in 1975 by Hammarström, Hamberg, and Samuelsson in collaboration with the dermatology group of Voorhees in Michigan [59]. In 1986 Pat Woollard presented the unexpected finding that the psoriatic 12-HETE is predominantly of the 12*R* configuration [60]. At the time, only cytochrome P450s were known to synthesize 12*R*-HETE, albeit as one of a complex mixture of epoxy and HETE products [61], and the concept of P450 involvement in mammalian 12*R*-HETE synthesis (mistakenly) dominated the thinking for the next decade.

5.1.8 *Purification and Cloning of the First* R-*Lipoxygenase*

Bill Boeglin, in my laboratory, succeeded in purifying an enzyme responsible for 8*R*-HPETE synthesis from the coral *P. homomalla*, and after obtaining peptide sequences we cloned a partial cDNA by homology-based PCR and cloned the full length by 5′- and 3′-RACE; the deduced sequence matched all the known peptides [62]. The encoded 79-kDa protein exhibited all the characteristic features of a typical LOX family member [1], establishing that closely related LOX proteins can synthesize either *S* or *R* (mirror image) fatty acid hydroperoxides.

As outlined in an earlier review [1], there are compelling reasons to conclude that fatty acid orientation in the LOX active site is a key determinant of reaction

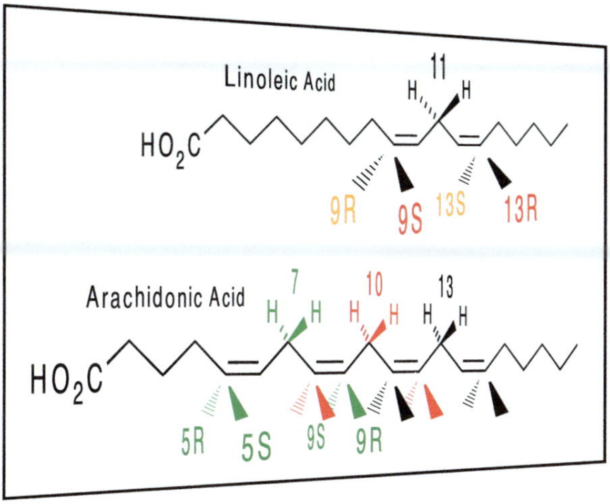

Fig. 5.2 Perspective view of the CH_2 hydrogens and available positions for oxygenation in linoleic and arachidonic acids. As shown on linoleate, the 9*S* and 13*R* positions are on the same face of the substrate (front face in this view) at opposite ends of the pentadiene, with 9*R*/13*S* on the back face. The same principles apply to arachidonate: for example, the *green* pentadiene with CH_2 hydrogens at C7 has the 5*S* and 9*R* positions facing towards the viewer with 5*R*/9*S* at the back. For clarity, the positions around the CH_2 at C10 and C13 are not labeled

specificity. Also, it is well established that both *R* and *S* lipoxygenases catalyze an initial hydrogen abstraction from the CH_2 group between the two *cis* double bonds followed by insertion of oxygen on the opposite face of the substrate (antarafacial relationship between H-abstraction and oxygenation) [63, 64]. To understand the relationship between the *R* and *S* positions in oxygenated polyunsaturated fatty acids, consider the simplest example, linoleic acid (Fig. 5.2, top). Linoleate has only one reacting pentadiene and therefore two available ends; one face presents the 9*S* and 13*R* positions, with 9*R* and 13*S* on the other side. The same applies to arachidonic acid (Fig. 5.2), which has three pentadienes available for reaction. A switch in the head-to-tail orientation of the fatty acid will present the correct face to the activating non-heme iron. Putting these concepts together, and knowing the sequence relatedness of *R*- and *S*-LOX, in the 1996 *Journal of Biological Chemistry* paper we proposed a conceptual model of how LOX enzymes of similar structure can form *R* or *S* HPETEs (Fig. 5.3) [62].

5.1.9 Identification of a Key Structural Determinant of Stereo Control in R- *and* S-*LOX*

By the early 2000s there were many *S*-LOX sequences available, although only four from *R*-LOX. Gianguido Coffa in my laboratory made extensive *R*/*S* LOX enzyme chimeras and selected mutations in attempts to identify amino acid residue(s) that

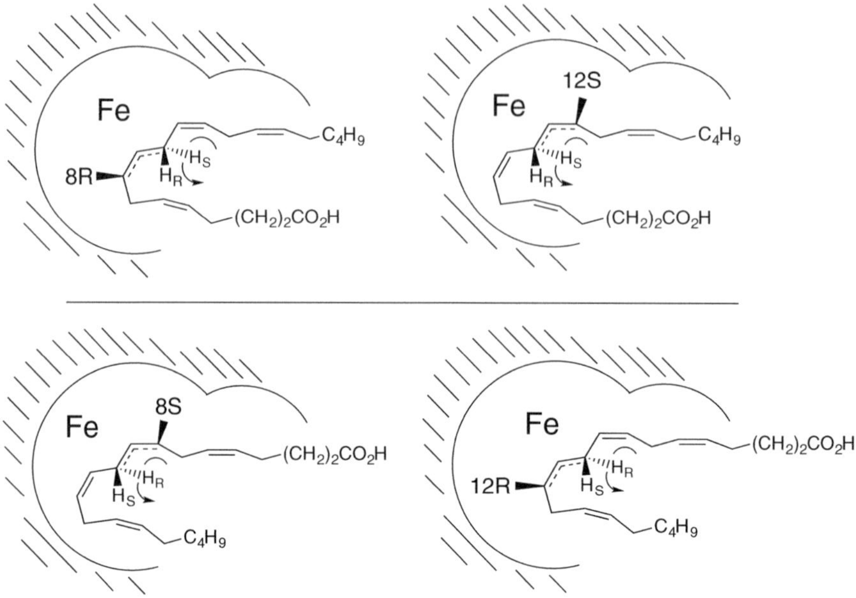

Fig. 5.3 The principle of *R* or *S* stereo control in lipoxygenases. In this example, formation of four different products is represented in lipoxygenase active sites of related structure. *Top panels:* The LOX non-heme iron removes the pro-*S* hydrogen from C10 (*curved arrow*), and O_2 reacts on the opposite face of the substrate (antarafacial to H-abstraction) to form either the 8*R*-hydroperoxide (*left side*) or 12*S*-hydroperoxide (*right side*). *Lower panels:* The fatty acid binds in the reverse head-to-tail orientation, allowing removal of the pro-*R* hydrogen from C10, followed by oxygenation on the opposite face of the substrate to form the 8*S*-hydroperoxide (*right*) or 12*R*-hydroperoxide (*left*) (Reproduced from Brash et al. [62] with permission)

controlled *R* or *S* specificity. One of the conserved differences he spotted was an Ala residue in *S*-LOX replaced by Gly in the four available *R*-LOX, and by making the Ala/Gly switch in two *S*-LOX and two *R*-LOX, he identified this Ala-or-Gly as a major determinant of stereospecificity [65]. With Ala in this critical position the oxygenation always favors reaction of O_2 at the end of the activated pentadiene deep in the substrate-binding pocket and *S* stereochemistry, whereas the Gly residue allows reaction at the other end and gives *R* stereochemistry [66].

In principle there are a number of ways in which the LOX enzyme could control the reaction of O_2 and thus direct formation of an *R* or *S* configuration product [67]. The most likely is shielding of the reacting pentadiene to prevent reaction of O_2 at one end or the other, and it appears that this is accomplished through the Ala/Gly switch by changing the conformation of a nearby leucine [2, 68, 69].

Only a few exceptions to this Ala/Gly stereo determinant have been identified [70], usually with a slight size modification of the amino acid; for example, *Anabaena* linoleate 9*R*-LOX has Ala in the key position (instead of the expected Gly), and its substitution with Val or Ile gives 13*S*-HPODE as a new product [71, 72]. Interestingly, there is a natural substitution of the expected Ala residue to Gly in a LOX from *Olea europaea* (common olive; accession no. EU678670); the expressed enzyme synthesizes 9*S*-HPODE along with 13*R*-HPODE [73], exactly as predicted by the model. The recent report from the Newcomer laboratory of the first crystal structure of a lipoxygenase–substrate complex [68] paves the way to a detailed understanding of stereo control in the LOX enzyme superfamily [2].

5.2 12*R*-LOX and eLOX3 and the Epidermal Water Barrier

5.2.1 12R-LOX in Human Skin and Its Role in Sealing the Epidermal Permeability Barrier

The eicosanoid group in the German Cancer Centre in Heidelberg led by Gerhard Fürstenberger and Peter Krieg has a long-standing interest in skin LOX enzymes and their role in carcinogenesis and epidermal proliferation. By the late 1990s they had cloned three epidermal LOX genes distinct from the mammalian 15-LOX-1, 12*S*-LOX, and 5-LOX known at the time [74–77]. The first of these epidermal LOX enzymes was identified in my laboratory by Mitsuo Jisaka and in Heidelberg as 8S-LOX in the mouse [74, 78] (its human homologue being 15-LOX-2, which we had cloned a couple of years beforehand [79]). The second epidermal LOX was identified by Bill Boeglin in my laboratory as (human) 12*R*-LOX [80]. Colin Funk and colleagues later described both the human and murine enzymes [81] and Peter Krieg and colleagues found that the murine 12R-LOX reacts only with fatty acid esters [82], an intriguing and biologically significant property in relation to its proposed role in the epidermis (see following). The third epidermal enzyme described by Fürstenberger and Krieg was also an enigma—at the time no catalytic activity could be ascribed [75]—and its name has endured as epidermal lipoxygenase-3, or eLOX3 (gene ALOXE3).

5.2.2 The Connection to Ichthyosis

Congenital ichthyosis is a scaly skin condition associated with an inactivation mutation in one of several genes involved in construction of the water permeability barrier in the outer epidermis. The discovery by geneticists led by Judith Fischer that mutations in 12*R*-LOX and eLOX3 are linked to congenital ichthyosis has set the stage for unraveling the role of these enzymes in epidermal barrier function [83]. Mutation of either LOX enzyme was associated with a similar scaly skin phenotype, leading to their suggestion that the two enzymes are involved in the same metabolic pathway. This proposal prompted our enzymological studies to test the hypothesis. Zheyong Yu in my laboratory showed that eLOX3 isomerizes HPETEs to epoxyalcohol products, with a preference for *R*-configuration fatty acid hydroperoxides [84, 85]; he thus characterized this apparently inactive enzyme as a hydroperoxide isomerase. Subsequently, LOX gene knockout studies in mice established the role in barrier function for both 12*R*-LOX and eLOX3 [86, 87]; in each case the knockout is neonatal lethal because of severe water loss within 5–10 h of birth. Human babies with inactivating mutations in 12*R*-LOX or eLOX3 survive and develop the hyperproliferative scaly skin of ichthyosis, presumably in attempts to counteract the defective barrier.

5.2.3 The Connection to the Scaly Skin of Essential Fatty Acid Deficiency

When animals [88] or humans [89] are nutritionally deficient in essential fatty acids, there is a permeability barrier defect, and they develop a scaly skin. Various topically applied fatty acids have the ability to reverse this skin defect, and the structural requirements for active fatty acids have been extensively examined. By testing twenty different fatty acids, Houtsmuller and van der Beek identified the minimal structural requirement as including a pair of *cis* double bonds in the ω6 and ω9 positions, as in linoleic acid [90]. Changing the position of this pair of double bonds in the fatty acid, removal of one of them or substitution with a *trans* double bond, or adding a CH_2 between the bonds eliminated activity [90]. The strong inference from these studies is that having double bonds alone is insufficient to relieve the skin symptoms of essential fatty acid deficiency. By all appearances the key structural features of an active fatty acid match the substrate specificity of a LOX enzyme that utilizes the ω9–ω6 pair of double bonds.

5.2.4 The Connection Between EFA Deficiency and LOX-Dependent Congenital Ichthyosis

A conceptual model of how 12*R*-LOX, eLOX3, and essential fatty acids are involved in sealing the epidermal barrier was developed by Yuxiang Zheng in my laboratory. The only EFA in the mammalian outer epidermis where the barrier resides is linoleic acid [91], and it is esterified to the omega-hydroxy group of a unique, epidermal-specific ceramide, EOS (esterified omega-hydroxy sphingosine) [92]. Yuxiang showed that the linoleate moiety in EOS can be oxidized by 12*R*-LOX and further transformed by eLOX3 to a specific epoxy-alcohol derivative [93]. She detected traces of these oxidized ceramides in pig epidermis and in the epidermis of neonatal mice, and the products were absent in the epidermis of 12*R*-LOX knockout mice [93]. Most significantly, the knockout mice also had an almost complete absence of omega-hydroxy ceramide covalently bound to protein. It is known that a fraction of EOS is de-esterified and the free omega-OH is coupled to a layer of polymerized protein on the outer surface of the corneocytes, replacing the original plasma membrane, and forming a covalently bound lipid coating named the corneocyte lipid envelope or CLE [94]. The CLE lies between the polymerized protein and extracellular lamellar lipids, and potentially functions as a scaffold that holds the two other parts together as a complete and intact barrier. Taking all the facts together, we propose that LOX-catalyzed oxidation of the linoleate ester in EOS is required to facilitate enzymatic hydrolysis of the ester, thus freeing OS for coupling to protein (Fig. 5.4) [93, 95].

In EFA deficiency, oleate (not a LOX substrate) replaces linoleate as the fatty acid ester in EOS [91]. Because it is not a LOX substrate it cannot be oxidized and therefore it cannot be hydrolyzed, resulting in a deficiency in free OS; the prediction, therefore, is that this should lead to decreased levels of covalently bound ceramides in the barrier layer, which is exactly what is found [96]. On the other hand, in LOX deficiency, in either knockout mice or afflicted human families, the

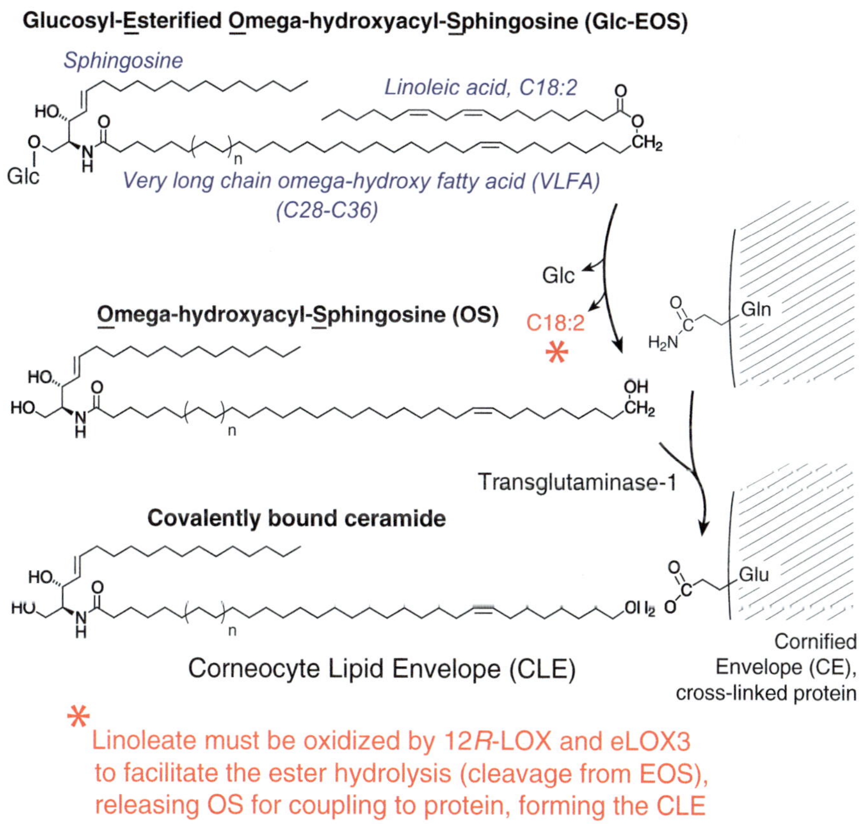

Fig. 5.4 Role of 12*R*-LOX and eLOX3 in forming the epidermal permeability barrier. In the water barrier of the outer epidermis the corneocyte lipid envelope (*CLE*, a lipid layer based upon ceramides covalently bonded to the protein coating of the corneocytes) is formed after initial cleavage of the linoleate moiety of EOS ceramide, releasing the free omega-hydroxyl of OS for coupling to protein by transglutaminase (TGase). The ester cleavage is facilitated by prior oxidation of the linoleate by 12*R*-LOX and eLOX3. Inactivation of either LOX or substitution of linoleate with oleate (in EFA deficiency) prevents the oxidation and results in a deficient or absent CLE and a defective water barrier

oxidizing enzymes are missing, with ultimately the same consequences: no oxidation, therefore no hydrolysis of the linoleate ester, therefore a lack of covalently bound ceramides, an epidermal barrier defect, and the resulting phenotype [93, 95].

There is a good chance that understanding the basis of the role of 12*R*-LOX and eLOX in epidermal function will lead to improved treatments for patients afflicted with LOX-dependent ichthyoses. In these patients it is predicted that the fundamental problem with the skin barrier is a defect in the covalently bound lipid envelope resulting from the lack of available free omega-hydroxy ceramide. As topical applications can be effective, the prediction is that topical OS should provide the missing substrate and ameliorate the skin symptoms. In the near future, this may emerge as another example of basic research resulting in a translational impact.

Acknowledgments This work is supported by grants from the NIH (GM-15431, GM-74888, and AR-51968).

References

1. Brash AR (1999) Lipoxygenases: occurrence, functions, catalysis, and acquisition of substrate. J Biol Chem 274:23679–23682
2. Newcomer ME, Brash AR (2015) The structural basis for specificity in lipoxygenase catalysis. Protein Sci 24:298–309
3. Haas LW, Bohn RM (1934) Bleaching agent and process of utilizing same for bleaching flour. In U.S. patent, editor. 1,957,333-957
4. André E, Hou K (1932) Sur la présence d'une oxydase des lipides ou lipoxydase dans la graine de soja, *Glycine soja* Lieb. Comptes Rendus 194:645–647
5. Tauber H (1940) Unsaturated fat oxidase. J Am Chem Soc 62:2251
6. Sumner JB, Sumner RJ (1940) The coupled oxidation of carotene and fat by carotene oxidase. J Biol Chem 134:531–533
7. Theorell H, Holman RT, Åkeson Å (1947) Crystalline lipoxidase. Acta Chem Scand 1:571–576
8. Bergström S, Holman RT (1948) Total conjugation of linoleic acid in oxidation with lipoxidase. Nature 161:55
9. Bergström S (1945) Autoxidation of linoleic acid. Nature 156:717–718
10. Hammond EG, White PJ (2011) A brief history of lipid oxidation. J Am Oil Chem Soc 88:891–897
11. Tappel AL, Boyer PD, Lundberg WO (1952) The reaction mechanism of soy bean lipoxidase. J Biol Chem 199:267–281
12. Smith WL, Lands WE (1972) Oxygenation of unsaturated fatty acids by soybean lipoxygenase. J Biol Chem 247:1038–1047
13. Cook HW, Lands WE (1975) Further studies of the kinetics of oxygenation of arachidonic acid by soybean lipoxygenase. Can J Biochem 53:1220–1231
14. Privett OS, Nickell C, Lundberg WO, Boyer PD (1955) Products of the lipoxidase-catalyzed oxidation of sodium linoleate. J Am Oil Chem Soc 32:505–511
15. Hamberg M, Samuelsson B (1965) On the specificity of the lipoxidase catalyzed oxygenation of unsaturated fatty acids. Biochem Biophys Res Commun 21:531–536
16. Hamberg M, Samuelsson B (1967) On the specificity of the oxygenation of unsaturated fatty acids catalyzed by soybean lipoxidase. J Biol Chem 242:5329–5335
17. Gardner HW, Weisleder D (1970) Lipoxygenase from *Zea mays*: 9-D-hydroperoxy-*trans*-10, *cis*-12-octadecadienoic acid from linoleic acid. Lipids 5:678–683

18. Hamberg M (1971) Steric analysis of hydroperoxides formed by lipoxygenase oxygenation of linoleic acid. Anal Biochem 43:515–526
19. Galliard T, Phillips DR (1971) Lipoxygenase from potato tubers. Partial purification and properties of an enzyme that specifically oxygenates the 9-position of linoleic acid. Biochem J 124:431–438
20. Veldink GA, Vliegenthart JFG, Boldingh J (1970) Proof of the enzymatic formation of 9-hydroperoxy-10-*trans*, 12-*cis*-octadecadienoic acid from linoleic acid by soya lipoxygenase. Biochim Biophys Acta 202:198–199
21. Gardner HW (1989) Soybean lipoxygenase-1 enzymically forms both (9*S*)- and (13*S*)-hydroperoxides from linoleic acid by a pH-dependent mechanism. Biochim Biophys Acta 1001:274–281
22. Smith JB, Willis AL (1971) Aspirin selectively inhibits prostaglandin production in human platelets. Nature (Lond) 231:235–237
23. Hamberg M, Svensson J, Samuelsson B (1975) Thromboxanes: a new group of biologically active compounds derived from prostaglandin endoperoxides. Proc Natl Acad Sci U S A 72:2994–2998
24. Hamberg M, Samuelsson B (1974) Prostaglandin endoperoxides. Novel transformations of arachidonic acid in human platelets. Proc Natl Acad Sci U S A 71:3400–3404
25. Nugteren DH (1975) Arachidonate lipoxygenase in blood platelets. Biochim Biophys Acta 380:299–307
26. Yamamoto S (1992) Mammalian lipoxygenases: molecular structures and functions. Biochim Biophys Acta 1128:117–131
27. Yoshimoto T, Yamamoto S (1995) Arachidonate 12-lipoxygenase. J Lipid Mediat Cell Signal 12:195–212
28. Schewe T, Halangk W, Hiebsch C, Rapoport SM (1975) A lipoxygenase in rabbit reticulocytes which attacks phospholipids and intact mitochondria. FEBS Lett 60:149–153
29. Rapoport SM, Schewe T, Wiesner R, Halangk W, Ludwig P, Janicke-Höhne M, Tannert C, Hiebsch C, Klatt D (1979) The lipoxygenase of reticulocytes. Purification, characterization, and biological dynamics of the lipoxygenase; its identity with the respiratory inhibitors of the reticulocyte. Eur J Biochem 96:545–561
30. Bryant RW, Bailey JM, Schewe T, Rapoport SM (1982) Positional specificity of a reticulocyte lipoxygenase. Conversion of arachidonic acid to 15*S*-hydroperoxy-eicosatetraenoic acid. J Biol Chem 257:6050–6055
31. Borgeat P, Hamberg M, Samuelsson B (1976) Transformation of arachidonic acid and homo-γ-linolenic acid by rabbit polymorphonuclear leukocytes. Monohydroxy acids from novel lipoxygenases. J Biol Chem 251:7816–7820
32. Borgeat P, Samuelsson B (1979) Transformation of arachidonic acid by rabbit polymorphonuclear leukocytes. Formation of a novel dihydroxyeicosatetraenoic acid. J Biol Chem 254:2643–2646
33. Borgeat P, Samuelsson B (1979) Arachidonic acid metabolism in polymorphonuclear leukocytes: unstable intermediate in formation of dihydroxy acids. Proc Natl Acad Sci U S A 76:3213–3217
34. Murphy RC, Hammarstrom S, Samuelsson B (1979) Leukotriene C: a slow-reacting substance from murine mastocytoma cells. Proc Natl Acad Sci U S A 76:4275–4279
35. Drazen J (1998) Clinical pharmacology of leukotriene receptor antagonists and 5-lipoxygenase inhibitors. Am J Respir Crit Care Med 157(S233–S237):S247–S248
36. Serhan CN (2010) Novel lipid mediators and resolution mechanisms in acute inflammation. To resolve or not? Am J Pathol 177:1576–1591
37. Samuelsson B, Dahlen SE, Lindgren JA, Rouzer CA, Serhan CN (1987) Leukotrienes and lipoxins: structures, biosynthesis, and biological effects. Science 237:1171–1176
38. Rouzer CA, Marnett LJ (2003) Mechanism of free radical oxygenation of polyunsaturated fatty acids by cyclooxygenases. Chem Rev 103:2239–2304
39. Bundy GL, Nidy EG, Epps DE, Mizsak SA, Wnuk RJ (1986) Discovery of an arachidonic acid C-8 lipoxygenase in the gorgonian coral *Pseudoplexaura porosa*. J Biol Chem 261:747–751

40. Meijer L, Guerrier P, Maclouf J (1984) Arachidonic acid, 12- and 15-hydroxyeicosatetraenoic acids, eicosapentaenoic acid, and phospholipase A_2 induce starfish oocyte maturation. Dev Biol 106:368–378
41. Brash AR, Porter AT, Maas RL (1985) Investigation of the selectivity of hydrogen abstraction in the non-enzymatic formation of hydroxyeicosatetraenoic acids and leukotrienes by autoxidation. J Biol Chem 260:4210–4216
42. Meijer L, Brash AR, Bryant RW, Ng K, Maclouf J, Sprecher H (1986) Stereospecific induction of starfish oocyte maturation by (8*R*)-hydroxyeicosatetraenoic acid. J Biol Chem 261:17040–17047
43. Hughes MA (1993) Role of 8(R)-hydroxyeicosatetraenoic acid (8(*R*)-HETE) in induction of starfish oocyte maturation. Ph.D. Vanderbilt University, Nashville
44. Corey EJ, Lansbury PT Jr, Yamada Y (1985) Identification of a new eicosanoid from in vitro biosynthetic experiments with *Clavularia viridis*. Implications for the biosynthesis of clavulones. Tetrahedron Lett 26:4171–4174
45. Weinheimer AJ, Spraggins RL (1969) The occurrence of two new prostaglandin derivatives (15-epi-PGA_2 and its acetate, methyl ester) in the gorgonian *Plexaura homomalla*. Tetrahedron Lett 59:5185–5188
46. Schneider WP, Hamilton RD, Rhuland LE (1972) Occurrence of esters of (15*S*)-prostaglandin A_2 and E_2 in coral. J Am Chem Soc 94:2122–2123
47. Corey EJ, Washburn WN, Chen JC (1973) Studies on the prostaglandin A_2 synthetase complex from *Plexaura homomalla*. J Am Chem Soc 95:2054–2055
48. Corey EJ, Ensley, HE, Hamberg M, Samuelsson B (1975) Disparate pathways of prostaglandin biosynthesis in coral and mammalian systems. J Chem Soc Chem Commun 277–278
49. Brash AR, Baertschi SW, Ingram CD, Harris TM (1987) On non-cyclooxygenase prostaglandin synthesis in the sea whip coral *Plexaura homomalla*: an 8(*R*)-lipoxygenase pathway leads to formation of an α-ketol and a racemic prostanoid. J Biol Chem 262:15829–15839
50. Corey EJ, Matsuda SPT, Nagata R, Cleaver MB (1988) Biosynthesis of 8-*R*-HPETE and preclavulone A from arachidonate in several species of Caribbean coral. A widespread route to marine prostanoids. Tetrahedron Lett 29:2555–2558
51. Hamberg M (1988) Biosynthesis of 12-oxo-10,15(Z)-phytodienoic acid: identification of an allene oxide cyclase. Biochem Biophys Res Commun 156:543–550
52. Feussner I, Wasternack C (2002) The lipoxygenase pathway. Annu Rev Plant Biol 53:275–297
53. Koljak R, Järving I, Kurg R, Boeglin WE, Varvas K, Valmsen K, Ustav M, Brash AR, Samel N (2001) The basis of prostaglandin synthesis in coral. Molecular cloning and expression of a cyclooxygenase from the Arctic soft coral *Gersemia fruticosa*. J Biol Chem 276:7033–7040
54. Valmsen K, Järving I, Boeglin WE, Varvas K, Koljak R, Pehk T, Brash AR, Samel N (2001) The origin of 15*R*-prostaglandins in the Caribbean coral *Plexaura homomalla*: molecular cloning and expression of a novel cyclooxygenase. Proc Natl Acad Sci U S A 98:7700–7705
55. Valmsen K, Boeglin WE, Järving J, Schneider C, Varvas K, Brash AR, Samel N (2004) Structural and functional comparison of 15*S*- and 15*R*-specific cyclooxygenases from the coral *Plexaura homomalla*. Eur J Biochem 271:3533–3538
56. Hawkins DJ, Brash AR (1987) Eggs of the sea urchin, *Strongylocentrotus purpuratus*, contain a prominent (11*R*) and (12*R*) lipoxygenase activity. J Biol Chem 262:7629–7634
57. Hawkins DJ, Brash AR (1989) Mechanism of biosynthesis of 11*R*-and 12*R*-hydroxyeicosatetraenoic acids by eggs of the sea urchin *Strongylocentrotus purpuratus*. FEBS Lett 247:9–12
58. Corey EJ, d'Alarcao M, Matsuda SPT, Lansbury PT Jr (1987) Intermediacy of 8-(*R*)-HPETE in the conversion of arachidonic acid to pre-clavulone A by *Clavularia viridis*. Implications for the biosynthesis of marine prostanoids. J Am Chem Soc 109:289–290
59. Hammarström S, Hamberg M, Samuelsson B, Duell EA, Stawiski M, Voorhees JJ (1975) Increased concentrations of nonesterified arachidonic acid, 12 L-hydroxy-5,8,10,14-eicosatetraenoic acid, prostaglandin E_2, and prostaglandin $F_{2\alpha}$ in epidermis of psoriasis. Proc Natl Acad Sci U S A 72:5130–5134

60. Woollard PM (1986) Stereochemical difference between 12-hydroxyeicosatetraenoic acid in platelets and psoriatic lesions. Biochem Biophys Res Commun 136:169–176
61. Capdevila J, Yadagiri P, Manna S, Falck JR (1986) Absolute configuration of the hydroxyeicosatetraenoic acids (HETEs) formed during catalytic oxygenation of arachidonic acid by microsomal cytochrome P-450. Biochem Biophys Res Commun 141:1007–1011
62. Brash AR, Boeglin WE, Chang MS, Shieh B-H (1996) Purification and molecular cloning of an 8*R*-lipoxygenase from the coral *Plexaura homomalla* reveal the related primary structures of *R*- and *S*-lipoxygenases. J Biol Chem 271:20949–20957
63. Coffa G, Imber AN, Maguire BC, Laxmikanthan G, Schneider C, Gaffney BJ, Brash AR (2005) On the relationships of substrate orientation, hydrogen abstraction and product stereochemistry in single and double dioxygenations by soybean lipoxygenase-1 and its Ala542Gly mutant. J Biol Chem 280:38756–38766
64. Brash AR, Schneider C, Hamberg M (2012) Applications of stereospecifically-labeled fatty acids in oxygenase and desaturase biochemistry. Lipids 47:101–116
65. Coffa G, Brash AR (2004) A single active site residue directs oxygenation stereospecificity in lipoxygenases: stereocontrol is linked to the position of oxygenation. Proc Natl Acad Sci U S A 101:15579–15584
66. Coffa G, Schneider C, Brash AR (2005) A comprehensive model of positional and stereo control in lipoxygenases. Biochem Biophys Res Commun 338:87–92
67. Schneider C, Pratt DA, Porter NA, Brash AR (2007) Control of oxygenation in lipoxygenase and cyclooxygenase catalysis. Chem Biol 14:473–488
68. Neau DB, Bender G, Boeglin WE, Bartlett SG, Brash AR, Newcomer ME (2014) Crystal structure of a lipoxygenase in complex with substrate: the arachidonic acid binding site of 8*R*-lipoxygenase. J Biol Chem 289:31905–31913
69. Neau DB, Gilbert NC, Bartlett SG, Boeglin W, Brash AR, Newcomer ME (2009) The 1.85 Å structure of an 8*R*-lipoxygenase suggests a general model for lipoxygenase product specificity. Biochemistry 48:7906–7915
70. Jansen C, Hofheinz K, Vogel R, Roffeis J, Anton M, Reddanna P, Kuhn H, Walther M (2011) Stereocontrol of arachidonic acid oxygenation by vertebrate lipoxygenases: newly cloned zebrafish lipoxygenase 1 does not follow the Ala-versus-Gly concept. J Biol Chem 286:37804–37812
71. Zheng Y, Boeglin WE, Schneider C, Brash AR (2008) A 49-kDa mini-lipoxygenase from *Anabaena* sp. PCC 7120 retains catalytically complete functionality. J Biol Chem 283:5138–5147
72. Andreou AZ, Vanko M, Bezakova L, Feussner I (2008) Properties of a mini 9*R*-lipoxygenase from *Nostoc* sp. PCC 7120 and its mutant forms. Phytochemistry 69:1832–1837
73. Palmieri-Thiers C, Canaan S, Brunini V, Lorenzi V, Tomi F, Desseyn JL, Garscha U, Oliw EH, Berti L, Maury J (2009) A lipoxygenase with dual positional specificity is expressed in olives (*Olea europaea* L.) during ripening. Biochim Biophys Acta 1791:339–346
74. Krieg P, Kinzig A, Heidt M, Marks F, Fürstenberger G (1998) cDNA cloning of a 8-lipoxygenase and a novel epidermis-type lipoxygenase from phorbol ester-treated mouse skin. Biochim Biophys Acta 1391:7–12
75. Kinzig A, Heidt M, Fürstenberger G, Marks F, Krieg P (1999) cDNA cloning, genomic structure, and chromosomal localization of a novel murine epidermis-type lipoxygenase. Genomics 58:158–164
76. Heidt M, Fürstenberger G, Vogel S, Marks F, Krieg P (2000) Diversity of mouse lipoxygenases: identification of a subfamily of epidermal isozymes exhibiting a differentiation-dependent mRNA expression pattern. Lipids 35:701–707
77. Krieg P, Marks F, Fürstenberger G (2001) A gene cluster encoding human epidermis-type lipoxygenases at chromosome 17p13.1: cloning, physical mapping, and expression. Genomics 73:323–330
78. Jisaka M, Kim RB, Boeglin WE, Nanney LB, Brash AR (1997) Molecular cloning and functional expression of a phorbol ester-inducible 8*S*-lipoxygenase from mouse skin. J Biol Chem 272:24410–24416

79. Brash AR, Jisaka M, Boeglin WE, Chang MS (1997) Molecular cloning of a second human 15*S*-lipoxygenase and its murine homologue, an 8*S*-lipoxygenase: their relationship to other mammalian lipoxygenases. In: Pace-Asciak CR, Nigam S (eds) Lipoxygenases and their products: biological functions. Plenum Press, New York, pp 29–36
80. Boeglin WE, Kim RB, Brash AR (1998) A 12*R*-lipoxygenase in human skin: mechanistic evidence, molecular cloning and expression. Proc Natl Acad Sci U S A 95:6744–6749
81. Sun D, McDonnell M, Chen X-S, Lakkis MM, Li H, Isaacs SN, Elsea SH, Patel PI, Funk CD (1998) Human 12(*R*)-lipoxygenase and the mouse ortholog. Molecular cloning, expression, and gene chromosomal assignment. J Biol Chem 273:33540–33547
82. Krieg P, Siebert M, Kinzig A, Bettenhausen R, Marks F, Fürstenberger G (1999) Murine 12(*R*)-lipoxygenase: functional expression, genomic structure and chromosomal localization. FEBS Lett 446:142–148
83. Jobard F, Lefèvre C, Karaduman A, Blanchet-Bardon C, Emre S, Weissenbach J, Özgüc M, Lathrop M, Prud'homme JF, Fischer J (2002) Lipoxygenase-3 (*ALOXE3*) and 12(*R*)-lipoxygenase (*ALOX12B*) are mutated in non-bullous congenital ichthyosiform erythroderma (NCIE) linked to chromosome 17p13.1. Hum Mol Genet 11:107–113
84. Yu Z, Schneider C, Boeglin WE, Marnett LJ, Brash AR (2003) The lipoxygenase gene *ALOXE3* implicated in skin differentiation encodes a hydroperoxide isomerase. Proc Natl Acad Sci U S A 100:9162–9167
85. Yu Z, Schneider C, Boeglin WE, Brash AR (2006) Human and mouse eLOX3 have distinct substrate specificities: implications for their linkage with lipoxygenases in skin. Arch Biochem Biophys 455:188–196
86. Moran JL, Qiu H, Turbe-Doan A, Yun Y, Boeglin WE, Brash AR, Beier DR (2007) A mouse mutation in the 12(*R*)-lipoxygenase, Alox12b, disrupts formation of the epidermal permeability barrier. J Invest Dermatol 127:1893–1897
87. Epp N, Furstenberger G, Muller K, de Juanes S, Leitges M, Hausser I, Thieme F, Liebisch G, Schmitz G, Krieg P (2007) 12*R*-lipoxygenase deficiency disrupts epidermal barrier function. J Cell Biol 177:173–182
88. Burr GO, Burr MM (1929) A new deficiency disease produced by the rigid exclusion of fat from the diet. J Biol Chem 82:345–367
89. Hansen AE, Haggard ME, Boelsche AN, Adam DJ, Wiese HF (1958) Essential fatty acids in infant nutrition. III. Clinical manifestations of linoleic acid deficiency. J Nutr 66:565–576
90. Houtsmuller UMT, van der Beek A (1981) Effects of topical application of fatty acids. Prog Lipid Res 20:219–224
91. Hansen HS (1986) The essential nature of linoleic acid in mammals. Trends Biochem Sci 11:263–265
92. Uchida Y, Holleran WM (2008) Omega-*O*-acylceramide, a lipid essential for mammalian survival. J Dermatol Sci 51:77–87
93. Zheng Y, Yin H, Boeglin WE, Elias PM, Crumrine D, Beier DR, Brash AR (2011) Lipoxygenases mediate the effect of essential fatty acid in skin barrier formation: a proposed role in releasing omega-hydroxyceramide for construction of the corneocyte lipid envelope. J Biol Chem 286:24046–24056
94. Elias PM, Gruber R, Crumrine D, Menon G, Williams ML, Wakefield JS, Holleran WM, Uchida Y (2014) Formation and functions of the corneocyte lipid envelope (CLE). Biochim Biophys Acta 1841:314–318
95. Muñoz-Garcia A, Thomas CP, Keeney DS, Zheng Y, Brash AR (2014) The importance of the lipoxygenase-hepoxilin pathway in the mammalian epidermal barrier. Biochim Biophys Acta 1841:401–408
96. Meguro S, Arai Y, Masukawa Y, Uie K, Tokimitsu I (2000) Relationship between covalently bound ceramides and transepidermal water loss (TEWL). Arch Dermatol Res 292:463–468

Chapter 6
Leukotriene B_4 Receptors

Tomoaki Koga and Takehiko Yokomizo

Abstract Leukotriene B_4 (LTB_4) is a classical pro-inflammatory lipid mediator that activates and recruits neutrophils into inflammatory areas. LTB_4 is an arachidonic acid-derived metabolite produced by 5-lipoxygenase and LTA_4 hydrolase. To date, two leukotriene B_4 receptors, the high-affinity receptor BLT1 and the low-affinity receptor BLT2, have been cloned, and a systemic knockout mouse for each receptor has been generated. BLT1 is mainly expressed in leukocytes, but BLT2 is expressed in epithelial cells. Based on many knockout mouse studies, BLT1 is now known to play important roles in acute and chronic inflammation and immune diseases. On the other hand, BLT2 protects against colitis, accelerates epidermal wound healing, and promotes cancer progression. Thus, BLT1 and BLT2 have totally different features with respect to their expression patterns and their physiological and pathological roles. In this review, we summarize the fundamental characteristics of BLT1 and BLT2, as well as recent advances in our understanding of the biosynthesis and degradation of LTB_4 and the LTB_4 receptors.

Keywords Leukotriene B_4 • BLT1 • Leukocyte • BLT2 • 12-HHT • Epithelial cells • Barrier function

6.1 Leukotriene B_4: Biosynthesis, Degradation, and Physiological Roles

Leukotriene B_4 (LTB_4) is a classical lipid mediator that activates and attracts neutrophils [1, 2]. LTB_4 is produced mainly in leukocytes, including neutrophils and macrophages, and it contains three conjugated double bonds (a triene), explaining the origin of the name "leukotriene." LTB_4 is an arachidonic acid metabolite that is produced by two enzymes, 5-lipoxygenase (5-LO) and leukotriene A_4 hydrolase (LTA_4H) [3]. Arachidonic acid is converted into 5-hydroperoxyeicosatetraenoic

T. Koga • T. Yokomizo (✉)
Department of Biochemistry, Juntendo University Graduate School of Medicine,
2-1-1 Hongo, Bunkyo-ku, Tokyo 113-8421, Japan
e-mail: yokomizo-tky@umin.ac.jp

T. Yokomizo, M. Murakami (eds.), *Bioactive Lipid Mediators*,
DOI 10.1007/978-4-431-55669-5_6

acids (5-HpETE) and subsequently into LTA_4. 5-LO mediates these reactions in functional cooperation with 5-lipoxygenase-activating protein (FLAP) [3]. When cells are exposed to certain stimuli (*e.g.*, Ca^{2+} influx), cytoplasmic 5-LO is activated and translocates to the nuclear membrane. FLAP binds to 5-LO to form an active complex on the nuclear membrane, and this complex catabolizes arachidonic acid into 5-HpETE and subsequently to LTA_4. LTA_4 is either converted to LTB_4 by LTA_4H or conjugated to glutathione to generate leukotriene C_4 (LTC_4) by LTC_4 synthase (LTC_4S). LTB_4 is exported from the cytoplasm to the extracellular space in an energy-dependent manner. In humans, LTB_4 export is mediated by the ATP-binding cassette (ABC) family transporter ABCC4/MRP4 [4]. The biosynthesis and transport pathway of LTB_4 is summarized in Fig. 6.1.

Degradation of LTB_4 is catalyzed by ω-oxidation and subsequent β-oxidation, or by the 12-hydroxydehydrogenase/15-oxo-prostaglandin-13-reductase (12HDH/15oPGR) pathway [3]. In the ω-oxidation pathway, LTB_4 is metabolized into 20-hydroxy-LTB_4 (20-OH-LTB_4), and subsequently into 20-carboxy-LTB_4 (20-COOH-LTB_4) [5]. The cytochrome P450 family protein CYP4F3A (CYP4F18 in mouse) is a human neutrophil LTB_4-ω-hydroxylase that is induced by retinoic acid and phorbol ester stimulation. This enzyme efficiently oxidizes the methyl moiety of LTB_4 at the ω-terminus. In neutrophils, 20-OH-LTB_4 is further oxidized by CYP4F3A into 20-COOH-LTB_4 through the 20-oxo-LTB_4 intermediate. In other tissues (*e.g.*, liver), an alternative pathway produces 20-COOH-LTB_4 from 20-OH-LTB_4; in the liver, the enzymes involved in LTB_4 oxidation are alcohol dehydrogenase (ADH) and aldehyde dehydrogenase (ALDH). Although 20-OH-LTB_4 binds fairly well to leukotriene B_4 receptor 1 (BLT1), 20-COOH-LTB_4 is not an active ligand of BLT1. A recent report showed that patients with Sjögren–Larsson syndrome, who have mutations in the fatty aldehyde dehydrogenase (*FALDH*) gene, exhibited high urinary levels of LTB_4 and 20-OH-LTB_4, indicating the importance of the ADH/ALDH pathway in LTB_4 metabolism. In some tissues, LTB_4 can be converted into 12-oxo-LTB_4, a reaction catalyzed by the enzyme 12HDH/15oPGR [6]. 12HDH/15oPGR recognizes the structural motif [R-CH(OH)-(*trans*)-CH=CH-R′] of LTB_4, and oxidizes the 12(*R*) hydroxyl group into the 12-oxo moiety. Biosynthesized 12-oxo-LTB_4 is unable to activate BLT1 signaling. A recent report showed that LTB_4 is extensively metabolized by the 12HDH/15oPGR pathway in human keratinocytes. Because CYP4F activity is very low in the human keratinocytes, ω-oxidized LTB_4 metabolite is rarely observed in these cells. The degradation pathway of LTB_4 is summarized in Fig. 6.2.

6.2 Leukotriene B_4 Receptor 1: BLT1

LTB_4 is recognized by two G protein-coupled receptors, BLT1 and BLT2, as shown in Fig. 6.3 [7, 8]. BLT1 is expressed in various cell types, including neutrophils, macrophages, osteoclasts, dendritic cells, and differentiated T cells [9–14]. Once BLT1 is activated by LTB_4, it cooperates with Gi and Gq proteins, inducing

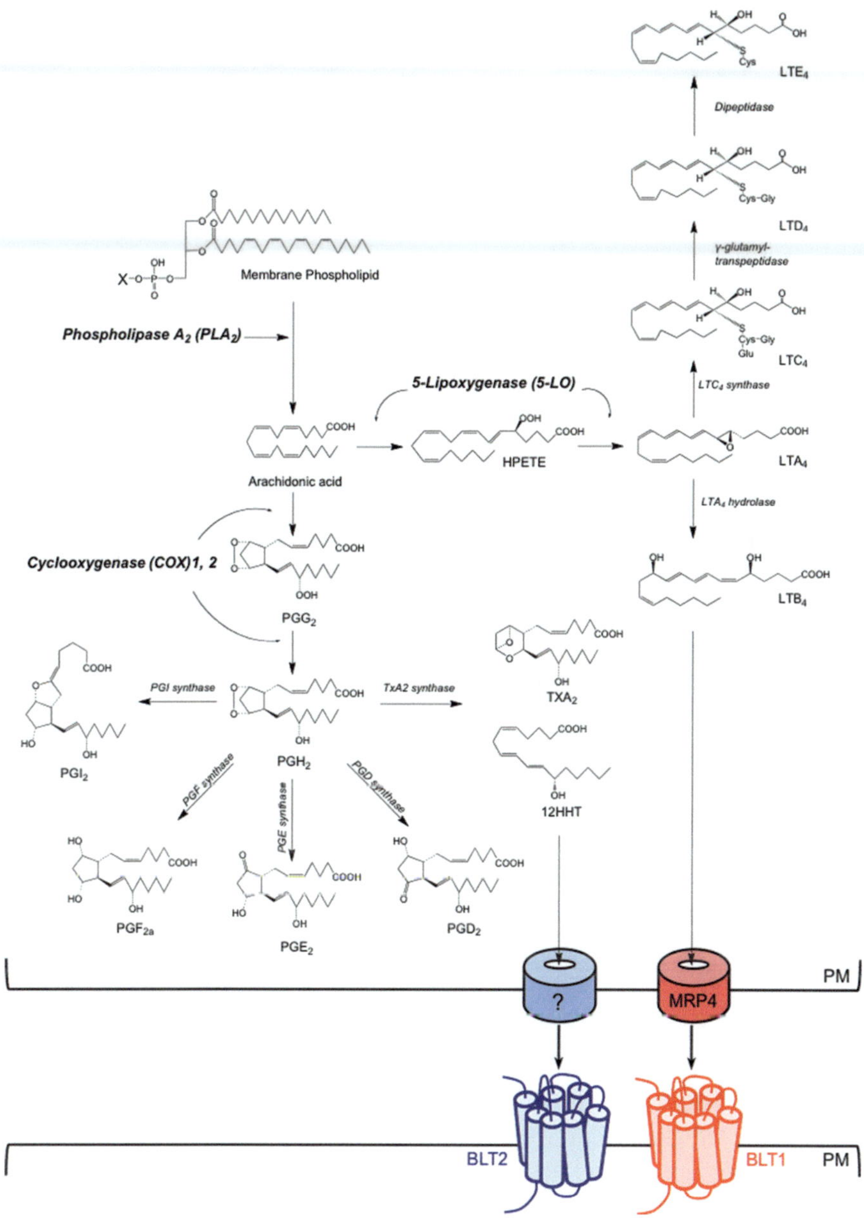

Fig. 6.1 Biosynthesis and export pathway of LTB_4 and 12-HHT. LTB_4 is synthesized from arachidonic acid by 5-LO, FLAP, and LTA_4H. LTB_4 is exported by MRP4, at least in humans, and recognized by high-affinity receptor BLT1. On the other hand, production of 12-HHT is catalyzed by COX-1, COX-2, and TxA_2S. 12-HHT is exported and recognized by BLT2. *PM* plasma membrane

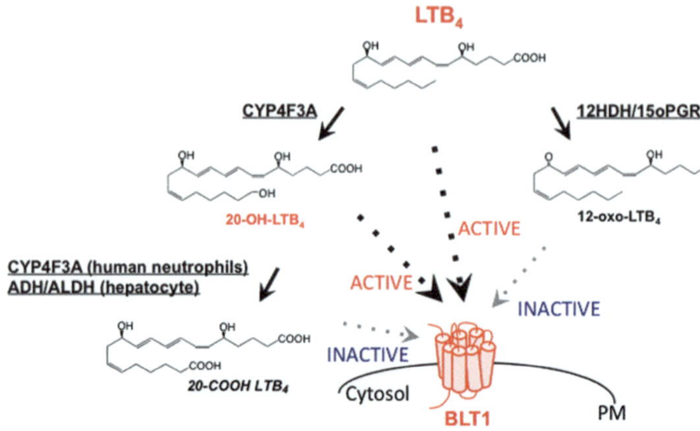

Fig. 6.2 Degradation pathway of LTB_4. Two pathways inactivate LTB_4. The ω-oxidation pathway is catalyzed by CYP4F3A. Its product, 20-hydroxy-LTB_4, still has the ability to activate BLT1. To inactivate 20-hydroxy-LTB_4, 20-carboxylation by either CYP4F3A (in human neutrophils) or ADH/ALDH (in hepatocytes) is required. Alternatively, the 12HDH/15oPGR pathway is also important for inactivation of LTB_4. Its metabolite, 12-oxo-LTB_4, is inactive. *PM* plasma membrane

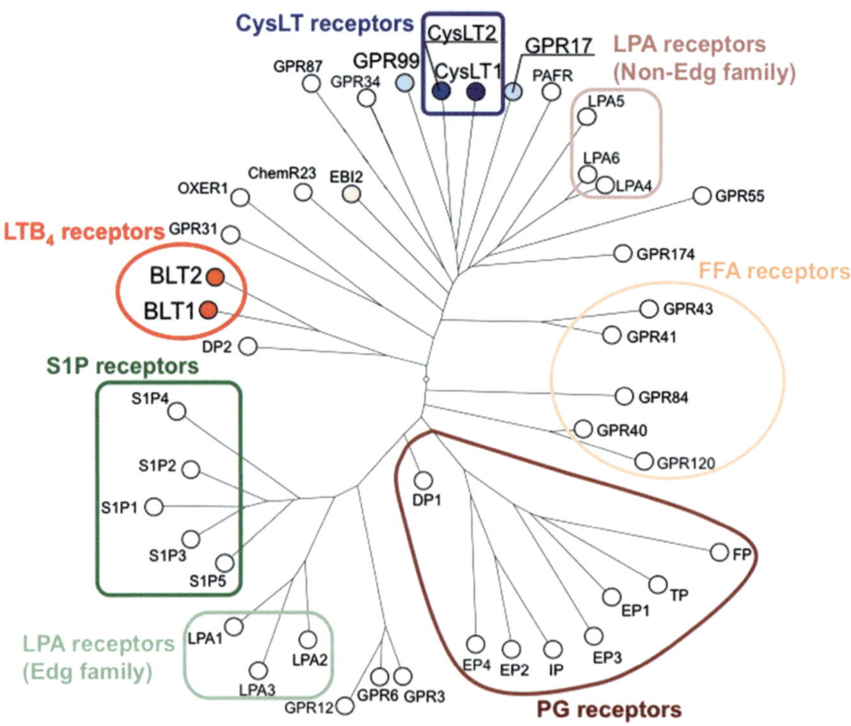

Fig. 6.3 Phylogenic tree of G protein-coupled receptors (GPCRs) for lipid mediators. Among all GPCRs, BLT1 and BLT2 receptors are the most closely related to each other

phospholipase Cβ-dependent intracellular Ca^{2+} mobilization and activating the phosphatidyl inositol-3-OH kinase (PI3K)-Akt pathway, mitogen-activated protein kinase (MAPK)/extracellular signal-regulated kinase (ERK) kinase (MEK) 1/2-ERK1/2 pathway, *etc*. Activation of downstream signaling causes chemotaxis, phagocytosis, and gene transcription [15].

Because BLT1 was originally identified as a receptor for LTB_4 expressed in neutrophils, it is thought to have important roles in inflammatory diseases. Consistent with this, neutrophil activation, called "swarming," in a mouse model of *Pseudomonas aeruginosa* infection is attenuated in BLT1-deficient mice [9]. The LTB_4–BLT1 axis is also required for neutrophils to secrete interleukin-1 (IL-1), which initiates joint inflammation. IL-1-producing neutrophils activate synovial cells and induce chemokines to recruit more inflammatory cells into joints [16]. In that arthritis model, LTB_4 release from neutrophils requires C5a receptor activation by C5a, indicating that the LTB_4–BLT1 axis in neutrophils plays an important role in the vicious cycle of arthritis [17]. A recent report demonstrated that BLT1 facilitates neutrophil infiltration in a mouse model of psoriasis *via* CXCR2-induced LTB_4 production, suggesting the existence of crosstalk between chemokine and lipid mediator [18]. As in neutrophils, BLT1 is also expressed in eosinophils, and it prevents nematode infection [19]. In macrophages, BLT1 plays a crucial role in phagocytosis *via* functional cooperation with the FcγR signaling pathway [20]. LTB_4–BLT1 signaling also enhances the expression of MyD88, an important adapter molecule for toll-like receptor (TLR) signaling molecules such as TLR2, TLR4, TLR5, TLR7/8, and TLR9 *via* suppression of suppressor of cytokine signaling (SOCS) 1 expression in macrophages [21–23].

Although BLT1 has important roles in inflammatory cells, recent reports showed that BLT1 is also expressed in immune cells, such as dendritic cells and differentiated T lymphocytes, including Th1 and Th2 [12–14]. BLT1 in dendritic cells upregulates CCR7 expression, which enhances the migration of dendritic cells into lymph nodes, and Th1-dependent contact hypersensitivity is attenuated by BLT1 deficiency [11, 24]. In addition, BLT1 knockout (KO) mice exhibit a less severe phenotype in an ovalbumin-induced Th2-dependent murine asthma model [25]. BLT1 is also crucial for the development of Th17-dependent experimental autoimmune encephalomyelitis (EAE), a murine model of multiple sclerosis [26]. Furthermore, a recent report showed that BLT1 is also expressed in $CD4^+$ $CD25^+$ $Foxp3^+$ Tregs [27]. In a lipopolysaccharide (LPS)-induced model of acute lung injury (ALI), the LTB_4–BLT1 axis is critical to promote the resolution from ALI through the recruitment of Tregs into the inflammatory area, suggesting that BLT1 is important in various immune responses.

In addition to acute inflammation, BLT1 is also involved in chronic inflammation. BLT1 is expressed in vascular smooth muscle cells, and BLT1 deficiency attenuates atherogenesis in the *Apoe* KO murine model of atherosclerosis [28, 29]. In addition, a recent report showed that BLT1 provides a potent signal that accelerates the differentiation from preadipocytes to adipocytes *in vitro* [30]. Moreover, BLT1 deficiency protects against systemic insulin resistance in high-fat-induced obesity, which arises from monocyte trafficking to adipose tissue [10]. Stimulation of osteoclasts with LTB_4 increases osteoclastic activity, and BLT1 deficiency attenuates osteoporosis after ovariectomy in mice [31]. In summary, although BLT1 was initially identified as a LTB_4 receptor expressed in neutrophils, recent reports reveal that it is expressed in various cell types, including neutrophils, eosinophils, macrophages, dendritic cells, differentiated T cells, smooth muscle cells, osteoclasts, and adipocytes, and that it plays important roles in acute and chronic inflammation and in immune diseases.

6.3 Leukotriene B_4 Receptor 2: BLT2

BLT2 was identified as a low-affinity receptor for LTB_4; the gene that encodes this protein is localized in the proximal promoter region of *BLT1* on human chromosome 14 [15]. Similar to BLT1, BLT2 is also coupled to two classes of G proteins (Gi and Gq), through which it activates the MAPK (ERK and p38), NADP oxidase (NOX)-reactive oxygen species (ROS), and NF-κB signaling pathways [32]. We recently identified 12(*S*)-hydroxyheptadeca-5*Z*,8*E*,10*E*-trienoic acid (12-HHT) as an endogenous ligand for BLT2. 12-HHT is a fatty acid with 17 carbons, and its physiological role has not been elucidated. 12-HHT is also a metabolite of arachidonic acid, which is biosynthesized during blood coagulation primarily by two enzymes, cyclooxygenase (COX) and thromboxane A_2 synthase (TxA_2S). In addition to this pathway, we recently identified a nonenzymatic production of 12-HHT from PGH_2 by using pharmacological and genetic inhibition of TxA_2S *in vitro* and *in vivo* [33]. 12-HHT is present in intestine, and BLT2 is highly expressed in intestinal epithelial cells [34, 35]. BLT2 plays protective roles against dextran sulfate-induced colitis by enhancing barrier function in intestinal epithelial cells [36]. BLT2 is also expressed in human and mouse skin. BLT2 deficiency delays epidermal wound healing by slowing keratinocyte migration without affecting fibroblast differentiation or keratinocyte proliferation [37]. 12-HHT is mainly produced by platelets in wound areas, and it activates BLT2 expressed in epidermal keratinocytes. Through the NF-κB pathway, activated keratinocytes induce tumor necrosis factor-α, which upregulates matrix metalloproteinase (MMP) 9 in an autocrine or paracrine manner. Although BLT2 is primarily expressed in epithelial cells, a recent report demonstrated that BLT2 is also expressed in Th2 cells; furthermore, BLT2-deficient mice exhibit enhanced eosinophil accumulation and IL-13 and IL-4 production in a Th2-dependent murine asthma model [38]. Furthermore, several studies showed that BLT2 is involved in progression of various cancers, including bladder, ovarian, prostate, and breast cancers [32]. BLT2 is upregulated in human bladder cancer, ovarian carcinoma, breast cancer, renal cell carcinoma, lung carcinoma, and other tumors [39]. In addition, cell transformation induced by constitutively active mutant Ras might be BLT2 dependent [39]. BLT2-mediated cancer progression is associated with the NOX-ROS-NF-κB signaling pathway, which promotes cancer cell survival and invasion through induction of androgen receptor and MMP9, respectively.

6.4 Perspectives and Future Directions

The biosynthesis and degradation pathways of LTB_4, with the exception of the export mechanism of LTB_4 to the extracellular space, have been clarified in detail. Because transporters are promising therapeutic targets, detailed studies of MRP4 by MRP4-deficient mice are required. On the other hand, the roles of leukotriene B_4 receptors have been extensively investigated in both BLT1 and BLT2 systemically deficient mice. However, because many studies found that BLT1 and BLT2 are expressed in various tissues and cell types, the cell-specific role of those receptors should be elucidated using conditional-knockout mice. Future studies should focus on determining which cells produce LTB_4, which cells express BLT1 *in vivo*, and the clinical relevance of the LTB_4–BLT1 and 12-HHT-BLT2 pathways.

References

1. Samuelsson B (1983) Leukotrienes: mediators of immediate hypersensitivity reactions and inflammation. Science 220:568–575
2. Ford-Hutchinson AW (1981) Leukotriene B4 and neutrophil function: a review. J R Soc Med 74:831–833
3. Murphy RC, Gijon MA (2007) Biosynthesis and metabolism of leukotrienes. Biochem J 405:379–395
4. Rius M, Hummel-Eisenbeiss J, Keppler D (2008) ATP-dependent transport of leukotrienes B4 and C4 by the multidrug resistance protein ABCC4 (MRP4). J Pharmacol Exp Ther 324:86–94
5. Soberman RJ, Sutyak JP, Okita RT, Wendelborn DF, Roberts LJ 2nd, Austen KF (1988) The identification and formation of 20-aldehyde leukotriene B4. J Biol Chem 263:7996–8002
6. Yokomizo T, Izumi T, Takahashi T, Kasama T, Kobayashi Y, Sato F, Taketani Y, Shimizu T (1993) Enzymatic inactivation of leukotriene B4 by a novel enzyme found in the porcine kidney. Purification and properties of leukotriene B4 12-hydroxydehydrogenase. J Biol Chem 268:18128–18135
7. Yokomizo T, Izumi T, Chang K, Takuwa Y, Shimizu T (1997) A G-protein-coupled receptor for leukotriene B4 that mediates chemotaxis. Nature 387:620–624
8. Yokomizo T, Kato K, Terawaki K, Izumi T, Shimizu T (2000) A second leukotriene B(4) receptor, BLT2. A new therapeutic target in inflammation and immunological disorders. J Exp Med 192:421–432
9. Lammermann T, Afonso PV, Angermann BR, Wang JM, Kastenmuller W, Parent CA, Germain RN (2013) Neutrophil swarms require LTB4 and integrins at sites of cell death in vivo. Nature 498:371–375
10. Spite M, Hellmann J, Tang Y, Mathis SP, Kosuri M, Bhatnagar A, Jala VR, Haribabu B (2011) Deficiency of the leukotriene B4 receptor, BLT-1, protects against systemic insulin resistance in diet-induced obesity. J Immunol 187:1942–1949
11. Toda A, Terawaki K, Yamazaki S, Saeki K, Shimizu T, Yokomizo T (2010) Attenuated Th1 induction by dendritic cells from mice deficient in the leukotriene B4 receptor 1. Biochimie 92:682–691
12. Tager AM, Bromley SK, Medoff BD, Islam SA, Bercury SD, Friedrich EB, Carafone AD, Gerszten RE, Luster AD (2003) Leukotriene B4 receptor BLT1 mediates early effector T cell recruitment. Nat Immunol 4:982–990

13. Goodarzi K, Goodarzi M, Tager AM, Luster AD, von Andrian UH (2003) Leukotriene B4 and BLT1 control cytotoxic effector T cell recruitment to inflamed tissues. Nat Immunol 4:965–973
14. Tager AM, Luster AD (2003) BLT1 and BLT2: the leukotriene B(4) receptors. Prostaglandins Leukot Essent Fatty Acids 69:123–134
15. Back M, Dahlen S-E, Drazen JM, Evans JF, Serhan CN, Shimizu T, Yokomizo T, Rovati GE (2011) International Union of Basic and Clinical Pharmacology. LXXXIV: leukotriene receptor nomenclature, distribution, and pathophysiological functions. Pharmacol Rev 63:539–584
16. Chou RC, Kim ND, Sadik CD, Seung E, Lan Y, Byrne MH, Haribabu B, Iwakura Y, Luster AD (2010) Lipid-cytokine-chemokine cascade drives neutrophil recruitment in a murine model of inflammatory arthritis. Immunity 33:266–278
17. Sadik CD, Kim ND, Iwakura Y, Luster AD (2012) Neutrophils orchestrate their own recruitment in murine arthritis through C5aR and Fc-gamma-R signaling. Proc Natl Acad Sci U S A 109:E3177–E3185
18. Sumida H, Yanagida K, Kita Y, Abe J, Matsushima K, Nakamura M, Ishii S, Sato S, Shimizu T (2014) Interplay between CXCR2 and BLT1 facilitates neutrophil infiltration and resultant keratinocyte activation in a murine model of imiquimod-induced psoriasis. J Immunol 192:4361–4369
19. Patnode ML, Bando JK, Krummel MF, Locksley RM, Rosen SD (2014) Leukotriene B4 amplifies eosinophil accumulation in response to nematodes. J Exp Med 211:1281–1288
20. Okamoto F, Saeki K, Sumimoto H, Yamasaki S, Yokomizo T (2010) Leukotriene B4 augments and restores Fc gamma-Rs-dependent phagocytosis in macrophages. J Biol Chem 285:41113–41121
21. Serezani CH, Lewis C, Jancar S, Peters-Golden M (2011) Leukotriene B4 amplifies NF-kappaB activation in mouse macrophages by reducing SOCS1 inhibition of MyD88 expression. J Clin Invest 121:671–682
22. Wang Z, Filgueiras LR, Wang S, Serezani APM, Peters-Golden M, Jancar S, Serezani CH (2014) Leukotriene B4 enhances the generation of proinflammatory microRNAs to promote MyD88-dependent macrophage activation. J Immunol 192:2349–2356
23. O'Neill LAJ, Bowie AG (2007) The family of five: TIR-domain-containing adaptors in Toll-like receptor signalling. Nat Rev Immunol 7:353–364
24. Del Prete A, Shao W-H, Mitola S, Santoro G, Sozzani S, Haribabu B (2007) Regulation of dendritic cell migration and adaptive immune response by leukotriene B4 receptors: a role for LTB4 in up-regulation of CCR7 expression and function. Blood 109:626–631
25. Terawaki K, Yokomizo T, Nagase T, Toda A, Taniguchi M, Hashizume K, Yagi T, Shimizu T (2005) Absence of leukotriene B4 receptor 1 confers resistance to airway hyperresponsiveness and Th2-type immune responses. J Immunol 175:4217–4225
26. Kihara Y, Yokomizo T, Kunita A, Morishita Y, Fukayama M, Ishii S, Shimizu T (2010) The leukotriene B4 receptor, BLT1, is required for the induction of experimental autoimmune encephalomyelitis. Biochem Biophys Res Commun 394:673–678
27. Wang L, Zhao L, Lv J, Yin Q, Liang X, Chu Y, He R (2012) BLT1-dependent alveolar recruitment of CD4(+)CD25(+) Foxp3(+) regulatory T cells is important for resolution of acute lung injury. Am J Respir Crit Care Med 186:989–998
28. Heller EA, Liu E, Tager AM, Sinha S, Roberts JD, Koehn SL, Libby P, Aikawa ER, Chen JQ, Huang P, Freeman MW, Moore KJ, Luster AD, Gerszten RE (2005) Inhibition of atherogenesis in BLT1-deficient mice reveals a role for LTB4 and BLT1 in smooth muscle cell recruitment. Circulation 112:578–586
29. Back M, Bu D-X, Branstrom R, Sheikine Y, Yan Z-Q, Hansson GK (2005) Leukotriene B4 signaling through NF-kappaB-dependent BLT1 receptors on vascular smooth muscle cells in atherosclerosis and intimal hyperplasia. Proc Natl Acad Sci U S A 102:17501–17506
30. Hirata K, Wada K, Murata Y, Nakajima A, Yamashiro T, Kamisaki Y (2013) Critical role of leukotriene B4 receptor signaling in mouse 3 T3-L1 preadipocyte differentiation. Lipids Health Dis 12:122

31. Hikiji H, Ishii S, Yokomizo T, Takato T, Shimizu T (2009) A distinctive role of the leukotriene B4 receptor BLT1 in osteoclastic activity during bone loss. Proc Natl Acad Sci U S A 106:21294–21299
32. Cho N-K, Joo Y-C, Wei JD, Park JI, Kim J-H (2013) BLT2 is a pro-tumorigenic mediator during cancer progression and a therapeutic target for anti-cancer drug development. Am J Cancer Res 3:347–355
33. Matsunobu T, Okuno T, Yokoyama C, Yokomizo T (2013) Thromboxane A synthase-independent production of 12-hydroxyheptadecatrienoic acid, a BLT2 ligand. J Lipid Res 54:2979–2987
34. Iizuka Y, Yokomizo T, Terawaki K, Komine M, Tamaki K, Shimizu T (2005) Characterization of a mouse second leukotriene B4 receptor, mBLT2: BLT2-dependent ERK activation and cell migration of primary mouse keratinocytes. J Biol Chem 280:24816–24823
35. Okuno T, Iizuka Y, Okazaki H, Yokomizo T, Taguchi R, Shimizu T (2008) 12(*S*)-Hydroxyheptadeca-5*Z*, 8*E*, 10*E*-trienoic acid is a natural ligand for leukotriene B4 receptor 2. J Exp Med 205:759–766
36. Iizuka Y, Okuno T, Saeki K, Uozaki H, Okada S, Misaka T, Sato T, Toh H, Fukayama M, Takeda N, Kita Y, Shimizu T, Nakamura M, Yokomizo T (2010) Protective role of the leukotriene B4 receptor BLT2 in murine inflammatory colitis. FASEB J 24:4678–4690
37. Liu M, Saeki K, Matsunobu T, Okuno T, Koga T, Sugimoto Y, Yokoyama C, Nakamizo S, Kabashima K, Narumiya S, Shimizu T, Yokomizo T (2014) 12-Hydroxyheptadecatrienoic acid promotes epidermal wound healing by accelerating keratinocyte migration via the BLT2 receptor. J Exp Med 211:1063–1078
38. Matsunaga Y, Fukuyama S, Okuno T, Sasaki F, Matsunobu T, Asai Y, Matsumoto K, Saeki K, Oike M, Sadamura Y, Machida K, Nakanishi Y, Kubo M, Yokomizo T, Inoue H (2013) Leukotriene B4 receptor BLT2 negatively regulates allergic airway eosinophilia. FASEB J 27:3306–3314
39. Yoo M-H, Song H, Woo C-H, Kim H, Kim J-H (2004) Role of the BLT2, a leukotriene B4 receptor, in Ras transformation. Oncogene 23:9259–9268

Chapter 7
Platelet-Activating Factor (PAF) in Infectious Diseases

Satoshi Ishii

Abstract Platelet-activating factor (PAF) is a phospholipid that was originally discovered as an IgE-sensitized rabbit basophil-derived substance responsible for platelet aggregation. The chemical structure was determined to be 1-*O*-alkyl-2-acetyl-sn-glycero-3-phosphocholine. The cellular effects of PAF are mediated by a specific G protein-coupled receptor (PAFR). When PAF is administered to laboratory animals, versatile pharmacological responses including platelet aggregation occur. In some studies, the pathophysiological roles of PAF have been deduced from observations that PAF production and PAFR expression are increased in the affected tissues or organs. Furthermore, mimicking of pathophysiological conditions by PAF and suppressive effects of PAF antagonists also suggest significant roles for PAF in some diseases. Under such experimental circumstances, PAFR-knockout (KO) mice were first reported in 1998. Since then, numerous studies using PAFR-KO mice have provided insight into multiple pathophysiological roles of PAF. Some of the studies used experimental infection models with various pathogens. This chapter reviews the current understanding of the PAF/PAFR axis and its protective or deleterious roles in infectious diseases.

Keywords *Klebsiella pneumoniae* • *Pseudomonas aeruginosa* • *Streptococcus pneumoniae* • *Haemophilus influenzae* • *Mycobacterium tuberculosis* • *Aggregatibacter actinomycetemcomitans* • Dengue virus • Influenza virus • *Leishmania amazonensis* • *Trypanosoma cruzi* • *Plasmodium berghei* • *Strongyloides venezuelensis*

7.1 PAF Biosynthesis and Metabolism

Phospholipids are major components of the biomembrane. Some phospholipids, when metabolized, are converted into intercellular signaling molecules, including lysophosphatidic acid [1], sphingosine-1-phosphate [2], and platelet-activating factor (PAF) [3, 4].

S. Ishii (✉)
Department of Immunology, Akita University Graduate School of Medicine,
Akita 010-8543, Japan
e-mail: satishii@med.akita-u.ac.jp

T. Yokomizo, M. Murakami (eds.), *Bioactive Lipid Mediators*,
DOI 10.1007/978-4-431-55669-5_7

PAF is synthesized intracellularly by either the *de novo* or the remodeling pathway [5, 6] (Fig. 7.1). The remodeling pathway is regulated by extracellular stimuli and is responsible for the bulk of PAF synthesis under inflammatory conditions. Stimulus-coupled PAF biosynthesis is initiated by the activation of cytosolic PLA_2 ($cPLA_2$), which hydrolyzes 1-*O*-alkyl-phosphatidylcholine in biomembranes to 1-*O*-alkyl-*sn*-glycero-3-phosphocholine (lyso-PAF) [7, 8]. This biologically inactive phospholipid is then acetylated at the *sn*-2 position to form PAF by the activity of LPCAT2, an acetyl-CoA:lyso-PAF acetyltransferase [9]. LPCAT2 expression is observed mainly in inflammatory cells such as peritoneal macrophages. In response to inflammatory signals that lead to the activation of p38 mitogen-activated protein kinase (MAPK) and protein kinase C (PKC)α, LPCAT2 is serine-phosphorylated and activated [10, 11]. Considering that the catalytic activity of $cPLA_2$ is stimulated in a Ca^{2+}- and ERK MAPK-dependent manner [12, 13], a mechanism in which

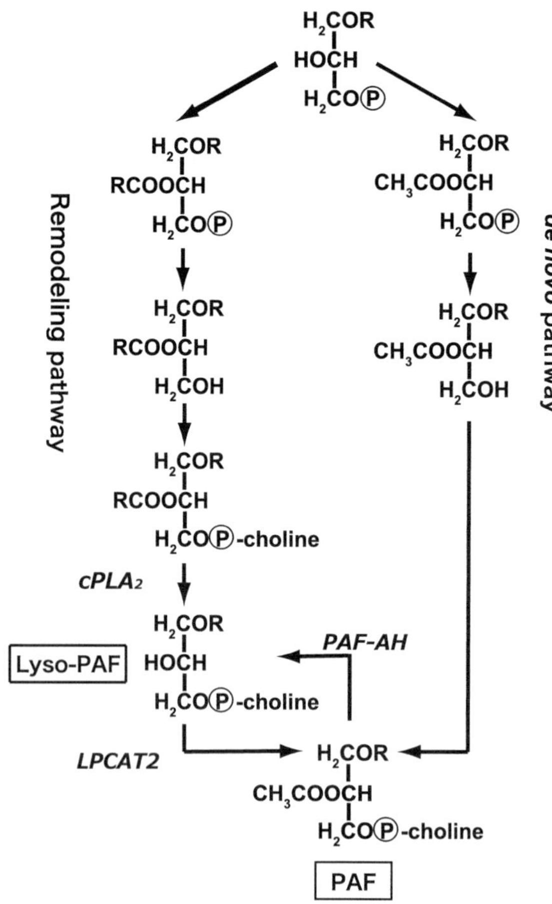

Fig. 7.1 Synthetic pathways of platelet-activating factor (PAF). PAF is synthesized via two distinct pathways, the *de novo* and the remodeling pathways. LPCAT2 is an enzyme with acetyl-CoA:lyso-PAF acetyltransferase activity that catalyzes the final reaction for PAF synthesis in the remodeling pathway. PAF is degraded by PAF acetylhydrolase (PAF-AH) to lyso-PAF. *P* in a *circle* indicates phosphate group

$cPLA_2$ and LPCAT2 act sequentially seems reasonable for the effective production of PAF in response to extracellular stimuli. Following secretion into the extracellular space, PAF exerts its bioactivity on a variety of cells through a specific G protein-coupled receptor (see following).

Deacetylation of PAF is catalyzed by plasma PAF acetylhydrolase, which results in its conversion into lyso-PAF [14]. This enzyme appears to be secreted mainly by macrophages [15, 16] and circulates in association with LDL and HDL particles [17]. Thus, PAF released into the extracellular space can be rapidly inactivated to prevent excessive and uncontrolled activation of the specific receptor for PAF (PAFR).

7.2 PAFR

Using an expression cloning method with *Xenopus* oocytes, Prof. Takao Shimizu and his research team isolated a cDNA clone encoding PAFR from a guinea pig lung cDNA library in 1991 [18]. Subsequent cross-hybridization screening studies identified PAFR orthologues from humans [19], rats [20], and mice [21]. Consistently, these PAFRs were found to be single polypeptides composed of approximately 340 amino acids. Similar to other G protein-coupled receptors, PAFR possesses seven transmembrane domains. A single receptor subtype is currently thought to mediate all the actions of PAF. Of note, PAFR can couple to both Gq/11 and Gi/o proteins, which initiate distinct signals (Fig. 7.2). This trait enables the receptor to simultaneously activate various kinases and phospholipases. The former involves MAPK, PKC, phosphatidylinositol 3-kinase (PI3K), and protein tyrosine kinase (PTK), and the latter includes phospholipase C (PLC)β and $cPLA_2$.

Northern hybridization experiments have revealed the distribution of PAFR mRNA expression in several species. In guinea pigs [18], PAFR mRNA is most abundant in leukocytes. Furthermore, the spleen and lung contain detectable amounts of PAFR mRNA. In humans, the PAFR message has been detected in neutrophils [19], monocytes [22], monocyte/$CD34^+$ cell-derived dendritic cells [23], and umbilical vein endothelial cells (HUVECs) [24]. As in guinea pigs, the human lung has been shown to express high levels of PAFR mRNA [25, 26]. Rat PAFR mRNA can be detected in the spleen, small intestine, and lung [20], as well as in microglia [27] and mesangial cells [28]. Mouse peritoneal macrophages express a significant amount of PAFR mRNA. In mouse tissues, abundant expression of PAFR mRNA has been observed in the spleen, lung, and small intestine [21]. Altogether, these observations indicate that tissues rich in myeloid cells commonly contain high levels of PAFR mRNA.

Specific receptors for PAF have been identified in numerous tissues and cells through the use of ^{3}H-labeled PAFR agonists and antagonists. The first binding experiment utilizing [^{3}H]-PAF was conducted in washed human platelets [29]. Consistent with the data obtained by Northern hybridization, specific receptors for PAF have been detected on neutrophils, macrophages, mononuclear leukocytes, eosinophils, and Kupffer cells (see [30] for review). In addition to these myeloid

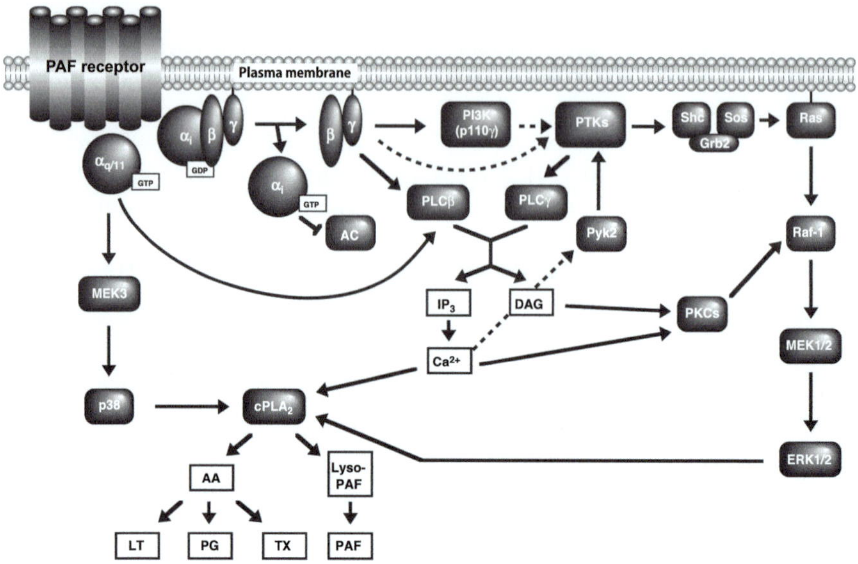

Fig. 7.2 Schematic representation of intracellular signaling events coupled to PAF receptor. This scheme represents an overlay of signaling pathways leading to intracellular Ca^{2+} increase and mitogen-activated protein kinase (MAPK) activation. Some of the signaling events may be cell-type specific. *Dashed lines* indicate the pathways that need to be further characterized. PKCs include conventional and novel PKC isozymes, which are activated by Ca^{2+} and DAG. PTKs include Src and Tec family tyrosine kinases as well as receptor tyrosine kinases. *AA* arachidonic acid, *AC* adenylate cyclase, *DAG* diacylglycerol, *IP3* inositol 1,4,5-trisphosphate, *MEK* MAP kinase kinase, *LT* leukotriene, *PG* prostaglandin, *TX* thromboxane. See text for other abbreviations

cells, tracheal epithelial cells [31] and HUVECs [32] also have shown specific binding activity to PAF. Although these cultured cells were assayed intact, membrane fractions were used to demonstrate specific PAFRs in tissues or to eliminate the intracellular degradation of PAF. Such specimens were prepared from the lung [33] and spleen [34].

7.3 Role of the PAF/PAFR Axis in Infectious Diseases

A variety of microorganisms can cause disease in mammals. These microorganisms are referred to as pathogenic microorganisms or pathogens, composed of four categories: viruses, bacteria, fungi, and parasites. Immune responses are induced against infection by pathogens. Since PAFR-knockout (KO) mice were first reported in 1998 [35], there have been several studies examining whether PAF contributes to host defense against or pathogenesis of infectious diseases. In the following, studies published during the past dozen years or so will be reviewed that offer insight into the protective or deleterious roles of the PAF/PAFR axis in infectious disease.

7.3.1 *Bacterial Infection*

7.3.1.1 *Klebsiella pneumoniae* and *Pseudomonas aeruginosa*

Soares *et al.* investigated the relevance of PAFR in a model of pulmonary infection in mice caused by the gram-negative bacterium *Klebsiella pneumoniae* [36]. Infection of PAFR-KO mice with *K. pneumoniae* resulted in significantly earlier lethality when compared with wild-type (WT) controls. This phenomenon was associated with a significant increase in the number of bacteria and a marked suppression of *K. pneumoniae* uptake by neutrophils in the lungs of PAFR-KO mice. Notably, there was little effect of PAFR deficiency on the number of neutrophils in the mouse lung. Thus, PAF is likely to exert a protective role against *K. pneumoniae* infection by activating infiltrating neutrophils in the lung. This study exemplifies how the innate immune system utilizes lipid mediators to protect the host against pathogens.

Pseudomonas pneumonia is a critical clinical problem with a high mortality rate. Another experimental pulmonary infection study was performed with *Pseudomonas aeruginosa* [37]. When compared with WT mice, PAFR-KO mice also show reduced resistance against *P. aeruginosa*, most likely the result of a reduced phagocytic capacity of PAFR-KO neutrophils in the lung. PAFR seems to contribute to host defense against *P. aeruginosa* infection in a manner similar to that of *K. pneumoniae* infection.

7.3.1.2 *Streptococcus pneumoniae* and *Haemophilus influenzae*

The characteristic features of each bacterial pathogen are diverse in terms of the mode of infection transmission, colonization procedure, and mechanism of pathogenesis. Phosphorylcholine is a critical moiety of PAF and is also a component of the cell walls of *Streptococcus pneumoniae* [38] and nontypeable *Haemophilus influenzae* [39].

Susceptibility to *S. pneumoniae* infection is reportedly observed after inflammatory activation of A549 alveolar epithelial cells and HUVECs *in vitro* [40], likely caused by upregulation of PAFR expression at the cell surface. The *in vivo* relevance of the interaction between pneumococcal phosphorylcholine and PAFR was evident in a model of pneumococcal pneumonia showing that PAFR-KO mice had an attenuated inflammatory response, reduced bacterial numbers, and an improved outcome [41].

It should be noted that the interaction between phosphorylcholine and PAFR does not affect murine host defense during nontypeable *H. influenzae* pneumonia [42]; PAFR-KO and WT mice display similar bacterial counts, myeloperoxidase activity, and histopathology in the lung. These results suggest that receptors other than PAFR are responsible for the regulation of innate immune responses to non-typeable *H. influenzae* infection despite *in vitro* data indicating a significant role of the phosphorylcholine–PAFR interaction in the invasion of pulmonary epithelial cells [43].

7.3.1.3 *Mycobacterium tuberculosis*

Mycobacteria such as *Mycobacterium tuberculosis* are intracellular pathogens that grow primarily in the phagosomes of macrophages [44]. Th1 cells can activate macrophages to eliminate these pathogens. To examine the role of PAF in host Th1 responses during pulmonary tuberculosis, PAFR-KO and WT mice were infected with *M. tuberculosis* [45]. Mycobacterial outgrowth as well as $CD4^+$ lymphocyte recruitment/activation and Th1 cytokine [interferon (IFN)-γ] production in the lungs did not differ significantly between PAFR-KO and WT mice, as determined during the early phase of infection when all mice were still alive. Subsequently, similar mortality rates in the two genotypes were observed. These data argue against a role for PAF in T-cell-mediated immune responses to *M. tuberculosis* infection and seem to be in contrast to a role for PAF in neutrophil-mediated innate immune responses to *K. pneumoniae* and *P. aeruginosa* infections. Indeed, little is known regarding the effects of PAF on lymphocyte activation [3].

7.3.1.4 *Aggregatibacter actinomycetemcomitans*

Periodontitis is a chronic inflammatory and alveolar bone destructive disease that affects the supporting structures of the tooth. This disease is caused by oral biofilm-producing bacteria, such as *Aggregatibacter actinomycetemcomitans*. There is a report that the mRNA expression level of the PAF biosynthetic enzyme LPCAT2 was increased after *A. actinomycetemcomitans* lipopolysaccharide (LPS) injection into mouse periodontal tissues [46]. Indeed, PAF has been detected in inflamed human gingival tissues [47]. After experimental oral infection with *A. actinomycetemcomitans*, PAFR-KO mice show reduced alveolar bone loss [46]. In line with this, the number of tartrate-resistant acid phosphatase-positive cells, which include osteoclasts, is reduced within the alveolar bone of PAFR-KO mice. We previously reported that LPCAT2 is expressed in osteoclasts [48]. The produced PAF is considered to activate PAFR on osteoclasts in an autocrine/paracrine manner and exert positive effects on cell survival and bone-resorptive activity. Altogether, it is likely that PAF, in concert with LPS, enhances the differentiation and activity of osteoclasts in the periodontal tissues during *A. actinomycetemcomitans* infection.

7.3.2 *Viral Infection*

7.3.2.1 Dengue Virus

Dengue fever is a mosquito-borne disease caused by dengue virus. In a small percentage of cases, individuals develop a severe syndrome, referred to as dengue hemorrhagic fever or dengue shock syndrome, which is characterized by increased vascular permeability, an altered number of leukocytes, increased hematocrit, thrombocytopenia, and varying amounts of hemorrhage [49]. When administered systemically to animals, PAF can mimic several responses that are commonly observed in the context of severe dengue infection. Consistently, a previous study showed that PAF production was greater in macrophages obtained from patients who were previously infected with dengue virus than those from healthy controls [50]. Therefore, it was hypothesized that excessive activation of PAFR during dengue infection could account, in part, for the symptoms of dengue hemorrhagic fever/dengue shock syndrome. Indeed, inoculation of PAFR-KO mice with dengue virus caused milder thrombocytopenia, hyperpermeability, and inflammatory cytokine production (including IFN-γ) and delay in death when compared with WT infected mice [51]. It is notable that viral titers in PAFR-KO mice were similar to those found in WT mice. These results are consistent with the hypothesis that PAF plays an important role in inflammation during viral infections without affecting viral clearance by the host.

7.3.2.2 Influenza Virus

In a model of experimental infection with influenza A virus, PAFR-KO mice had reduced pulmonary inflammation, Th1 cytokine production, and lethality rates [52], suggesting the importance of PAF in the pathogenesis of influenza A virus infection. Consistently, the infection induced the mRNA expression level of LPCAT2. Again, immune responses against the virus, as assessed by viral titers and specific antibodies, were unaffected in PAFR-KO mice.

7.3.3 *Parasite Infection*

7.3.3.1 *Leishmania amazonensis* and *Trypanosoma cruzi*

The intracellular protozoans *Leishmania* and *Trypanosoma* are medically important parasites [53]. Several *Leishmania* species are responsible for cutaneous, mucocutaneous, or visceral leishmaniasis, and *Trypanosoma cruzi* causes Chagas' disease. The roles of PAFR in the outcome of infection with these protozoan parasites were assessed in experimental models using PAFR-KO mice.

Leishmania preferentially infects macrophages and drives Th1 immune responses in the host, which are effective for elimination of this parasite [53]. The inflammation observed in infected tissues is characterized by infiltration of lymphocytes, neutrophils, and macrophages, which appear to secrete a range of inflammatory mediators, including PAF. PAFR-KO mice are more susceptible to *Leishmania amazonensis* infection than WT controls, as observed both by the inflammatory area and parasite number at the site of infection [54]. Consistently, IFN-γ production is impaired in PAFR-KO mice by unknown mechanisms. Expression of nitric oxide (NO) synthase-2 mRNA is also impaired in PAFR-KO mice. Thus, in concert with IFN-γ, PAF may activate macrophages to eliminate infecting *L. amazonensis*, most likely through NO production.

Trypanosoma cruzi infects many cell types [53]. In addition to parasite-driven inflammation, an autoimmune component may also be involved in the tissue damage that occurs during chronic stages of the disease, especially in cardiac tissues. Effective immune responses to *T. cruzi* consist of macrophage activation that favors phagocytosis and NO production. Additionally, the establishment of antigen-specific Th1 response is important for host protection against the parasite. The hearts of *T. cruzi*-infected PAFR-KO mice have an increased number of parasite nests associated with a more intense inflammatory infiltrate [55]. This pathology is accompanied by greater parasitemia and lethality, although there is no significant change in the expression of IFN-γ in the hearts of infected PAFR-KO mice. *In vitro*, macrophages from PAFR-KO mice fail to phagocytose *T. cruzi* in response to leukotriene B4 and MCP-1 as well as PAF and produce reduced amounts of NO following infection and activation with IFN-γ. These data suggest that PAF is involved in host resistance to *T. cruzi* infection by activating macrophages to phagocytose the parasite and subsequently produce NO.

7.3.3.2 *Plasmodium berghei*

Plasmodium falciparum is an intracellular protozoan parasite that causes malaria in humans. Cerebral malaria is a severe form of the disease that may result, in part, from an overt inflammatory response during infection by *P. falciparum* [56]. Although the exact pathogenesis of cerebral malaria remains unclear, several mechanisms have been proposed to explain it, including breakdown of the blood–brain barrier in the central nervous system and inflammatory adaptive immune responses [56, 57]. Infection with *P. berghei* is the main mouse model for cerebral malaria [57]. PAFR deletion in mice causes a delay to death in experimental cerebral malaria [58]. In addition, brain inflammation is significantly reduced in PAFR-KO mice, as demonstrated by histology, vascular permeability, and recruitment/activation of $CD8^+$ T cells. In contrast, PAFR-KO mice display normal parasitemia, which suggests that PAFR is not relevant to parasite replication and does not appear to have any significant antimalarial properties. Therefore, PAFR appears to contribute to the inflammatory events that lead to premature death of the host.

7.3.3.3 *Strongyloides venezuelensis*

Gastrointestinal nematode species with a pulmonary migration component of their life cycles are the most prevalent parasites in humans [59]. Experimental murine infections with gastrointestinal nematodes such as *Strongyloides venezuelensis* induce predominantly Th2 immune responses for host protection. Following their obligatory migration through the lungs of mice, *S. venezuelensis* larvae become established in the duodenal mucosa [60]. Arrival of the parasite in the intestine is accompanied by intestinal eosinophilia and mastocytosis, which may be involved in the process of worm elimination by approximately 10 days after infection. Although larval migration and intestinal establishment are not affected by the absence of the PAFR, *S. venezuelensis*-infected PAFR-KO mice show a delay in the elimination of worms compared to infected WT mice [61]. Histopathological changes in the small intestine are reduced in PAFR-KO mice. Indeed, PAFR-KO mice express lower levels of Th2 cytokines. Although the mechanisms underlying PAFR-driven Th2 immune responses remain to be established, these results suggest that PAFR activation is associated with the elimination of *S. venezuelensis* from the small intestine.

7.3.4 Conclusions

In this review, the involvement of the PAF/PAFR axis in several models of infectious diseases is described. These experimental infection studies revealed several distinctive functions of PAFR in host defense, depending on the pathogenic mechanisms of infection. Outcomes for each study are summarized in Table 7.1.

By facilitating phagocytosis and killing of bacteria, PAF appears to affect the ability of phagocytes (likely mainly neutrophils) to address pulmonary infections with *K. pneumoniae* [36] and *P. aeruginosa* [37]. Most likely because of different characteristic features of pathogens, the other pulmonary infections with extracellular bacteria, that is, *S. pneumoniae* [41] and *H. influenzae* [42], appear to develop independently of PAFR-activated phagocytes. It is of interest that PAFR mediated activation of phagocytes (likely mainly macrophages) is also responsible for host resistance to the intracellular protozoan parasites *L. amazonensis* [54] and *T. cruzi* [55]. NO produced in PAFR-activated macrophages may be a key effector molecule of the intracellular parasite killing. Alveolar macrophages serve as the major host cell niche for the growth and survival of *M. tuberculosis*, and NO also has mycobactericidal activity in mice [62]. However, PAFR-KO mice have normal protective immune responses to *M. tuberculosis* [45], the reason for which remains unclear.

PAF exerts different functions during intracellular infection with dengue and influenza A viruses [51, 52]. PAF-producing cells, such as phagocytes, in the infected tissues appear to release a large amount of PAF, which aberrantly activates PAFR in an autocrine/paracrine manner and causes tissue injury and, in some instances, death. A similar mechanism may underlie the pathogenesis of cerebral malaria [58], pneumococcal pneumonia [41], and periodontitis [46], although the

Table 7.1 Characteristics of the experimental infection studies using platelet-activating receptor (PAFR)-knockout (KO) mice

Pathogens	Effects of PAF/PAFR on		Responsible immune mechanisms	References
	Pathogen burden	Disease symptoms		
Klebsiella pneumoniae	⇓	⇓	PAF activates neutrophil phagocytosis.	[36]
Pseudomonas aeruginosa	⇓	⇓	PAF activates neutrophil phagocytosis.	[37]
Streptococcus pneumoniae	⇑	⇑	PAFR anchors *S. pneumoniae* to epithelial and endothelial cells.	[41]
Haemophilus influenzae	⇒	⇒	N/A[a]	[42]
Mycobacterium tuberculosis	⇒	⇒	N/A	[45]
Aggregatibacter actinomycetemcomitans	⇒	⇑	PAF enhances the differentiation and activity of osteoclasts in inflamed periodontal tissues.	[46]
Dengue virus	⇒	⇑	PAF induces thrombocytopenia, hyperpermeability, hemoconcentration, and Th1 cytokine production.	[51]
Influenza virus	⇒	⇑	PAF induces lung injury with increased neutrophil recruitment and Th1 cytokine production.	[52]
Leishmania amazonensis	⇓	⇓	PAF stimulates macrophages and Th1 polarization.	[54]
Trypanosoma cruzi	⇓	⇓	PAF stimulates macrophages.	[55]
Plasmodium berghei	⇒	⇑	PAF contributes to vascular permeability and $CD8^+$ T cell activation in the brain.	[58]
Strongyloides venezuelensis	⇓	⇓	PAF stimulates Th2-predominant intestinal inflammation.	[61]

[a]*N/A*, not applicable

latter two are caused by infection with extracellular pathogens. Thus, PAFR could be a useful therapeutic target to interfere with the aggravation of these diseases. It is worth noting that treatment of mice with PAFR antagonists prevented, at least in part, the progression of infection with these pathogens [51, 52, 58, 40, 63]. Equivalent studies in humans are required to understand the potential role of PAFR blockade in the amelioration of disease symptoms.

In many of the experimental infection models mentioned earlier, adaptive immune responses are induced: Th1 immune responses during infections with dengue and influenza A viruses [51, 52] and *L. amazonensis* [54], Th2 responses during infection with *S. venezuelensis* [61], and CD8$^+$ T-cell responses with *P. berghei* [58]. PAFR-KO mice are deficient in the ability to induce these adaptive immune responses. Considering that PAFR activation is likely to be important for phagocytosis [64, 65], the ensuing antigen presentation by phagocytes may be impaired in PAFR-KO mice. Future research will further characterize the role of the PAF/PAFR axis in the modulation of adaptive immunity.

References

1. Nakanaga K, Hama K, Aoki J (2010) Autotaxin–an LPA producing enzyme with diverse functions. J Biochem 148(1):13–24
2. Maceyka M, Harikumar KB, Milstien S, Spiegel S (2012) Sphingosine-1-phosphate signaling and its role in disease. Trends Cell Biol 22(1):50–60
3. Ishii S, Shimizu T (2000) Platelet-activating factor (PAF) receptor and genetically engineered PAF receptor mutant mice. Prog Lipid Res 39(1):41–82
4. Shimizu T (2009) Lipid mediators in health and disease: enzymes and receptors as therapeutic targets for the regulation of immunity and inflammation. Annu Rev Pharmacol Toxicol 49:123–150
5. Snyder F (1995) Platelet-activating factor: the biosynthetic and catabolic enzymes. Biochem J 305:689–705
6. Prescott SM, Zimmerman GA, Stafforini DM, McIntyre TM (2000) Platelet-activating factor and related lipid mediators. Annu Rev Biochem 69:419–445
7. Uozumi N, Kume K, Nagase T, Nakatani N, Ishii S, Tashiro F, Komagata Y, Maki K, Ikuta K, Ouchi Y, Miyazaki J, Shimizu T (1997) Role of cytosolic phospholipase A2 in allergic response and parturition. Nature 390(6660):618–622
8. Shindou H, Ishii S, Uozumi N, Shimizu T (2000) Roles of cytosolic phospholipase A(2) and platelet-activating factor receptor in the Ca-induced biosynthesis of PAF. Biochem Biophys Res Commun 271(3):812–817
9. Shindou H, Hishikawa D, Nakanishi H, Harayama T, Ishii S, Taguchi R, Shimizu T (2007) A single enzyme catalyzes both platelet-activating factor production and membrane biogenesis of inflammatory cells. Cloning and characterization of acetyl-CoA:LYSO-PAF acetyltransferase. J Biol Chem 282(9):6532–6539
10. Morimoto R, Shindou H, Oda Y, Shimizu T (2010) Phosphorylation of lysophosphatidylcholine acyltransferase 2 at Ser34 enhances platelet-activating factor production in endotoxin-stimulated macrophages. J Biol Chem 285(39):29857–29862
11. Morimoto R, Shindou H, Tarui M, Shimizu T (2014) Rapid production of platelet-activating factor is induced by protein kinase c alpha-mediated phosphorylation of lysophosphatidylcholine acyltransferase 2 protein. J Biol Chem 289(22):15566–15576
12. Nalefski EA, Sultzman LA, Martin DM, Kriz RW, Towler PS, Knopf JL, Clark JD (1994) Delineation of two functionally distinct domains of cytosolic phospholipase A2, a regulatory Ca(2+)-dependent lipid-binding domain and a Ca(2+)-independent catalytic domain. J Biol Chem 269(27):18239–18249
13. Lin LL, Wartmann M, Lin AY, Knopf JL, Seth A, Davis RJ (1993) cPLA2 is phosphorylated and activated by MAP kinase. Cell 72(2):269–278
14. Tjoelker LW, Wilder C, Eberhardt C, Stafforini DM, Dietsch G, Schimpf B, Hooper S, Le Trong H, Cousens LS, Zimmerman GA, Yamada Y, McIntyre TM, Prescott SM, Gray PW

(1995) Anti-inflammatory properties of a platelet-activating factor acetylhydrolase. Nature 374(6522):549–553
15. Elstad MR, Stafforini DM, McIntyre TM, Prescott SM, Zimmerman GA (1989) Platelet-activating factor acetylhydrolase increases during macrophage differentiation. A novel mechanism that regulates accumulation of platelet-activating factor. J Biol Chem 264(15):8467–8470
16. Asano K, Okamoto S, Fukunaga K, Shiomi T, Mori T, Iwata M, Ikeda Y, Yamaguchi K (1999) Cellular source(s) of platelet-activating-factor acetylhydrolase activity in plasma. Biochem Biophys Res Commun 261(2):511–514
17. Stafforini DM, McIntyre TM, Carter ME, Prescott SM (1987) Human plasma platelet-activating factor acetylhydrolase. Association with lipoprotein particles and role in the degradation of platelet-activating factor. J Biol Chem 262(9):4215–4222
18. Z H, Nakamura M, Miki I, Minami M, Watanabe T, Seyama Y, Okado H, Toh H, Ito K, Miyamoto T, Shimizu T (1991) Cloning by functional expression of platelet-activating factor receptor from guinea-pig lung. Nature 349(6307):342–346
19. Nakamura M, Honda Z, Izumi T, Sakanaka C, Mutoh H, Minami M, Bito H, Seyama Y, Matsumoto T, Noma M, Shimizu T (1991) Molecular cloning and expression of platelet-activating factor receptor from human leukocytes. J Biol Chem 266(30):20400–20405
20. Bito H, Honda Z, Nakamura M, Shimizu T (1994) Cloning, expression and tissue distribution of rat platelet-activating-factor-receptor cDNA. Eur J Biochem 221(1):211–218
21. Ishii S, Matsuda Y, Nakamura M, Waga I, Kume K, Izumi T, Shimizu T (1996) A murine platelet-activating factor receptor gene: cloning, chromosomal localization and up-regulation of expression by lipopolysaccharide in peritoneal resident macrophages. Biochem J 314:671–678
22. Thivierge M, Alami N, Muller E, de Brum-Fernandes AJ, Rola-Pleszczynski M (1993) Transcriptional modulation of platelet-activating factor receptor gene expression by cyclic AMP. J Biol Chem 268(23):17457–17462
23. Sozzani S, Longoni D, Bonecchi R, Luini W, Bersani L, D'Amico G, Borsatti A, Bussolino F, Allavena P, Mantovani A (1997) Human monocyte-derived and $CD34^+$ cell-derived dendritic cells express functional receptors for platelet activating factor. FEBS Lett 418(1-2):98–100
24. Shen Y, Sultana C, Arditi M, Kim KS, Kalra VK (1998) Endotoxin-induced migration of monocytes and PECAM-1 phosphorylation are abrogated by PAF receptor antagonists. Am J Physiol 275:E479–E486
25. Ye RD, Prossnitz ER, Zou AH, Cochrane CG (1991) Characterization of a human cDNA that encodes a functional receptor for platelet activating factor. Biochem Biophys Res Commun 180(1):105–111
26. Kunz D, Gerard NP, Gerard C (1992) The human leukocyte platelet-activating factor receptor. cDNA cloning, cell surface expression, and construction of a novel epitope-bearing analog. J Biol Chem 267(13):9101–9106
27. Mori M, Aihara M, Kume K, Hamanoue M, Kohsaka S, Shimizu T (1996) Predominant expression of platelet-activating factor receptor in the rat brain microglia. J Neurosci 16(11):3590–3600
28. Nakao A, Watanabe T, Bitoh H, Imaki H, Suzuki T, Asano K, Taniguchi S, Nosaka K, Shimizu T, Kurokawa K (1997) cAMP mediates homologous downregulation of PAF receptor mRNA expression in mesangial cells. Am J Physiol 273:F445–F450
29. Valone FH, Coles E, Reinhold VR, Goetzl EJ (1982) Specific binding of phospholipid platelet-activating factor by human platelets. J Immunol 129(4):1637–1641
30. Chao W, Olson MS (1993) Platelet-activating factor: receptors and signal transduction. Biochem J 292:617–629
31. Herbert JM (1992) Characterization of specific binding sites of 3H-labelled platelet-activating factor ([^{3}H]PAF) and a new antagonist, [^{3}H]SR 27417, on guinea-pig tracheal epithelial cells. Biochem J 284:201–206

32. Korth RM, Hirafuji M, Benveniste J, Russo MF (1995) Human umbilical vein endothelial cells: specific binding of platelet-activating factor and cytosolic calcium flux. Biochem Pharmacol 49(12):1793–1799
33. Hwang SB, Lam MH, Shen TY (1985) Specific binding sites for platelet activating factor in human lung tissues. Biochem Biophys Res Commun 128(2):972–979
34. Bito H, Nakamura M, Honda Z-I, Izumi T, Iwatsubo T, Seyama Y, Ogura A, Kudo Y, Shimizu T (1992) Platelet-activating factor (PAF) receptor in rat brain: PAF mobilizes intracellular Ca^{2+} in hippocampal neurons. Neuron 9(2):285–294
35. Ishii S, Kuwaki T, Nagase T, Maki K, Tashiro F, Sunaga S, Cao WH, Kume K, Fukuchi Y, Ikuta K, Miyazaki J, Kumada M, Shimizu T (1998) Impaired anaphylactic responses with intact sensitivity to endotoxin in mice lacking a platelet-activating factor receptor. J Exp Med 187(11):1779–1788
36. Soares AC, Pinho VS, Souza DG, Shimizu T, Ishii S, Nicoli JR, Teixeira MM (2002) Role of the platelet-activating factor (PAF) receptor during pulmonary infection with gram negative bacteria. Br J Pharmacol 137(5):621–628
37. van Zoelen MAD, Florquin S, Meijers JCM, De Beer R, De Vos AF, De Boer OJ, Van Der Poll T (2008) Platelet-activating factor receptor contributes to host defense against *Pseudomonas aeruginosa* pneumonia but is not essential for the accompanying inflammatory and procoagulant response. J Immunol 180(5):3357–3365
38. Fischer W, Behr T, Hartmann R, Peter-Katalinic J, Egge H (1993) Teichoic acid and lipoteichoic acid of *Streptococcus pneumoniae* possess identical chain structures. A reinvestigation of teichoid acid (C polysaccharide). Eur J Biochem 215(3):851–857
39. Risberg A, Masoud H, Martin A, Richards JC, Moxon ER, Schweda EK (1999) Structural analysis of the lipopolysaccharide oligosaccharide epitopes expressed by a capsule-deficient strain of *Haemophilus influenzae* Rd. Eur J Biochem 261(1):171–180
40. Cundell DR, Gerard NP, Gerard C, Idanpaan-Heikkila I, Tuomanen EI (1995) *Streptococcus pneumoniae* anchor to activated human cells by the receptor for platelet-activating factor. Nature 377(6548):435–438
41. Rijneveld AW, Weijer S, Florquin S, Speelman P, Shimizu T, Ishii S, Van Der Poll T (2004) Improved host defense against pneumococcal pneumonia in platelet-activating factor receptor-deficient mice. J Infect Dis 189(4):711–716
42. Branger J, Wieland CW, Florquin S, Maris NA, Pater JM, Speelman P, Shimizu T, Ishii S, Van Der Poll T (2004) Platelet-activating factor receptor-deficient mice show an unaltered clearance of nontypeable *Haemophilus influenzae* from their respiratory tract. Shock 22(6):543–547
43. Swords WE, Buscher BA, Ver Steeg Ii K, Preston A, Nichols WA, Weiser JN, Gibson BW, Apicella MA (2000) Non-typeable *Haemophilus influenzae* adhere to and invade human bronchial epithelial cells via an interaction of lipooligosaccharide with the PAF receptor. Mol Microbiol 37(1):13–27
44. van Crevel R, Ottenhoff TH, van der Meer JW (2002) Innate immunity to *Mycobacterium tuberculosis*. Clin Microbiol Rev 15(2):294–309
45. Weijer S, Leemans JC, Florquin S, Shimizu T, Ishii S, Van Der Poll T (2003) Host response of platelet-activating factor receptor-deficient mice during pulmonary tuberculosis. Immunology 109(4):552–556
46. Madeira MF, Queiroz-Junior CM, Costa GM, Werneck SM, Cisalpino D, Garlet GP, Teixeira MM, Silva TA, Souza DG (2013) Platelet-activating factor receptor blockade ameliorates *Aggregatibacter actinomycetemcomitans*-induced periodontal disease in mice. Infect Immun 81(11):4244–4251
47. Emingil G, Cinarcik S, Baylas H, Huseyinov A (2001) Levels of platelet-activating factor in gingival crevicular fluid and gingival tissue in specific periodontal diseases. J Periodontol 72(8):1032–1037
48. Hikiji H, Ishii S, Shindou H, Takato T, Shimizu T (2004) Absence of platelet-activating factor receptor protects mice from osteoporosis following ovariectomy. J Clin Invest 114(1):85–93

49. Rothman AL (2004) Dengue: defining protective versus pathologic immunity. J Clin Invest 113(7):946–951
50. Yang KD, Lee CS, Shaio MF (1995) A higher production of platelet activating factor in ex vivo heterologously secondary dengue-2 virus infections. Acta Microbiol Immunol Hung 42(4):403–407
51. Souza DG, Fagundes CT, Sousa LP, Amaral FA, Souza RS, Souza AL, Kroon EG, Sachs D, Cunha FQ, Bukin E, Atrasheuskaya A, Ignatyev G, Teixeira MM (2009) Essential role of platelet-activating factor receptor in the pathogenesis of dengue virus infection. Proc Natl Acad Sci U S A 106(33):14138–14143
52. Garcia CC, Russo RC, Guabiraba R, Fagundes CT, Polidoro RB, Tavares LP, Salgado APC, Cassali GD, Sousa LP, Machado AV, Teixeira MM (2010) Platelet-activating factor receptor plays a role in lung injury and death caused by influenza A in mice. PLoS Pathog 6(11):e1001171
53. Sibley LD (2011) Invasion and intracellular survival by protozoan parasites. Immunol Rev 240(1):72–91
54. Santiago HC, Braga Pires MF, Souza DG, Roffê E, Côrtes DF, Tafuri WL, Teixeira MM, Vieira LQ (2006) Platelet activating factor receptor-deficient mice present delayed interferon-γ upregulation and high susceptibility to *Leishmania amazonensis* infection. Microbes Infect 8(11):2569–2577
55. Talvani A, Santana G, Barcelos LS, Ishii S, Shimizu T, Romanha ÁJ, Silva JS, Soares MBP, Teixeira MM (2003) Experimental *Trypanosoma cruzi* infection in platelet-activating factor receptor-deficient mice. Microbes Infect 5(9):789–796
56. van der Heyde HC, Nolan J, Combes V, Gramaglia I, Grau GE (2006) A unified hypothesis for the genesis of cerebral malaria: sequestration, inflammation and hemostasis leading to microcirculatory dysfunction. Trends Parasitol 22(11):503–508
57. Lou J, Lucas R, Grau GE (2001) Pathogenesis of cerebral malaria: recent experimental data and possible applications for humans. Clin Microbiol Rev 14(4):810–820
58. Lacerda-Queiroz N, Rodrigues DH, Vilela MC, Rachid MA, Soriani FM, Sousa LP, Campos RDL, Quesniaux VFJ, Teixeira MM, Teixeira AL (2012) Platelet-activating factor receptor is essential for the development of experimental cerebral malaria. Am J Pathol 180(1):246–255
59. Brooker S (2010) Estimating the global distribution and disease burden of intestinal nematode infections: adding up the numbers–a review. Int J Parasitol 40(10):1137–1144
60. Sato Y, Toma H (1990) *Strongyloides venezuelensis* infections in mice. Int J Parasitol 20(1):57–62
61. Negrão-Corrêa D, Souza DG, Pinho V, Barsante MM, Souza ALS, Teixeira MM (2004) Platelet-activating factor receptor deficiency delays elimination of adult worms but reduces fecundity in *Strongyloides venezuelensis*-infected mice. Infect Immun 72(2):1135–1142
62. MacMicking JD, North RJ, LaCourse R, Mudgett JS, Shah SK, Nathan CF (1997) Identification of nitric oxide synthase as a protective locus against tuberculosis. Proc Natl Acad Sci U S A 94(10):5243–5248
63. Cabellos C, MacIntyre DE, Forrest M, Burroughs M, Prasad S, Tuomanen E (1992) Differing roles for platelet-activating factor during inflammation of the lung and subarachnoid space. The special case of *Streptococcus pneumoniae*. J Clin Invest 90(2):612–618
64. Ichinose M, Hara N, Sawada M, Maeno T (1994) A flow cytometric assay reveals an enhancement of phagocytosis by platelet activating factor in murine peritoneal macrophages. Cell Immunol 156(2):508–518
65. Au BT, Teixeira MM, Collins PD, Williams TJ (2001) Blockade of PAF receptors controls interleukin-8 production by regulating the activation of neutrophil CD11/CD18. Eur J Pharmacol 425(1):65–71

Chapter 8
Lysophospholipid Mediators: Their Receptors and Synthetic Pathways

Kuniyuki Kano, Kumiko Makide, Jun Ishiguro, Hiroshi Yukiura, Shizu Aikawa, Akiharu Uwamizu, Yuji Shinjo, Kahori Namiki, Hiroki Kawana, Saki Nemoto, Hirotaka Matsumoto, Ryoji Kise, Asuka Inoue, and Junken Aoki

Abstract It is now widely accepted that lysophospholipids (LPLs), a product of the phospholipase A reaction, function as mediators through G protein-coupled receptors (GPCRs). Notably, recent studies of lysophosphatidic acid (LPA) and sphingosine 1-phosphate (S1P) have revealed their essential roles *in vivo*. In addition, recent studies have identified several GPCRs for other lysophospholipids such as lysophosphatidylserine (LysoPS) and lysophosphatidylinositol (LPI). In this chapter, we summarize the actions and production of these LPLs as lipid mediators including LysoPS and LPI.

Keywords Lysophospholipids • Lysophosphatidic acid • Lysophosphatidylserine • G protein-coupled receptors (GPCRs) • Autotaxin • Phosphatidic acid-selective phospholipase A_1

8.1 Introduction

Lysophospholipids (LPLs) are deacylated forms of phospholipids. They include lysophosphatidic acid (LPA), sphingosine 1-phosphate (S1P), lysophosphatidylcholine (LPC), lysophosphatidylethanolamine (LPE), lysophosphatidylinositol (LPI), lysophosphatidylserine (LysoPS), and lysophosphatidylglycerol (LPG). LPLs had long been thought as merely by-products of the arachidonate cascade and cholesterol metabolism. However, they have been shown to induce multiple cellular responses such as stimulation of proliferation, migration, and mast cell degranulation, and suppression of apoptosis. Moreover, in the past two decades, a number of

---

K. Kano • K. Makide • J. Ishiguro • H. Yukiura • S. Aikawa • A. Uwamizu • Y. Shinjo • K. Namiki • H. Kawana • S. Nemoto • H. Matsumoto • R. Kise • A. Inoue • J. Aoki (✉)
Graduate School of Pharmaceutical Sciences, Tohoku University,
Sendai, Miyagi 980-8578, Japan
e-mail: jaoki@m.tohoku.ac.jp

T. Yokomizo, M. Murakami (eds.), *Bioactive Lipid Mediators*,
DOI 10.1007/978-4-431-55669-5_8

studies have identified multiple receptors and synthetic enzymes specific to each LPL. From the studies on their receptors and producing enzymes together with studies on LPLs themselves, the pathophysiological roles have been identified especially for LPA, S1P, LysoPS, and LPI. Thus, these LPLs are now known as lipid mediators. In this chapter, we review the receptors and synthetic enzymes for each LPL and summarize their pathophysiological function.

8.2 Receptors for lysophospholipids

Most lysophospholipids exert their functions via G protein-coupled receptors (GPCRs). More than 20 GPCRs have been identified. They respond to LPA, S1P, LPI, LysoPS, LPC, and sphingosylphosphorylcholine (SPC). However, for some of them, including G2A (for LPC) and OGR1 (for SPC), it remains controversial whether they are the genuine LPLs receptor, and these need further study. In this chapter, we review only legitimate LPLs receptors that are supported by independent studies (Table 8.1). LPLs also act as ligands for other types of receptors such as peroxisome proliferator-activated receptor-γ (PPARγ) and transient receptor

Table 8.1 Lysophospholipid receptors and their functions

Receptor	Ligand	G protein	mRNA localization	Functions
LPA_1 (Edg2)	LPA	G_i, G_q, $G_{12/13}$	Ubiquitous	Development of central nervous system, osteogenesis, neuropathic pain, fibrosis
LPA_2 (Edg4)	LPA	G_i, G_q, $G_{12/13}$	Spleen, testis	Suppression of dendritic cell activation
LPA_3 (Edg7)	LPA	G_i, G_q	Uterus, testis, lung, kidney	Embryo implantation
LPA_4	LPA	G_s, G_i, G_q, $G_{12/13}$	Ovary, endothelium	Vascular development
LPA_5 (GPR92)	LPA, FPP	G_i, $G_{12/13}$	Thymus, small intestine, dorsal root ganglion, platelets	Neuropathic pain
LPA_6 (P2Y5)	LPA	$G_{12/13}$	Hair follicle	Differentiation and maturation of hair follicles
GPR55	LPI	$G_{i/o}$, G_q, $G_{12/13}$	Immune cell, osteoclast	Nociceptive pain, bone morphogenesis
LPS_1 (GPR34)	LysoPS	G_i	Immune cell, microglia	Suppression of cytokine release
LPS_2 (P2Y10)	LysoPS	$G_{12/13}$	Thymus, spleen	Unknown
LPS_{2L} (A630033H20)	LysoPS	$G_{12/13}$	Thymus, spleen	Unknown
LPS_3 (GPR174)	LysoPS	G_{13}, G_s	Thymus, spleen	Unknown

potential V1 (TRPV1), but we shall not consider these in this chapter. After their discovery, multiple nomenclatures have been proposed. In this chapter, we principally follow the nomenclature proposed recently by Kihara et al. [1], except that LysoPS receptors are noted as LPS_x. In all cases, receptor proteins are noted as $LP(X)(_y)$, whereas receptor genes are $LP(X)R(_y)$, where X and *X* denote the head group of LPLs and y and *y* denote the order of discovery.

8.2.1 Receptors for LPA

In 1989, van Corven et al. reported that LPA-induced cell proliferation is G protein ($G\alpha_i$) dependent and indicated that the LPA receptor was GPCR [2]. In 1996, Hecht et al. first reported that EDG2 was a cellular receptor for LPA [3]. More recently, Noguchi et al. reported that P2Y9 reacted specifically with LPA [4]. To date there are six GPCRs for LPA belonging to either of two clusters: one is the EDG family members consisting of LPA_1/EDG2, LPA_2/EDG4, and LPA_3/EDG7, and the other is the P2Y family members consisting of LPA_4/P2Y9, LPA_5/GPR92, and LPA_6/P2Y5. These LPA receptors are expressed by particular cells in different organs and activate multiple intracellular signaling pathways, which accounts for their many biological functions. Here, we summarize recent advances in the studies of LPA receptors and their physiological function obtained mainly from analyses of knockout (KO) mice and human genetic diseases.

8.2.1.1 LPA_1 (*LPAR1*)

LPA_1 is the first identified LPA receptor. It was identified by Chun and colleagues in 1996 [3]. LPA_1 is expressed ubiquitously in various tissues. It couples to multiple G proteins ($G\alpha_i$, $G\alpha_q$, $G\alpha_{12/13}$) to activate various intracellular signaling pathways.

Several studies have shown the pathophysiological functions of LPA_1. (1) LPA_1 KO mice have defects in the development of parts of the central nervous system such as the cerebral cortex and olfactory neurons. (2) LPA_1 has a role in the adult peripheral nervous system. LPA_1 KO mice are resistant to injury-induced demyelination of the spinal cord and less sensitive to neuropathic pain [5, 6]. (3) LPA_1 has been linked to fibrotic disease in multiple organs. LPA_1 promotes pulmonary fibrosis and renal fibrosis by mediating fibroblast migration and vascular permeability, suggesting that LPA_1 could be a promising new therapeutic target in fibrosis [7]. LPA_1 KO mice were also reported to be resistant to bleomycin-induced dermal fibrosis [8]. (4) LPA_1 KO mice show cranial and rib cage deformities and low bone mass [9]. We found that LPA_1 KO mice exhibited abnormal chondrocyte arrangement in cartilage tissue throughout the body, including the cranial bone (unpublished).

8.2.1.2 LPA$_2$ (*LPAR2*)

Two years after identification of LPA$_1$, the second LPA receptor LPA$_2$ was discovered by virtue of its homology to LPA$_1$ [10]. LPA$_2$ is coupled with G proteins similar to LPA$_1$ (Gα_i, Gα_q, G$\alpha_{12/13}$). LPA$_2$ KO mice exhibit no obvious phenotypic abnormalities under normal conditions, and it is speculated that LPA$_2$ acts redundantly with LPA$_1$ [11]. However, colon cancer induced by azoxymethane and dextran sulfate sodium was dramatically reduced in LPA$_2$ KO mice [12]. In addition, dendritic cells from LPA$_2$ KO mice were found to induce T-cell proliferation and interleukin (IL)-13 production more efficiently than their wild-type (WT) counterparts, and LPA$_2$ KO mice developed greater allergen-induced lung inflammation in ovalbumin-challenged asthma [13], revealing the immunomodulatory roles of LPA$_2$.

8.2.1.3 LPA$_3$ (*LPAR3*)

LPA$_3$ is the third LPA receptor identified by our group in 1999 [14]. Interestingly, LPA$_3$ appears to discriminate among different LPA species (classes and positions of fatty acid), and LPA$_3$ is fully activated by 2-acyl-LPA with unsaturated fatty acids such as 18:1, 18:2, and 20:4. LPA$_3$ is known to couple to mainly Gα_q and Gα_i.

Female LPA$_3$ KO mice exhibited implantation failure, as revealed by delayed implantation and crowded implantation sites [15]. As a result, LPA$_3$ KO mice showed decreased litter size and shared placenta. In the mouse uterus, LPA$_3$ is expressed specifically in the endometrial epithelium in a progesterone-dependent fashion during the peri-implantation period or menstrual cycle. This manner of expression is also observed in humans, sheep, and pigs [16, 17]. On the other hand, clinical observations revealed that LPA$_3$ expression is low in the secretory endometrium of patients with recurrent implantation failure [18]. These findings suggest that LPA$_3$ is essential for normal pregnancy in a wide range of mammals.

8.2.1.4 LPA$_4$ (*LPAR4*)

LPA$_4$ is a novel type of LPA receptor classified in the P2Y receptor family that is very distant from Edg family LPA receptors. Noguchi et al. identified this receptor by screening for orphan GPCRs using a large number of lipids in 2003 [4]. LPA$_4$ is coupled with various G proteins including Gα_s, Gα_q, Gα_i, and G$\alpha_{12/13}$.

LPA$_4$ deficiency in C57BL/6 mice was reported to be partially embryonically lethal [19]. LPA$_4$-deficient embryo displayed hemorrhage or edema in many organs as a result of abnormal development of blood and lymphatic vessels. Mouse embryos deficient in autotaxin (ATX), an LPA-producing enzyme, showed profound vascular defects and died around embryonic day 9.5–10.5 [20]. Although the phenotype of ATX-deficient mice was more severe than that of LPA$_4$ KO mice, some of the phenotypes observed in ATX-deficient mice might be attributed to LPA$_4$. Recently, we found that zebrafish embryos showed aberrant segmental vascular development upon knockdown of both LPA_4 and LPA_1 [21], again suggesting that LPA_4 is one of the candidates in vascular development by ATX-LPA signaling.

8.2.1.5 LPA_5 (*LPAR5*)

GPR92/LPA_5 was shown to be an LPA receptor by two groups in 2006 [22, 23]. LPA_5 was more strongly activated by alkyl-LPA than by acyl-LPA. Because alkyl-LPA robustly aggregates human platelets, it has been speculated that LPA-induced platelet activation and aggregation are mediated through LPA_5 [24]. Although LPA_5 did react with LPA, it also responds to a dietary protein hydrolysate (peptone) at nanomolar (nM) concentrations [25] and to farnesyl pyrophosphate (FPP) [26]. So, further studies are needed to reveal the real endogenous ligand for LPA_5 and how LPA_5 distinguishes different ligands.

LPA_5 is expressed in platelets, small intestine, and dorsal root ganglia. LPA_5 KO mice, which are used as a sciatic nerve ligation model, are apparently normal, but not sensitive to injury-induced neuropathic pain [27]. As already described, LPA_1 KO mice were also insensitive to neuropathic pain. However, LPA_5 seems to have a downstream signaling pathway that is mechanistically distinct from the LPA_1 pathway, because nerve injury induced demyelination in LPA_5 KO mice but not in LPA_1 KO mice.

8.2.1.6 LPA_6 (*LPAR6*)

LPA_6 (P2Y5) is also a member of the P2Y family. It has high sequence homology to LPA_4. Patients who have mutations in *LPAR6* or *LIPH*, a gene coding an LPA-producing enzyme (PA-$PLA_1\alpha$), show the same congenital hypotrichia phenotypes [28, 29], which strongly suggests that P2Y5 is an LPA receptor (and PA-$PLA_1\alpha$ is an LPA-producing enzyme). However, conventional GPCR assays including Ca^{2+} and cAMP assays failed to detect LPA-induced activation of P2Y5. In 2009, Yanagida et al. found that in B103 stably expressing P2Y5, LPA induced significant neurite regression, and the LPA effect was blocked by ROCK inhibitor [30]. On the other hand, we found that PA-$PLA_1\alpha$ KO mice exhibited wavy hairs, which resembled hairs in mutant mice defective in tumor necrosis factor-α-converting enzyme (TACE), transforming growth factor-α (TGF-α), and epidermal growth factor receptor (EGFR). In addition, we showed that LPA as well as recombinant PA-$PLA_1\alpha$ itself induced P2Y5- and TACE-mediated ectodomain shedding of TGF-α in HEK293 cells [31]. These findings clearly showed that P2Y5 is a receptor for LPA. Now, P2Y5 is recognized as the sixth LPA receptor, LPA_6.

LPA_6 showed a marked preference for 2-acyl LPA with unsaturated fatty acid [30] and primarily coupled with $G\alpha_{12/13}$. In addition to being expressed in hair follicles, LPA_6 is highly expressed in endothelial cells of blood vessels [32], which suggests that LPA_6 is involved in endothelial functions such as blood vessel formation.

8.2.1.7 Other LPA receptors

LPA receptors other than those described here have been proposed: these include GPR87 and P2Y10 because cells stably expressing these GPCRs responded to LPA. However, further studies are needed to confirm these findings.

8.2.2 Receptors for LysoPS

Several orphan GPCRs have been recently identified as LysoPS receptors, including GPR34, P2Y10, A630033H20, and GPR174, all of which are P2Y family members. According to the nomenclature of lysophospholipid receptors, we propose that GPR34, P2Y10, A630033H20, and GPR174 be designated as LPS_1, LPS_2, LPS2-like (LPS_{2L}), and LPS_3, respectively. These four LysoPS receptors are encoded by the X chromosome and are highly expressed in immune cells, but their functions are largely unknown.

8.2.2.1 LPS_1/GPR34 (*LPSR1*)

In 2006, Sugo et al. identified GPR34 as the first receptor that is specific to LysoPS [33]. They showed that LysoPS inhibited forskolin-stimulated cAMP accumulation, promoted incorporation of [^{35}S]-GTPγS, and induced phosphorylation of ERK in GPR34-expressing cells. The response was completely abolished by treatment of pertussis toxin (PTX), indicating that GPR34 couples to Gi/Go-type G proteins.

Under normal breeding condition, GPR34 KO mice show no remarkable abnormalities in development, reproductive potential, size of organs, histology, and blood parameters, although they were slightly less active in the open field test and more active in the light arena in a light–dark test [34]. GPR34 is highly expressed in mononuclear cells of the immune system, suggesting that it has a role in the immune response. Accordingly, GPR34 KO mice (1) showed an increased delayed-type hypersensitivity response, (2) were more susceptible toward a disseminating *Cryptococcus neoformans* infection, and (3) accumulated significantly fewer granulocytes/macrophages in the spleen after methylated bovine serum albumin (BSA) immunization. They produced more cytokines (TNF-α, GM-CSF, interferon (IFN)-γ). These results show that GPR34 is important in appropriate immune response. Further studies are needed to elucidate what types of cells are involved.

8.2.2.2 LPS_2/P2Y10 (*LPSR2*)

In 2012, we identified three LysoPS receptors (P2Y10, A630033H20, and GPR174) by using a TGF-α shedding assay, which is a novel method for detecting GPCR activation [35] (see Chapter 28). These three LysoPS receptors are clustered in a range of about 200,000 bp on the X chromosomes. P2Y10 is coupled with $G\alpha_{12/13}$

but not with other G proteins. In 2008, it was reported that a fusion protein of P2Y10 and $G\alpha_{16}$ was activated by LPA and S1P [36]. However, our TGF-α shedding assay showed that P2Y10 is a LysoPS-specific receptor [35]. Neither LPA nor S1P induced P2Y10-dependent response, even in the presence of $G\alpha_{16}$. P2Y10 is highly expressed in thymus and spleen, suggesting that it has a role in the immune system. However, the *in vivo* role of P2Y10 is unknown.

8.2.2.3 LPS_{2L}/A630033H20 (*LPSR2L*)

As mentioned earlier, in 2012 we found that A630033H20 also reacted specifically with LysoPS [35]. Human A630033H20 has a truncated open reading frame, causing it to become a pseudogene (*P2RY10P2*). We confirmed that A630033H20 derived from mouse and rat species responded to LysoPS. A630033H20 is coupled with $G\alpha_{12/13}$. As A630033H20 is the closest homologue of P2Y10/LPS_2 (75 % homology to P2Y10 at the amino acid level) and has a similar expression pattern, we propose to name A630033H20 as LPS_{2L} (LPS_2-like).

8.2.2.4 LPS_3/GPR174 (*LPSR3*)

In 2012, we identified GPR174 as a LysoPS receptor [35]. In contrast to P2Y10 and A630033H20, GPR174 is coupled with $G\alpha_{13}$, but not $G\alpha_{12}$. GPR174 is also coupled with $G\alpha_s$ [37]. As is P2Y10, GPR174 is highly expressed in thymus and spleen, but its *in vivo* role is unknown.

8.2.2.5 Other LysoPS Receptors

It is reported that G2A expressed on macrophages was activated by LysoPS and enhanced apoptotic cell phagocytosis [38]. On the other hand, G2A is known to be the receptor for protons [39] and the oxidized fatty acid, 9-HODE [40]. In G2A-expressing cells, 9-HODE was found to trigger a TGF-α shedding response, but LysoPS did not, which suggests that G2A is not a LysoPS receptor.

LysoPS is known to promote degranulation of mast cells. Degranulation is triggered by aggregation of high-affinity IgE receptor (FcεRI). When LysoPS is added, degranulation is dramatically increased. This event required strictly LysoPS structure because conversion of the L-serine of LysoPS to D-serine abolishes the activity [41, 42]. Degranulation is also partially blocked by PTX pretreatment [33]. These results suggest that mast cells have a LysoPS-specific receptor that promotes degranulation. In the course of synthesis of an LysoPS structural analogue, we identified lysophosphatidylthreonine (LysoPT) as a strong promoter of degranulation (about tenfold more potent than LysoPS) [41]. However, LysoPT does not activate GPR34, P2Y10, A630033H20, or GPR174. Moreover, LysoPS promotes degranulation of mast cells from GPR34 KO, P2Y10 KO, and GPR174 KO mice. These results indicate that the LysoPS receptor that promotes degranulation is different from identified LysoPS receptors.

8.2.3 *Receptors for LPI*

The role of LPI was first demonstrated in 1986, when it was shown to stimulate the release of insulin from pancreatic cells [43]. Subsequent studies found that LPI is produced in various cell systems and that it induces a number of cellular events [44]. LPI is known as an agonist for two GPCRs (GPR55 and GPR119). However, GPR119 is activated by not only LPI, but also by other lysophospholipids including lysophosphatidylcholine [45, 46], so that in this section, we focus on GPR55 as an LPI-specific receptor.

8.2.3.1 GPR55

Human GPR55 is 319 amino acids in length, and its gene maps to human chromosome 2q37. A database search for sequences similar to human GPR55 revealed that GPR55 is conserved among vertebrates from fish to mammals. The closest homologues to GPR55, as judged by amino acid homology, are LPA_6/P2Y5 (29 %), LPA_4/GPR23 (30 %), GPR35 (27 %), and the chemokine receptor CCR4 (23 %). GPR55 was once proposed as the third receptor for cannabinoid. However, to date, the most potent ligand identified for GPR55 is LPI [47]. Although GPR55 clearly interacts with certain cannabinoid ligands that interact with cannabinoid receptors such as CB1 and CB2, it is not clear whether CB1/CB2-independent cannabinoid actions are mediated by GPR55.

GPR55 appears primarily to be coupled with $G\alpha_{13}$. GPR55 is expressed in osteoclasts, and GPR55 activation in the cells results in osteoclastogenesis, cell polarization, and bone resorption [48]. Knockout of GPR55 in male mice significantly increased the numbers of osteoclasts. These findings indicate that the LPI–GPR55 axis affects differentiation and proliferation of osteoclasts and thus regulates bone metabolism. GPR55 KO mice were also resistant to mechanical hyperalgesia associated with Freund's complete adjuvant (FCA)-induced inflammation or partial nerve ligation [49]. In GPR55 KO female mice, the onset of experimentally autoimmune encephalomyelitis was delayed, and the symptoms were less severe than those in WT mice [50].

8.3 Synthetic Pathways and Enzymes for Lysophospholipids

LPLs are mainly produced by deacylation mediated by phospholipase A_1 (PLA_1) or phospholipase A_2 (PLA_2). For LPA, phospholipase D (PLD) is involved in addition to PLA_1/PLA_2 (Fig. 8.1). Although a number of enzymes that are able to produce LPLs *in vitro* have been identified, a few of them were proved to be producing enzymes for LPL mediators. Many PLA_1/PLA_2 enzymes appear to be involved in the release of arachidonic acid from phospholipids to be used in the arachidonate cascade or fatty acid remodeling of phospholipids.

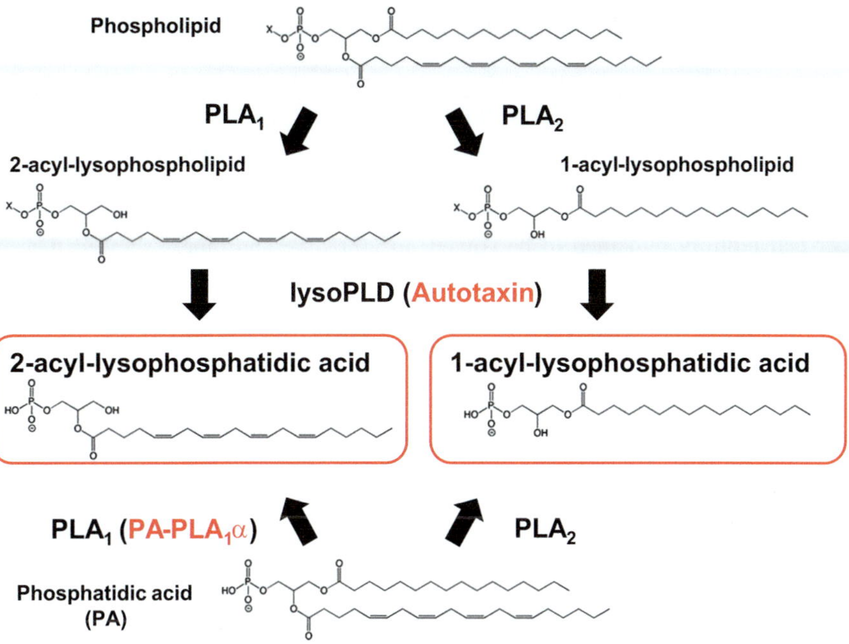

Fig. 8.1 Lysophosphatidic acid (LPA)-producing pathway. There are two pathways of LPA production. First, lysophospholipids generated by phospholipase A_1 (PLA_1) or PLA_2 reaction are subsequently converted to LPA by lysophospholipase D (lysoPLD) reaction. Second, phosphatidic acid (PA) is converted to LPA by the PLA_1 or PLA_2 reaction. In extracellular LPA production, mainly two phospholipases, autotaxin (*ATX*)/lysoPLD and phosphatidic acid-specific phospholipase $A_1\alpha$ (PA-$PLA_1\alpha$), are crucial

8.3.1 Synthetic Pathways and Enzymes for LPA

At least two pathways have been postulated for bioactive lipid LPA production. In the first pathway, lysophospholipids are produced by phospholipase A_1 (PLA_1) or phospholipase A_2 (PLA_2) and then converted to LPA by lysophospholipase D (lysoPLD). In the other pathway, phosphatidic acid is hydrolyzed by PLA_1 or PLA_2, generating LPA. Extracellular LPA production mainly involves two phospholipases, autotaxin (ATX)/lysoPLD and phosphatidic acid-selective phospholipase $A_1\alpha$ (PA-$PLA_1\alpha$). Inside the cell, LPA is generated as intermediates by addition of fatty acyl-CoA to glycerol-3-phosphate in the course of de novo phospholipid biosynthesis. This LPA-producing pathway is conserved in lower organisms such as *Escherichia coli* that do not have LPA receptors. Currently, there is no evidence that this pathway contributes to the extracellular functions of LPA production. In this section, we describe the latest findings in LPA-producing enzymes, especially ATX and PA-$PLA_1\alpha$.

8.3.1.1 LPA Production by ATX

ATX is a member of the nucleotide pyrophosphate phosphodiesterase (NPP) family and is also referred to as NPP2. Some NPP family members such as NPP2 and NPP3 hydrolyze nucleotides such as ATP. ATX was originally found as an autocrine motility-stimulating factor identified in the culture cell supernatant of human malignant melanoma cells [51]. However, the mechanism by which ATX stimulated cancer cell motility remained to be solved. In 2002, we and others showed that an enzyme (lysoPLD) that acts on various LPLs and produces LPA by its phosphodiesterase activity is identical to ATX [52, 53]. ATX is a secretory glycoprotein with a molecular weight about 100 kDa. Multiple ATX isoforms are present, generated by alternative splicing sites. ATX activity depends on divalent cations such as Co^{2+} or Mn^{2+}. The preferred substrates of ATX are LPLs such as LPC with short fatty acids (12:0, 14:0, 16:0) or unsaturated fatty acids (18:2, 20:4). Recently, the crystal structure of ATX was determined, showing that these substrate specificities can be explained by the structure of the catalytic pocket [54, 55]. ATX has a deep hydrophobic pocket for substrates that other NPP family members do not have: LPA acyl chains are accommodated in this hydrophobic pocket. Unsaturated fatty acids bend sharply at the unsaturated bonds, which also confirmed that highly unsaturated fatty acids bind more stably.

In mice, ATX knockout (KO) is lethal at around embryonic day 10.5 because of impaired vessel formation in the yolk sac and embryo, suggesting that ATX-LPA signaling has an essential role in vessel formation during embryonic development [20, 56]. Because ATX KO mice are embryonically lethal, the pathophysiological functions of ATX in the adult stage are largely unknown. ATX is present in various biological fluids such as blood, seminal fluid, and cerebrospinal fluid [57]. Thus, by measuring lysoPLD activity or directly measuring ATX protein using specific antibodies, the ATX level in such biological fluids can be determined, providing important insights into ATX function in human pathophysiology. ATX concentrations are significantly higher in serum from patients with liver cirrhosis and follicular lymphoma and from pregnant women [58–60]. Interestingly, serum ATX levels in patients with pregnancy-induced hypertension were found to be significantly lower than those in normal pregnancy controls [60]. In cholestatic pruritus, the intensity of itching is significantly correlated with ATX concentration [61], which raises the possibility that the itching is caused by ATX-LPA signaling. ATX is also highly expressed in particular endothelial cells such as cells of the choroid plexus and high endothelial venules (HEV) [62, 63]. The HEV is where lymphocytes pass from the bloodstream into lymphoid tissue. Experiments with an ATX inhibitor revealed that ATX promotes the adhesion of lymphocytes to intercellular adhesion molecule (ICAM) on the surface of HEV endothelial cells and promotes their transmigration across the basal lamina of HEV [64]. ATX have been implicated in the progression of arthritis [65], lung fibrosis [66], and kidney fibrosis. ATX promotes cancer cell motility via the LPA_1 receptor *in vitro* [67]. Some malignant tumors such as glioblastoma, non-small cell lung cancer, and breast cancer strongly express

ATX, although the role of ATX in these tumors is unknown. Mice overexpressing ATX exhibit a higher rate of the initiation and metastasis of breast cancer, which suggests that ATX is related to the initiation and progression of cancer [68].

8.3.1.2 LPA Production by PA-PLA$_1\alpha$

Phosphatidic acid-selective phospholipase A$_1$ (PA-PLA$_1\alpha$) has sequence similarity to phosphatidylserine-specific phospholipase A$_1$ (PS-PLA$_1$). These enzymes, both with molecular weight about 55 kDa, are classified to the pancreatic lipase family, including hepatic lipase and lipoprotein lipase [69]. Although PA-PLA$_1\alpha$ has a signal sequence at the N-terminus as do other lipases, it is mainly detected in the membrane fraction in both tissues and cultured cells, indicating that PA-PLA$_1\alpha$ is a membrane-binding protein rather than a secreted protein. PA-PLA$_1\alpha$-expressing cells constantly produce LPA with an unsaturated fatty acid at the *sn*-2 position of the glycerol backbone from phosphatidic acid (PA). Interestingly, LPA$_3$ and LPA$_6$ receptors are more strongly activated by 2-acyl LPA than by 1-acyl LPA. Therefore, PA-PLA$_1\alpha$ might supply the ligand for these LPA receptors.

PA-PLA$_1\alpha$, which is encoded by *LIPH*, is the gene responsible for autosomal recessive hypotrichosis [28]. Many mutations in *LIPH* have been identified in some family lineages with hypotrichosis [70–72]. The LPA$_6$ receptor has also been identified as a causative gene for the same disorder [29, 73], which suggests that LPA produced by PA-PLA$_1\alpha$ promotes the formation of hair follicles. PA-PLA$_1\alpha$ KO mice showed a hair abnormality (wooly hair) as was observed in humans [31]. That study also demonstrated that, in wild-type mice, PA-PLA$_1\alpha$ and LPA$_6$ receptor are highly expressed on epidermal cells in the inner root sheath of hair follicles. Together, these findings show that PA-PLA$_1\alpha$ and LPA$_6$ receptor regulates growth or differentiation of epidermal cells in the follicles. Interestingly, similar abnormal hair (wooly hair) was also found in knockout mice of TACE or TGF-α and in mice with mutations in epidermal growth factor receptor genes. The activation of LPA$_6$ induces the TACE-dependent-ectodomain shedding of TGF-α, followed by activation of EGFR, described earlier. It is now accepted that a PA-PLA$_1\alpha$–LPA–LPA$_6$ axis regulates differentiation and maturation of hair follicles via a TACE–TGF-α–EGFR pathway.

8.3.2 *Synthetic Pathways for Lysophosphatidylserine and Lysophosphatidylinositol*

Although the synthetic pathway of LPA is well known, the synthetic pathways for LysoPS and LPI remained to be solved. Analyses of these LPLs in tissues and cells by a recently developed LC-MS/MS-based method revealed that they are mixtures of 1-acyl-LPLs and 2-acyl-LPLs, indicating that both PLA$_1$ and PLA$_2$ are involved.

8.3.2.1 Production of LysoPS

Rat platelets express two extracellular PLA enzymes, secretory PLA_2 group IIA ($sPLA_2IIA$) and PS-specific PLA_1 ($PS\text{-}PLA_1$), and secrete them upon activation (Fig. 8.2). In the course of activation of rat platelets, $sPLA_2$-IIA and $PS\text{-}PLA_1$ are thought to participate in production of 1-acyl LysoPS and 2-acyl LysoPS, respectively.

$PS\text{-}PLA_1$ is stored in granules in rat platelets and is secreted into the medium when activated. Although $PS\text{-}PLA_1$ is structurally homologous to triglyceride (TG) lipase, it selectively hydrolyzes PS and does not have lipase activity for TG [74]. The amino acid sequence of $PS\text{-}PLA_1$ showed about 30 % identity to an LPA-producing enzyme, $PA\text{-}PLA_1\alpha$, and $PS\text{-}PLA_1$ expression is dramatically induced at the mRNA and protein levels by various inflammatory stimuli. For these reasons, $PS\text{-}PLA_1$ is the most likely candidate for an *in vivo* LysoPS-producing enzyme.

Because secretory PLA_2 and $PS\text{-}PLA_1$ are secreted proteins, they should act on PS extracellularly. However, PS exists in the inner leaflet of the plasma membrane. Recently, TMEM16F was identified as a scramblase that triggers exposure of PS in activated platelets [75]. Although LysoPS exists at only several nanomoles (nM) in mouse plasma, it exists at about 100 nM in mouse serum, and especially LysoPS species with unsaturated fatty acids are abundant. These observations suggest that LysoPS is produced as follows: PS is exposed outside the activated platelets by TMEM16F during serum preparation or in wound sites and then deacylated by a PLA_1 such as $PS\text{-}PLA_1$.

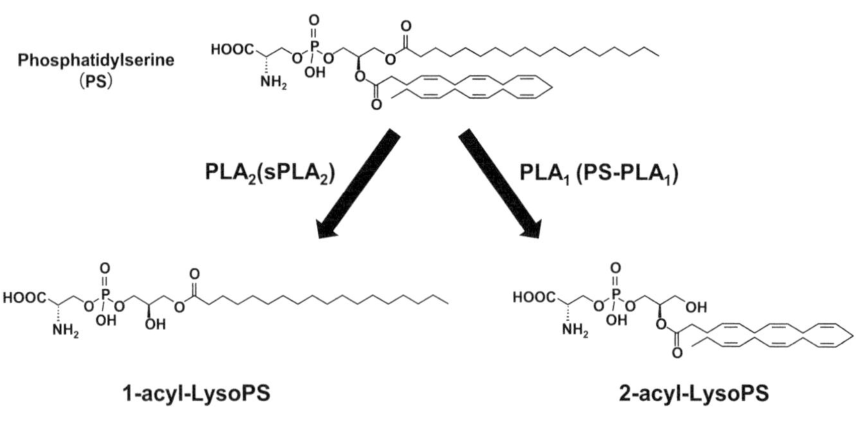

Fig. 8.2 Lysophosphatidylserine (LysoPS)-producing pathway. LysoPS is produced from phosphatidylserine (PS) via phospholipase A. PLA_2 enzymes hydrolyze the acyl chain at *sn*-2 position and produce 1-acyl-LysoPS with saturated fatty acids; PLA_1 enzymes hydrolyze the acyl chain at *sn*-1 position and produce 2-acyl-LysoPS with unsaturated fatty acids

8.3.2.2 Production of LPI

Although the tissue distribution of lysophosphatidylinositol (LPI) is mostly unknown, a mass spectrometer analysis showed that LPI exists inside and outside of cells. Although LPI is thought to be produced inside and outside of cells by PLA_1 and/or PLA_2, this has not been demonstrated *in vivo*. In this section, we review candidate enzymes involved in the production of LPI.

The *sn*-1 and *sn*-2 positions of phospholipids usually have saturated and unsaturated fatty acids, respectively. In phosphatidylinositol (PI), the *sn*-1 and *sn*-2 positions predominantly have stearic acid (18:0) and arachidonic acid (20:4), respectively.

Caenorhabditis elegans synthesizes PI (with oleic acid (18:1) in the *sn*-1 position) by a *de novo* pathway. The *sn*-1 linkage is hydrolyzed by an intracellular PLA_1, producing LPI. Subsequently, stearic acid is incorporated into the *sn*-1 position [76], which means that the PLA_1 is involved in remodeling the *sn*-1 position of PI.

Mammals have three intracellular isoforms of PLA_1: $iPLA_1\alpha$ (DDHD1/PA-PLA_1), $iPLA_1\beta$ (p125), and $iPLA_1\gamma$ (DDHD2/KIAA0725). $iPLA_1\alpha$ purified from the cytosolic fraction of bovine testis was reported to predominantly hydrolyze phosphatidic acid (PA) [77], but it was later found to hydrolyze other phospholipids including PE and PI. Although a PLA_2 that specifically cleaves PI has not been reported currently, $cPLA_2$ (a type of cytosolic PLA_2) and $iPLA_2$ may be involved in the production of LPI. In thyroid PCCL-3 cells, $cPLA_2\alpha$ was found to cleave PI and produce 1-acyl-LPI and GPI [78]. There is no report that cells expressing LPI-producing enzymes stimulate cells expressing GPR55. Furthermore, it is unclear how LPI is released from inside the cell to outside and then activates GPR55. Recent advances in mass spectrometry make it possible to quantify LPI easily and can help to reveal the production pathway of LPI.

8.4 Conclusion

In contrast to LPA, which is relatively well characterized, LysoPS and LPI are just beginning to be investigated. Receptors for LPI and LysoPS have only recently been identified. In the future, physiological and pathophysiological roles of LPI and LysoPS as novel lysophospholipid mediators will be elucidated through functional analysis of these receptors.

References

1. Kihara Y, Maceyka M, Spiegel S, Chun J (2014) Lysophospholipid receptor nomenclature review: IUPHAR Review 8. Br J Pharmacol 171:3575–3594
2. Moolenaar WH, van Corven EJ (1990) Growth factor-like action of lysophosphatidic acid: mitogenic signalling mediated by G proteins. Ciba Found Symp 150:99–106, discussion 106–111

3. Hecht JH, Weiner JA, Post SR, Chun J (1996) Ventricular zone gene-1 (vzg-1) encodes a lysophosphatidic acid receptor expressed in neurogenic regions of the developing cerebral cortex. J Cell Biol 135:1071–1083
4. Noguchi K, Ishii S, Shimizu T (2003) Identification of p2y9/GPR23 as a novel G protein-coupled receptor for lysophosphatidic acid, structurally distant from the Edg family. J Biol Chem 278:25600–25606
5. Estivill-Torrus G, Llebrez-Zayas P, Matas-Rico E, Santin L, Pedraza C, De Diego I, Del Arco I, Fernandez-Llebrez P, Chun J, De Fonseca FR (2008) Absence of LPA1 signaling results in defective cortical development. Cereb Cortex 18:938–950
6. Inoue M, Rashid MH, Fujita R, Contos JJ, Chun J, Ueda H (2004) Initiation of neuropathic pain requires lysophosphatidic acid receptor signaling. Nat Med 10:712–718
7. Tager AM, LaCamera P, Shea BS, Campanella GS, Selman M, Zhao Z, Polosukhin V, Wain J, Karimi-Shah BA, Kim ND et al (2008) The lysophosphatidic acid receptor LPA1 links pulmonary fibrosis to lung injury by mediating fibroblast recruitment and vascular leak. Nat Med 14:45–54
8. Castelino FV, Seiders J, Bain G, Brooks SF, King CD, Swaney JS, Lorrain DS, Chun J, Luster AD, Tager AM (2011) Amelioration of dermal fibrosis by genetic deletion or pharmacologic antagonism of lysophosphatidic acid receptor 1 in a mouse model of scleroderma. Arthritis Rheum 63:1405–1415
9. Gennero I, Laurencin-Dalicieux S, Conte-Auriol F, Briand-Mesange F, Laurencin D, Rue J, Beton N, Malet N, Mus M, Tokumura A et al (2011) Absence of the lysophosphatidic acid receptor LPA1 results in abnormal bone development and decreased bone mass. Bone 49:395–403
10. An S, Bleu T, Hallmark OG, Goetzl EJ (1998) Characterization of a novel subtype of human G protein-coupled receptor for lysophosphatidic acid. J Biol Chem 273:7906–7910
11. Contos JJ, Ishii I, Fukushima N, Kingsbury MA, Ye X, Kawamura S, Brown JH, Chun J (2002) Characterization of lpa(2) (Edg4) and lpa(1)/lpa(2) (Edg2/Edg4) lysophosphatidic acid receptor knockout mice: signaling deficits without obvious phenotypic abnormality attributable to lpa(2). Mol Cell Biol 22:6921–6929
12. Lin S, Wang D, Iyer S, Ghaleb AM, Shim H, Yang VW, Chun J, Yun CC (2009) The absence of LPA2 attenuates tumor formation in an experimental model of colitis-associated cancer. Gastroenterology 136:1711–1720
13. Emo J, Meednu N, Chapman TJ, Rezaee F, Balys M, Randall T, Rangasamy T, Georas SN (2012) Lpa2 is a negative regulator of both dendritic cell activation and murine models of allergic lung inflammation. J Immunol 188:3784–3790
14. Bandoh K, Aoki J, Hosono H, Kobayashi S, Kobayashi T, Murakami-Murofushi K, Tsujimoto M, Arai H, Inoue K (1999) Molecular cloning and characterization of a novel human G-protein-coupled receptor, EDG7, for lysophosphatidic acid. J Biol Chem 274:27776–27785
15. Ye X, Hama K, Contos JJ, Anliker B, Inoue A, Skinner MK, Suzuki H, Amano T, Kennedy G, Arai H et al (2005) LPA3-mediated lysophosphatidic acid signalling in embryo implantation and spacing. Nature 435:104–108
16. Ziecik AJ, Waclawik A, Bogacki M (2008) Conceptus signals for establishment and maintenance of pregnancy in pigs: lipid signaling system. Exp Clin Endocrinol Diabetes 116:443–449
17. Liszewska E, Reinaud P, Dubois O, Charpigny G (2012) Lysophosphatidic acid receptors in ovine uterus during estrous cycle and early pregnancy and their regulation by progesterone. Domest Anim Endocrinol 42:31–42
18. Achache H, Tsafrir A, Prus D, Reich R, Revel A (2010) Defective endometrial prostaglandin synthesis identified in patients with repeated implantation failure undergoing in vitro fertilization. Fertil Steril 94:1271–1278
19. Sumida H, Noguchi K, Kihara Y, Abe M, Yanagida K, Hamano F, Sato S, Tamaki K, Morishita Y, Kano MR et al (2010) LPA4 regulates blood and lymphatic vessel formation during mouse embryogenesis. Blood 116:5060–5070

20. Tanaka M, Okudaira S, Kishi Y, Ohkawa R, Iseki S, Ota M, Noji S, Yatomi Y, Aoki J, Arai H (2006) Autotaxin stabilizes blood vessels and is required for embryonic vasculature by producing lysophosphatidic acid. J Biol Chem 281:25822–25830
21. Yukiura H, Hama K, Nakanaga K, Tanaka M, Asaoka Y, Okudaira S, Arima N, Inoue A, Hashimoto T, Arai H et al (2011) Autotaxin regulates vascular development via multiple lysophosphatidic acid (LPA) receptors in zebrafish. J Biol Chem 286:43972–43983
22. Lee CW, Rivera R, Gardell S, Dubin AE, Chun J (2006) GPR92 as a new G12/13- and Gq-coupled lysophosphatidic acid receptor that increases cAMP, LPA5. J Biol Chem 281:23589–23597
23. Kotarsky K, Boketoft A, Bristulf J, Nilsson NE, Norberg A, Hansson S, Owman C, Sillard R, Leeb-Lundberg LM, Olde B (2006) Lysophosphatidic acid binds to and activates GPR92, a G protein-coupled receptor highly expressed in gastrointestinal lymphocytes. J Pharmacol Exp Ther 318:619–628
24. Williams JR, Khandoga AL, Goyal P, Fells JI, Perygin DH, Siess W, Parrill AL, Tigyi G, Fujiwara Y (2009) Unique ligand selectivity of the GPR92/LPA5 lysophosphatidate receptor indicates role in human platelet activation. J Biol Chem 284:17304–17319
25. Choi S, Lee M, Shiu AL, Yo SJ, Aponte GW (2007) Identification of a protein hydrolysate responsive G protein-coupled receptor in enterocytes. Am J Physiol Gastrointest Liver Physiol 292:G98–G112
26. Oh DY, Yoon JM, Moon MJ, Hwang JI, Choe H, Lee JY, Kim JI, Kim S, Rhim H, O'Dell DK et al (2008) Identification of farnesyl pyrophosphate and N-arachidonylglycine as endogenous ligands for GPR92. J Biol Chem 283:21054–21064
27. Lin ME, Rivera RR, Chun J (2012) Targeted deletion of LPA5 identifies novel roles for lysophosphatidic acid signaling in development of neuropathic pain. J Biol Chem 287:17608–17617
28. Kazantseva A, Goltsov A, Zinchenko R, Grigorenko AP, Abrukova AV, Moliaka YK, Kirillov AG, Guo Z, Lyle S, Ginter EK et al (2006) Human hair growth deficiency is linked to a genetic defect in the phospholipase gene LIPH. Science 314:982–985
29. Shimomura Y, Wajid M, Ishii Y, Shapiro L, Petukhova L, Gordon D, Christiano AM (2008) Disruption of P2RY5, an orphan G protein-coupled receptor, underlies autosomal recessive woolly hair. Nat Genet 40:335–339
30. Yanagida K, Masago K, Nakanishi H, Kihara Y, Hamano F, Tajima Y, Taguchi R, Shimizu T, Ishii S (2009) Identification and characterization of a novel lysophosphatidic acid receptor, p2y5/LPA(6). J Biol Chem 284:17731–17741
31. Inoue A, Arima N, Ishiguro J, Prestwich GD, Arai H, Aoki J (2011) LPA-producing enzyme PA-PLA(1)alpha regulates hair follicle development by modulating EGFR signalling. EMBO J 30:4248–4260
32. Ren Y, Guo L, Tang X, Apparsundaram S, Kitson C, Deguzman J, Fuentes ME, Coyle L, Majmudar R, Allard J et al (2013) Comparing the differential effects of LPA on the barrier function of human pulmonary endothelial cells. Microvasc Res 85:59–67
33. Sugo T, Tachimoto H, Chikatsu T, Murakami Y, Kikukawa Y, Sato S, Kikuchi K, Nagi T, Harada M, Ogi K et al (2006) Identification of a lysophosphatidylserine receptor on mast cells. Biochem Biophys Res Commun 341:1078–1087
34. Liebscher I, Muller U, Teupser D, Engemaier E, Engel KM, Ritscher L, Thor D, Sangkuhl K, Ricken A, Wurm A et al (2011) Altered immune response in mice deficient for the G protein-coupled receptor GPR34. J Biol Chem 286:2101–2110
35. Inoue A, Ishiguro J, Kitamura H, Arima N, Okutani M, Shuto A, Higashiyama S, Ohwada T, Arai H, Makide K et al (2012) TGFalpha shedding assay: an accurate and versatile method for detecting GPCR activation. Nat Methods 9:1021–1029
36. Murakami M, Shiraishi A, Tabata K, Fujita N (2008) Identification of the orphan GPCR, P2Y(10) receptor as the sphingosine-1-phosphate and lysophosphatidic acid receptor. Biochem Biophys Res Commun 371:707–712

37. Sugita K, Yamamura C, Tabata K, Fujita N (2013) Expression of orphan G-protein coupled receptor GPR174 in CHO cells induced morphological changes and proliferation delay via increasing intracellular cAMP. Biochem Biophys Res Commun 430:190–195
38. Frasch SC, Fernandez-Boyanapalli RF, Berry KZ, Leslie CC, Bonventre JV, Murphy RC, Henson PM, Bratton DL (2011) Signaling via macrophage G2A enhances efferocytosis of dying neutrophils by augmentation of Rac activity. J Biol Chem 286:12108–12122
39. Murakami N, Yokomizo T, Okuno T, Shimizu T (2004) G2A is a proton-sensing G-protein-coupled receptor antagonized by lysophosphatidylcholine. J Biol Chem 279:42484–42491
40. Obinata H, Hattori T, Nakane S, Tatei K, Izumi T (2005) Identification of 9-hydroxyoctadecadienoic acid and other oxidized free fatty acids as ligands of the G protein-coupled receptor G2A. J Biol Chem 280:40676–40683
41. Iwashita M, Makide K, Nonomura T, Misumi Y, Otani Y, Ishida M, Taguchi R, Tsujimoto M, Aoki J, Arai H et al (2009) Synthesis and evaluation of lysophosphatidylserine analogues as inducers of mast cell degranulation. Potent activities of lysophosphatidylthreonine and its 2-deoxy derivative. J Med Chem 52:5837–5863
42. Chang HW, Inoue K, Bruni A, Boarato E, Toffano G (1988) Stereoselective effects of lysophosphatidylserine in rodents. Br J Pharmacol 93:647–653
43. Metz SA (1986) Lysophosphatidylinositol, but not lysophosphatidic acid, stimulates insulin release. A possible role for phospholipase A2 but not de novo synthesis of lysophospholipid in pancreatic islet function. Biochem Biophys Res Commun 138:720–727
44. Pineiro R, Falasca M (2012) Lysophosphatidylinositol signalling: new wine from an old bottle. Biochim Biophys Acta 1821:694–705
45. Andersson DA, Nash M, Bevan S (2007) Modulation of the cold-activated channel TRPM8 by lysophospholipids and polyunsaturated fatty acids. J Neurosci 27:3347–3355
46. Soga T, Ohishi T, Matsui T, Saito T, Matsumoto M, Takasaki J, Matsumoto S, Kamohara M, Hiyama H, Yoshida S et al (2005) Lysophosphatidylcholine enhances glucose-dependent insulin secretion via an orphan G-protein-coupled receptor. Biochem Biophys Res Commun 326:744–751
47. Yamashita A, Oka S, Tanikawa T, Hayashi Y, Nemoto-Sasaki Y, Sugiura T (2013) The actions and metabolism of lysophosphatidylinositol, an endogenous agonist for GPR55. Prostaglandins Other Lipid Mediat 107:103–116
48. Whyte LS, Ryberg E, Sims NA, Ridge SA, Mackie K, Greasley PJ, Ross RA, Rogers MJ (2009) The putative cannabinoid receptor GPR55 affects osteoclast function in vitro and bone mass in vivo. Proc Natl Acad Sci U S A 106:16511–16516
49. Staton PC, Hatcher JP, Walker DJ, Morrison AD, Shapland EM, Hughes JP, Chong E, Mander PK, Green PJ, Billinton A et al (2008) The putative cannabinoid receptor GPR55 plays a role in mechanical hyperalgesia associated with inflammatory and neuropathic pain. Pain 139:225–236
50. Sisay S, Pryce G, Jackson SJ, Tanner C, Ross RA, Michael GJ, Selwood DL, Giovannoni G, Baker D (2013) Genetic background can result in a marked or minimal effect of gene knockout (GPR55 and CB2 receptor) in experimental autoimmune encephalomyelitis models of multiple sclerosis. PLoS One 8, e76907
51. Stracke ML, Krutzsch HC, Unsworth EJ, Arestad A, Cioce V, Schiffmann E, Liotta LA (1992) Identification, purification, and partial sequence analysis of autotaxin, a novel motility-stimulating protein. J Biol Chem 267:2524–2529
52. Tokumura A, Majima E, Kariya Y, Tominaga K, Kogure K, Yasuda K, Fukuzawa K (2002) Identification of human plasma lysophospholipase D, a lysophosphatidic acid-producing enzyme, as autotaxin, a multifunctional phosphodiesterase. J Biol Chem 277:39436–39442
53. Umezu-Goto M, Kishi Y, Taira A, Hama K, Dohmae N, Takio K, Yamori T, Mills GB, Inoue K, Aoki J et al (2002) Autotaxin has lysophospholipase D activity leading to tumor cell growth and motility by lysophosphatidic acid production. J Cell Biol 158:227–233

54. Hausmann J, Kamtekar S, Christodoulou E, Day JE, Wu T, Fulkerson Z, Albers HM, van Meeteren LA, Houben AJ, van Zeijl L et al (2011) Structural basis of substrate discrimination and integrin binding by autotaxin. Nat Struct Mol Biol 18:198–204
55. Nishimasu H, Okudaira S, Hama K, Mihara E, Dohmae N, Inoue A, Ishitani R, Takagi J, Aoki J, Nureki O (2011) Crystal structure of autotaxin and insight into GPCR activation by lipid mediators. Nat Struct Mol Biol 18:205–212
56. van Meeteren LA, Ruurs P, Stortelers C, Bouwman P, van Rooijen MA, Pradere JP, Pettit TR, Wakelam MJ, Saulnier-Blache JS, Mummery CL et al (2006) Autotaxin, a secreted lysophospholipase D, is essential for blood vessel formation during development. Mol Cell Biol 26:5015–5022
57. Tanaka M, Kishi Y, Takanezawa Y, Kakehi Y, Aoki J, Arai H (2004) Prostatic acid phosphatase degrades lysophosphatidic acid in seminal plasma. FEBS Lett 571:197–204
58. Masuda A, Nakamura K, Izutsu K, Igarashi K, Ohkawa R, Jona M, Higashi K, Yokota H, Okudaira S, Kishimoto T et al (2008) Serum autotaxin measurement in haematological malignancies: a promising marker for follicular lymphoma. Br J Haematol 143:60–70
59. Nakamura K, Igarashi K, Ide K, Ohkawa R, Okubo S, Yokota H, Masuda A, Oshima N, Takeuchi T, Nangaku M et al (2008) Validation of an autotaxin enzyme immunoassay in human serum samples and its application to hypoalbuminemia differentiation. Clin Chim Acta 388:51–58
60. Masuda A, Fujii T, Iwasawa Y, Nakamura K, Ohkawa R, Igarashi K, Okudaira S, Ikeda H, Kozuma S, Aoki J et al (2011) Serum autotaxin measurements in pregnant women: application for the differentiation of normal pregnancy and pregnancy-induced hypertension. Clin Chim Acta 412:1944–1950
61. Kremer AE, Martens JJWW, Kulik W, Rueff F, Kuiper EMM, van Buuren HR, van Erpecum KJ, Kondrackiene J, Prieto J, Rust C et al (2010) Lysophosphatidic acid is a potential mediator of cholestatic pruritus. Gastroenterology 139:1008–1018
62. Kanda H, Newton R, Klein R, Morita Y, Gunn MD, Rosen SD (2008) Autotaxin, an ectoenzyme that produces lysophosphatidic acid, promotes the entry of lymphocytes into secondary lymphoid organs. Nat Immunol 9:415–423
63. Nakasaki T, Tanaka T, Okudaira S, Hirosawa M, Umemoto E, Otani K, Jin S, Bai Z, Hayasaka H, Fukui Y et al (2008) Involvement of the lysophosphatidic acid-generating enzyme autotaxin in lymphocyte-endothelial cell interactions. Am J Pathol 173:1566–1576
64. Zhang Y, Chen YC, Krummel MF, Rosen SD (2012) Autotaxin through lysophosphatidic acid stimulates polarization, motility, and transendothelial migration of naive T cells. J Immunol 189:3914–3924
65. Inoue M, Ma L, Aoki J, Chun J, Ueda H (2008) Autotaxin, a synthetic enzyme of lysophosphatidic acid (LPA), mediates the induction of nerve-injured neuropathic pain. Mol Pain 4:6
66. Oikonomou N, Thanasopoulou A, Tzouvelekis A, Harokopos V, Paparountas T, Nikitopoulou I, Witke W, Karameris A, Kotanidou A, Bouros D et al (2009) Gelsolin expression is necessary for the development of modelled pulmonary inflammation and fibrosis. Thorax 64:467–475
67. Hama K, Aoki J, Fukaya M, Kishi Y, Sakai T, Suzuki R (2004) Lysophosphatidic acid and autotaxin stimulate cell motility of neoplastic and non-neoplastic cells through LPA(1). J Biol Chem 279:17634–17639
68. Liu SY, Umezu-Goto M, Murph M, Lu YL, Liu WB, Zhang F, Yu SX, Stephens LC, Cui XJ, Murrow G et al (2009) Expression of autotaxin and lysophosphatidic acid receptors increases mammary tumorigenesis, invasion, and metastases. Cancer Cell 15:539–550
69. Sonoda H, Aoki J, Hiramatsu T, Ishida M, Bandoh K, Nagai Y, Taguchi R, Inoue K, Arai H (2002) A novel phosphatidic acid-selective phospholipase A1 that produces lysophosphatidic acid. J Biol Chem 277:34254–34263
70. Shinkuma S, Akiyama M, Inoue A, Aoki J, Natsuga K, Nomura T, Arita K, Abe R, Ito K, Nakamura H et al (2010) Prevalent LIPH founder mutations lead to loss of P2Y5 activation ability of PA-PLA1alpha in autosomal recessive hypotrichosis. Hum Mutat 31:602–610

71. Ali G, Chishti MS, Raza SI, John P, Ahmad W (2007) A mutation in the lipase H (LIPH) gene underlie autosomal recessive hypotrichosis. Hum Genet 121:319–325
72. Tariq M, Azhar A, Baig SM, Dahl N, Klar J (2012) A novel mutation in the lipase H gene underlies autosomal recessive hypotrichosis and woolly hair. Sci Rep 2:730
73. Pasternack SM, von Kugelgen I, Al Aboud K, Lee YA, Ruschendorf F, Voss K, Hillmer AM, Molderings GJ, Franz T, Ramirez A et al (2008) G protein-coupled receptor P2Y5 and its ligand LPA are involved in maintenance of human hair growth. Nat Genet 40:329–334
74. Sato T, Aoki J, Nagai Y, Dohmae N, Takio K, Doi T, Arai H, Inoue K (1997) Serine phospholipid-specific phospholipase A that is secreted from activated platelets. A new member of the lipase family. J Biol Chem 272:2192–2198
75. Suzuki J, Umeda M, Sims PJ, Nagata S (2010) Calcium-dependent phospholipid scrambling by TMEM16F. Nature 468:834–838
76. Imae R, Inoue T, Kimura M, Kanamori T, Tomioka NH, Kage-Nakadai E, Mitani S, Arai H (2010) Intracellular phospholipase A1 and acyltransferase, which are involved in *Caenorhabditis elegans* stem cell divisions, determine the sn-1 fatty acyl chain of phosphatidylinositol. Mol Biol Cell 21:3114–3124
77. Higgs HN, Han MH, Johnson GE, Glomset JA (1998) Cloning of a phosphatidic acid-preferring phospholipase A1 from bovine testis. J Biol Chem 273:5468–5477
78. Mariggio S, Sebastia J, Filippi BM, Iurisci C, Volonte C, Amadio S, De Falco V, Santoro M, Corda D (2006) A novel pathway of cell growth regulation mediated by a PLA2alpha-derived phosphoinositide metabolite. FASEB J 20:2567–2569

Chapter 9
Sphingolipid Metabolism via Sphingosine 1-Phosphate and Its Role in Physiology, Pathology, and Nutrition

Akio Kihara

Abstract A delicate balance between the synthesis and degradation of sphingolipids must be kept to maintain cellular sphingolipid levels. Otherwise, cellular functions are impaired, leading to various disorders. Complex sphingolipids are degraded to sphingosine by the actions of a series of lysosomal hydrolases, mutations in the genes of which are known to be responsible for approximately ten sphingolipid storage diseases to date. The resultant sphingosine is either recycled to sphingolipids or metabolized to glycerophospholipids via sphingosine 1-phosphate (S1P). Extracellular S1P is well known to be a lipid mediator, whereas intracellular S1P is a key intermediate of the sphingolipid metabolic pathway linking sphingolipids to glycerophospholipids. This pathway is important for sphingolipid homeostasis, and its impairment results in several harmful effects on cells and tissues. We have recently identified and described in detail the downstream metabolic pathway of S1P. S1P is metabolized to palmitoyl-CoA via *trans*-2-hexadecenal, trans-2-hexadecenoic acid, and *trans*-2-hexadecenoyl-CoA, and then incorporated into glycerophospholipids. One of the genes involved in this pathway is *ALDH3A2*, the causative gene of Sjögren–Larsson syndrome. This review focuses on the physiological, pathological, and nutritional aspects of S1P as an intermediate of the sphingolipid-metabolic pathway.

Keywords Sphingolipid • Sphingosine 1-phosphate • Metabolism • Glycerophospholipid • Sphingolipid storage disease • Sjögren–Larsson syndrome • Fatty aldehyde

9.1 Sphingolipid Structure, Function, and Diversity

Sphingolipids are one of the major eukaryotic lipid species and have a role in a variety of physiological functions, including embryogenesis, organogenesis, skin barrier formation, neural functions, cell adhesion, recognition of bacterial toxins

A. Kihara (✉)
Laboratory of Biochemistry, Faculty of Pharmaceutical Sciences, Hokkaido University, Kita 12-jo, Nishi 6-chome, Kita-ku, Sapporo 060-0812, Japan
e-mail: kihara@pharm.hokudai.ac.jp

T. Yokomizo, M. Murakami (eds.), *Bioactive Lipid Mediators*,
DOI 10.1007/978-4-431-55669-5_9

and viruses, spermatogenesis, immunity, and glucose metabolism [1–5]. Sphingolipids are particularly enriched in the plasma membrane. Sphingolipids account for 28 % [sphingomyelin, 18 %; glycosphingolipids (GSLs), 10 %] of total human erythrocyte lipids [6]; this is thought to represent a rough estimate of the sphingolipid percentage in the plasma membrane of mammalian cells because the cellular membranes of erythrocytes are solely composed of the plasma membrane.

Sphingolipids consist of two parts: the hydrophobic ceramide (Cer) backbone and a polar head group. Sphingolipid synthesis occurs in the endoplasmic reticulum (ER) until the Cer synthesis step. In contrast, the synthesis of complex sphingolipids (sphingomyelin and GSLs) occurs in the Golgi complex, after which they are transported to the plasma membrane. Sphingolipid synthesis starts with the condensation of serine and palmitoyl-CoA to produce 3-ketosphinganine (3-ketodihydrosphingosine), a reaction catalyzed by serine palmitoyltransferases (SPTs) [7]. Mammals contain three SPT subunits (SPTLC1, -2, -3), which function as heterodimers (SPTLC1/SPTLC2 and SPTLC1/SPTLC3) [7]. Particular mutations in the *SPTLC1* or *SPTLC2* genes affect the substrate specificities of the SPTs. In addition to the physiological substrate serine, these mutated SPTs are able to use alanine and glycine as substrates, producing 1-deoxysphinganine and 1-deoxymethylsphinganine, respectively. These compounds are neurotoxic and can cause hereditary sensory neuropathy type 1 [8].

The 3-ketosphinganine reductase FVT1/KDSR catalyzes the second reaction of sphingolipid synthesis, namely, the conversion of 3-ketosphinganine to sphinganine (dihydrosphingosine) [9]. Sphinganine is then amide-linked to a fatty acid, generating dihydroceramide. This reaction is catalyzed by one of the six Cer synthases (CERS1-6) [2]. Sphinganine is classified as a long-chain base (LCB). Although sphinganine-containing sphingolipids exist in mammalian cells at low levels, the major LCB in mammals is sphingosine, which contains a *trans* double bond between the C4 and C5 positions of sphinganine. The sphinganine moiety of dihydroceramide is converted to sphingosine by the dihydroceramide desaturase DES1 [10], producing Cer. Therefore, free sphingosine is not generated de novo through the sphingolipid-biosynthetic pathway but is instead produced via the degradation pathway, that is, through the hydrolysis of Cer. Sphingosine can also be used as a substrate for Cer synthases, in which case Cer is produced directly. Each CERS exhibits characteristic substrate specificity toward the acyl-CoA substrate. CERS3 is involved in the formation of Cer with extremely long fatty acids (≥C26), which is important for skin barrier function [11]. Accordingly, mutations in the *CERS3* gene cause the cutaneous disorder ichthyosis [12].

The head group of mammalian sphingolipids is either phosphocholine (in sphingomyelin) or a sugar chain (in GSLs). GSLs are unique in that they are highly diverse molecules. In mammals, hundreds of GSLs exist, with differing sugar composition and linkage positions among the sugar residues. The sugar residues found in GSLs include glucose, galactose, *N*-acetylglucosamine, *N*-acetylgalactosamine, sialic acid, fucose, and mannose. The simplest GSLs are glucosylceramide (GlcCer) and galactosylceramide (GalCer), which contain glucose and galactose, respectively, at the C1 position of Cers (Fig. 9.1). These compounds are referred to as

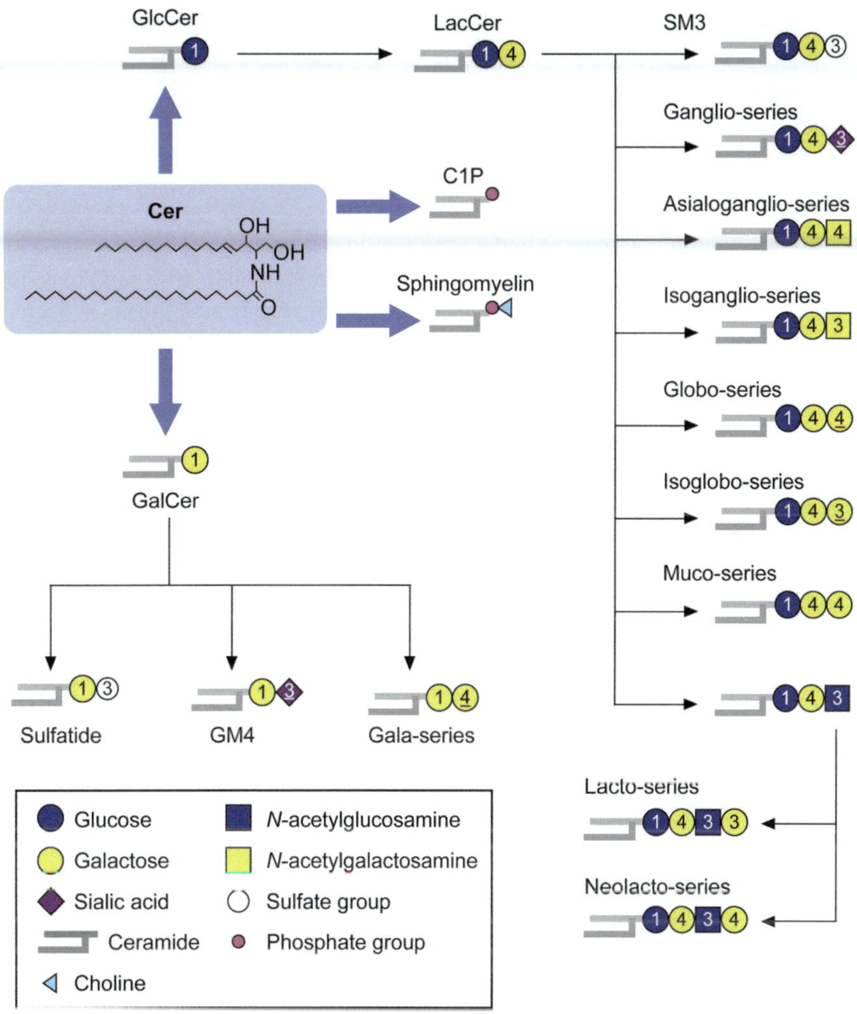

Fig. 9.1 Synthetic pathways of complex sphingolipids. In the sphingolipid-synthetic pathway, ceramide (Cer) is converted to sphingomyelin, GlcCer, GalCer, or C1P. The *number* indicates the position of each bond, in which the C1 hydroxy group of the numbered sugar residue links to the numbered position of its left sugar residue. *Underlined numbers* represent an α-linkage and *non-underlined numbers* indicate a β-linkage

monohexosylceramides or cerebrosides. Synthesis of GlcCer is catalyzed by the GlcCer synthase (Cer glucosyltransferase) UGCG. *Ugcg*-knockout mice die during early embryogenesis between embryonic day (E)7.5 and E9.5 [13]. The addition of galactose to the glucose residue of GlcCer produces lactosylceramide (LacCer), which is a starting material of a variety of GSL series. Deficient LacCer synthesis also causes embryonic lethality [14]. The LacCer-derived GSL series include the ganglio-series, asialoganglio-series, isoganglio-series, globo-series, isoglobo-series, lacto-series, and neolacto-series (Fig. 9.1). SM3, which contains a sulfate group at the 3-position of the galactose residue, is also produced from LacCer.

GalCer is produced by the Cer galactosyltransferase CGT/UGT8. Some GalCer is further converted to sulfatide, GM4, or the gala-series GSLs (Fig. 9.1). GalCer and sulfatide are enriched in myelin. Accordingly, *Cgt*-knockout mice, which produce neither GalCer nor sulfatide, exhibit neural symptoms such as severe generalized tremors and mild ataxia [15].

Sphingomyelin, the most abundant sphingolipid in mammals, can interact with cholesterol to form lipid microdomains in the plasma membrane [16]. Sphingomyelin synthases catalyze the formation of sphingomyelin by transferring the phosphocholine moiety of phosphatidylcholine to Cer. Two sphingomyelin synthases with differing intracellular localizations exist in mammals; SMS1 is localized to the *trans*-Golgi network whereas SMS2 is localized to the plasma membrane [17]. *Sms1*-knockout mice exhibit impaired insulin secretion, immune function, and hearing [18–20], whereas *Sms2*-knockout mice are resistant to high-fat diet-induced obesity and atherosclerosis [21, 22].

Cer can also be metabolized to Cer 1-phosphate (C1P) by the Cer kinase CerK (Fig. 9.1). C1P is involved in the regulation of inflammation, stimulation of phagocytosis by neutrophils, degranulation of mast cells, and obesity [1, 23]. However, significant amounts of C1P are still present in *CerK*-knockout mice, strongly suggesting the existence of another C1P-generating pathway [24].

9.2 The Sphingolipid Degradation Pathway and Sphingolipid Storage Diseases

The polar head groups of sphingolipids are removed by a series of hydrolases, producing Cer that is then degraded to sphingosine and a fatty acid by ceramidase. Mutations in the genes encoding these hydrolases result in sphingolipid storage diseases (SLSDs), also known as sphingolipidoses. Because the hydrolases responsible for SLSs are localized to the lysosomes, SLSDs are classified as lysosomal diseases. To date, about 10 SLSDs with about 40 genetically distinct forms have been reported [25, 26]. For example, mutations in the lysosomal forms of ceramidase (acid ceramidase) and sphingomyelinase (acid sphingomyelinase) genes are the underlying cause of Farber disease and Niemann–Pick disease (type A and B), respectively [26, 27] (Fig. 9.2). Another form of Niemann–Pick disease (type C) is caused by mutations in *NPC-1* or *NPC-2* [28]. NPC-1 and NPC-2 are thought to be involved in the removal of cholesterol from endosomes, in their roles as a cholesterol transporter and a cholesterol-binding protein, respectively [26, 28]. Furthermore, it has been suggested that the interaction of sphingomyelin with cholesterol obstructs the removal of cholesterol from endosomes [26]. Accordingly, patients with Niemann–Pick disease type A or B exhibit accumulation of cholesterol as well as sphingomyelin [29].

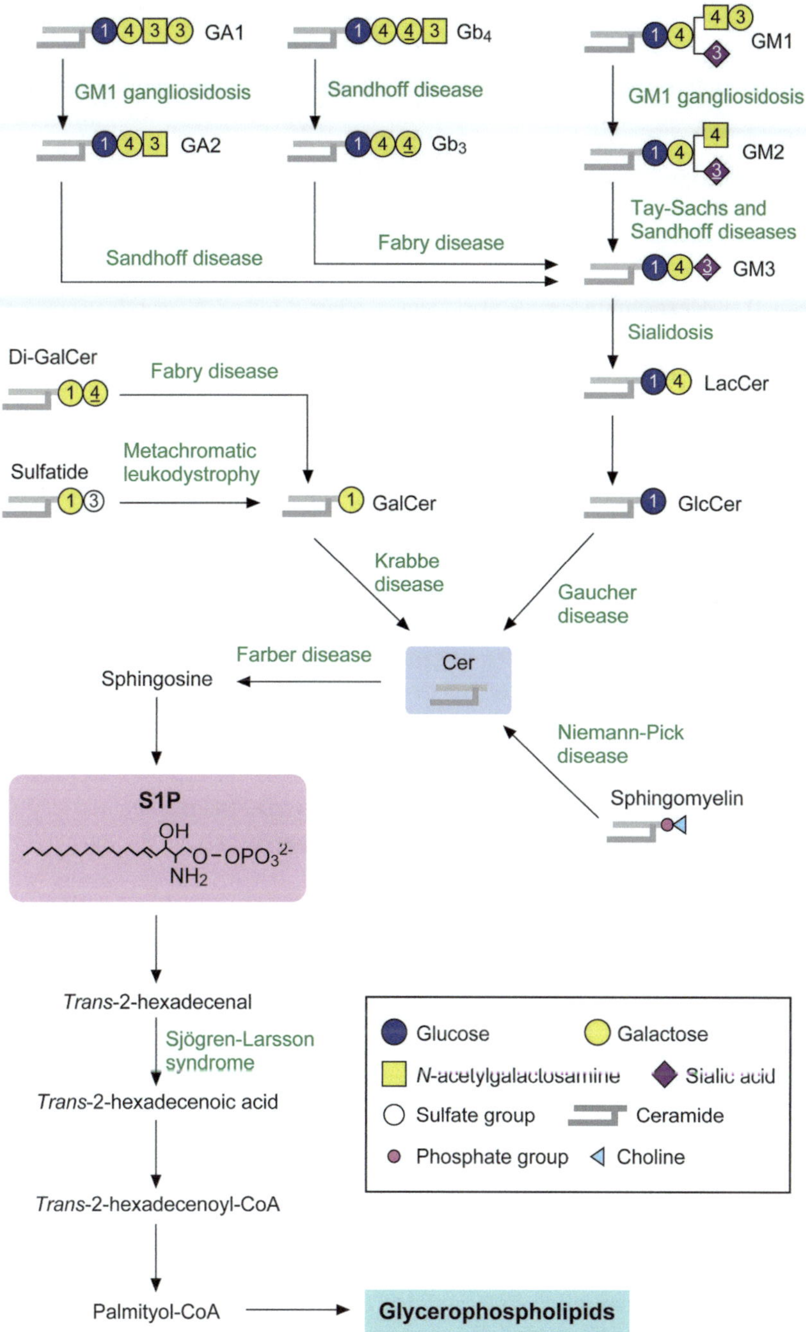

Fig. 9.2 Sphingolipid degradation pathways and related disorders. In the sphingolipid degradation pathway, the polar head groups of complex sphingolipids are removed by the successive actions of lysosomal hydrolases. The resulting Cer is then converted to sphingosine, which is either recycled for sphingolipid synthesis or converted to S1P. When S1P is irreversibly degraded by S1P lyase to *trans*-2-hexadecenal, it is metabolized to glycerophospholipids via *trans*-2-hexadecenoic acid, *trans*-2-hexadecenoyl-CoA, and palmitoyl-CoA

Sphingolipids containing sialic acid are known as gangliosides. The ganglioside GM2 accumulates in the neurons of patients with GM2 gangliosidoses, which can be divided into the B variant (Tay–Sachs disease; *HEXA* mutation), the AB variant (*GM2A* mutation), and the O variant (Sandhoff disease; *HEXB* mutation) [26] (Fig. 9.2). The *N*-acetylgalactosamine residue of GM2 is removed by β-hexosaminidase A, which consists of an α-subunit (the *HEXA* gene product) and a β-subunit (the *HEXB* gene product), in cooperation with the GM2 activator protein (the *GM2A* gene product). β-Hexosaminidase A also cleaves other β-linked *N*-acetylgalactosamine and *N*-acetylglucosamine residues from negatively charged and uncharged glycoconjugates [26]. β-Hexosaminidase B is composed of two β-subunits and exhibits activity toward uncharged substrates such as the GSLs GA2 and Gb_4.

The GM2 activator protein and saposins (saposin-A, -B, -C, -D) comprise the class of sphingolipid activator proteins (SAPs). Saposins are small glycoproteins that are produced from the common precursor protein (prosaposin) by processing [30]. These saposins activate several hydrolases involved in sphingolipid degradation. For example, saposin-C is involved in the degradation of GlcCer together with β-glucocerebrosidase. Mutations in the β-glucocerebrosidase or saposin-C genes cause Gaucher disease [25, 26, 30]. Saposin-D is required for the optimal enzyme activity of acid ceramidase, and a mutation in the saposin-D gene leads to clinical symptoms that are reminiscent of Farber disease [27].

9.3 Sphingosine 1-Phosphate Is a Key Intermediate of the Sphingolipid Degradation Pathway

After the degradation of Cer by ceramidase, the resultant fatty acids are converted to acyl-CoAs and used in the synthesis of a variety of lipids such as glycerophospholipids, triglycerides, and sphingolipids or in energy production via β-oxidation. On the other hand, sphingosine, another degradation product of Cer, is recycled for sphingolipid synthesis or metabolized to glycerophospholipids via sphingosine 1-phosphate (S1P). The latter pathway, referred to as the S1P-metabolic pathway hereafter, is the sole route for the removal of the sphingosine (LCB) component of sphingolipids. Therefore, the S1P-metabolic pathway is important for the maintenance of sphingolipid homeostasis: indeed, this pathway is conserved in eukaryotes.

Sphingosine/LCB is phosphorylated to S1P/LCB 1-phosphate (LCBP) by sphingosine/LCB kinases. Mammals contain two sphingosine kinases: SPHK1 and SPHK2 [31, 32]. Extracellular S1P acts as a lipid mediator by binding to one of the five S1P receptors ($S1P_1$–$S1P_5$). The functions of S1P as a lipid mediator are particularly important in the immune and vascular systems. Taking advantage of its role in immune system, the sphingosine analogue fingolimod (FTY720) has been developed to treat multiple sclerosis [33]. Fingolimod is a pro-drug: its functional form is fingolimod phosphate, which binds to all S1P receptors with the exception of $S1P_2$ [34]. Only vertebrates and chordates possess S1P receptors, yet all eukaryotes produce S1P/LCBP. Thus, in evolutionary terms, the role of S1P/LCBP in sphingolipid metabolism greatly precedes its role as a lipid mediator [35].

The first irreversible (i.e., committed) step of the S1P-metabolic pathway is the cleavage of S1P between the C2 and C3 positions by the S1P lyase SPL/SGPL1 [36]. This reaction produces the fatty aldehyde *trans*-2-hexadecenal and phosphoethanolamine. *Spl*-knockout mice die approximately 1 month after birth, and exhibit abnormalities in the lung, heart, urinary tract, and bone, as well as displaying enhanced pro-inflammatory responses, myeloid cell hyperplasia, and aberrant lipid homeostasis in the liver and brain [37–40], thus clearly indicating the importance of the S1P-metabolic pathway. The S1P metabolite phosphoethanolamine is converted to CDP-ethanolamine, which is then used for synthesis of the glycerophospholipid phosphatidylethanolamine. On the other hand, *trans*-2-hexadecenal is converted to palmitoyl-CoA via *trans*-2-hexadecenoic acid and *trans*-2-hexadecenoyl-CoA [41, 42]. The majority of palmitoyl-CoA is then incorporated into glycerophospholipids [35, 41–43]. All these reactions of the S1P-metabolic pathway occur in the ER [35].

The conversion of *trans*-2-hexadecenal to *trans*-2-haxadecenoic acid is catalyzed by the fatty aldehyde dehydrogenase ALDH3A2 [41]. Mutations in the *ALDH3A2* gene cause the neurocutaneous disorder Sjögren–Larsson syndrome (SLS). ALDH3A2 exhibits activity toward a broad range of aliphatic aldehydes with medium- and long-chain lengths in vitro [44]. A variety of ALDH3A2 substrates are generated via the metabolism of lipids such as fatty alcohol, leukotriene B_4, and phytol (other than S1P) and through lipid peroxidation [45]. Aldehyde molecules are reactive and can form Schiff bases with the amino groups of biomolecules. Furthermore, α,β-unsaturated aldehydes, to which *trans*-2-hexadecenal belongs, can react with general nucleophiles (e.g., lysine, cysteine, and histidine side chains) via a 1,4-Michael addition [46]. Therefore, accumulated aldehydes are thought to cause SLS by reacting with important cellular biomolecules and impairing their functions. However, the primary aldehyde responsible for the pathology of SLS is still unknown. Both the activities of the S1P-producing enzyme (sphingosine kinase) and the degradation enzyme (S1P lyase) are high in most tissues [47, 48], suggesting that S1P metabolism is an active pathway that occurs ubiquitously and continuously in cells. Therefore, the S1P metabolite *trans*-2-hexadecenal appears to be produced constitutively at a relatively high level. In addition, both ALDH3A2 and *trans*-2-hexadecenal coexist in the ER [35], suggesting that *trans-2-hexadecenal* may be a major ALDH3A2 substrate. Thus, it is possible that impairment of the S1P-metabolic pathway is one of the major causes of SLS pathology.

Fatty acids must be converted to their active acyl-CoA form by acyl-CoA synthetases (ACSs) for further metabolism. Humans possess 26 ACSs, which are classified into six subfamilies—ACSS, ACSM, ACSL, ACSVL, ACSF, and ACSBG—based on their substrate specificity and sequence similarity [49]. Among them, the ACSL subfamily members (ACSL1, -3, -4, -5, -6), which exhibit activity toward long-chain fatty acids, have central roles in the conversion of *trans*-2-hexadecanoic acid to *trans*-2-hexadecenoyl-CoA in the S1P-metabolic pathway [41, 43].

9.4 The Importance of the S1P-Metabolic Pathway from a Nutritional Perspective

Foods contain abundant sphingolipids: it is estimated that humans consume 300–400 mg sphingolipids per day [50]. Dietary sphingolipids have several beneficial effects, including reducing serum LDL cholesterol, inhibiting colon carcinogenesis and inflammation, and reinforcing the skin barrier [51–56]. Because complex sphingolipids and Cer do not readily enter cells, they must be first converted to sphingosine for intestinal absorption. The successive actions of alkaline sphingomyelinase and neutral ceramidase, both of which are localized to the intestinal lumen, produce sphingosine from sphingomyelin [57, 58]. The sphingosine absorbed by small intestinal epithelial cells is converted either to sphingolipids or to glycerophospholipids via the S1P-metabolic pathway. Because the quantity of sphingosine obtained from the diet is much higher than required for sphingolipid synthesis in the small intestine, the majority of the sphingosine is converted to glycerophospholipids [42]. Consistent with this notion, the protein level and activity of the S1P lyase SPL, which catalyzes the committed step of the S1P-metabolic pathway, have been reported as the highest in the small intestine among 12 mouse tissues examined [48].

Plasma sphingolipids can also enter cells in the vascular system that are in contact with the blood, such as endothelial cells, after conversion to sphingosine. In addition to sphingomyelin and Cer, which are present mainly in LDL and VLDL, respectively [50], plasma S1P can also be a source of extracellular sphingosine. The concentration of S1P in plasma is several hundreds of nanomoles [1]. S1P is first dephosphorylated by the cell-surface lipid phosphate phosphatase LPP1 before entering endothelial cells [59]. Imported sphingosine is metabolized to sphingolipids or glycerophospholipids via the S1P-metabolic pathway.

9.5 Perspective and Future Directions

The function of S1P as a lipid mediator has attracted significant attention and has been analyzed extensively. Although research on the function of S1P as a metabolic intermediate has slowed since the identification of S1P as an intermediate in the sphingolipid degradation pathway in the late 1960s [60], the recent elucidation of the entire S1P-metabolic pathway provides welcome information allowing further analysis of the pathophysiological functions of this pathway. It is possible that impairment of the S1P-metabolic pathway is responsible for the pathology of SLS and other disorders. Recent analyses of *Spl*-knockout mice also emphasize the physiological importance of the S1P-metabolic pathway. Future studies will doubtless reveal additional physiological, pathological, and nutritional functions of the S1P-metabolic pathway.

References

1. Kihara A, Mitsutake S, Mizutani Y, Igarashi Y (2007) Metabolism and biological functions of two phosphorylated sphingolipids, sphingosine 1-phosphate and ceramide 1-phosphate. Prog Lipid Res 46:126–144
2. Mizutani Y, Mitsutake S, Tsuji K, Kihara A, Igarashi Y (2009) Ceramide biosynthesis in keratinocyte and its role in skin function. Biochimie 91:784–790
3. Kihara A (2012) Very long-chain fatty acids: elongation, physiology and related disorders. J Biochem 152:387–395
4. Pontier SM, Schweisguth F (2012) Glycosphingolipids in signaling and development: from liposomes to model organisms. Dev Dyn 241:92–106
5. Mitsutake S, Igarashi Y (2013) Sphingolipids in lipid microdomains and obesity. Vitam Horm 91:271–284
6. Tanford C (1980) The hydrophobic effect: formation of micelles and biological membranes, 2nd edn. Wiley, New York
7. Lowther J, Naismith JH, Dunn TM, Campopiano DJ (2012) Structural, mechanistic and regulatory studies of serine palmitoyltransferase. Biochem Soc Trans 40:547–554
8. Penno A, Reilly MM, Houlden H, Laura M, Rentsch K, Niederkofler V, Stoeckli ET, Nicholson G, Eichler F, Brown RH Jr, von Eckardstein A, Hornemann T (2010) Hereditary sensory neuropathy type 1 is caused by the accumulation of two neurotoxic sphingolipids. J Biol Chem 285:11178–11187
9. Kihara A, Igarashi Y (2004) FVT-1 is a mammalian 3-ketodihydrosphingosine reductase with an active site that faces the cytosolic side of the endoplasmic reticulum membrane. J Biol Chem 279:49243–49250
10. Ternes P, Franke S, Zahringer U, Sperling P, Heinz E (2002) Identification and characterization of a sphingolipid Δ4-desaturase family. J Biol Chem 277:25512–25518
11. Jennemann R, Rabionet M, Gorgas K, Epstein S, Dalpke A, Rothermel U, Bayerle A, van der Hoeven F, Imgrund S, Kirsch J, Nickel W, Willecke K, Riezman H, Grone HJ, Sandhoff R (2012) Loss of ceramide synthase 3 causes lethal skin barrier disruption. Hum Mol Genet 21:586–608
12. Eckl KM, Tidhar R, Thiele H, Oji V, Hausser I, Brodesser S, Preil ML, Onal-Akan A, Stock F, Muller D, Becker K, Casper R, Nurnberg G, Altmuller J, Nurnberg P, Traupe H, Futerman AH, Hennies HC (2013) Impaired epidermal ceramide synthesis causes autosomal recessive congenital ichthyosis and reveals the importance of ceramide acyl chain length. J Invest Dermatol 133:2202–2211
13. Yamashita T, Wada R, Sasaki T, Deng C, Bierfreund U, Sandhoff K, Proia RL (1999) A vital role for glycosphingolipid synthesis during development and differentiation. Proc Natl Acad Sci U S A 96:9142–9147
14. Nishie T, Hikimochi Y, Zama K, Fukusumi Y, Ito M, Yokoyama H, Naruse C, Asano M (2010) β4-Galactosyltransferase-5 is a lactosylceramide synthase essential for mouse extra-embryonic development. Glycobiology 20:1311–1322
15. Coetzee T, Fujita N, Dupree J, Shi R, Blight A, Suzuki K, Popko B (1996) Myelination in the absence of galactocerebroside and sulfatide: normal structure with abnormal function and regional instability. Cell 86:209–219
16. Slotte JP (2013) Biological functions of sphingomyelins. Prog Lipid Res 52:424–437
17. Huitema K, van den Dikkenberg J, Brouwers JF, Holthuis JC (2004) Identification of a family of animal sphingomyelin synthases. EMBO J 23:33–44
18. Yano M, Watanabe K, Yamamoto T, Ikeda K, Senokuchi T, Lu M, Kadomatsu T, Tsukano H, Ikawa M, Okabe M, Yamaoka S, Okazaki T, Umehara H, Gotoh T, Song WJ, Node K, Taguchi R, Yamagata K, Oike Y (2011) Mitochondrial dysfunction and increased reactive oxygen species impair insulin secretion in sphingomyelin synthase 1-null mice. J Biol Chem 286:3992–4002

19. Dong L, Watanabe K, Itoh M, Huan CR, Tong XP, Nakamura T, Miki M, Iwao H, Nakajima A, Sakai T, Kawanami T, Sawaki T, Masaki Y, Fukushima T, Fujita Y, Tanaka M, Yano M, Okazaki T, Umehara H (2012) $CD4^+$ T-cell dysfunctions through the impaired lipid rafts ameliorate concanavalin A-induced hepatitis in sphingomyelin synthase 1-knockout mice. Int Immunol 24:327–337
20. Lu MH, Takemoto M, Watanabe K, Luo H, Nishimura M, Yano M, Tomimoto H, Okazaki T, Oike Y, Song WJ (2012) Deficiency of sphingomyelin synthase-1 but not sphingomyelin synthase-2 causes hearing impairments in mice. J Physiol 590:4029–4044
21. Mitsutake S, Zama K, Yokota H, Yoshida T, Tanaka M, Mitsui M, Ikawa M, Okabe M, Tanaka Y, Yamashita T, Takemoto H, Okazaki T, Watanabe K, Igarashi Y (2011) Dynamic modification of sphingomyelin in lipid microdomains controls development of obesity, fatty liver, and type 2 diabetes. J Biol Chem 286:28544–28555
22. Liu J, Huan C, Chakraborty M, Zhang H, Lu D, Kuo MS, Cao G, Jiang XC (2009) Macrophage sphingomyelin synthase 2 deficiency decreases atherosclerosis in mice. Circ Res 105:295–303
23. Gomez-Muñoz A, Gangoiti P, Arana L, Ouro A, Rivera IG, Ordoñez M, Trueba M (2013) New insights on the role of ceramide 1-phosphate in inflammation. Biochim Biophys Acta 1831:1060–1066
24. Boath A, Graf C, Lidome E, Ullrich T, Nussbaumer P, Bornancin F (2008) Regulation and traffic of ceramide 1-phosphate produced by ceramide kinase: comparative analysis to glucosylceramide and sphingomyelin. J Biol Chem 283:8517–8526
25. Sillence DJ, Platt FM (2003) Storage diseases: new insights into sphingolipid functions. Trends Cell Biol 13:195–203
26. Schulze H, Sandhoff K (2014) Sphingolipids and lysosomal pathologies. Biochim Biophys Acta 1841:799–810
27. Park JH, Schuchman EH (2006) Acid ceramidase and human disease. Biochim Biophys Acta 1758:2133–2138
28. Storch J, Xu Z (2009) Niemann-Pick C2 (NPC2) and intracellular cholesterol trafficking. Biochim Biophys Acta 1791:671–678
29. Vanier MT (1983) Biochemical studies in Niemann-Pick disease. I. Major sphingolipids of liver and spleen. Biochim Biophys Acta 750:178–184
30. Tamargo RJ, Velayati A, Goldin E, Sidransky E (2012) The role of saposin C in Gaucher disease. Mol Genet Metab 106:257–263
31. Kohama T, Olivera A, Edsall L, Nagiec MM, Dickson R, Spiegel S (1998) Molecular cloning and functional characterization of murine sphingosine kinase. J Biol Chem 273:23722–23728
32. Liu H, Sugiura M, Nava VE, Edsall LC, Kono K, Poulton S, Milstien S, Kohama T, Spiegel S (2000) Molecular cloning and functional characterization of a novel mammalian sphingosine kinase type 2 isoform. J Biol Chem 275:19513–19520
33. Bigaud M, Guerini D, Billich A, Bassilana F, Brinkmann V (2014) Second generation S1P pathway modulators: research strategies and clinical developments. Biochim Biophys Acta 1841:745–758
34. Mandala S, Hajdu R, Bergstrom J, Quackenbush E, Xie J, Milligan J, Thornton R, Shei GJ, Card D, Keohane C, Rosenbach M, Hale J, Lynch CL, Rupprecht K, Parsons W, Rosen H (2002) Alteration of lymphocyte trafficking by sphingosine-1-phosphate receptor agonists. Science 296:346–349
35. Kihara A (2014) Sphingosine 1-phosphate is a key metabolite linking sphingolipids to glycerophospholipids. Biochim Biophys Acta 1841:766–772
36. Zhou J, Saba JD (1998) Identification of the first mammalian sphingosine phosphate lyase gene and its functional expression in yeast. Biochem Biophys Res Commun 242:502–507
37. Vogel P, Donoviel MS, Read R, Hansen GM, Hazlewood J, Anderson SJ, Sun W, Swaffield J, Oravecz T (2009) Incomplete inhibition of sphingosine 1-phosphate lyase modulates immune system function yet prevents early lethality and non-lymphoid lesions. PLoS One 4:e4112

38. Bektas M, Allende ML, Lee BG, Chen W, Amar MJ, Remaley AT, Saba JD, Proia RL (2010) Sphingosine 1-phosphate lyase deficiency disrupts lipid homeostasis in liver. J Biol Chem 285:10880–10889
39. Allende ML, Bektas M, Lee BG, Bonifacino E, Kang J, Tuymetova G, Chen W, Saba JD, Proia RL (2011) Sphingosine-1-phosphate lyase deficiency produces a pro-inflammatory response while impairing neutrophil trafficking. J Biol Chem 286:7348–7358
40. Hagen-Euteneuer N, Lutjohann D, Park H, Merrill AH Jr, van Echten-Deckert G (2012) Sphingosine 1-phosphate (S1P) lyase deficiency increases sphingolipid formation via recycling at the expense of *de novo* biosynthesis in neurons. J Biol Chem 287:9128–9136
41. Nakahara K, Ohkuni A, Kitamura T, Abe K, Naganuma T, Ohno Y, Zoeller RA, Kihara A (2012) The Sjögren-Larsson syndrome gene encodes a hexadecenal dehydrogenase of the sphingosine 1-phosphate degradation pathway. Mol Cell 46:461–471
42. Wakashima T, Abe K, Kihara A (2014) Dual functions of the *trans*-2-enoyl-CoA reductase TER in the sphingosine 1-phosphate metabolic pathway and in fatty acid elongation. J Biol Chem 289:24736–24748
43. Ohkuni A, Ohno Y, Kihara A (2013) Identification of acyl-CoA synthetases involved in the mammalian sphingosine 1-phosphate metabolic pathway. Biochem Biophys Res Commun 442:195–201
44. Kelson TL, Secor McVoy JR, Rizzo WB (1997) Human liver fatty aldehyde dehydrogenase: microsomal localization, purification, and biochemical characterization. Biochim Biophys Acta 1335:99–110
45. Rizzo WB (2011) The role of fatty aldehyde dehydrogenase in epidermal structure and function. Dermatoendocrinology 3:91–99
46. Catalá A (2009) Lipid peroxidation of membrane phospholipids generates hydroxy-alkenals and oxidized phospholipids active in physiological and/or pathological conditions. Chem Phys Lipids 157:1–11
47. Fukuda Y, Kihara A, Igarashi Y (2003) Distribution of sphingosine kinase activity in mouse tissues: contribution of SPHK1. Biochem Biophys Res Commun 309:155–160
48. Ikeda M, Kihara A, Igarashi Y (2004) Sphingosine-1-phosphate lyase SPL is an endoplasmic reticulum-resident, integral membrane protein with the pyridoxal 5′-phosphate binding domain exposed to the cytosol. Biochem Biophys Res Commun 325:338–343
49. Watkins PA, Maiguel D, Jia Z, Pevsner J (2007) Evidence for 26 distinct acyl-coenzyme A synthetase genes in the human genome. J Lipid Res 48:2736–2750
50. Vesper H, Schmelz EM, Nikolova-Karakashian MN, Dillehay DL, Lynch DV, Merrill AH Jr (1999) Sphingolipids in food and the emerging importance of sphingolipids to nutrition. J Nutr 129:1239–1250
51. Imaizumi K, Tominaga A, Sato M, Sugano M (1992) Effects of dietary sphingolipids on levels of serum and liver lipids in rats. Nutr Res 12:543–548
52. Dillehay DL, Webb SK, Schmelz EM, Merrill AH Jr (1994) Dietary sphingomyelin inhibits 1,2-dimethylhydrazine-induced colon cancer in CF1 mice. J Nutr 124:615–620
53. Schmelz EM, Dillehay DL, Webb SK, Reiter A, Adams J, Merrill AH Jr (1996) Sphingomyelin consumption suppresses aberrant colonic crypt foci and increases the proportion of adenomas versus adenocarcinomas in CF1 mice treated with 1,2-dimethylhydrazine: implications for dietary sphingolipids and colon carcinogenesis. Cancer Res 56:4936–4941
54. Kobayashi T, Shimizugawa T, Osakabe T, Watanabe S, Okuyama H (1997) A long-term feeding of sphingolipids affected the levels of plasma cholesterol and hepatic triacylglycerol but tissue phospholipids and sphingolipids. Nutr Res 17:111–114
55. Duan J, Sugawara T, Sakai S, Aida K, Hirata T (2011) Oral glucosylceramide reduces 2,4-dinitrofluorobenzene induced inflammatory response in mice by reducing TNF-α levels and leukocyte infiltration. Lipids 46:505–512
56. Duan J, Sugawara T, Hirose M, Aida K, Sakai S, Fujii A, Hirata T (2012) Dietary sphingolipids improve skin barrier functions via the upregulation of ceramide synthases in the epidermis. Exp Dermatol 21:448–452

57. Zhang Y, Cheng Y, Hansen GH, Niels-Christiansen LL, Koentgen F, Ohlsson L, Nilsson A, Duan RD (2011) Crucial role of alkaline sphingomyelinase in sphingomyelin digestion: a study on enzyme knockout mice. J Lipid Res 52:771–781
58. Kono M, Dreier JL, Ellis JM, Allende ML, Kalkofen DN, Sanders KM, Bielawski J, Bielawska A, Hannun YA, Proia RL (2006) Neutral ceramidase encoded by the *Asah2* gene is essential for the intestinal degradation of sphingolipids. J Biol Chem 281:7324–7331
59. Zhao Y, Kalari SK, Usatyuk PV, Gorshkova I, He D, Watkins T, Brindley DN, Sun C, Bittman R, Garcia JG, Berdyshev EV, Natarajan V (2007) Intracellular generation of sphingosine 1-phosphate in human lung endothelial cells: role of lipid phosphate phosphatase-1 and sphingosine kinase 1. J Biol Chem 282:14165–14177
60. Stoffel W, Sticht G (1967) Metabolism of sphingosine bases. I. Degradation and incorporation of [3-^{14}C]*erythro*-DL-dihydrosphingosine and [7-$^{3}H_2$]*erythro*-DL-sphingosine into sphingolipids of rat liver. Hoppe Seylers Z Physiol Chem 348:941–943

Chapter 10
Fatty Acids Receptors

Akira Hirasawa, Masato Takeuchi, Takafumi Hara, Ayako Hirata, Soshi Tanabe, and Naoya Umeda

Abstract In the past decade, a strategy to deorphanize G protein-coupled receptors (GPCRs) has identified a series of receptors for free fatty acids (FFAs) that play significant roles in nutrition regulation. In this free fatty acid receptor family, FFAR1 (GPR40) and FFAR4 (GPR120) are activated by medium- and long-chain FFAs. FFAR1 regulates insulin secretion in pancreatic β-cells, whereas FFAR4 promotes the secretion of glucagon-like peptide-1 (GLP-1) in the intestine and also act as the lipid sensor in the adipose tissue to sense dietary fat and control energy balance. In this chapter, we discuss recent advances in the identification of ligands and the pharmacological characterization of FFAR1 and FFAR4, and we present a summary of the current understanding of their physiological roles and potential as drug targets.

Keywords GPCR • Fatty acid receptor • FFAR1 • FFAR4 • GPR40 • GPR120

10.1 Introduction

Free fatty acids (FFAs) are not only essential dietary nutrients but they also act as signaling molecules in various physiological functions. The nuclear receptors peroxisome proliferator-activated receptors (PPARs) and fatty acid-binding proteins (FABPs) are known to act as 'sensors' of FFAs. They maintain homeostasis under

A. Hirasawa (✉)
Department of Genomic Drug Discovery Science, Graduate School of Pharmaceutical Sciences, Kyoto University, Kyoto 606-8501, Japan

Institute for Integrated Medical Sciences, Tokyo Women's Medical University, Tokyo 162-8666, Japan
e-mail: akira_h@pharm.kyoto-u.ac.jp

M. Takeuchi • T. Hara • A. Hirata • S. Tanabe • N. Umeda
Department of Genomic Drug Discovery Science, Graduate School of Pharmaceutical Sciences, Kyoto University, Kyoto 606-8501, Japan

T. Yokomizo, M. Murakami (eds.), *Bioactive Lipid Mediators*,
DOI 10.1007/978-4-431-55669-5_10

physiological and pathophysiological conditions by coordinating the expression of proteins involved in lipid uptake, synthesis, transport, storage, degradation, and elimination [5]. However, these mechanisms could not explain all biological effects of FFAs. Some effects were thought to be mediated by other mechanisms such as signaling through cell-surface receptors [28, 40]. In the last decade, a strategy to deorphanize G protein-coupled receptors (GPCRs) has identified a series of such receptors for FFAs that have significant roles in nutrition regulation (Table 10.1). Among FFARs, FFAR2 (GPR43) and FFAR3 (GPR41) are activated by short-chain FFAs (SCFAs) such as acetate, propionate, and butyrate. On the other hand, GPR84 is activated by medium-chain fatty acids, and FFAR1 (GPR40) and FFAR4 (GPR120) are activated by medium- and long-chain saturated and unsaturated FFAs. The importance of the characterization of these GPCRs is emphasized by the fact that 30 % of all prescription drugs target GPCRs, and many groups have reported that these FFARs are expressed in the gastrointestinal tract and have several important roles involved in energy homeostasis. These FFARs are also widely conserved among vertebrates, which suggests that they have common important physiological functions. Therefore, FFARs have received considerable attention as potential therapeutic targets for metabolic disorders. In this chapter, we focus on recent advances in our understanding of FFARs, especially FFAR1 and FFAR4, and their roles in energy homeostasis.

10.2 FFAR1

10.2.1 Ligand and Tissue Distribution

FFAR1 is activated by medium- and long-chain saturated and unsaturated FFAs, as reported by three independent groups almost simultaneously [1, 24, 27]. A variety of FFAs has been found to act as agonists of FFAR1 in the micromolar concentration range, with eicosatrienoic acid being the most potent. FFAR1 is enriched 2- to 100-fold in pancreatic islets as compared with the whole pancreas. FFAR1 is also expressed in the intestine. FFAR2 and FFAR3, on the other hand, are both activated by short-chain FFAs, such as formate, acetate, propionate, butyrate, and pentanoate [7, 8, 13]. FFAR3 is activated equally by propionate, butyrate, and pentanoate, whereas FFAR2 prefers propionate over the other short-chain FFAs [7, 8]. Short-chain fatty acids activate FFAR2 or FFAR3 in a relatively high submillimolar concentration range. FFAR2 and FFAR3 are expressed in the adipose tissue and sympathetic ganglions, respectively [26]. There are also several reports suggesting that fermentation end products, especially short-chain fatty acids produced by gut microbiota, affect inflammation via FFARs [29, 38].

Table 10.1 Free fatty acids receptors family

Nomenclature	FFAR1 GPR40	FFAR2 GPR43	FFAR3 GPR41	FFAR4 GPR120	GPR84
Agonist(FFA)	Medium-long	Short Chain C3∼C4∼C2	Short Chain C3>C4>>C2	Medium-long	Medium- 2-OH- or 3-OH MCFA
(Other)	TAK-875,Thiazolidinedione			NCG21	6-OAU
G protein coupling	Gq/11	Gq/11,Gi/o	Gi/o	Gq/11	—
Gene/chromosomal	GPR40	GPR43	GPR41	GPR120	GPR84
Localization	19q13,1	19q13.1	19q13.1	10q23.33	12q13.13
Protein(human)	NP_005294, 300a.a	NP_005297, 330a.a	NP_005295, 346a.a	NP_859529, 377a.a	NP_065103 396a.a.
Expression	Pancreatic β-cell intestine	Adipose tissue intestine	Adipose tissue Sympathetic ganglions	Colon/Adipose tissue	Spleen(B cell & T cell)
Physiological role	Insulin secretion	Lipid and energy metabolism anti-inflammatory effect	Energy regulation	GLP-1 secretion	IL-12 p40

Table modified from references [32].

10.2.2 Genomic Structure and Evolution

FFAR1, FFAR2, FFAR3, and GPR42, which is thought to be a pseudogene, are all GPCRs of the rhodopsin family located within a gene cluster on the human 19q13 chromosome. GPR42 only exists in the family *Hominidae* and cannot be detected in species below gibbons. The members of this subfamily share approximately 30–40 % identity, with the exception that human GPR42 (hGPR42) differs from human FFAR3 (hFFAR3) at only six amino acid positions [3]. These findings suggest that hGPR42 arose as the result of a gene duplication of hFFAR3 that occurred after the gibbons branched off from the superfamily *Hominoidea* [3]. Recent advances in genomic analysis of various species have revealed that FFAR1, FFAR2, and FFAR3 are widely conserved throughout vertebrates from fishes to mammals (with the exception of birds). In amphibians, reptiles, and mammals, FFAR1, FFAR2, and FFAR3 create a family of genes in tandem sequence with shared synteny (Fig. 10.1). However, similar genomic structures cannot be found in birds, at least among genomically analyzed species such as pigeons and chickens. On the other hand, we have found multiple clusters of these genes existing in *Teleostei*, which suggests that multiplication of these genes occurred after the divergence of amphibians. We have also found that only one homologous gene exists in cartilaginous fishes. To date, the function of FFARs have only been investigated in mammals, and we lack information about their expression and physiological function in other species. Thus, the connection between these receptors and physiological functions are not fully understood and should become a matter of great interest in the upcoming years.

10.2.3 Signal Transduction

In Chinese hamster ovary (CHO) cells, exogenously expressed FFAR1 is coupled to the formation of inositol 1,4,5-trisphosphate, intracellular Ca^{2+} mobilization, and the activation of extracellular signal-regulated kinase (ERK) 1/2 [24]. These results suggest that FFAR1 is coupled to Gαq and/or $G_{i/o}$ protein. Fujiwara et al. showed, in rat islet β-cells, that oleic acid (OA) interacts with FFAR1 to increase intracellular Ca^{2+}, via the phospholipase (PL) C- and L-type Ca^{2+} channel-mediated pathway, which links to insulin release [10]. Feng et al. showed that linolenic acid reduces the voltage-gated K^+ current in rat pancreatic β-cells through the FFAR1-mediated regulation of cAMP levels and protein kinase A activity; the reduction in K^+ current leads in turn to enhanced β-cell excitability and insulin secretion [9].

10.2.4 Protein Structure and New Chemicals

The first report identifying a series of FFAR1 agonists based on 3-(4-[*N*-alkyl]aminophenyl)propanoic acid was by Garrido et al. [12]. In particular, the physiological and pharmacological properties of GW9508 have been studied in detail, as it has potential as an agonist for not only FFAR1 but also FFAR4 (Fig. 10.2). Furthermore,

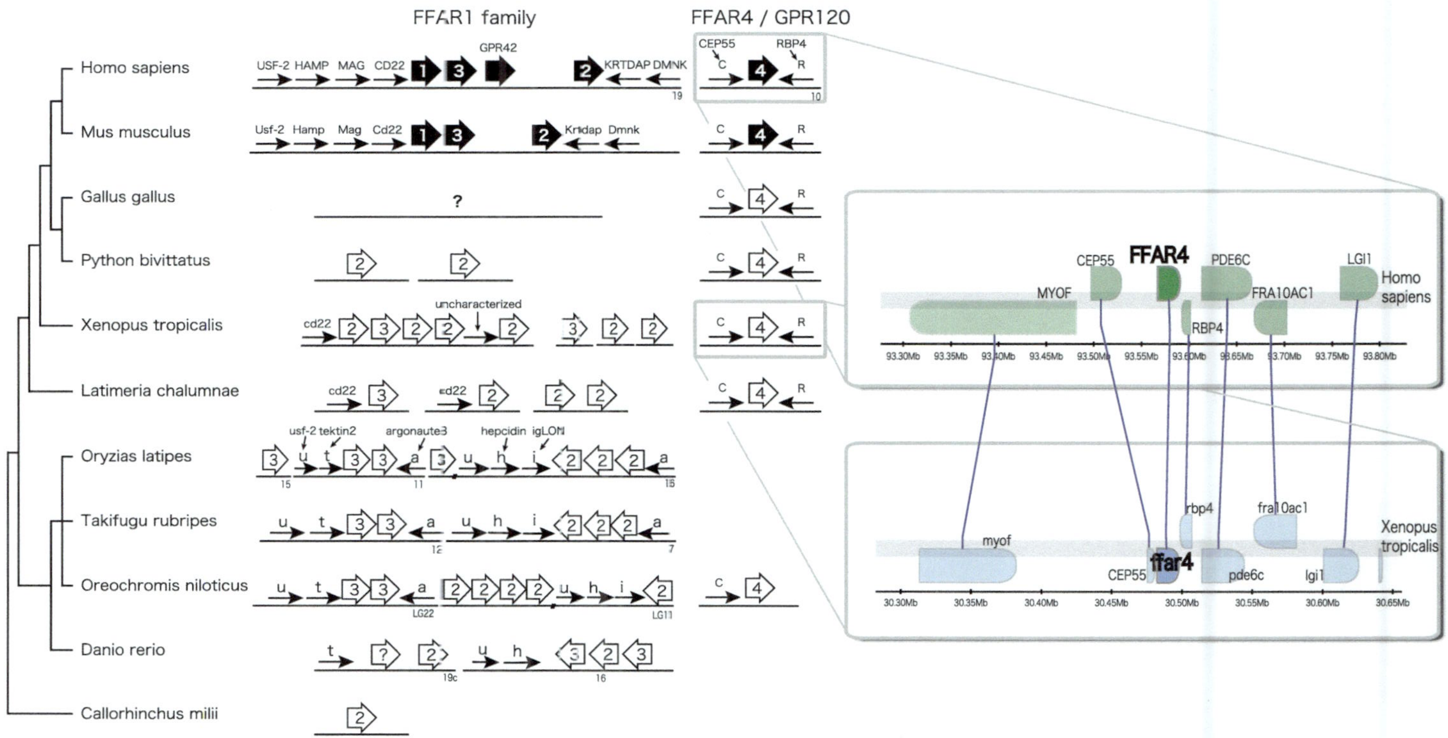

Fig. 10.1 Schematic representation of genomic structures around free fatty acids receptors (FFAR)1–4 of multiple animal species, and synteny analysis of the FFAR4 locus between *Homo sapiens* and *Xenopus tropicalis* genomes. The homology genes of FFAR1 and FFAR4 are compared and aligned in each species. *Black* or *white boxes* indicate genes that have been studied or not, respectively. Synteny analysis using Synteny Database [4] reveals syntenic conservation of many neighbor genes around FFAR4 between human and *Xenopus* (cluster ID#1424337 according to the Synteny Database)

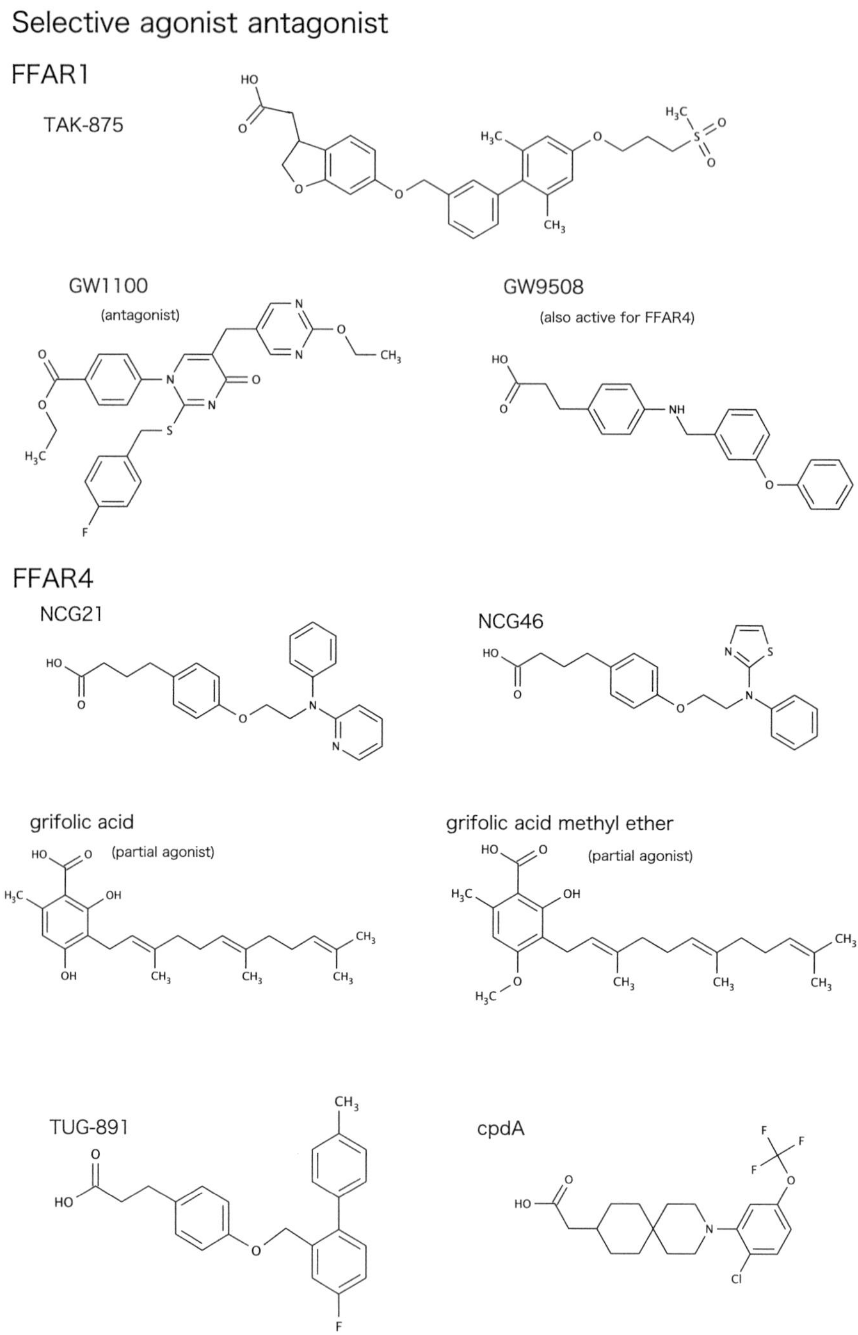

Fig. 10.2 Chemical structures of FFAR1 and FFAR4 ligands

a synthetic FFAR1 antagonist, GW1100, has been identified and its antagonistic activities were examined via in both in vitro and in vivo studies [1, 19, 22, 53]. The antidiabetic thiazolidinediones troglitazone and rosiglitazone, and the experimental anti-obesity compound MEDICA16, also activate FFAR1 [2, 16]. Very recently, the X-ray crystallography of co-crystals of a novel compound, TAK-875 (fasiflifam), and FFAR1 was performed, and the state of the ligand binding to the receptor was revealed [42]. It was reported that binding sites other than those predicted in recent models exist and that TAK-875 could act as a partial agonist or an allosteric ligand. In the same paper, a mass spectrometry-based ligand-binding assay system for FFAR1 was also established, and the FFAR–ligand interaction has been studied in great detail.

10.2.5 Physiological Roles of FFAR1

Free fatty acids have so far been known to exert versatile effects on pancreatic β-cells. Chronic exposure to high levels of FFAs results in the impairment of β-cell function and secretory capacity, whereas acute administration of FFAs stimulates insulin release. FFAs are considered to be important for maintaining basal insulin secretion as well as increasing glucose-stimulated insulin secretion when fasting [1, 7, 8, 13, 43, 44], but the mechanisms of these phenomena have not been explained. Itoh et al. showed that long-chain free fatty acids amplified glucose-stimulated insulin secretion from pancreatic β-cells via activation of FFAR1 [24]. Because inhibition of FFAR1 expression with small interfering RNA (siRNA) resulted in quenching of FFA-stimulated insulin secretion, FFAR was presumed to be involved in this pathway. In contrast to the decrease in FFA-stimulated insulin secretion observed in FFAR1-deficient β-cells, FFAR1-deficient mice actually exhibited resistance to obesity-induced hyperinsulinemia, hepatic steatosis, hypertriglyceridemia, increased hepatic glucose output, hyperglycemia, and glucose intolerance. Steneberg et al. have indicated that both acute and chronic effects of FFAs were mediated by FFAR1 [44]. In contrast, overexpression of FFAR1 in β-cells by methods via the mouse Ipf1/Pdx1 promoter impaired β-cell function and resulted in hyperinsulinemia and diabetes [44]. Moreover, FFAR1 was found to regulate glucose-stimulated insulin secretion by overexpressing FFAR1 under the control of mouse insulin II promoter [33]. On the basis of these studies, we speculate that FFAR1 is involved in an essential pathway that connects obesity and type 2 diabetes.

Furthermore, there have been several studies reporting genomic polymorphisms in human FFAR1. One type of polymorphisms, D175N, has the same EC_{50} but lower maximum response compared to the wild-type receptor. This polymorphism, however, has no relevance to changes in insulin secretion [15]. Another polymorphism, R211H, shows no difference in primary response, but results from laboratory data comparison suggests its involvement in insulin secretion [34]. Moreover, β-cell response to FFAs is quenched in the polymorphism G180S because of impaired mechanisms in increasing intracellular Ca^{2+} concentration [52].

These results have raised a great deal of interest in FFAR1 as a potential target for novel drugs in metabolic diseases such as type 2 diabetes. Various experimental models have identified chemical compounds that display agonistic or antagonistic activity, and their physiological and pharmacological functions are being examined. One such compound was TAK-875 (fasiglifam, mentioned earlier), an orally available, potent, and selective agonist of FFAR1 [39]. This agent was tested in a phase III clinical trial for the potential treatment of type 2 diabetes mellitus, but the trial was cancelled because of an undesired side effect. Another FFAR1 agonist, JTT-851, has completed phase II clinical trials and is anticipated to become the first therapeutic drug to target FFAR1.

10.3 FFAR4

10.3.1 Ligand and Tissue Distribution

Using a receptor internalization assay [11], medium- to long-chain FFAs were identified as endogenous ligands of FFAR4. Saturated FFAs (C14-18) and unsaturated FFAs (C16-22) activate FFAR4. Although some have claimed that FFAR4 was a selective receptor for ω-3 polyunsaturated fatty acids (PUFAs), it is now widely accepted that both ω-3 and ω-6 PUFAs act as agonists. A variety of PUFAs, regardless of ω-3 or ω-6 species, can activate FFAR4 in the micromolar concentration range [18]. The ligand profiles for FFAR4 are similar to those for FFAR1; however, the amino acid homology between FFAR4 and FFAR1 is only 10 %.

10.3.2 Genomic Structure and Evolution

FFAR1 and FFAR4 have no homology in structure, although some ligands activate both receptors. FFAR4 has been experimentally proved to function as a receptor in only mammals. Nonetheless, the FFAR4 gene is conserved in vertebrates from *Coelacanthiformes* to mammals, and the neighboring genomic structures are also quite similar, as shown in the comparison of the human and *Silurana tropicalis* genomes (Fig. 10.1). However, differing from FFAR1, FFAR2, and FFAR3, which prevail throughout vertebrates, the FFAR4 gene cannot be seen in teleost fish, with an exception of the family *Cichlidae*. Sequences similar to FFAR4 are detected in *Cichlidae* genomes, but the reason for this exception is unknown. Multiplication of genes such as those in FFAR1, FFAR2, and FFAR3 are also not observed in FFAR4. Comparing the FFAR4 orthologues including *Cichlidae*, we found that the chief amino acid sequences are well conserved among species. The amino acid residue equivalent to human R99 (described later in this chapter), which is important for the interaction of FFAR4 and FFAs, is conserved in all species, strongly suggesting that the function is conserved as well (Fig. 10.3).

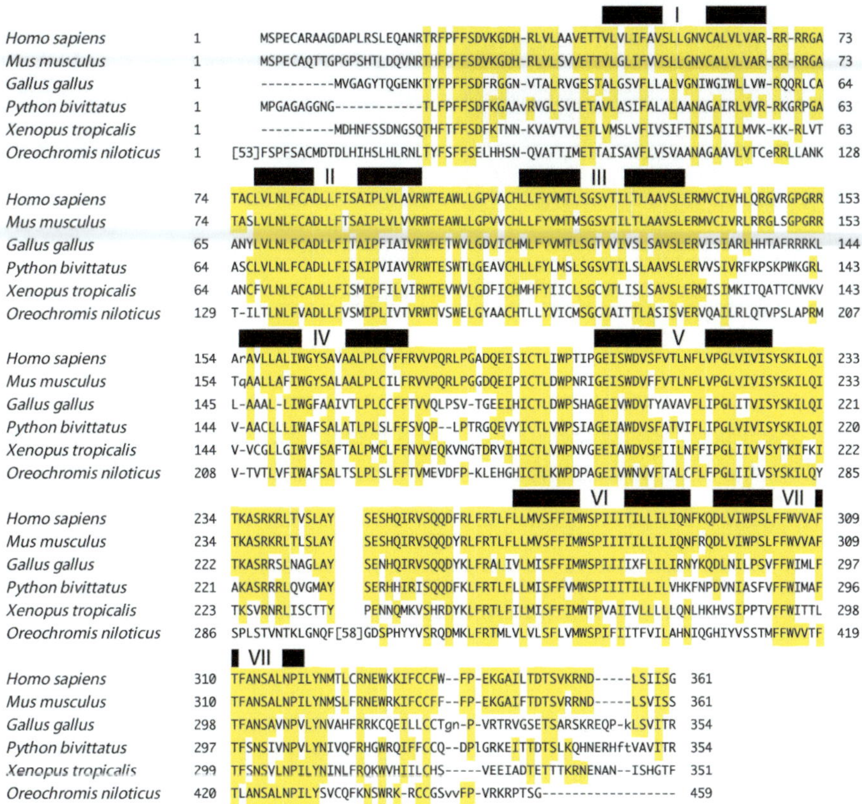

Fig. 10.3 Sequence alignment of FFAR4 in multiple animal species. The alignment is obtained by multiple sequence alignment for six homology protein sequences. Amino acid sequences corresponding to FFAR4 from different animal species were aligned using the NCBI COBALT algorithm [37]. Transmembrane ™ helix regions are shown between *black bars*. Highly conserved residues are highlighted by *yellow*

10.3.3 Signal Transduction

Both PUFAs and synthetic ligands induced a rise in cytosolic free Ca^{2+} in FFAR4-overexpressing HEK293 cells, suggesting that FFAR4 is coupled with the Gαq protein family. Recently, Shah et al. showed that PUFA-induced depolarization induced by the monovalent cation-specific transient receptor potential channel type M5 (TRPM5) is related to intracellular Ca^{2+} rise as well as CCK secretion from STC-1 cells, suggesting that TRPM5 plays a crucial role in FFAR4 signaling in STC-1 cells [41]. Oh et al. showed that a FFAR4 agonist exerts anti-inflammatory effects through β-arrestin 2 signaling in monocytic RAW 264.7 cells and primary macrophages [36]. FFAR4 can also induce the activation of ERK1/2 under certain conditions and activation of PI3-kinase and the serine/threonine protein kinase Akt in FFAR4-expressing cells [17, 25].

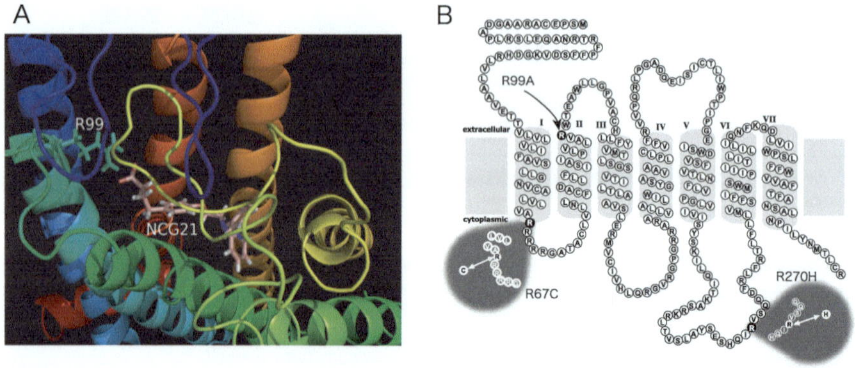

Fig. 10.4 Single-nucleotide polymorphisms (SNPs) in human FFAR4 and homology model of FFAR4. (**a**) Representation of secondary structure of FFAR4. Locus of its SNPs (R67C, R270H) and an interaction site (R99) are marked. (**b**) FFAR4 homology model docked with NCG21

10.3.4 Structure–Activity Relationships of FFAR4 Ligands

Because the three-dimensional structure of FFAR4 has not yet been elucidated by X-ray crystallography, structure–activity relationship studies are being conducted by combining site-directed mutagenesis and homology modeling. The results of the site-directed mutagenesis showed that the R99 residue is significant for the ligand binding for FFAR4, and the amino acid sequences around R99 are well conserved in aforementioned orthologues [47] (Figs. 10.3 and 10.4a). The calculation of the docking simulation and homology model of FFAR4 revealed significant correlation between the calculated value of the hydrogen bond energy and ligand-induced activity in many compounds, which led us to predict the activity of novel compounds [47, 49]. To identify other natural ligands of FFAR4, we screened and identified a selective partial agonist among a series of natural compounds derived from fruiting bodies of *Albatrellus ovinus* [17]. Depending on the experimental conditions, this compound is also useful as an antagonist selective for FFAR4. In addition, based on the structure of the PPARγ agonist thiazolidinediones, we synthesized a series of compounds containing carboxylic acids and developed a selective agonist using a homology model of FFAR4 [48] (Fig. 10.2). Hudson et al. have also reported the synthesis of compounds selective for FFAR4 [20], and many patents of compounds have been claimed [14]. The structure–activity relationship studies combining site-directed mutagenesis and homology modeling of FFAR4 showed that hydrophobic amino acid residues facing the ligand-binding pocket play an important role in the binding of FFAR4 ligand [21]. These compounds might be useful tools to monitor the physiological effects of FFAR4, and they might be potentially useful in the development of novel drug candidates for the treatment of type 2 diabetes, obesity, and metabolic diseases.

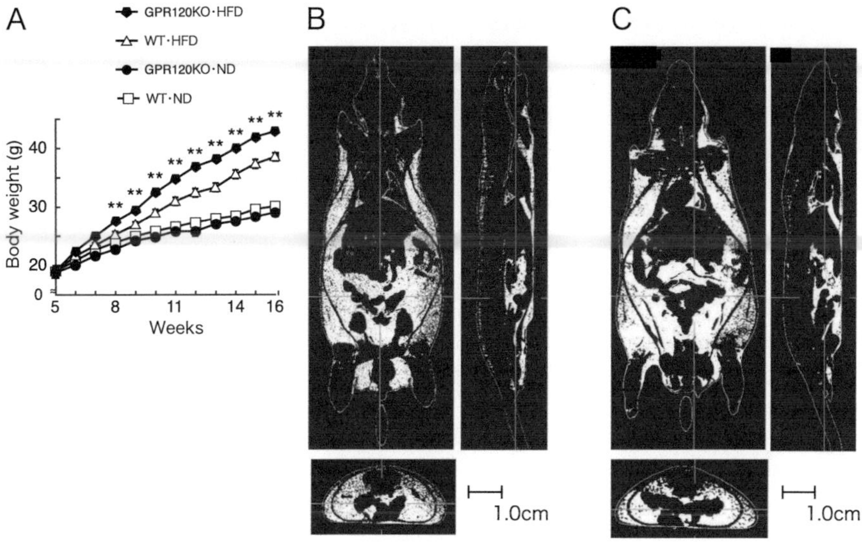

Fig. 10.5 Obesity in FFAR4-deficient mice fed a high-fat diet. (**a**) Body weight changes of wild-type and FFAR4-deficient mice fed a normal diet (ND) or a high-fat diet (HFD). All data represent mean + SEM. $^{**}P<0.01$ versus the corresponding wild-type value. (**b**, **c**) Computed tomography images of fat accumulation in wild-type (**b**) and FFAR4-deficient (**c**) male mice fed a high-fat diet. Fat depots are demarcated for illustration

10.3.5 Physiological Roles of FFAR4

We recently reported that dysfunctional FFAR4 led to obesity in both mice and humans [23]. We found that FFAR4-deficient mice fed a high-fat diet developed obesity, glucose intolerance, and fatty liver along with decreased adipocyte differentiation and lipogenesis and enhanced hepatic lipogenesis (Fig. 10.5). Insulin resistance in these mice was associated with reduced insulin signaling and enhanced inflammation in adipose tissue. FFAR4 exon sequencing in obese subjects revealed a deleterious nonsynonymous mutation (R270H) that inhibited FFAR4 signaling activity (Fig. 10.4b). Furthermore, the R270H variant increases the risk of obesity in European populations. Overall, this study demonstrates that the lipid sensor FFAR4 has a key role in sensing dietary fat and, therefore, in the control of energy balance in both humans and rodents.

Endogenous expression of FFAR4 has been identified in the intestine of humans and mice. Our previous study showed that FFAR4-expressing cells were located in the GLP-1-expressing enteroendocrine cells in the large intestine [18, 31]. Furthermore, the enteroendocrine cell line STC-1 also expressed FFAR4 endogenously, and PUFA or synthetic ligand stimulation induced the secretion of GLP-1 and cholecystokinin (CCK) as well as the $[Ca^{2+}]_i$ response [50]. These studies led us to speculate the physiological function of FFAR4 in incretin secretion in vivo.

FFAR4 is also expressed in other cells and tissues. Oh et al. found endogenous expression of FFAR4 in monocytic RAW 264.7 cells and in primary pro-inflammatory, M1-like macrophages. Matsumura et al. found the expression of FFAR4 in taste buds [30]. In the lung, we found that FFAR4 protein was colocalized with the Clara cell 10-kDa protein, a marker of Clara cells [31]. Further studies are needed to reveal the physiological function of FFAR4 in the lung.

Taneera et al. performed a systems genomics approach to identify genes for type 2 diabetes, and FFAR4 was ranked among the top 20 possibly associated genes [51]. In this report, FFAR4 expression in human islets was positively correlated with insulin secretion and insulin content and with lower HbA1c. Although inconsistent with previous reports that FFAR1 is dominantly detected in mouse pancreatic islets [33], these data suggest that FFAR4 can protect pancreatic islets from lipotoxicity in humans.

Recently, it has become clear that FFAR4 is expressed in the pancreas and contributes to glucagon secretion [46]. According to a report by Oh et al., continued administration of an agonist selective for FFAR4 improved glucose tolerance and insulin sensitivity in mice fed a high-fat diet [35]. However, they attributed their results to a macrophage-mediated pathway and did not mention its relationship to obesity; this is in disagreement with our results.

From the analyses of human and mice described here, there is no doubt that FFAR4 is strongly involved in diet-induced obesity and acts as a lipid sensor that maintains the balance of energy metabolism by controlling lipid biosynthesis [23]. Further investigation is anticipated to develop therapeutic drugs targeting FFAR4 for the treatment of obesity-related metabolic disorders.

10.4 Conclusion

Because of the low binding affinity of FFAs to FFARs, there was skepticism toward FFARs when originally discovered. However, the experimental facts described here confirmed that FFAs were indeed ligands for FFARs; thus, the names of the fatty acid receptor family, which were originally the numbers of the orphan receptors (GPR40, GPR120, etc.), are now officially changed to FFAR1, FFAR4, etc, respectively. [6, 45]. The ligands for some nutrient-sensing GPCRs bind with lower affinity (in the micromolar or millimolar range) than that of classic high-affinity ligands, such as hormones or growth factors, for their receptors. The fatty acid receptor family is considered to be a group of sensor molecules that detect FFAs of various lengths and structures as natural ligands with binding constants comparable with their vivo concentrations. Among the FFARs, FFAR1 and FFAR4 are considered to be potential drug targets for the treatment of metabolic diseases such as type 2 diabetes, because their physiological functions are related to energy homeostasis. Further analysis of FFARs may also be important to better understand the nutrient-sensing process and to develop novel therapeutic compounds to treat metabolic diseases.

References

1. Briscoe CP, Tadayyon M, Andrews JL, Benson WG, Chambers JK, Eilert MM, Ellis C, Elshourbagy NA, Goetz AS, Minnick DT, Murdock PR, Sauls HR, Shabon U, Spinage LD, Strum JC, Szekeres PG, Tan KB, Way JM, Ignar DM, Wilson S, Muir AI (2003) The orphan G protein-coupled receptor GPR40 is activated by medium and long chain fatty acids. J Biol Chem 278(13):11303–11311
2. Briscoe CP, Peat AJ, McKeown SC, Corbett DF, Goetz AS, Littleton TR, McCoy DC, Kenakin TP, Andrews JL, Ammala C, Fornwald JA, Ignar DM, Jenkinson S (2006) Pharmacological regulation of insulin secretion in MIN6 cells through the fatty acid receptor GPR40: identification of agonist and antagonist small molecules. Br J Pharmacol 148(5):619–628
3. Brown AJ, Goldsworthy SM, Barnes AA, Eilert MM, Tcheang L, Daniels D, Muir AI, Wigglesworth MJ, Kinghorn I, Fraser NJ, Pike NB, Sturm JC, Steplewski KM, Murdock PR, Holder JC, Marshall FH, Szekeres PG, Wilson S, Ignar DM, Foord SM, Wise A, Dowell SJ (2003) The orphan G protein-coupled receptors GPR41 and GPR43 are activated by propionate and other short chain carboxylic acids. J Biol Chem 278:11312–11319
4. Catchen JM, Conery JS, Postlethwait JH (2009) Automated identification of conserved synteny after whole-genome duplication. Genome Res 19(8):1497–1505
5. Chawla A, Repa JJ, Evans RM, Mangelsdorf DJ (2001) Nuclear receptors and lipid physiology: opening the X-files. Science 294(5548):1866–1870
6. Davenport AP, Harmar AJ (2013) Evolving pharmacology of orphan GPCRs: IUPHAR commentary. Br J Pharmacol 170(4):693–695
7. Dobbins RL, Chester MW, Daniels MB, McGarry JD, Stein DT (1998) Circulating fatty acids are essential for efficient glucose-stimulated insulin secretion after prolonged fasting in humans. Diabetes 47(10):1613–1618
8. Dobbins RL, Chester MW, Stevenson BE, Daniels MB, Stein DT, McGarry JD (1998) A fatty acid-dependent step is critically important for both glucose- and non-glucose-stimulated insulin secretion. J Clin Invest 101(11):2370–2376
9. Feng DD, Luo Z, Roh SG, Hernandez M, Tawadros N, Keating DJ, Chen C (2006) Reduction in voltage-gated K^+ currents in primary cultured rat pancreatic beta-cells by linoleic acids. Endocrinology 147(2):674–682
10. Fujiwara K, Maekawa F, Yada T (2005) Oleic acid interacts with GPR40 to induce Ca^{2+} signaling in rat islet beta-cells: mediation by PLC and L-type Ca^{2+} channel and link to insulin release. Am J Physiol Endocrinol Metab 289(4):E670–E677
11. Fukunaga S, Setoguchi S, Hirasawa A, Tsujimoto G (2006) Monitoring ligand-mediated internalization of G protein-coupled receptor as a novel pharmacological approach. Life Sci 80(1):17–23
12. Garrido DM, Corbett DF, Dwornik KA, Goetz AS, Littleton TR, McKeown SC, Mills WY, Smalley TL, Briscoe CP, Peat AJ (2006) Synthesis and activity of small molecule GPR40 agonists. Bioorg Med Chem Lett 16(7):1840–1845
13. Gravena C, Mathias PC, Ashcroft SJ (2002) Acute effects of fatty acids on insulin secretion from rat and human islets of Langerhans. J Endocrinol 173(1):73–80
14. Halder S, Kumar S, Sharma R (2013) The therapeutic potential of GPR120: a patent review. Expert Opin Ther Pat 23(12):1581–1590
15. Hamid YH, Vissing H, Holst B, Urhammer SA, Pyke C, Hansen SK, Glumer C, Borch-Johnsen K, Jørgensen T, Schwartz TW, Pedersen O, Hansen T (2005) Studies of relationships between variation of the human G protein-coupled receptor 40 gene and type 2 diabetes and insulin release. Diabet Med 22(1):74–80
16. Hara T, Hirasawa A, Sun Q, Koshimizu TA, Itsubo C, Sadakane K, Awaji T, Tsujimoto G (2009) Flow cytometry-based binding assay for GPR40 (FFAR1; free fatty acid receptor 1). Mol Pharmacol 75(1):85–91

17. Hara T, Hirasawa A, Sun Q, Sadakane K, Itsubo C, Iga T, Adachi T, Koshimizu TA, Hashimoto T, Asakawa Y, Tsujimoto G (2009) Novel selective ligands for free fatty acid receptors GPR120 and GPR40. Naunyn Schmiedebergs Arch Pharmacol 380(3):247–255
18. Hirasawa A, Tsumaya K, Awaji T, Katsuma S, Adachi T, Yamada M, Sugimoto Y, Miyazaki S, Tsujimoto G (2005) Free fatty acids regulate gut incretin glucagon-like peptide-1 secretion through GPR120. Nat Med 11(1):90–94
19. Hu H, He LY, Gong Z, Li N, Lu YN, Zhai QW, Liu H, Jiang HL, Zhu WL, Wang HY (2009) A novel class of antagonists for the FFAs receptor GPR40. Biochem Biophys Res Commun 390(3):557–563
20. Hudson BD, Shimpukade B, Mackenzie AE, Butcher AJ, Pediani JD, Christiansen E, Heathcote H, Tobin AB, Ulven T, Milligan G (2013) The pharmacology of TUG-891, a potent and selective agonist of the free fatty acid receptor 4 (FFA4/GPR120), demonstrates both potential opportunity and possible challenges to therapeutic agonism. Mol Pharmacol 84(5):710–725
21. Hudson BD, Shimpukade B, Milligan G, Ulven T (2014) The molecular basis of ligand interaction at free fatty acid receptor 4 (FFA4/GPR120). J Biol Chem 289(29):20345–20358
22. Humphries PS, Benbow JW, Bonin PD, Boyer D, Doran SD, Frisbie RK, Piotrowski DW, Balan G, Bechle BM, Conn EL, Dirico KJ, Oliver RM, Soeller WC, Southers JA, Yang X (2009) Synthesis and SAR of 1,2,3,4-tetrahydroisoquinolin-1-ones as novel G-protein-coupled receptor 40 (GPR40) antagonists. Bioorg Med Chem Lett 19(9):2400–2403
23. Ichimura A, Hirasawa A, Poulain-Godefroy O, Bonnefond A, Hara T, Yengo L, Kimura I, Leloire A, Liu N, Iida K, Choquet H, Besnard P, Lecoeur C, Vivequin S, Ayukawa K, Takeuchi M, Ozawa K, Tauber M, Maffeis C, Morandi A, Buzzetti R, Elliott P, Pouta A, Jarvelin MR, Korner A, Kiess W, Pigeyre M, Caiazzo R, Van Hul W, Van Gaal L, Horber F, Balkau B, Levy-Marchal C, Rouskas K, Kouvatsi A, Hebebrand J, Hinney A, Scherag A, Pattou F, Meyre D, Koshimizu TA, Wolowczuk I, Tsujimoto G, Froguel P (2012) Dysfunction of lipid sensor GPR120 leads to obesity in both mouse and human. Nature 483(7389):350–354
24. Itoh Y, Kawamata Y, Harada M, Kobayashi M, Fujii R, Fukusumi S, Ogi K, Hosoya M, Tanaka Y, Uejima H, Tanaka H, Maruyama M, Satoh R, Okubo S, Kizawa H, Komatsu H, Matsumura F, Noguchi Y, Shinohara T, Hinuma S, Fujisawa Y, Fujino M (2003) Free fatty acids regulate insulin secretion from pancreatic beta cells through GPR40. Nature 422(6928):173–176
25. Katsuma S, Hatae N, Yano T, Ruike Y, Kimura M, Hirasawa A, Tsujimoto G (2005) Free fatty acids inhibit serum deprivation-induced apoptosis through GPR120 in a murine enteroendocrine cell line STC-1. J Biol Chem 280(20):19507–19515
26. Kimura I, Inoue D, Maeda T, Hara T, Ichimura A, Miyauchi S, Kobayashi M, Hirasawa A, Tsujimoto G (2011) Short-chain fatty acids and ketones directly regulate sympathetic nervous system via G protein-coupled receptor 41 (GPR41). Proc Natl Acad Sci U S A 108(19):8030–8035
27. Kotarsky K, Nilsson NE, Flodgren E, Owman C, Olde B (2003) A human cell surface receptor activated by free fatty acids and thiazolidinedione drugs. Biochem Biophys Res Commun 301(2):406–410
28. Louet JF, Chatelain F, Decaux JF, Park EA, Kohl C, Pineau T, Girard J, Pegorier JP (2001) Long-chain fatty acids regulate liver carnitine palmitoyl-transferase I gene (L-CPT I) expression through a peroxisome-proliferator-activated receptor alpha (PPARalpha)-independent pathway. Biochem J 354(pt 1):189–197
29. Maslowski KM, Vieira AT, Ng A, Kranich J, Sierro F, Yu D, Schilter HC, Rolph MS, Mackay F, Artis D, Xavier RJ, Teixeira MM, Mackay CR (2009) Regulation of inflammatory responses by gut microbiota and chemoattractant receptor GPR43. Nature 461(7268):1282–1286
30. Matsumura S, Mizushige T, Yoneda T, Iwanaga T, Inoue K, Tsuzuki S, Fushiki T (2007) GPR expression in the rat taste bud relating to fatty acid sensing. Biomed Res 28(1):49–55
31. Miyauchi S, Hirasawa A, Iga T, Liu N, Itsubo C, Sadakane K, Hara T, Tsujimoto G (2009) Distribution and regulation of protein expression of the free fatty acid receptor GPR120. Naunyn Schmiedebergs Arch Pharmacol 379(4):427–434

32. Miyauchi S, Hirasawa A, Ichimura A, Hara T, Tsujimoto G (2010) New frontiers in gut nutrient sensor research: free fatty acid sensing in the gastrointestinal tract. J Pharmacol Sci 112(1):19–24
33. Nagasumi K, Esaki R, Iwachidow K, Yasuhara Y, Ogi K, Tanaka H, Nakata M, Yano T, Shimakawa K, Taketomi S, Takeuchi K, Odaka H, Kaisho Y (2009) Overexpression of GPR40 in pancreatic beta-cells augments glucose-stimulated insulin secretion and improves glucose tolerance in normal and diabetic mice. Diabetes 58(5):1067–1076
34. Ogawa T, Hirose H, Miyashita K, Saito I, Saruta T (2005) GPR40 gene Arg211His polymorphism may contribute to the variation of insulin secretory capacity in Japanese men. Metab Clin Exp 54(3):296–299
35. Ohda Y, Walenta E, Akiyama TE, Lagakos WS, Lackey D, Pessentheiner AR, Sasik R, Hah N, Chi TJ, Cox JM, Powels MA, Di Salvo J, Sinz C, Watkins SM, Armando AM, Chung H, Evans RM, Quehenberger O, McNelis J, Bogner-Strauss JG, Olefsky JM (2014) A Gpr120-selective agonist improves insulin resistance and chronic inflammation in obese mice. Nat Med 20(8):942–947
36. Oh DY, Talukdar S, Bae EJ, Imamura T, Morinaga H, Fan W, Li P, Lu WJ, Watkins SM, Olefsky JM (2010) GPR120 is an omega-3 fatty acid receptor mediating potent anti-inflammatory and insulin-sensitizing effects. Cell 142(5):687–698
37. Papadopoulos JS, Agarwala R (2007) COBALT: constraint-based alignment tool for multiple protein sequences. Bioinformatics 23(9):1073–1079
38. Remely M, Aumueller E, Merold C, Dworzak S, Hippe B, Zanner J, Pointner A, Brath H, Haslberger AG (2014) Effects of short chain fatty acid producing bacteria on epigenetic regulation of FFAR3 in type 2 diabetes and obesity. Gene 537(1):85–92
39. Sasaki S, Kitamura S, Negoro N, Suzuki M, Tsujihata Y, Suzuki N, Santou T, Kanzaki N, Harada M, Tanaka Y, Kobayashi M, Tada N, Funami M, Tanaka T, Yamamoto Y, Fukatsu K, Yasuma T, Momose Y (2011) Design, synthesis, and biological activity of potent and orally available G protein-coupled receptor 40 agonists. J Med Chem 54(5):1365–1378
40. Sauer LA, Dauchy RT, Blask DE (2000) Mechanism for the antitumor and anticachectic effects of n-3 fatty acids. Cancer Res 60(18):5289–5295
41. Shah BP, Liu P, Yu T, Hansen DR, Gilbertson TA (2012) TRPM5 is critical for linoleic acid-induced CCK secretion from the enteroendocrine cell line, STC-1. Am J Physiol Cell Physiol 302(1):C210–C219
42. Srivastava A, Yano J, Hirozane Y, Kefala G, Gruswitz F, Snell G, Lane W, Ivetac A, Aertgeerts K, Nguyen J, Jennings A, Okada K (2014) High-resolution structure of the human GPR40 receptor bound to allosteric agonist TAK-875. Nature 513(7516):124–127
43. Stein DT, Esser V, Stevenson BE, Lane KE, Whiteside JH, Daniels MB, Chen S, McGarry JD (1996) Essentiality of circulating fatty acids for glucose- stimulated insulin secretion in the fasted rat. J Clin Invest 97(12):2728–2735
44. Steneberg P, Rubins N, Bartoov-Shifman R, Walker MD, Edlund H (2005) The FFA receptor GPR40 links hyperinsulinemia, hepatic steatosis, and impaired glucose homeostasis in mouse. Cell Metab 1(4):245–258
45. Stoddart LA, Smith NJ, Milligan G (2008) International Union of Pharmacology. LXXI. Free fatty acid receptors FFA1, -2, and -3: pharmacology and pathophysiological functions. Pharmacol Rev 60(4):405–417
46. Suckow AT, Polidori D, Yan W, Chon S, Ma JY, Leonard J, Briscoe CP (2014) Alteration of the glucagon axis in GPR120 (FFAR4) knockout mice: a role for GPR120 in glucagon secretion. J Biol Chem 289(22):15751–15763
47. Sun Q, Hirasawa A, Hara T, Kimura I, Adachi T, Awaji T, Ishiguro M, Suzuki T, Miyata N, Tsujimoto G (2010) Structure–activity relationships of GPR120 agonists based on a docking simulation. Mol Pharmacol 78(5):804–810
48. Suzuki T, Igari S, Hirasawa A, Hata M, Ishiguro M, Fujieda H, Itoh Y, Hirano T, Nakagawa H, Ogura M, Makishima M, Tsujimoto G, Miyata N (2008) Identification of G protein-coupled receptor 120-selective agonists derived from PPARgamma agonists. J Med Chem 51(23):7640–7644

49. Takeuchi M, Hirasawa A, Hara T, Kimura I, Hirano T, Suzuki T, Miyata N, Awaji T, Ishiguro M, Tsujimoto G (2013) FFA1-selective agonistic activity based on docking simulation using FFA1 and GPR120 homology models. Br J Pharmacol 168(7):1570–1583
50. Tanaka T, Katsuma S, Adachi T, Koshimizu TA, Hirasawa A, Tsujimoto G (2008) Free fatty acids induce cholecystokinin secretion through GPR120. Naunyn Schmiedebergs Arch Pharmacol 377(4-6):523–527
51. Taneera J, Lang S, Sharma A, Fadista J, Zhou Y, Ahlqvist E, Jonsson A, Lyssenko V, Vikman P, Hansson O, Parikh H, Korsgren O, Soni A, Krus U, Zhang E, Jing XJ, Esguerra JL, Wollheim CB, Salehi A, Rosengren A, Renstrom E, Groop L (2012) A systems genetics approach identifies genes and pathways for type 2 diabetes in human islets. Cell Metab 16(1):122–134
52. Vettor R, Granzotto M, De Stefani D, Trevellin E, Rossato M, Farina MG, Milan G, Pilon C, Nigro A, Federspil G, Vigneri R, Vitiello L, Rizzuto R, Baratta R, Frittitta L (2008) Loss-of-function mutation of the GPR40 gene associates with abnormal stimulated insulin secretion by acting on intracellular calcium mobilization. J Clin Endocrinol Metab 93(9):3541–3550
53. Zhang X, Yan G, Li Y, Zhu W, Wang H (2010) DC2601 in obese Zucker rats. Biomed Pharmacother 64(9):647–651

Chapter 11
Omega-3 Fatty Acid Metabolism and Regulation of Inflammation

Yosuke Isobe and Makoto Arita

Abstract Increasing evidence from both human and animal studies has demonstrated that omega-3 polyunsaturated fatty acids (PUFAs), primarily eicosapentaenoic acid (EPA) and docosahexaenoic acid (DHA), can suppress inflammation and play a beneficial role in a variety of inflammation-related human diseases, such as inflammatory bowel disease, rheumatoid arthritis, asthma, cancer, and cardiovascular diseases. Omega-3 PUFAs serve as substrates for the production of potent bioactive anti-inflammatory lipid mediators such as resolvins. Herein we review recent advances in understanding the formation and action of these mediators, especially focusing on the LC-ESI-MS/MS-based lipidomics approach and on recently identified bioactive compounds with potent anti-inflammatory properties.

Keywords Omega-3 PUFA • Anti-inflammation • Lipidomics • Lipid mediator • Lipoxygenase • LC-MS/MS

11.1 Introduction

Polyunsaturated fatty acids (PUFAs) are essential in human nutrition and can be divided into two subcategories, termed omega-3 and omega-6, based on the location of their first double bond relative to the tail (omega) of the carbon chain. Omega-6 arachidonic acid is a common precursor of many eicosanoids, a family of bioactive lipid mediators important in controlling inflammatory responses (Fig. 11.1). Omega-3 PUFAs such as eicosapentaenoic acid (EPA) and docosahexaenoic acid

Y. Isobe
Laboratory for Metabolomics, RIKEN Center for Integrative Medical Sciences, Yokohama, Japan

M. Arita (✉)
Laboratory for Metabolomics, RIKEN Center for Integrative Medical Sciences, Yokohama, Japan

Graduate School of Medical Life Science, Yokohama City University, Yokohama, Japan

PRESTO, Japan Science and Technology Agency, Kawaguchi, Saitama, Japan
e-mail: makoto.arita@riken.jp

T. Yokomizo, M. Murakami (eds.), *Bioactive Lipid Mediators*,
DOI 10.1007/978-4-431-55669-5_11

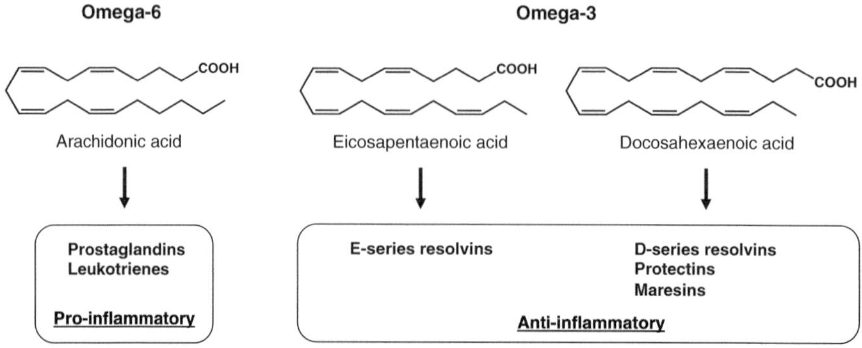

Fig. 11.1 Polyunsaturated fatty acid (PUFA)-derived mediators. Arachidonic acid is a metabolic precursor of eicosanoids [e.g., prostaglandins (PGs) and leukotrienes (LTs)], which have distinct roles as pro-inflammatory mediators. Omega-3 PUFAs prevent conversion of arachidonic acid into pro-inflammatory eicosanoids via substrate competition. In addition, omega-3 PUFAs are converted to bioactive metabolites such as resolvins and protectins with anti-inflammatory and pro-resolving properties

(DHA), which are enriched in some fish oils, are believed to exert beneficial effects on a wide range of human inflammatory disorders, including inflammatory bowel disease, rheumatoid arthritis, and cardiovascular diseases [1–3]. In addition, studies using omega-3 desaturase (*fat-1*) transgenic mice, which have abundant endogenous omega-3 PUFAs, strongly support the idea that omega-3 PUFAs are protective in inflammatory pathology [4, 5]. Omega-3 PUFAs prevent conversion of arachidonic acid (AA) to the pro-inflammatory eicosanoids by serving as an alternative substrate for cyclooxygenase (COX) or lipoxygenase (LOX), resulting in the production of less potent products. In addition, a number of enzymatically oxygenated metabolites derived from omega-3 PUFAs were recently identified as anti-inflammatory mediators. These compounds may contribute to the beneficial actions that have been attributed to omega-3 PUFAs in human diseases.

11.2 LC-ESI-MS/MS-Based Lipidomics

Liquid chromatography-tandem mass spectrometry (LC-ESI-MS/MS) is the most reliable and sensitive method to identify and quantify PUFA metabolites (Fig. 11.2). The development of electrospray ionization (ESI) technology provided an ideal interface between LC and MS because it avoids unwanted degradation and generates primarily molecular ions for MS/MS analysis. ESI is a soft ionization technology that forms either positive or negative ions through the addition of a proton to form $[M+H]^+$ or the removal of a proton to form $[M-H]^-$. In case of PUFA-derived mediators, ESI results in $[M-H]^-$ carboxylate ions that can be detected with relatively high sensitivity. The triple quadrupole mass spectrometer is capable of

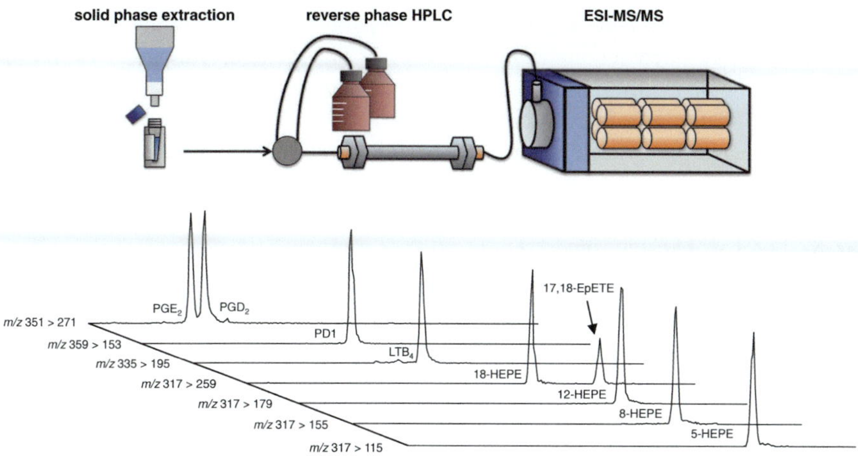

Fig. 11.2 Liquid chromatography-tandem mass spectrometry (LC-MS/MS)-based lipidomics. After solid-phase extraction, samples are separated by reverse-phase HPLC, and PUFA metabolites are detected and quantified by multiple reaction monitoring (MRM) using triple quadrupole MS/MS. Representative MRM chromatograms of PUFA metabolites are shown

operating an MS/MS mode called multiple reaction monitoring (MRM). A specified precursor ion is selected according to its mass-to-charge ratio (m/z) in the first quadrupole mass filter, and it is then fragmented into product ions in the second chamber by collision-induced dissociation (CID). Then, the third quadrupole mass filter is locked onto its specified product ion. Representative chromatograms of PUFA metabolites are depicted in Fig. 11.2. Among HEPE isomers, 8-HEPE and 12-HEPE do not resolve well using a C18 column. However, the choice of structure-specific product ions allows their differentiation (8-HEPE m/z 317>155, and 12-HEPE m/z 317>179). Also, the MRM transition used for 18-HEPE (m/z 317>215) showed cross-reactivity with 17,18-EpETE, but these hydroxy- and epoxy-fatty acids could be resolved well by C18 column chromatography. This MRM mode leads to further improvement in detection and quantification limits when combined with high-resolution LC separations [6].

11.3 E-Series Resolvins: Novel EPA-Derived Mediators with Potent Anti-inflammatory Properties

Using an unbiased lipidomics approach, Serhan's group identified resolvin E1 (RvE1) and RvE2 as novel anti-inflammatory and pro-resolving mediators derived from EPA [7–9]. RvE1 (5*S*,12*R*,18*R*-trihydroxy-6*Z*,8*E*,10*E*,14*Z*,16E-eicosapentaenoic acid) and RvE2 (5*S*,18*R*-dihydroxy-6*E*,8*Z*,11*Z*,14*Z*,16E-eicosapentaenoic acid) are biosynthesized by human polymorphonuclear leukocytes

(PMNs) through the 5-lipoxygenase (5-LOX) pathway from a common precursor, 18-hydroxyeicosapentaenoic acid (18-HEPE). The general functions of RvE1 include reducing neutrophil transmigration, attenuating inflammation and pain, and promoting resolution of acute inflammation [8–14]. RvE2 is the second E-series member to inhibit zymosan-induced PMN infiltration and display potent anti-inflammatory properties in murine peritonitis [15, 16].

Using a similar strategy, we recently identified RvE3 (17*R*,18-dihydroxy-5*Z*,8*Z*,11*Z*,13*E*,15*E*-eicosapentaenoic acid) as a potent and stereoselective antiinflammatory metabolite [17]. RvE3 blocked PMN infiltration in acute peritonitis, and inhibited LTB_4-induced PMN chemotaxis in vitro at low nanomolar concentrations. Of interest, the unnatural stereoisomers (i.e., 17*S*,18-dihydroxy-5*Z*,8*Z*,11*Z*,13*E*,15*E*-eicosapentaenoic acid obtained via chemical synthesis [18]) were much less active compared with the natural isomers, demonstrating the stereoselective action of RvE3 [19]. A recent study showed that the administration of RvE3 to LPS-exposed pregnant mice lowered the incidence of preterm birth [20].

In contrast to RvE1 and E2 biosynthesized by PMNs via the 5-LOX pathway, RvE3 is biosynthesized by eosinophils via the 12/15-LOX pathway [17]. Eosinophils are circulating granulocytes known to be involved in the pathogenesis of allergic diseases such as asthma. On the other hand, several studies have identified novel beneficial roles for eosinophils in maintaining tissue homeostasis [21–23]. Also, our recent studies suggest that eosinophils are recruited to the inflammatory site during the resolution phase, where they locally produce pro-resolving lipid mediators via the 12/15-LOX-initiated biosynthetic route [24, 25]. Therefore, RvE3 may contribute, at least in part, to the eosinophil functions of regulating acute inflammation and resolution. 12/15-LOX is also expressed by tissue resident macrophages, dendritic cells, mast cells, and airway epithelial cells [26]. 12/15-LOX deficiency leads to progressive atherosclerosis [27], exacerbation of arthritis, and inflammatory joint destruction [28], reduced corneal re-epithelialization [29], and a decline in immune self-tolerance [30]. Cells expressing 12/15-LOX may be involved in regulating inflammatory responses by locally producing anti-inflammatory lipid mediators such as RvE3.

11.4 Novel Bioactive Metabolites Formed via Omega-3 Epoxygenation of EPA

In addition to E-series resolvins, which are formed via a common precursor 18-HEPE, we recently discovered a new EPA-metabolic pathway via omega-3 epoxygenation, and identified a novel bioactive metabolite, 12-OH-17,18-EpETE, with potent and stereoselective anti-inflammatory properties [31]. Intravenous administration of 12-OH-17,18-EpETE dose-dependently blocked acute PMN infiltration in zymosan-induced peritonitis. This compound also significantly reduced PMN migration speed toward an LTB_4 chemotactic gradient at low nanomolar concentrations. PMNs treated with 12-OH-17,18-EpETE exhibited less directed migration toward the chemotactic gradient of LTB_4 and extended multiple pseudopodia during migration. Interestingly, when compared with RvE3, both compounds reduced speed of PMN migration toward the chemotactic gradient of LTB_4, but only 12-OH-17,18-EpETE treatment resulted in less direct PMN migration. These results indicate that 12-OH-17,18-EpETE has a structure-specific effect and a mechanism of action different from that of RvE3.

12-OH-17,18-EpETE is presumably biosynthesized in vivo through 12-hydroxylation of 17,18-EpETE or 17,18-epoxygenation of 12-HEPE. Although 12-OH-17,18-EpETE has potent anti-inflammatory activity, the structurally related 17,18-EpETE, 12-HEPE or 12,17,18-triHETE were essentially devoid of activity in zymosan-induced peritonitis and LTB_4-induced PMN polarization. Because the epoxide moiety of 12-OH-17,18-EpETE can be rapidly hydrolyzed by epoxide hydrolase to form 12,17,18-triHETE, this may be the physiological enzymatic pathway used to inactivate this endogenous autacoid.

Using chemically synthesized stereoisomers, the complete structures of two natural isomers were assigned as 12*S*-OH-17*R*,18*S*-EpETE and 12*S*-OH-17*S*,18R-EpETE. These natural isomers both displayed potent anti-inflammatory action, whereas the unnatural stereoisomers were essentially devoid of activity. The low nanomolar activity levels of these compounds and their stereospecificity activity requirements support the existence of a high-affinity receptor for 12-OH-17,18-EpETE on PMNs.

11.5 Perspectives

The formation of endogenous autacoids derived from omega-3 PUFA may explain in part the well-known, essential roles of the omega-3 PUFA in health and disease. E-series reolvins are biosynthesized from a common precursor, 18-HEPE, and 12-OH-17,18-EpETE is formed from 17,18-EpETE. These oxidation reactions target the omega-3 double bond of EPA, which distinguishes it from omega 6 arachidonic acid. Also, as in EPA, our recent study has identified novel anti-inflammatory metabolites generated via omega-3 oxidation of DHA [32]. Thus, we propose that omega-3 oxidation is an important structural and metabolic feature for the anti-inflammatory actions of omega-3 PUFA (Fig. 11.3). EPA is converted to 18-HEPE by aspirin-acetylated COX-2 [7] or cytochrome P450 monooxygenase (CYP) [33]. Several CYPs, including the CYP1A, CYP2C, and CYP2J subfamily members, can preferentially introduce a *cis*-epoxide at the omega-3 double bond of EPA to form 17,18-EpETE [34–36]. Recent studies have demonstrated contributions of CYP1 enzymes to the resolvin-biosynthetic pathway, and there was increased neutrophil recruitment in zymosan-induced peritoneal exudates of *Cyp1a1/1a2/1b1* triple-knockout mice [37]. Thus, cells expressing these enzymes are likely to be involved in regulating inflammatory responses via local production of anti-inflammatory metabolites. Further studies to understand lipid mediator biosynthesis and the

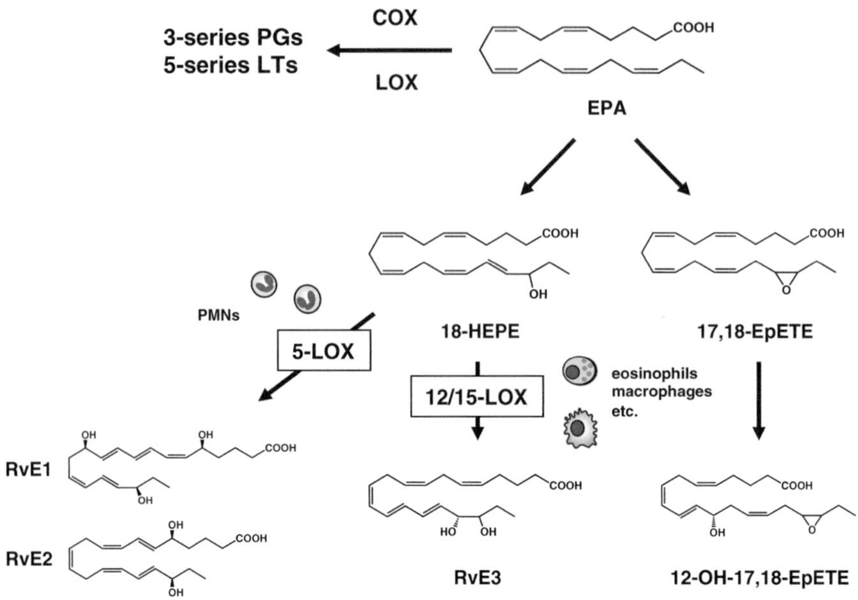

Fig. 11.3 Novel anti-inflammatory metabolic pathways of eicosapentaenoic acid (EPA). EPA is converted by COX or LOX to form 3-series PGs or 5-series LTs. In addition, recent studies uncovered a novel EPA metabolic pathway via omega-3 hydroxylation or epoxygenation in vivo and identified potent anti-inflammatory metabolites such as E-series resolvins and 12-OH-17,18-EpETE. These metabolic pathways may contribute to the anti-inflammatory actions of omega-3 PUFAs in vivo

structure–activity relationship of these mediators will provide insight into endogenous protection against aberrant or uncontrolled innate inflammatory responses as well as a molecular basis for the beneficial actions of omega-3 PUFAs noted in humans.

References

1. Simopoulos AP (2002) Omega-3 fatty acids in inflammation and autoimmune diseases. J Am Coll Nutr 21:495–505
2. Kremer JM (2000) N-3 Fatty acid supplements in rheumatoid arthritis. Am J Clin Nutr 71(1 suppl):349S–351S
3. De Caterina R, Endres S, Kristensen SD, Schmidt EB (eds) (1993) N-3 fatty acids and vascular disease. Springer, London
4. Kang JX (2007) Fat-1 transgenic mice: a new model for omega-3 research. Prostaglandins Leukot Eessent Fatty Acids 77:263–267
5. Endo J, Sano M, Isobe Y et al (2014) 18-HEPE, an n-3 fatty acid metabolite released by macrophages, prevents pressure overload-induced cardiac remodeling. J Exp Med 211(8):1673–1687
6. Arita M (2012) Mediator lipidomics in acute inflammation and resolution. J Biochem 152:313–319

7. Serhan CN, Clish CB, Brannon J et al (2000) Novel functional sets of lipid-derived mediators with anti-inflammatory actions generated from omega-3 fatty acids via cyclooxygenase 2-nonsteroidal anti-inflammatory drugs and transcellular processing. J Exp Med 192:1197–1204
8. Arita M, Bianchini F, Aliberti J et al (2005) Stereochemical assignment, anti-inflammatory properties, and receptor for the omega-3 lipid mediator resolvin E1. J Exp Med 201:713–722
9. Arita M, Yoshida M, Hong S et al (2005) Resolvin E1, an endogenous lipid mediator derived from omega-3 eicosapentaenoic acid, protects against 2,4,6-trinitrobenzene sulfonic acid-induced colitis. Proc Natl Acad Sci U S A 102:7671–7676
10. Schwab JM, Chiang N, Arita M et al (2007) Resolvin E1 and protectin D1 activate inflammation-resolution programmes. Nature 447:869–874
11. Ohira T, Arita M, Omori K et al (2010) Resolvin E1 receptor activation signals phosphorylation and phagocytosis. J Biol Chem 285:3451–3461
12. Connor KM, SanGiovanni JP, Lofqvist C et al (2007) Increased dietary intake of omega-3 polyunsaturated fatty acids reduces pathological retinal angiogenesis. Nat Med 13:868–873
13. Campbell EL, MacManus CF, Kominsky DJ et al (2010) Resolvin E1-induced intestinal alkaline phosphatase promotes resolution of inflammation through LPS detoxification. Proc Natl Acad Sci U S A 107:14298–14303
14. Xu ZZ, Zhang L, Liu T et al (2010) Resolvins RvE1 and RvD1 attenuate inflammatory pain via central and peripheral actions. Nat Med 16:592–597
15. Tjonahen E, Oh SF, Siegelman J et al (2006) Resolvin E2: identification and anti-inflammatory actions. Pivotal role of human 5-lipoxygenase in resolvin E series biosynthesis. Chem Biol 13:1193–1202
16. Oh SF, Dona M, Fredman G et al (2012) Resolvin E2 formation and impact in inflammation resolution. J Immunol 188:4527–4534
17. Isobe Y, Arita M, Matsueda S et al (2012) Identification and structure determination of novel anti-inflammatory mediator resolvin E3, 17,18-dihydroxyeicosapentaenoic acid. J Biol Chem 287:10525–10534
18. Urabe D, Todoroki H, Masuda K, Inoue M (2012) Total syntheses of four possible stereoisomers of resolvin E3. Tetrahedron 68:3210–3219
19. Isobe Y, Arita M, Iwamoto R et al (2013) Stereochemical assignment and anti-inflammatory properties of the omega-3 lipid mediator resolvin E3. J Biochem 153:355–360
20. Yamashita A, Kawana K, Tomio K et al (2013) Increased tissue levels of omega-3 polyunsaturated fatty acids prevents pathological preterm birth. Sci Rep 3:3113
21. Wu D, Molofsky AB, Liang HE et al (2011) Eosinophils sustain adipose alternatively activated macrophages associated with glucose homeostasis. Science 332:243–247
22. Molofsky AB, Nussbaum JC, Liang HE et al (2013) Innate lymphoid type 2 cells sustain visceral adipose tissue eosinophils and alternatively activated macrophages. J Exp Med 210:535–549
23. Heredia JE, Mukundan L, Chen FM et al (2013) Type 2 innate signals stimulate fibro/adipogenic progenitors to facilitate muscle regeneration. Cell 153:376–388
24. Yamada T, Tani Y, Nakanishi H et al (2011) Eosinophils promote resolution of acute peritonitis by producing proresolving mediators in mice. FASEB J 25:561–568
25. Tani Y, Isobe Y, Imoto Y et al (2014) Eosinophils control the resolution of inflammation and draining lymph node hypertrophy through the proresolving mediators and CXCL13 pathway in mice. FASEB J 28(9):4036–4043
26. Kühn H, O'Donnell VB (2006) Inflammation and immune regulation by 12/15-lipoxygenases. Prog Lipid Res 45:334–356
27. Merched AJ, Ko K, Gotlinger KH et al (2008) Atherosclerosis. Evidence for impairment of resolution of vascular inflammation governed by specific lipid mediators. FASEB J 22:3595–3606
28. Krönke G, Katzenbeisser J, Uderhardt S et al (2009) 12/15-Lipoxygenase counteracts inflammation and tissue damage in arthritis. J Immunol 183:3383–3389

29. Gronert K, Maheshwari N, Khan N et al (2005) A role for the mouse 12/15-lipoxygenase pathway in promoting epithelial wound healing and host defense. J Biol Chem 280:15267–15278
30. Uderhardt S, Herrmann M, Oskolkova OV et al (2012) 12/15-Lipoxygenase orchestrates the clearance of apoptotic cells and maintains immunologic tolerance. Immunity 36:1–13
31. Kubota T, Arita M, Isobe Y et al (2014) Eicosapentaenoic acid is converted via ω-3 epoxygenation to the anti-inflammatory metabolite 12-hydroxy-17,18-epoxyeicosatetraenoic acid. FASEB J 28:586–593
32. Yokokura Y, Isobe Y, Matsueda S et al (2014) Identification of 14,20-dihydroxy-docosahexaenoic acid as a novel anti-inflammatory metabolite. J Biochem 156(6):315–321
33. Arita M, Clish CB, Serhan CN (2005) The contributions of aspirin and microbial oxygenase to the biosynthesis of anti-inflammatory resolvins. Novel oxygenase products from omega-3 polyunsaturated fatty acids. Biochem Biophys Res Commun 338:149–157
34. Schwarz D (2004) Arachidonic and eicosapentaenoic acid metabolism by human CYP1A1: highly stereoselective formation of 17(*R*),18(*S*)-epoxyeicosatetraenoic acid. Biochem Pharmacol 67:1445–1457
35. Lucas D, Goulitquer S, Marienhagen J et al (2010) Stereoselective epoxidation of the last double bond of polyunsaturated fatty acids by human cytochromes P450. J Lipid Res 51:1125–1133
36. Arnold C, Markovic M, Blossey K et al (2010) Arachidonic acid-metabolizing cytochrome p450 enzymes are targets of n-3 fatty acids. J Biol Chem 285:32720–32733
37. Divanovic S, Dalli J, Jorge-Nebert LF et al (2013) Contributions of the three CYP1 monooxygenases to pro-inflammatory and inflammation-resolution lipid mediator pathways. J Immunol 191:3347–3357

Part II
Lipid Mediators in *Drosophila* and Zebrafish

Chapter 12
Membrane Lipid Transporters in *Drosophila melanogaster*

Kohjiro Nagao, Naoto Juni, and Masato Umeda

Abstract Membrane lipid transport within and across the membrane is mediated by lipid transport machineries known as flippase, floppase, and scramblase. Flippase translocates lipids from the exocytoplasmic to the cytoplasmic leaflet of cellular membranes, floppase mediates the translocation of lipids in the opposite direction, and scramblase facilitates bidirectional translocation of lipids. These specialized lipid transport machineries are now demonstrated to have crucial roles in a variety of biological processes, including lipid metabolism, immune response, apoptosis, and neural function, in many mammalian species. The *Drosophila melanogaster* genome contains orthologues to about 70 % of all human disease-associated genes, and thus both traditional genetic approaches and more recent genome-wide screening techniques in *Drosophila* have been powerful tools for the study of lipid-related processes. There are, however, many open questions about the structure and function of lipids and their transport machineries in *Drosophila*. In this review, we summarize the functions of flippase, floppase, and scramblase from several species, and discuss the roles of these lipid transporters in *D. melanogaster*.

Keywords *Drosophila melanogaster* • Flippase • Floppase • Scramblase • P4-ATPase • CDC50 • ABC transporter • PLSCR • TMEM16F • XKR8

12.1 Introduction

The fruit fly *Drosophila melanogaster* has played a central role in establishing the link between genetics and embryology and has provided a useful model system for cell biology, physiology, immunology, social biology, and population genetics. Although evolutionarily distant from humans, the *D. melanogaster* genome contains orthologues to around 70 % of all human disease-associated genes [1],

K. Nagao • N. Juni • M. Umeda (✉)
Department of Synthetic Chemistry and Biological Chemistry, Graduate School of Engineering, Kyoto University, Kyoto 615-8510, Japan
e-mail: umeda@sbchem.kyoto-u.ac.jp

T. Yokomizo, M. Murakami (eds.), *Bioactive Lipid Mediators*,
DOI 10.1007/978-4-431-55669-5_12

including genes whose disruption are responsible for a variety of human diseases such as developmental disorders, cancer, neurological diseases, and metabolic disorders [1–3]. Recent advances in *Drosophila* genetics have provided a powerful tool for genome-wide screening of disease-associated genes and a pharmacological approach for the development of anticancer drugs [4, 5]. Despite the continued accumulation of knowledge about gene and protein function in *Drosophila,* however, there are many open questions about the structure and function of lipids in this fly species. *Drosophila* has been shown to have a unique phospholipid composition in which phosphatidylethanolamine (PE) is the dominant component, in contrast to other animals and plants in which phosphatidylcholine (PC) is the major phospholipid component [6]. Because of its sterol auxotrophy and facile genetic manipulation, *Drosophila* also provides a unique system for the study of dietary sterol uptake and the genetic basis of diet-induced metabolic disorders [7, 8]. A recent lipidomics study identified more than 500 molecular species of lipids during *Drosophila* development, and showed the dynamic remodeling of lipid species at specific stages during development [9].

It is now known that lipid molecules are not tethered to the site of their synthesis, but rather are actively transported and assembled into specific sites of cellular membranes [10]. Specialized lipid transport machineries are now demonstrated to have crucial roles in the formation of distinct membrane domains, which are involved in highly localized remodeling of membrane structures as well as recruitment and activation of membrane proteins [11]. It is well established that phospholipids in biological membranes are distributed asymmetrically between the inner and outer leaflets of the lipid bilayer. In many eukaryotic plasma membranes, aminophospholipids such as PE and phosphatidylserine (PS) reside in the inner leaflet, whereas choline-containing phospholipids, such as PC and sphingomyelin (SM), are localized mainly in the outer leaflet [12]. This phospholipid asymmetry is generated and maintained in part by the three putative phospholipid transport machineries known as phospholipid flippase, floppase, and scramblase [13, 14] (Fig. 12.1). Phospholipid flippase translocates phospholipids from the exocytoplasmic (extracellular/luminal) to the cytoplasmic leaflet of cellular membranes and is now known to belong to a subfamily of P-type ATPases known as type IV P-type ATPases (P4-ATPases). Some members of the ATP-binding cassette (ABC) transporters function as phospholipid floppases, which translocate phospholipids from the cytoplasmic to the

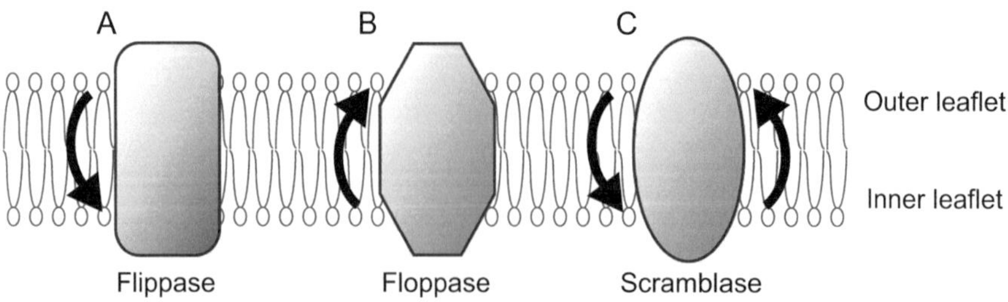

Fig. 12.1 Lipid-transport machineries: flippase (**a**), floppase (**b**), and scramblase (**c**)

exocytoplasmic leaflet of the membrane bilayer. Phospholipid scramblase causes a rapid disruption of the phospholipid asymmetry by bidirectional translocation of phospholipids, and several candidates for the scramblase, such as PLSCR [15], TMEM16F [16], and XKR8 [17], have been identified. In this review, we summarize the function of flippases, floppases, and scramblases from several species, and discuss the roles of these lipid transporters in *D. melanogaster.*

12.2 Flippase

12.2.1 A Complex of P4-ATPase and CDC50 as a Phospholipid Flippase

Bovine ATPase II (currently designated as ATP8A1), which mediates the net transfer of PS and PE from the outer to inner leaflets of the plasma membrane bilayer, was identified as the first candidate for a phospholipid flippase [18]. A subsequent genome search revealed that ATP8A1 and its closest yeast homologue, Drs2p, were founding members of a novel subfamily of P-type ATPases known as P4-ATPases [19]. P4-ATPases share structural similarity with other P-type ATPases, including ten transmembrane-spanning segments and a P-type ATPase-specific sequence motif, most notably the DKTGTLT sequence motif in which the canonical aspartic acid is phosphorylated during the reaction cycle, and a nucleotide (ATP)-binding site [20]. The unique structural feature of P4-ATPases is that the conserved charged and polar amino acids in the transmembrane domains 4 and 6 of cation-transport ATPases are replaced with hydrophobic and aromatic amino acid residues. Although it is not known how the P4-ATPases couple ATP hydrolysis and phospholipid translocation, recent studies by Graham's group proposed a structural model for their phospholipid specificity and transport mechanism [21, 22]. It is also shown that the association of P4-ATPases with CDC50 family proteins is required for their exit from the endoplasmic reticulum (ER) and for their proper cellular localization [23, 24].

P4-ATPases represent the largest subfamily of P-type ATPases and are present only in eukaryotic cells. In mammals, at least 14 members of P4-ATPases, designated ATP8A1 through ATP11C, and three CDC50 proteins (CDC50A, CDC50B, and CDC50C) have been identified. Among the P4-ATPases expressed in mammalian cells, ATP8A1, ATP8A2, ATP8B1, ATP8B3, ATP8B5, and ATP11C have been implicated in the translocation of phospholipids. With regard to the biological functions of phospholipid flippases, studies using the budding yeast *Saccharomyces cerevisiae* have demonstrated that P4-ATPases interact genetically and directly with the components involved in clathrin-dependent vesicular transport [25]. The translocation of aminophospholipids such as PS and PE creates an imbalance in the numbers of phospholipids between the two leaflets, which increases the membrane curvature and anionic phospholipid content of the cytosolic leaflet, thereby driving the recruitment of functional molecules and the resulting vesicular formation [26, 27]. Recent studies of mammalian P4-ATPases have also shown their involvement in various cellular and pathophysiological events such as neural function, bile salt secretion, acrosome reaction, B-cell differentiation, cell migration, and apoptosis [24, 28].

12.2.2 **Drosophila** ***P4-ATPases and CDC50 Protein***

In *D. melanogaster*, 6 P4-ATPase genes (*CG4301, CG9981, CG14741, CG31729, CG33298*, and *CG42321*) and a single CDC50 gene (*CG9947*) have been identified (Fig. 12.2a, b). Based on sequence similarity, P4-ATPase could be divided into several classes [28]. CG42321 has high homology to human ATP8A1 (55 % identical) and ATP8A2 (54 % identical), which mediate the flipping of PE and PS [24, 29–32]. CG14741 is mainly expressed in the brain (reported in Flybase [33]), and is categorized into a group containing human ATP8B1, ATP8B2, ATP8B3, and ATP8B4. The ubiquitously expressed CG31729 shows high homology with human ATP9A (52 % identical) and ATP9B (53 % identical), and CG33298 is classified into a group containing ATP10A, ATP10B, and ATP10D. Both CG4301 and CG9981 are specifically expressed in midgut, hindgut, and Malpighian tubules and belong to a group with human ATP11A, ATP11B, and ATP11C. Although flippase activity of *Drosophila* P4-ATPase has not been demonstrated in vitro and little information on the substrate specificities and cellular functions of *Drosophila* phospholipid flippases is available, recent genetic analyses have revealed some unique functions of these transporters.

12.2.3 **A Drosophila** ***P4-ATPase Involved in Odorant Receptor Function***

Recently, two groups have independently shown that dATP8B (CG14741), a *Drosophila* homologue of the mammalian ATP8B subclass, modulates odorant receptor function [34, 35]. By immunocytochemical analysis, dATP8B was shown to be enriched in a subset of olfactory neuron cilia, the site of odorant transduction. In dATP8B mutant flies, responses of odorant receptor 67d (OR67d)-expressing neurons to a male-specific pheromone, 11-*cis*-vaccenyl acetate (cVA), are severely impaired, whereas responses mediated by other receptor families (ionotropic-like and gustatory receptors) are not affected. Because the amount of OR67d localized in dendritic cilia is decreased in dATP8B mutants, the olfactory defects of the dATP8B mutant are likely caused by the mislocalization of odorant receptors. In addition, the cVA responses of dATP8B mutants are restored by expressing wild-type dATP8B or bovine ATP8A2 in pheromone-sensing neurons. On the other hand, defects in the

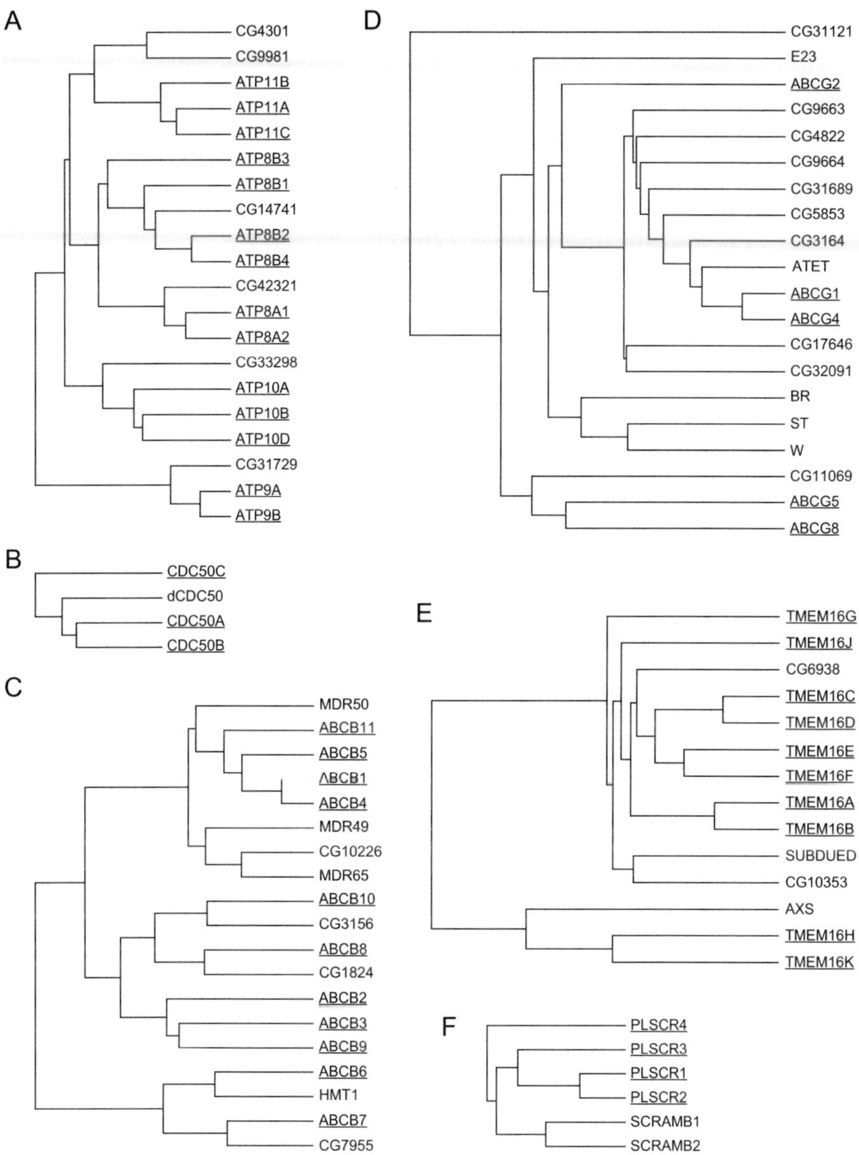

Fig. 12.2 Phylogenic trees of lipid transporters in *Homo sapiens* and *Drosophila melanogaster*: P4-ATPase (**a**), CDC50 (**b**), ABCB subfamily (**c**), ABCG subfamily (**d**), TMEM16 (**e**), PLSCR (**f**). *H. sapiens* proteins are *underlined*

dATP8B mutant are not rescued by expression of a dATP8B protein harboring a mutation in the DGETN motif, which is important for the phosphatase activity of P-type ATPase. Because proper lipid localization is required for vesicle docking and receptor recycling, dATP8B may regulate the odorant receptor trafficking by flipping phospholipids in olfactory neurons. Thus, the lipid-flipping activity of dATP8B is important for proper subcellular trafficking of odorant receptors and modulates odorant sensitivity in *Drosophila* [34, 35].

12.2.4 The Role of Drosophila *P4-ATPase in Secretory Vesicle Formation and Cholesterol Homeostasis*

Modulating the composition and the distribution of phospholipids and sterols is important for intracellular vesicle trafficking. It has been reported that overexpression of the *Drosophila* oxysterol-binding protein (OSBP) leads to sterol accumulation in the Golgi apparatus in salivary gland cells, and OSBP-overexpressing flies fail to perform post-eclosion behaviors including wing expansion [36]. The wing expansion defects in OSBP-overexpressing flies are the result of impaired function of crustacean cardioactive peptide (CCAP) neurons that release the neuropeptide bursicon, a regulator of post-eclosion behaviors. In OSBP-overexpressing CCAP neurons, the axon outgrowth and the synthesis of bursicon are normal, but the function and morphology of the Golgi apparatus and intracellular trafficking of bursicon are impaired. Ma et al. reported that the wing expansion defect in OSBP-overexpressing flies can be restored by disruption of the *CG33298* gene, a *Drosophila* homologue of ATP10D [37]. Mutation in the *CG33298* gene (truncation of amino acid residue 1-302) suppresses both defect in the bursicon trafficking and abnormality in Golgi apparatus morphology caused by overexpression of OSBP. However, the accumulation of sterol is not suppressed by the loss of one copy of *CG33298* gene. These results indicate that the mutation of the *CG33298* gene does not suppress the OSBP-overexpression-induced phenotype by reducing the sterol level in the Golgi apparatus. Therefore, it is likely that the reduction of phospholipid asymmetry by the *CG33298* mutation modulates the membrane trafficking by influencing the membrane curvature or fluidity in the sterol-overloaded Golgi apparatus [37].

12.2.5 The Role of Drosophila *CDC50 in Body Size Control*

In contrast to the variation among the P4-ATPases, only a single CDC50 family protein designated as dCDC50 (CG9947) has been identified in *D. melanogaster*. The expression profile of dCDC50 was analyzed by using an antibody that we raised against a synthetic peptide corresponding to the carboxyl terminus of the protein. The expression of dCDC50 was detected throughout the developmental stages with a prominent peak at the first-instar stage. In adult flies, dCDC50 was abundantly expressed in gut and brain tissues, with ubiquitous expression detected in all organs or body parts examined.

Ubiquitous gene silencing of the *dCdc50* gene using the GAL4/UAS system [38], in which a ubiquitous promoter from a *Drosophila αTubulin84B* gene and *dCdc50* double-strand RNA (dsRNA) were employed, resulted in larval/pupal lethality with no adult flies emerging. In the same study, the *dCdc50* gene silencing induced developmental retardation and a striking reduction of body size, but no apparent morphological defects. The body weights at the last-instar stage were half of normal. Internal organs, including the gut, fat body, and salivary glands, were also smaller and less developed and consisted of smaller-than-normal cells.

To examine whether the reduction in cell size was the consequence of a cell-autonomous effect of *dCdc50* silencing, we next employed a somatic mosaic technique combining site-specific recombination and GAL4/UAS-driven RNAi [39]. In this experiment, we established flies carrying four transgene constructs. The first transgene consists of a ubiquitous promoter of the *Drosophila Actin5C* gene and a coding sequence of the GAL4 transcription factor with an "FLPout cassette" that blocks GAL4 expression. The second transgene expresses a yeast site-specific recombinase, FLP, under the control of a heat shock promoter. Heat shock induction of FLP causes recombination between FRT sequences placed at both ends of the FLPout cassette in the first transgene, which cancels the suppression of GAL4 and induces the expression of *dCdc50* dsRNA and green fluorescence protein (GFP) under the control of UAS from the third and fourth transgenes. Because the recombination event occurs randomly in somatic cells, *dCdc50*-silenced cells appear embedded within wild-type cells. Thus, the somatic mosaic technique facilitates comparison of *dCdc50*-silenced cells with wild-type cells.

Organisms to be examined were subjected to a heat shock at the first-instar stage and allowed to grow to the last-instar stage. The resulting larvae appeared normal in size and morphology. In their internal tissues, *dCdc50* dsRNA-expressing cells as marked by GFP also appeared normal (Fig. 12.3). These results demonstrate that the reduction of cell size caused by ubiquitous *dCdc50* silencing is unlikely to be cell autonomous. This finding implies that dCDC50 in particular tissues/organs might affect systemic growth in a non-cell-autonomous fashion.

12.3 Floppase

12.3.1 The ABC Transporter as a Lipid Floppase

The ABC transporter binds and hydrolyzes ATP in nucleotide-binding domains (NBDs) to induce conformational change within the transmembrane domains (TMDs), which drive the transport of substrates [40]. Most ABC transporters consist of two NBDs and two TMDs. Although the individual domains are frequently expressed as separate polypeptides in prokaryotic ABC transporters, many eukaryotic ABC transporters have four domains in one or two peptides; the former transporter is called a full type and the latter a half type [41]. Each NBD contains three highly conserved sequence motifs, Walker A, Walker B, and the ABC signature sequence (also known as the C motif). Residues within the Walker A motif (G-X-X-G-X-G-K-S/T-S/T) interact with the phosphate group and magnesium ion of the

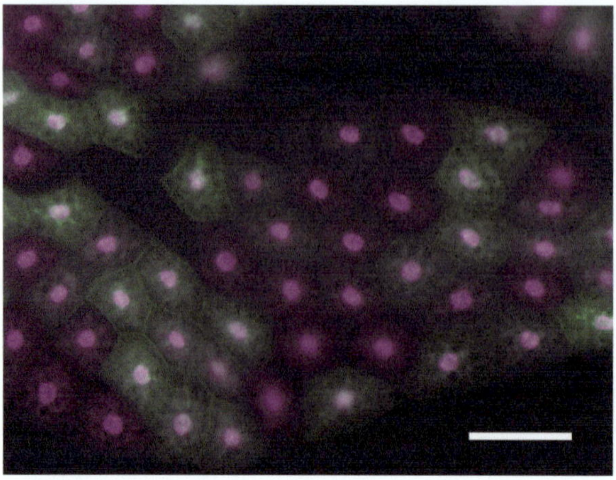

Fig. 12.3 Whole-mount preparation of a fat body from a last-instar larva with a *dCdc50*-silenced somatic mosaic. *dCdc50*-silenced cells were randomly generated at the first-instar stage. Fat bodies were dissected at the last-instar stage and observed with a Zeiss Axiovert 200 M fluorescent microscope. Nuclei are visualized with DAPI (*magenta*). Note that *dCdc50*-silenced cells [marked by green fluorescent protein (*GFP*), *green*] were indistinguishable from the neighboring wild-type cells in both size and morphology. *Bar* 100 μm

Mg^{2+}–nucleotide complex, and an amino acid substitution of the conserved lysine residue results in a loss of ATP hydrolysis activity in several ABC transporters [42, 43]. The Walker B motif is h-h-h-h-D, where 'h' is a hydrophobic residue, and the highly conserved aspartate residues are involved in the coordination of the catalytic Mg^{2+} ion [44, 45]. Although Walker A and B motifs are also widely conserved among the many nucleotide-binding proteins (e.g., Ras, adenylate kinase, myosin, F_1-ATPase), the ABC signature sequence (L-S-G-G-Q/R/K-Q-R) is unique to NBD of the ABC transporter. The ABC signature sequence is located immediately N terminal of the Walker B motif, and is involved in recognition of the γ-phosphate and ribose moiety of ATP.

Although most ABC transporters function as efflux pumps that secrete hydrophobic substrates from the membrane to the extracellular environment, some members of the ABC transporter family function as floppases that translocate lipids from the inner to the outer leaflet of the plasma membrane [40]. Thus, in contrast to inward translocation by P4-ATPases, ABC transporters function as outward translocases by flopping or secreting the various kinds of hydrophobic substrates, including phospholipids, sphingolipids, sterols, and xenobiotics. Although details of the mechanism for substrate recognition and transport by ABC transporters remain to be clarified, the processes for outward transport of hydrophobic substrates across the membrane can be classified as follows.

1. *Flopping from the inner to outer leaflet.* In this process, the substrate enters the transporter from the inner leaflet and exits from the transporter into the outer leaflet. Substrates could be moved from one substrate-binding site to another in the transporter in this "flopping" manner.

2. *Secretion from the inner leaflet to extracellular environment*. The substrate enters the transporter from the inner leaflet as in model 1, but then exits the transporter directly to an exogenous environment. Because the hydrophobic substrate readily returns to the outer leaflet of the lipid bilayer, an acceptor is required for the hydrophobic substrate to be solubilized in an exogenous environment.
3. *Secretion from the outer leaflet to extracellular environment*. In this model, the substrate enters the transporter from the outer leaflet and exits the transporter directly to an exogenous environment. An acceptor would also be required for the substrate to be solubilized in an exogenous environment in this case. The movement of substrates in this model can be called "projection" to the exogenous environment.

12.3.2 **Drosophila** ***ABC Transporters***

Based on the sequence similarity of NBDs, ABC transporters could be divided into eight subfamilies (ABCA to ABCH) [46, 47]. Although the *D. melanogaster* genome encodes 56 ABC transporter genes (10 for ABCA, 8 for ABCB, 14 for ABCC, 2 for ABCD, 1 for ABCE, 3 for ABCF, 15 for ABCG, 3 for ABCH), information about the function and substrate of *Drosophila* ABC transporters is limited. In this section, we summarize the functions of ABCG and ABCB subfamily members that have been reported to be involved in the transport of hydrophobic substrate.

12.3.3 ABCG Subfamily Members in **Drosophila**

Among the 15 ABCG subfamily members of *D. melanogaster* (Fig. 12.2d), the best characterized members are W, BW, and ST (the gene products of *white*, *brown*, and *scarlet*, respectively), which are involved in the determination of eye color [18]. Normally, the eye color of *D. melanogaster* is red-brown, because two color light-screening pigments, xanthommatin (brown pigment) and a class of drosopterins (red pigment), are deposited in membrane-bound granules in specialized pigment cells in each ommatidium of the compound eye. Xanthommatin and drosopterins are synthesized from tryptophan and guanosine triphosphate (GTP), respectively. Two *Drosophila* ABC transporters, W and ST, form a heterodimer and mediate the uptake of tryptophan and two intermediates of the xanthommatin pathway. W also forms a heterodimer with BW to mediate the uptake of the drosopterin precursor guanine. Therefore, W is required for the synthesis of both pigments, whereas ST and BW are involved in the synthesis of xanthommatin and drosopterins, respectively. Reflecting the role of each ABC transporter in pigment production, null mutation of the *white* gene causes white eye color because of the lack of deposition of either xanthommatin or drosopterins in the pigment cells, whereas flies with deletion of the *scarlet* gene have red eyes and the null mutation of the *brown* gene causes brown eyes.

Although three *Drosophila* ABCG subfamily proteins (W, BW, and ST) transport hydrophilic compounds in eye pigment synthesis, mammalian ABCG subfamily members mediate the secretion of hydrophobic molecules (e.g., ABCG1 and ABCG4 mediate cholesterol efflux into HDL particles; the ABCG5-ABCG8 heterodimer transports cholesterol and sitosterol). There are 12 *Drosophila* ABCG subfamily proteins in addition to the three transports for eye pigment synthesis. Some of these proteins show high homology with mammalian lipid transporters and are predicted to be lipid transporters. For example, ATET has high homology with human ABCG1 (44 % identical) and human ABCG4 (41 % identical), but the transport substrate of ATET is not clear [49]. E23 (CG3327) is involved in ecdyson-mediated gene activation [50]. Ecdyson is a steroid hormone that regulates developmental processes, and the concentration of ecdyson is drastically changed during the developmental course. The temporal pattern of E23 transcript accumulation is consistent with that of an ecdyson pulse during development, and the expression of E23 is robustly induced by ecdyson in cultured larval tissues, demonstrating that E23 is an ecdyson-inducible ABC transporter. Ectopic expression of E23 suppresses the ecdyson-mediated transcriptional activation of the *Eip74EF* and *Eip75B* genes, indicating that E23 may modulate the intracellular ecdyson concentration by secreting ecdyson into the extracellular environment. Interestingly, the system for induction of the putative ecdyson transporter E23 by ecdyson is similar to the induction mechanism of mammalian cholesterol transporters. For example, expression of the cholesterol transporters ABCG1 and ABCG4 is induced by oxidized cholesterol via activation of a nuclear receptor, liver X receptor (LXR) [51, 52]. Because ecdyson induces the transcription of several genes via binding to the nuclear receptor ecdyson receptor (EcR) [53], mammals and *Drosophila* utilize similar mechanisms for the modulation of cellular sterol levels. As E23 is a half-type transporter similar to *Drosophila* W and mammalian ABCG1, E23 requires the formation of a dimer to function as an active transporter. However, it is not clear whether E23 forms a homodimer or a heterodimer with the other half-type transporter. Identification of a dimer partner of E23 is important for understanding both the mechanism and regulation of ecdyson transport during developmental processes. Thus, the *Drosophila* ABCG subfamily transporter is also involved in the transport of hydrophobic compounds, and further research is needed to reveal the role of other ABCG subfamily members.

12.3.4 ABCB Subfamily Members in Drosophila

As already mentioned, the *D. melanogaster* genome encodes eight ABCB subfamily transporter genes (Fig. 12.2c). Four members (MDR49, MDR50, MDR65, CG10226) are full-type transporters in the manner of human ABCB1 (MDR1), whereas the others (CG1824, CG3156, CG7955, HMT1) are half-type transporters that form dimers to serve as active transporters. Among them, two transporters have been shown to mediate the transport of hydrophobic substrate. Because human ABCB1 and ABCB4 secrete lipophilic xenobiotics and PC, respectively [54, 55], it is plausible that *Drosophila* ABCB transporters mediate the transport of hydrophobic substrates.

The full-type transporter MDR65 is specifically localized at the humoral barrier of the *Drosophila* central nervous system (CNS) of subperineural glia and is required for protection against chemical attack [56]. Another full-type transporter, MDR49, has been shown to be involved in the secretion of chemoattractant [57]. The migration of primordial germ cells from the site of origin to the somatic part of the gonad is crucial for embryonic development. Chemoattractants are commonly secreted through a classic, signal peptide-dependent pathway, but the secretion of a geranyl-modified attractant requires MDR49. Among ABCB subfamily transporters of *Drosophila*, the expression pattern and phenotype of the *Mdr49* mutant are consistent with a role in germ cell migration. Further, the overexpression of MDR49 and HMGCR, an enzyme required for production of the geranylgeranyl moiety in insect cell lines, enhances the migration of germ cells, demonstrating the role of the MDR49 transporter in the secretion of the geranyl-modified attractant. Although the transport mechanisms of MDR49 and MDR65 are not clear, these functions strongly suggest the importance of *Drosophila* ABCB subfamily transporters in lipid transport, as seen in mammalian cells.

12.4 Scramblase

12.4.1 Phospholipid Scrambling by PLSCR, TMEM16 (Anoctamin), and XK Protein Family Members

Phospholipid scrambling is involved in several physiological processes. For instance, apoptotic cells expose PS on the cell surface as an "eat-me" signal to be removed by phagocytes, and PS exposure in platelets is required for triggering of the coagulation cascade. A single-pass membrane protein, PLSCR1, was originally identified as a protein enhancing the transbilayer movement of phospholipids [58], and humans and mice each have four PLSCRs (PLSCR1–4). However, it has been reported that mice lacking PLSCR1 show no defects in lipid scrambling [59]. Suzuki et al. found that TMEM16F, an eight-transmembrane protein, mediates a Ca^{2+}-dependent scramblase activity on the cell surface [16]. Calcium ionophore-dependent PS and PE exposure detected by annexin V and Ro09-0198 are enhanced in TMEM16F-overexpressing Ba/F3 cells. Overexpression of TMEM16F also enhances calcium ionophore-induced flipping of NBD-PC and NBD-SM. Among the ten members of the TMEM16 protein family, TMEM16C, 16D, 16F, 16G, and 16J show phospholipid scramblase activity with different preferences for lipid substrate [60]. In contrast to the Ca^{2+}-dependent scrambling activity of these five

TMEM16 family members, TMEM16A and -16B work as Ca^{2+}-dependent Cl^- channels. Because TMEM16C, -16D, -16F, -16G, and -16J cannot act as Ca^{2+}-dependent Cl^- channels, their Ca^{2+}-dependent phospholipid scramblase activities are independent of Cl^- channel activity. Recently, human and mouse XKR8, a member of the evolutionarily conserved XK protein family, was shown to be responsible for PS exposure in apoptotic cells [17]. It was also demonstrated that CED-8, the only *Caenorhabditis elegans* XK-family homologue, is involved in apoptotic PS exposure and cell-corpse engulfment.

12.4.2 Phospholipid Scramblase of **Drosophila**

The genome of *D. melanogaster* contains two homologues of PLSCR (SCRAMB1 and SCRAMB2), four homologues of the TMEM16 family (CG6938, CG10353, SUBDUED, AXS), and one homologue of the XK family (CG32579) (Fig. 12.2e, f). Flies lacking both SCRAMB1 and SCRAMB2 show no defects in developmentally regulated and ectopically induced apoptotic events in vivo, and overexpression and knockdown of *Scramb1* and *Scramb2* in S2 cells have no effect on PS exposure on the cell surface [61]. However, the recruitment of vesicles from a reserve pool of vesicles and the secretion of neurotransmitters are enhanced in SCRAMB1- and SCRAMB2-deficient flies, suggesting that these scramblases play a role in the process of neurotransmission [61]. SUBDUED, a TMEM16A homologue of *Drosophila*, is found to be a calcium-activated chloride channel upon heterologous expression in HEK293T cells [62]. Furthermore, *Subdued*-knockout flies more readily succumb to death caused by ingesting the pathogenic bacteria *Serratia marcescens*, indicating that SUBDUED is involved in the host defense of *Drosophila*. A TMEM16K homologue, Aberrant X segregation (AXS), has been shown to be localized on the meiotic spindle and to be involved in the regulation of meiotic chromosomal segregation [63]. Because mouse TMEM16A and TMEM16K do not function as scramblases, AXS and SUBDUED may not have phospholipid scramblase activity. *Drosophila* CG6938 is a homologue of TMEM16E, and human TMEM16E has the highest homology with human TMEM16F, although TMEM16F$^{-/-}$ immortalized fetal thymocytes overexpressing mouse TMEM16E cannot scramble phospholipids on the plasma membrane. However, because TMEM16E is localized in the intracellular compartment (64), it is likely that TMEM16E also possesses the scramblase activity, and CG6983 may be a scramblase of *Drosophila*. A *Drosophila* homologue of the XK family, CG32579, is ubiquitously expressed, and has low homology with human XKR8 (18 % identical) and *C. elegans* CED-8 (17 % identical). For an improved understanding of the roles of scramblase in *Drosophila*, biochemical and cell biological analyses of the scramblase activity of *Drosophila* homologues are needed.

12.5 Perspectives

As already discussed, *D. melanogaster* has a variety of lipid transport machineries that function as flippase, floppase, and scramblase. It has become clear that the lipid transport machineries play a crucial role in the multilayered regulatory networks of lipid metabolism, transport, and signal transduction at the cellular and whole organism levels. Recent advances in gene technologies have allowed us to manipulate gene expression in a cell- and tissue-specific manner in *D. melanogaster* more easily than any other multicellular organisms. This technique provides us a unique opportunity to explore the pathophysiological role of the lipid transporters in various lipid-related pathophysiological processes, such as development, energy homeostasis, cancer, and neurological disorders. Future research in this field using *D. melanogaster* may provide a novel concept for lipid transport and lipid-mediated signaling and contribute to a better understanding of the biological processes.

References

1. Bier E (2005) *Drosophila*, the golden bug, emerges as a tool for human genetics. Nat Rev Genet 6:9–23
2. Xu Y, Condell M, Plesken H, Edelman-Novemsky I, Ma J, Ren M, Schlame M (2006) A *Drosophila* model of Barth syndrome. Proc Natl Acad Sci U S A 103:11584–11588
3. Takeuchi K, Nakano Y, Kato U, Kaneda M, Aizu M, Awano W, Yonemura S, Kiyonaka S, Mori Y, Yamamoto D, Umeda M (2009) Changes in temperature preferences and energy homeostasis in dystroglycan mutants. Science 323:1740–1743
4. Dar AC, Das TK, Shokat KM, Cagan RL (2012) Chemical genetic discovery of targets and anti-targets for cancer polypharmacology. Nature 486:80–84
5. Lenz S, Karsten P, Schulz JB, Voigt A (2013) *Drosophila* as a screening tool to study human neurodegenerative diseases. J Neurochem 127:453–460
6. Pavlidis P, Ramaswami M, Tanouye MA (1994) The *Drosophila* easily shocked gene: a mutation in a phospholipid synthetic pathway causes seizure, neuronal failure, and paralysis. Cell 79:23–33
7. Pospisilik JA, Schramek D, Schnidar H, Cronin SJ, Nehme NT, Zhang X, Knauf C, Cani PD, Aumayr K, Todoric J, Bayer M, Haschemi A, Puviindran V, Tar K, Orthofer M, Neely GG, Dietzl G, Manoukian A, Funovics M, Prager G, Wagner O, Ferrandon D, Aberger F, Hui CC, Esterbauer H, Penninger JM (2010) *Drosophila* genome-wide obesity screen reveals hedgehog as a determinant of brown versus white adipose cell fate. Cell 140:148–160
8. Carvalho M, Sampaio JL, Palm W, Brankatschk M, Eaton S, Shevchenko A (2012) Effects of diet and development on the *Drosophila* lipidome. Mol Syst Biol 8:600
9. Guan XL, Cestra G, Shui G, Kuhrs A, Schittenhelm RB, Hafen E, van der Goot FG, Robinett CC, Gatti M, Gonzalez-Gaitan M, Wenk MR (2013) Biochemical membrane lipidomics during *Drosophila* development. Dev Cell 24:98–111
10. van Meer G, Voelker DR, Feigenson GW (2008) Membrane lipids: where they are and how they behave. Nat Rev Mol Cell Biol 9:112–124
11. Holthuis JC, Menon AK (2014) Lipid landscapes and pipelines in membrane homeostasis. Nature 510:48–57

12. Zachowski A (1993) Phospholipids in animal eukaryotic membranes: transverse asymmetry and movement. Biochem J 294(pt 1):1–14
13. Coleman JA, Quazi F, Molday RS (2013) Mammalian P4-ATPases and ABC transporters and their role in phospholipid transport. Biochim Biophys Acta 1831:555–574
14. Kunzelmann K, Nilius B, Owsianik G, Schreiber R, Ousingsawat J, Sirianant L, Wanitchakool P, Bevers EM, Heemskerk JW (2014) Molecular functions of anoctamin 6 (TMEM16F): a chloride channel, cation channel, or phospholipid scramblase? Pflugers Arch 466:407–414
15. Bevers EM, Williamson PL (2010) Phospholipid scramblase: an update. FEBS Lett 584:2724–2730
16. Suzuki J, Umeda M, Sims PJ, Nagata S (2010) Calcium-dependent phospholipid scrambling by TMEM16F. Nature 468:834–838
17. Suzuki J, Denning DP, Imanishi E, Horvitz HR, Nagata S (2013) Xk-related protein 8 and CED-8 promote phosphatidylserine exposure in apoptotic cells. Science 341:403–406
18. Zachowski A, Henry JP, Devaux PF (1989) Control of transmembrane lipid asymmetry in chromaffin granules by an ATP-dependent protein. Nature 340:75–76
19. Tang X, Halleck MS, Schlegel RA, Williamson P (1996) A subfamily of P-type ATPases with aminophospholipid transporting activity. Science 272:1495–1497
20. Tanaka K, Fujimura-Kamada K, Yamamoto T (2011) Functions of phospholipid flippases. J Biochem 149:131–143
21. Baldridge RD, Graham TR (2012) Identification of residues defining phospholipid flippase substrate specificity of type IV P-type ATPases. Proc Natl Acad Sci U S A 109:E290–E298
22. Baldridge RD, Graham TR (2013) Two-gate mechanism for phospholipid selection and transport by type IV P-type ATPases. Proc Natl Acad Sci U S A 110:E358–E367
23. Kato U, Emoto K, Fredriksson C, Nakamura H, Ohta A, Kobayashi T, Murakami-Murofushi K, Kobayashi T, Umeda M (2002) A novel membrane protein, Ros3p, is required for phospholipid translocation across the plasma membrane in *Saccharomyces cerevisiae*. J Biol Chem 277:37855–37862
24. Kato U, Inadome H, Yamamoto M, Emoto K, Kobayashi T, Umeda M (2013) Role for phospholipid flippase complex of ATP8A1 and CDC50A proteins in cell migration. J Biol Chem 288:4922–4934
25. Muthusamy BP, Natarajan P, Zhou X, Graham TR (2009) Linking phospholipid flippases to vesicle-mediated protein transport. Biochim Biophys Acta 1791:612–619
26. Sebastian TT, Baldridge RD, Xu P, Graham TR (2012) Phospholipid flippases: building asymmetric membranes and transport vesicles. Biochim Biophys Acta 1821:1068–1077
27. Xu P, Baldridge RD, Chi RJ, Burd CG, Graham TR (2013) Phosphatidylserine flipping enhances membrane curvature and negative charge required for vesicular transport. J Cell Biol 202:875–886
28. Lopez-Marques RL, Theorin L, Palmgren MG, Pomorski TG (2014) P4-ATPases: lipid flippases in cell membranes. Pflugers Arch 466:1227–1240
29. Paterson JK, Renkema K, Burden L, Halleck MS, Schlegel RA, Williamson P, Daleke DL (2006) Lipid-specific activation of the murine P4-ATPase Atp8a1 (ATPase II). Biochemistry 45:5367–5376
30. Coleman JA, Molday RS (2011) Critical role of the beta-subunit CDC50A in the stable expression, assembly, subcellular localization, and lipid transport activity of the P4-ATPase ATP8A2. J Biol Chem 286:17205–17216
31. Coleman JA, Kwok MC, Molday RS (2009) Localization, purification, and functional reconstitution of the P4-ATPase Atp8a2, a phosphatidylserine flippase in photoreceptor disc membranes. J Biol Chem 284:32670–32679
32. Coleman JA, Vestergaard AL, Molday RS, Vilsen B, Andersen JP (2012) Critical role of a transmembrane lysine in aminophospholipid transport by mammalian photoreceptor P4-ATPase ATP8A2. Proc Natl Acad Sci U S A 109:1449–1454
33. St. Pierre SE, Ponting L, Stefancsik R, McQuilton P, FlyBase C (2014) FlyBase 102: advanced approaches to interrogating FlyBase. Nucleic Acids Res 420:D780–D788

34. Ha TS, Xia R, Zhang H, Jin X, Smith DP (2014) Lipid flippase modulates olfactory receptor expression and odorant sensitivity in *Drosophila*. Proc Natl Acad Sci U S A 111:7831–7836
35. Liu YC, Pearce MW, Honda T, Johnson TK, Charlu S, Sharma KR, Imad M, Burke RE, Zinsmaier KE, Ray A, Dahanukar A, de Bruyne M, Warr CG (2014) The *Drosophila melanogaster* phospholipid flippase dATP8B is required for odorant receptor function. PLoS Genet 10, e1004209
36. Ma Z, Liu Z, Huang X (2010) OSBP- and FAN-mediated sterol requirement for spermatogenesis in *Drosophila*. Development 137:3775–3784
37. Ma Z, Liu Z, Huang X (2012) Membrane phospholipid asymmetry counters the adverse effects of sterol overloading in the Golgi membrane of *Drosophila*. Genetics 190:1299–1308
38. Elliott DA, Brand AH (2008) The GAL4 system: a versatile system for the expression of genes. Methods Mol Biol 420:79–95
39. Bischof J, Basler K (2008) Recombinases and their use in gene activation, gene inactivation, and transgenesis. Methods Mol Biol 420:175–195
40. Nagao K, Kimura Y, Mastuo M, Ueda K (2010) Lipid outward translocation by ABC proteins. FEBS Lett 584:2717–2723
41. Hyde SC, Emsley P, Hartshorn MJ, Mimmack MM, Gileadi U, Pearce SR, Gallagher MP, Gill DR, Hubbard RE, Higgins CF (1990) Structural model of ATP-binding proteins associated with cystic fibrosis, multidrug resistance and bacterial transport. Nature 346:362–365
42. Zhang DW, Graf GA, Gerard RD, Cohen JC, Hobbs HH (2006) Functional asymmetry of nucleotide-binding domains in ABCG5 and ABCG8. J Biol Chem 281:4507–4516
43. Urbatsch IL, Beaudet L, Carrier I, Gros P (1998) Mutations in either nucleotide-binding site of P-glycoprotein (Mdr3) prevent vanadate trapping of nucleotide at both sites. Biochemistry 37:4592–4602
44. Tombline G, Bartholomew LA, Urbatsch IL, Senior AE (2004) Combined mutation of catalytic glutamate residues in the two nucleotide binding domains of P-glycoprotein generates a conformation that binds ATP and ADP tightly. J Biol Chem 279:31212–31220
45. Hrycyna CA, Ramachandra M, Germann UA, Cheng PW, Pastan I, Gottesman MM (1999) Both ATP sites of human P-glycoprotein are essential but not symmetric. Biochemistry 38:13887–13899
46. Dean M, Rzhetsky A, Allikmets R (2001) The human ATP-binding cassette (ABC) transporter superfamily. Genome Res 11:1156–1166
47. Dermauw W, Van Leeuwen T (2014) The ABC gene family in arthropods: comparative genomics and role in insecticide transport and resistance. Insect Biochem Mol Biol 45:89–110
48. Ewart GD, Howells AJ (1998) ABC transporters involved in transport of eye pigment precursors in *Drosophila melanogaster*. Methods Enzymol 292:213–224
49. Kuwana H, Shimizu-Nishikawa K, Iwahana H, Yamamoto D (1996) Molecular cloning and characterization of the ABC transporter expressed in Trachea (ATET) gene from *Drosophila melanogaster*. Biochim Biophys Acta 1309:47–52
50. Hock T, Cottrill T, Keegan J, Garza D (2000) The E23 early gene of *Drosophila* encodes an ecdysone-inducible ATP-binding cassette transporter capable of repressing ecdysone-mediated gene activation. Proc Natl Acad Sci U S A 97:9519–9524
51. Kennedy MA, Venkateswaran A, Tarr PT, Xenarios I, Kudoh J, Shimizu N, Edwards PA (2001) Characterization of the human ABCG1 gene: liver X receptor activates an internal promoter that produces a novel transcript encoding an alternative form of the protein. J Biol Chem 276:39438–39447
52. Engel T, Lorkowski S, Lueken A, Rust S, Schluter B, Berger G, Cullen P, Assmann G (2001) The human ABCG4 gene is regulated by oxysterols and retinoids in monocyte-derived macrophages. Biochem Biophys Res Commun 288:483–488
53. Koelle MR, Talbot WS, Segraves WA, Bender MT, Cherbas P, Hogness DS (1991) The *Drosophila* EcR gene encodes an ecdysone receptor, a new member of the steroid receptor superfamily. Cell 67:59–77

54. Ueda K, Cardarelli C, Gottesman MM, Pastan I (1987) Expression of a full-length cDNA for the human "MDR1" gene confers resistance to colchicine, doxorubicin, and vinblastine. Proc Natl Acad Sci U S A 84:3004–3008
55. Morita SY, Kobayashi A, Takanezawa Y, Kioka N, Handa T, Arai H, Matsuo M, Ueda K (2007) Bile salt-dependent efflux of cellular phospholipids mediated by ATP binding cassette protein B4. Hepatology 46:188–199
56. Mayer F, Mayer N, Chinn L, Pinsonneault RL, Kroetz D, Bainton RJ (2009) Evolutionary conservation of vertebrate blood-brain barrier chemoprotective mechanisms in *Drosophila*. J Neurosci 29:3538–3550
57. Ricardo S, Lehmann R (2009) An ABC transporter controls export of a *Drosophila* germ cell attractant. Science 323:943–946
58. Zhou Q, Zhao J, Stout JG, Luhm RA, Wiedmer T, Sims PJ (1997) Molecular cloning of human plasma membrane phospholipid scramblase. A protein mediating transbilayer movement of plasma membrane phospholipids. J Biol Chem 272:18240–18244
59. Zhou Q, Zhao J, Wiedmer T, Sims PJ (2002) Normal hemostasis but defective hematopoietic response to growth factors in mice deficient in phospholipid scramblase 1. Blood 99:4030–4038
60. Suzuki J, Fujii T, Imao T, Ishihara K, Kuba H, Nagata S (2013) Calcium-dependent phospholipid scramblase activity of TMEM16 protein family members. J Biol Chem 288:13305–13316
61. Acharya U, Edwards MB, Jorquera RA, Silva H, Nagashima K, Labarca P, Acharya JK (2006) *Drosophila melanogaster* scramblases modulate synaptic transmission. J Cell Biol 173:69–82
62. Wong XM, Younger S, Peters CJ, Jan YN, Jan LY (2013) Subdued, a TMEM16 family Ca^{2+}-activated Cl^- channel in *Drosophila melanogaster* with an unexpected role in host defense. eLife 2, e00862
63. Kramer J, Hawley RS (2003) The spindle-associated transmembrane protein Axs identifies a membranous structure ensheathing the meiotic spindle. Nat Cell Biol 5:261–263
64. Mizuta K, Tsutsumi S, Inoue H, Sakamoto Y, Miyatake K, Miyawaki K, Noji S, Kamata N, Itakura M (2007) Molecular characterization of GDD1/TMEM16E, the gene product responsible for autosomal dominant gnathodiaphyseal dysplasia. Biochem Biophys Res Commun 357:126–132

Chapter 13
Drosophila: A Model for Studying Prostaglandin Signaling

Andrew J. Spracklen and Tina L. Tootle

Abstract Prostaglandin (PG) synthesis and signaling are conserved in *Drosophila melanogaster*. PGs are produced downstream of cyclooxygenase or COX enzymes, the targets of nonsteroidal anti-inflammatory drugs. Almost 20 years ago, biochemical studies suggested that *Drosophila* possess COX activity. Recent efforts utilizing a combination of pharmacological and genetic approaches revealed that PGs have critical functions in *Drosophila* oogenesis or follicle development. Pxt was identified as the COX-like enzyme and is required for multiple aspects of female fertility, including temporal regulation of both gene expression and actin cytoskeletal remodeling. Here we review the PG synthesis and signaling machinery, the evidence for PG activity in *Drosophila*, the roles of PGs in flies, primarily focused on oogenic activities, and the conservation of PG function in higher animals. We also point out how studies on PGs in a genetic model system, such as flies, can significantly advance our understanding of the molecular actions of PGs.

Keywords Prostaglandins • *Drosophila* • Oogenesis • Reproduction • Actin cytoskeleton • Fascin • Enabled • Cancer • Gene amplification

Prostaglandins (PGs) are lipid signaling molecules that mediate a wide range of physiological processes including reproduction, cardiovascular function and disease, pain and inflammation, and cancer development and progression (reviewed in [1, 2]). Although PGs were identified more than 60 years ago, the study of PGs in the genetic model system of *Drosophila melanogaster* (hereafter referred to as *Drosophila*) has largely been restricted to the past decade. Here we review (1) the

A.J. Spracklen
Anatomy and Cell Biology Department, University of Iowa Carver College of Medicine, Iowa City, IA 52242, USA

Current Address: Lineberger Comprehensive Cancer Center, Department of Biology, University of North Carolina at Chapel Hill, Chapel Hill, NC 27599, USA

T.L. Tootle (✉)
Anatomy and Cell Biology Department, University of Iowa Carver College of Medicine, Iowa City, IA 52242, USA
e-mail: tina-tootle@uiowa.edu

T. Yokomizo, M. Murakami (eds.), *Bioactive Lipid Mediators*,
DOI 10.1007/978-4-431-55669-5_13

PG synthesis and signaling machinery, (2) the evidence for PG signaling in *Drosophila*, (3) the currently known roles of PGs in flies, and (4) the extent to which the roles of PGs in *Drosophila* are conserved across organisms.

13.1 Prostaglandin Synthesis and Signaling

Prostaglandin (PG) synthesis is a multistep processes that begins with the release of arachidonic acid (AA) from the glycerol backbone of membrane phospholipids through the enzymatic activity of phospholipase A_2 (PLA_2). This free AA is then converted into the PG precursor, PGH_2, through the enzymatic activity of cyclooxygenase enzymes (in mammals, COX-1 and COX-2), which are the pharmacological targets of nonsteroidal anti-inflammatory drugs (reviewed in [3–6]). Downstream of COX enzymes, PGH_2 is processed into the biologically active PGs (PGD_2, PGE_2, $PGF_{2\alpha}$, PGI_2) and thromboxane (TXA_2) through the activity of specific synthases (PGD_2: H-PGDS, L-PGDS; PGE_2: mPGES-1, mPGES-2, cPGES; $PGF_{2\alpha}$: AKR1B1; PGI_2: PGIS; TXA_2: TXAS) [7]. These bioactive species then go on to serve as autocrine/paracrine signaling molecules.

Although PGs may induce MAPK signaling pathways [8–12], or serve as peroxisome proliferator-activated receptor-gamma (PPARγ) nuclear hormone receptor ligands [13–16] independently of G protein-coupled receptors (GPCRs) (reviewed in [17]), their most widely accepted and best understood mechanism of action is to serve as ligands for specific GPCRs [18]. Each bioactive species of PG can bind to and activate from one to four cognate GPCRs (PGD_2: DP, CRTH2; PGE_2: EP1, EP2, EP3, EP4; $PGF_{2\alpha}$: FP; PGI_2: IP), which elicit their downstream effects through activation of Gα [17, 19] and, in some cases, Gβγ [20].

13.2 Evidence for Prostaglandin Synthesis and Signaling in *Drosophila*

13.2.1 Evidence for COX Activity

PGs are derived from the COX-dependent oxygenation of three long-chain polyunsaturated fatty acids (PUFAs): arachidonic acid (AA, 20:4n-6), eicosapentaenoic acid (EPA, 20:5n-3), and dihomo-gamma-linolenic acid (DGLA, 20:3n-6). In many organisms, these long-chain PUFAs are acquired through the diet or through the elongation/desaturation of the essential fatty acid, linoleic acid (LA, 18:2n-6). Although some insects are capable of de novo synthesis of LA [21], there is little evidence that this occurs in *Drosophila* [22, 23].

The presence of long-chain PUFAs and their biological significance in *Drosophila* remains unclear. Early studies indicated that AA is not present in the fly, but its precursor, LA, is [23–25]. Interestingly, Shen et al. found that when 22-carbon PUFAs are supplied in the diet these lipids are readily converted to 20-carbon PUFAs [25]. Thus, flies possess the machinery to utilize very long chain PUFAs. Other more recent studies, conducted using liquid chromatography coupled to tandem mass spectrometry (LC-MS/MS), suggest that 20-carbon PUFAs are present in *Drosophila*. Using LC-MS/MS to analyze the fatty acid content of membrane phospholipids in both whole adults and isolated adult testes, Steinhauer et al. reported the presence of numerous phospholipid species containing AA in a wild-type fly strain, ranging from 0.2 to 1.5 % of the total phospholipid class [26]. Other groups also suggest 20-carbon PUFAs may be present in *Drosophila* [27–29]. Together these studies support the idea that the lipid precursors for PG synthesis are present at low levels in *Drosophila*.

In 1986, Pages et al. found that *Drosophila* extracts incubated with AA can generate PGE_2, PGF_1, and $PGF_{2\alpha}$, as detected by gas chromatography–mass spectrometry. Furthermore, endogenous PGE_2 was detected in untreated extracts using high performance liquid chromatography–radioimmunoassay [24]. These data were the first to suggest that a COX-like activity may be conserved in *Drosophila*.

Given the finding by Pages et al. [24] and the highly conserved roles of PGs in female reproduction [30, 31], we hypothesized that if PG synthesis and signaling were conserved in *Drosophila* it would regulate oogenesis or follicle development. Initially, we took advantage of the ability of mid-oogenesis stage 10B (S10B) follicles or eggs to mature in in vitro culture to ask whether COX enzyme activity was required to facilitate late-stage oogenesis [32]. These studies demonstrated that COX-1-like activity is required for follicle maturation as COX-1 inhibitors, but not COX-2 selective inhibitors, block follicle maturation in a dose-dependent manner [32]. Importantly, this COX inhibitor-dependent block in development is rescued by concomitant treatment with exogenous PGH_2, $PGF_{2\alpha}$, or fluprostenol, a stabilized $PGF_{2\alpha}$ analogue. These studies revealed that both COX-like activity and PGs are required for late-stage follicle development in *Drosophila* [32].

13.2.2 Identification the COX-Like Enzyme Pxt

The results of our pharmacological experiments [32], in combination with the previous findings of Pages et al. [24], strongly suggested the presence of a COX-like enzyme in *Drosophila*. BLAST analysis [33] revealed *Drosophila* Pxt as a candidate COX-like enzyme [32]. Sequence alignment using MUltiple Sequence Comparison by Log-Expectation (MUSCLE) [34, 35] reveals that Pxt is 26.76 % identical to ovine COX-1 and that a number of key residues [36] are conserved between COX-1 and Pxt. Most notably, the three critical residues for heme coordination in the peroxidase active site of COX-1 (Gln203, His207, and His388) are conserved in Pxt (Gln396, His402, and His590). Interestingly, Pxt possesses a candidate COX catalytic residue (Pxt Tyr564 vs. COX-1 Tyr385), although there is no

Table 13.1 Putative *Drosophila* homologues of prostaglandin (PG) synthesis and signaling proteins defined by BLASTp

PG pathway component	Function	Putative *Drosophila* homologue	Expression level during mid-to-late oogenesis[b]
COX	COX1-like	Pxt	High
	COX-like?	CG4009	Low, except high during S12
		CG10211	Below detection
PGD_2 synthases	H-PGDS	Gsts1	Low
	L-PGDS	n/a	n/a
PGE_2 synthases	mPGES1	Mgst1	Medium
		CG33178	Below detection
	mPGES2	Su(P) (CG4086)	Low
	cPGES	CG16817	High
		CG9267	Medium
$PGF_{2\alpha}$ synthases	AKR1C3	CG6084[a]	High
	AKR1B1		
PGI_2 synthase	PGIS	n/a	n/a
TXA_2 synthase	TBXAS	n/a	n/a
PG-like receptor	GPCR	CG7497	Below detection
15d-PGDH	Degrades PGs	Pdh	Below detection
		CG4086	Medium

[a]Indicates existence of numerous other similar proteins in *Drosophila*; however, CG6084 exhibits the highest homology

[b]Expression during oogenesis determined by microarray analysis of staged wild-type follicles [37]

clear conservation of the residues that have been shown to be critical for substrate binding through mutagenesis studies performed on mammalian COX enzymes. Additionally, the residue that is the target of aspirin-mediated acetylation (Ser530) is not clearly conserved.

Although the sequence homology between Pxt and COX-1 enzymes is not particularly striking, genetic loss of Pxt phenocopies the effects of COX inhibition. Specifically, similar to wild-type follicles treated with COX inhibitors, *pxt* mutant S10B follicles fail to complete maturation in vitro and exogenous PGs can restore development [32]. Additionally, *pxt* mutants are female sterile, and this sterility can be rescued by germline expression of mouse COX-1 [32]. Together, these data suggest that Pxt is the *Drosophila* COX-like enzyme and that both COX-like activity and PG signaling are required for *Drosophila* follicle maturation.

In addition to Pxt, *Drosophila* possesses putative homologues of other PG synthesis and signaling components (identified by BLAST [33]). Table 13.1 summarizes these candidates and their level of expression during mid-to-late oogenesis, as revealed by our microarray analysis [37].

13.3 The Roles of Prostaglandins in *Drosophila*

Given the widely conserved roles of PGs in reproduction [30, 31] and our finding that both COX activity and Pxt, the COX-like enzyme, mediate *Drosophila* follicle development [32], we have continued to exploit the system of *Drosophila* oogenesis to discover the specific activities of PGs. Thus, here we primarily discuss the oogenic roles of PGs. Additionally, we briefly discuss functions of PGs in *Drosophila* other than female reproduction.

13.3.1 Oogenic Roles of Prostaglandins

The *Drosophila* ovary is comprised of about 15 or 16 ovarioles or chains of sequentially developing follicles. Each follicle passes through 14 well-characterized, morphologically defined stages of development, termed Stages 1–14 (S1–S14) [38]. The germarium, a specialized structure housing both the germline and somatic follicle cell stem cells, is located at the anterior tip of each ovariole. Each developing follicle is a self-contained unit consisting of 16 interconnected germline-derived cells (15 nurse cells and a single oocyte) and approximately 1000 somatic epithelial cells, termed follicle cells.

Pxt, and thus PG signaling, are required for multiple aspects of *Drosophila* oogenesis [32]. Loss of Pxt results in age-dependent defects in follicle packaging/fusion, nurse cell cortical actin integrity during mid-to-late oogenesis (see following), border cell migration during Stage 9 (S9), germline stem cell division, and ovulation [32]. Here we focus on the characterized roles of PG signaling in eggshell formation [37] and temporal regulation of actin remodeling [32, 39, 40].

13.3.1.1 Prostaglandins Regulate Eggshell Formation

Pxt is required for the coordination of eggshell gene expression throughout the end of *Drosophila* oogenesis [37]. The somatic follicle cells secrete the eggshell. The *Drosophila* eggshell consists of five structural layers, which are sequentially synthesized in the following order: the vitelline membrane, the wax layer, and the chorion, which consists of the inner chorion layer, the endochorion, and the exochorion (Fig. 13.1a, d). Proper eggshell assembly requires tight temporal regulation of gene expression for the eggshell structural components. As such, the expression of vitelline membrane components begins during Stage 8 (S8), peaks during S10, and ceases by the end of S11, whereas the expression of chorion components occurs in three distinct phases, spanning from S10 to S14: early, middle, and late [41].

In *pxt* mutants, the expression of many vitelline membrane genes is prolonged while chorion gene expression is severely disrupted. The onset of expression of some chorion genes is early, although the expression of other chorion genes is

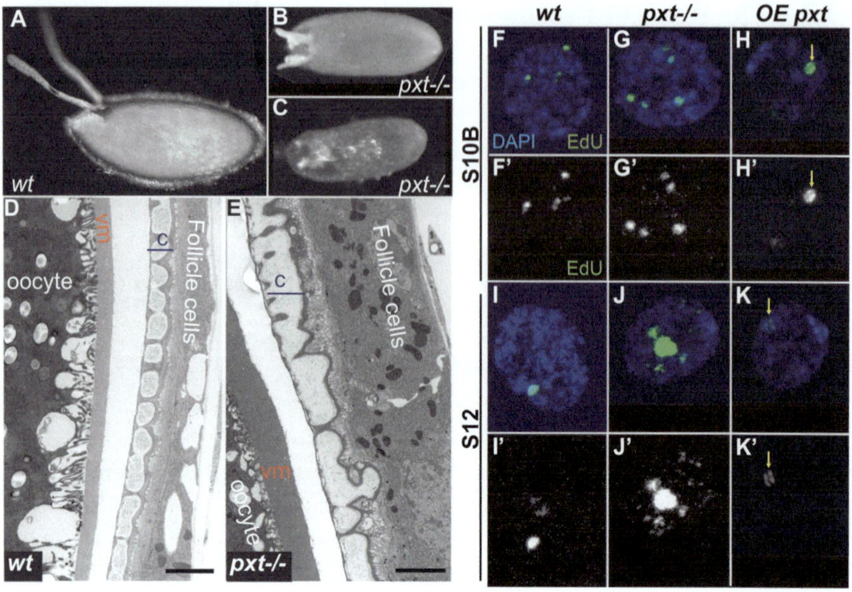

Fig. 13.1 Loss of Pxt in the soma results in defects in eggshell gene expression and eggshell formation. (**a**, **b**) Images of laid eggs of the indicated genotypes. (**d**, **e**) Transmission electron micrographs (Michael Sepanski) of the eggshell of S14 follicles (*c* chorion, *vm* vitelline membrane). *Bars* 5 μm. (**f**–**k**′) Confocal images of follicle cell nuclei from specified stages and genotypes labeled with DAPI (*blue* in **f**–**k**) to mark the nucleus and EdU (*green* in **f**–**k**; *white* in **f**′–**k**′) to label the number and size of gene amplification sites. Loss of Pxt results in short eggs with defective eggshells (**b**, **c** compared to **a**). In *pxt* mutants, the vitelline membrane is produced and fuses prematurely (not shown), and the chorion of S14 follicles exhibits structural defects (**e** compared to **d**). These eggshell defects are likely caused by the altered temporal regulation of eggshell gene expression during mid-to-late oogenesis (not shown), which may be caused by the altered gene amplification observed. When Pxt is lost, S10B follicle cells exhibit an increase in the number of sites of gene amplification observed by EdU labeling (**g**–**g**′ compared to **f**–**f**′), and during S12, when most of the amplification has ceased in wild-type follicles (**i**–**i**′), multiple and larger sites of amplification are observed in *pxt* mutants (**j**–**j**′). Conversely, overexpression of Pxt results in decreased sites of amplification and an increase in the rate of elongation as separated replication forks, indicated by double-bar EdU structures, are often observed (**h**–**h**′ and **k**–**k**′, *yellow arrows*)

delayed in *pxt* mutants [37]. Additionally, the expression of some chorion genes persists longer in *pxt* mutants than in wild-type [37]. These defects in temporal regulation of gene expression ultimately lead to numerous eggshell abnormalities, including a loss of vitelline membrane integrity and altered chorion production, resulting in short, uneven dorsal appendages and chorion patterning defects [37] (Fig. 13.1b, c compared to Fig. 13.1a, e compared to Fig. 13.1d). Although Pxt is required in the germline for nurse cell actin remodeling and nurse cell dumping (see following) [32], it is required in the soma (follicle cells) for temporal coordination of eggshell gene expression [37].

One means by which PG signaling could regulate the timing of eggshell gene expression is by affecting gene amplification of eggshell gene clusters. To promote the proper formation of the eggshell, the appropriate eggshell genes must be rapidly transcribed at high rates during a strict temporal window. The eggshell-encoding genes are organized into a few clusters throughout the *Drosophila* genome. These gene clusters undergo gene amplification [42]. During gene amplification, particular regions of the genome undergo multiple rounds of replication to increase the DNA copy number of those regions. In nondividing cells, such as the *Drosophila* follicle cells after S6, gene amplification can be visualized as spots of nucleotide analogue, such as EdU incorporation. The size of the EdU spot generally corresponds to the amount gene amplification. There are six characterized sites of gene amplification in the *Drosophila* follicle cells [43]. During S10B–S11, gene amplification is initiating and all the sites of amplification are visible (Fig. 13.1f–f′); by S12 it has shifted to elongation and only a subset of the sites are visible (Fig. 13.1i–i′) [44].

Both loss of and overexpression of Pxt affect gene amplification during follicle development. Follicles from *pxt* mutants exhibit an increase in both the visible number and size of amplification sites during both S10B and S12 (Fig. 13.1g–g′, j-j' compared to Fig. 13.1f–f'', i–i′, respectively). Conversely, ubiquitous expression of either Pxt or mouse COX-1 results in a reduction in the level of amplification as decreased EdU incorporation is observed during both S10B and S12 (Fig. 13.1h–h′, k–k′ compared to Fig. 13.1f–f′, i–i′, respectively). Additionally, double-bar EdU spots, indicative of replication forks [44], are easily observed in follicles overexpressing Pxt or mouse COX-1, suggesting increased elongation. These data indicate the Pxt and PG signaling regulate the sites, level, and extent of elongation of gene amplification (Tootle, Williams, and Spradling, unpublished observations).

13.3.1.2 Prostaglandins Regulate Actin Cytoskeletal Remodeling

Actin cytoskeletal remodeling has critical roles in *Drosophila* follicle development and female fertility [45]. Here we specifically focus on those events occurring during mid-to-late oogenesis (S9–S14).

During S9 (Fig. 13.2a), the developing follicle undertakes numerous processes that are essential for female fertility, including yolk uptake [46, 47], slow cytoplasmic streaming [48, 49], and border cell migration [50]. Aside from cortical actin deposits underlying the nurse cell membranes, the cytoplasm of the germline-derived nurse cells is largely devoid of actin filament bundles through the end of S10A (Fig. 13.2b–b′).

Loss of Pxt results in aberrant, early actin remodeling during S9. Specifically, the majority of *pxt* mutant follicles exhibit early actin filament or aggregate formation in the posterior nurse cells (Fig. 13.2c–c′) [40]. Wild-type follicles exhibit similar actin structures at a low frequency. Studies using immunofluorescence reveal that the actin elongation factor Enabled (Ena), the sole *Drosophila* Ena/VASP family member, preferentially localizes to the early actin structures in *pxt* mutants. Subsequent genetic interaction studies show that reduced Ena levels rescue the aberrant early actin remodeling when Pxt is lost [40]. Notably, Ena expression is not affected in *pxt* mutants. These data lead to the model that, during S9, PGs elicits a

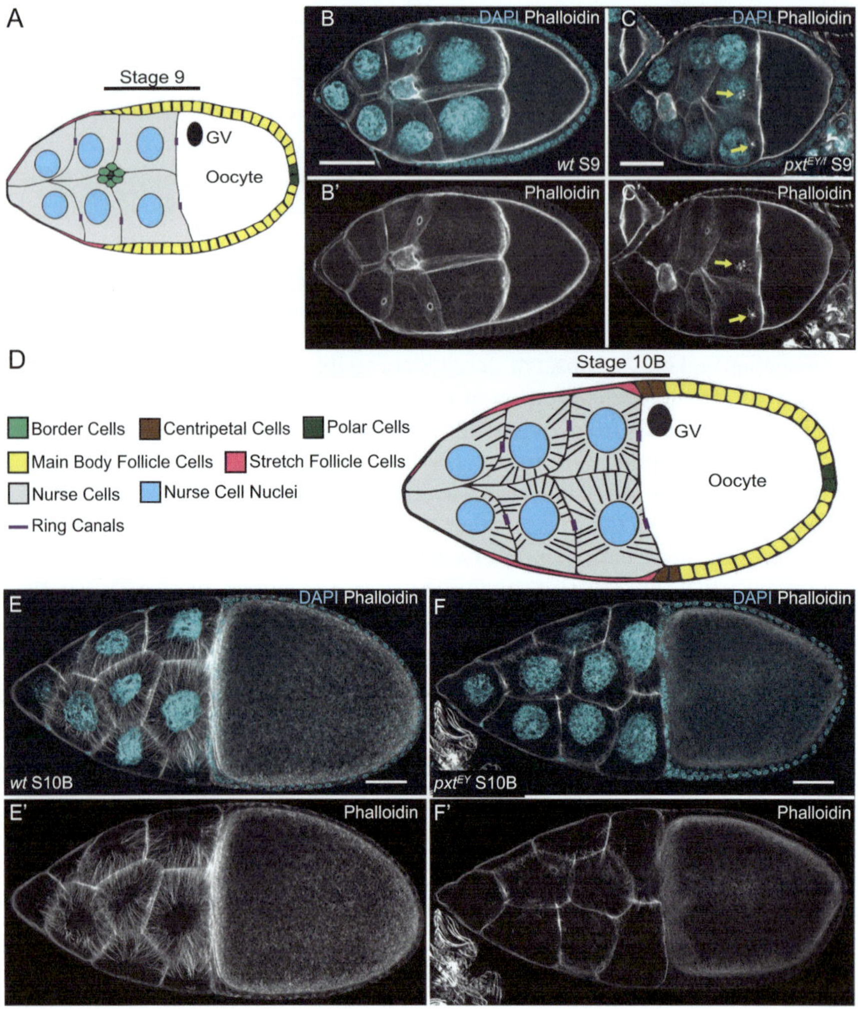

Fig. 13.2 Actin remodeling during *Drosophila* oogenesis requires the *Drosophila* COX-like enzyme, Pxt. (**a**) Schematic detailing the cellular composition of a S9 follicle (*GV* germinal vesicle). (**b**–**b**′, **d**–**f**′) Maximum projections of three confocal slices of follicles, staged as indicated, taken at 20×. (**c**–**c**′) Single confocal slice of S9 follicle, taken at 20×. Anterior is to the *left*. F-actin (phalloidin) *white*, DNA (DAPI) *cyan*. (**b**–**b**′, **e**–**e**′) Wild type *wt* (*yw*). (**c**–**c**′, **f**–**f**′) $pxt^{EY03052}$ mutant, pxt^{EY}. (**d**) Schematic detailing the cellular composition of a S10B follicle (*GV* germinal vesicle). S9 and S10B follicles consist of 16 germline-derived cells [1 oocyte (*white*) and 15 nurse cells (*gray*)] that are surrounded by a somatic epithelium (**a**, **c**). During S9, the nurse cell cytoplasm is largely devoid of actin filament structures aside from the cortical actin meshwork underlying the membranes (**b**–**b**′). *pxt* mutants exhibit a range of actin remodeling defects during S9, including cortical actin breakdown (not shown) and early actin remodeling, resulting in extensive early actin filaments and actin aggregate structures (*yellow arrows*). (**c**–**c**′) During S10B, wild-type follicles rapidly undergo actin remodeling to generate a network of parallel actin filament bundles extending from the nurse cell membranes toward the nuclei (**e**–**e**′). *pxt* mutants exhibit a range of actin remodeling defects at S10B, ranging from mild defects in the number and distribution of actin filament bundles to a near complete loss of actin filament bundles (**f**–**f**′). Images are representative, taken from multiple experiments. *Scale Bars* 50 μm

signaling cascade that normally inhibits Ena localization or activity to temporally restrict actin remodeling.

During S10B (Fig. 13.2d), the actin cytoskeleton within the nurse cells undergoes a rapid remodeling resulting in increased cortical actin deposition and the formation of a cage-like network of parallel actin filament bundles extending from the nurse cell membranes inward, toward the nurse cell nuclei (Fig. 13.2e–e′) [51, 52]. This dramatic actin remodeling is required to provide the contractile force necessary for the rapid transfer of nurse cell cytoplasm (nurse cell dumping) into the growing oocyte at S11 [53], while preventing the nurse cell nuclei from obstructing the ring canals, the specialized cytoplasmic bridges through which the nurse cell cytoplasm must flow [54, 55]. Importantly, nurse cell dumping, and thus actin remodeling, are required for female fertility. By S12, the nurse cells have completely transferred their cytoplasmic contents to the elongated oocyte, retaining only their nuclei, which will persist through S13. The mature follicle (S14) consists of only the oocyte, somatic follicle cells, and fully formed eggshell structures.

Initial studies revealed that numerous actin cytoskeletal defects are observed in *pxt* mutant S10B follicles. In contrast to wild-type follicles, *pxt* mutant follicles exhibit a significant reduction, if not complete loss, in the ability to form cytoplasmic actin filament bundles during S10B (Fig. 13.2f–f′) [32]. Additionally, nurse cell cortical actin levels are highly reduced in *pxt* mutants. As a result, *pxt* mutant follicles fail to generate the contractile force necessary for nurse cell dumping during S11 and exhibit multinucleate nurse cells resulting from cortical actin and nurse cell membrane breakdown [32, 39]. Expression of mouse COX-1 is able to almost completely suppress all the actin remodeling defects observed in *pxt* mutants [32]. Similarly, COX inhibitor treatment of cultured follicles results in the formation of large actin aggregates within the nurse cells and a reduction in nurse cell cortical actin deposits. Importantly, exogenous PGH_2 suppresses these COX inhibitor-induced actin cytoskeletal defects [32]. Together, these data suggest that both the COX-like activity of Pxt and PG signaling are required for appropriate actin cytoskeletal remodeling during S10B.

Although PGs have been implicated in regulating the actin cytoskeleton in other systems, the molecular mechanisms by which this occurs remain largely unknown [56–62]. Given our finding that PGs are required for the dynamic remodeling events occurring during *Drosophila* S10B [32] and the well-established roles of numerous actin-binding proteins in this process [45], we reasoned that we could exploit *Drosophila* genetics and our ability to mature S10B in culture to identify the actin-binding proteins whose activity is regulated by PG signaling [32, 63].

A pharmaco-interaction screen was performed to identify downstream targets of PG signaling based on their ability to dominantly modify follicle sensitivity to the effects of COX inhibition. Subsequently, using a combination of phenotypic analyses, pharmaco-genetic interactions with additional COX inhibitors, and genetic interactions, the actin bundling protein, Fascin was validated as a novel downstream target of PG signaling [39]. Indeed, overexpression of Fascin in the germline significantly restores cortical actin integrity and actin bundle formation in *pxt* mutants. Additionally, this screen uncovered an allele-dependent interaction that suggests Ena may be a downstream target of PG signaling (Spracklen, Meyer, and Tootle, unpublished observation). Supporting this idea, Ena localization to the sites of actin remodeling during S10B is reduced in *pxt* mutants [40]. As Ena [40] and Fascin [39] expression are unaffected by loss of Pxt, PGs likely regulate these actin-binding proteins via posttranslational mechanisms.

PGs are critical in regulating the actin cytoskeleton during *Drosophila* oogenesis. During S9, Pxt leads to the production of PGs that block Ena localization and activity to preclude aberrant actin remodeling [40]. Pharmacological data suggest PGE_2-dependent signaling restricts actin remodeling [32]. Conversely, during S10B, Pxt-dependent generation of PGs, likely $PGF_{2\alpha}$ [32], mediates actin remodeling by positively regulating Ena [40] and Fascin [39]. In future, it will be critical to define the molecular mechanisms and the signal transduction cascades by which PGs regulate the identified targets Fascin and Ena and to identify other novel downstream effectors of PG signaling.

13.3.2 Other Roles of Prostaglandins in Drosophila

The studies of the roles of PGs in *Drosophila* have largely been limited to oogenesis, but PGs are likely to have additional functions in this organism. Indeed, *pxt* mutant flies are sickly; they exhibit reduced viability, developmental delays, reduced lifespan, motility defects, and abnormal fluid retention. Furthermore, COX inhibitor studies have implicated PGs in Fascin-dependent neural morphogenesis and branching [64]. Additionally, it has recently been observed that loss of Pxt results in sperm individualization defects (Josefa Steinhauer, personal communication); notably, these defects may be caused by altered actin dynamics [65]. Thus, *Drosophila* provides a rich system to elucidate both functions and molecular mechanisms of PG action in a variety of contexts.

13.4 Conservation of Prostaglandin Function

13.4.1 Follicle Development

The female reproductive functions of PG signaling are highly conserved. PGs regulate egg development and ovulation from insects [31] to mammals [30]. Indeed, PG synthesis inhibitors cause fertility defects in women, likely the result of altered follicle maturation and ovulation [66–68]. Interestingly, COX2-dependent production of PGE_2 and $PGF_{2\alpha}$ are implicated in mediating follicle development in mammals [69–71], whereas COX-1-dependent $PGF_{2\alpha}$ mediates it in zebrafish [72], silkmoths [73], and *Drosophila* [32]. It is important to determine if the molecular targets of these PG signaling pathways are conserved across organisms.

13.4.2 Gene Amplification

Both PGs [74–81] and gene amplification [82, 83] are implicated in driving cancer development and progression and are independently associated with poor patient prognosis. We speculate that one means by which PGs may contribute to cancer is by modulating gene amplification. In breast cancer, when the oncogene HER-2/neu is highly expressed it is often caused by gene amplification [84]. Subbaramaiah et al. found that the majority of HER-2/neu-positive tumor samples tested exhibit high COX-2 levels [33]. It will be interesting to determine if there is mechanistic association between PG signaling and gene amplification in cancer.

13.4.3 Actin Cytoskeletal Remodeling

Numerous in vitro studies have implicated PG signaling in regulating the actin cytoskeleton. However, such studies have provided limited insight into the underlying mechanisms of PG action. Multiple studies indicate that PGs induce changes in cytoplasmic actin bundles by cAMP-dependent mechanisms. Indeed, PGE_2- and PGI_2-induced actin stress fiber disassembly in human pulmonary artery endothelial cells occurs via cAMP-dependent kinase (PKA) and nucleotide exchange proteins directly activated by cAMP (Epac1)/Ras-related protein 1 (Rap1)-dependent activation of Rac [57]. In human umbilical vein endothelial cells, TXA_2 inhibits, while PGE_2 promotes, $\alpha_v\beta_3$-dependent cell adhesion and cell spreading by both PKA-dependent Rac activation and Rac-independent activities [59]. PGE_2 mediates actin stress fiber disassembly in human aortic smooth muscle cells by PKA-dependent decreases in focal adhesion kinase (FAK) phosphorylation [58]. PGs can also modulate the actin cytoskeleton via Rho GTPases. Specifically, PGE_2 promotes actin stress fiber assembly in rat IMCD cells [85], and $PGF_{2\alpha}$ mediates filopodia retraction and actin stress fiber assembly in 293-EBNA cells via Rho activation [85, 86].

PGs also regulate actin cytoskeletal dynamics in vivo to control platelet activation and aggregation [87]. The major prostanoid produced in platelets is TXA_2, which serves as a potent activator of platelet aggregation [88]. Conversely, PGI_2 [89], PGE_1 [90], and PGD_2 [91] inhibit platelet aggregation, and PGE_2 may both potentiate [90, 92] and inhibit platelet aggregation [93, 94]. The main activity of these prostanoids is to regulate vasodilator-stimulated phosphoprotein (VASP), a member of the Ena/VASP family of actin elongation factors. VASP is activated by TXA_2 and is inhibited through cAMP/cGMP-dependent phosphorylation downstream of PGI_2 and PGE_1 [95, 96]. The opposing actions of distinct prostanoids in regulating VASP activity is strikingly similar to our findings on PG regulation of Ena, the sole *Drosophila* Ena/VASP family member. Indeed, we find that PG signaling inhibits Ena during S9, but likely promotes Ena activity during S10B [40].

Another context in which PGs likely regulate actin cytoskeletal dynamics is in cancer. Cytoskeletal remodeling is necessary for proliferation, adhesion, and migration—cellular properties altered in cancer [97]. Remodeling is achieved through the coordinated activities of actin-binding proteins, the misregulation of which is associated with cancer development and progression [98–104]. Given that actin remodeling is aberrant in cancer, PGs regulate actin cytoskeletal remodeling in multiple contexts, and PGs play critical roles in cancer development and progression [74–81], it is tempting to speculate that PG signaling controls actin-binding protein localization/activity by posttranslation mechanisms to drive tumorigenic and metastatic events. This idea is particularly appealing as our studies in *Drosophila* identified Fascin [39] and Ena [40], the homologues of known cancer-associated actin-binding proteins, as downstream effectors of PG-dependent actin cytoskeletal remodeling.

Fascin is required for the generation of filopodia [105, 106] and invadopodia [98, 107], structures required for cancer invasion and metastasis [108, 109]. Indeed, high fascin-1 levels correlate with highly aggressive cancers in patients [108, 110–114]. Notably, high PG levels are similarly associated with poor patient prognosis [74–77, 79]. Together these data lead us to postulate that PG signaling posttranslationally regulates Fascin to drive cancer aggressiveness.

Our data also implicate Ena/VASP family members as targets of PG signaling during *Drosophila* oogenesis [40]. Alternatively spliced invasive isoforms of mammalian Ena (Mena), which is the closest homologue to *Drosophila* Ena [115], are upregulated as tumors become more advanced and strongly correlate with increased invasiveness and metastatic potential [116, 117]. Given that PGs regulate another Ena/VASP family member, VASP, in platelets [87] and our data that Ena activity is tightly controlled by PGs in *Drosophila* [40], it seems likely that PG signaling is a conserved means of modulating this family of actin-binding proteins.

13.5 Concluding Remarks

The study of PG signaling in *Drosophila* is in its infancy, yet it has already provided key insights into how PGs regulate actin cytoskeletal remodeling. By taking advantage of the genetic tools in *Drosophila*, further insights into the developmental and homeostatic roles of PGs and their downstream signaling cascades can be uncovered. Thus, this system is poised to truly advance our understanding of PG signaling. In the future, it will be important to determine the extent to which the means of PG signaling in *Drosophila* are conserved in higher organisms, both during normal physiology, such as reproduction and tissue homeostasis, and disease, such as cardiovascular diseases and cancer.

References

1. Funk CD (2001) Prostaglandins and leukotrienes: advances in eicosanoid biology. Science 294(5548):1871–1875
2. Tootle TL (2013) Genetic insights into the in vivo functions of prostaglandin signaling. Int J Biochem Cell Biol 45(8):1629–1632
3. Garavito RM, DeWitt DL (1999) The cyclooxygenase isoforms: structural insights into the conversion of arachidonic acid to prostaglandins. Biochim Biophys Acta 1441(2-3):278–287
4. Smith WL, DeWitt DL, Garavito RM (2000) Cyclooxygenases: structural, cellular, and molecular biology. Annu Rev Biochem 69:145–182
5. Garavito RM, Mulichak AM (2003) The structure of mammalian cyclooxygenases. Annu Rev Biophys Biomol Struct 32:183–206
6. Simmons DL, Botting RM, Hla T (2004) Cyclooxygenase isozymes: the biology of prostaglandin synthesis and inhibition. Pharmacol Rev 56(3):387–437
7. Smith WL, Urade Y, Jakobsson PJ (2011) Enzymes of the cyclooxygenase pathways of prostanoid biosynthesis. Chem Rev 111(10):5821–5865
8. Buchanan FG et al (2003) Prostaglandin E_2 regulates cell migration via the intracellular activation of the epidermal growth factor receptor. J Biol Chem 278(37):35451–35457
9. Han C, Michalopoulos GK, Wu T (2006) Prostaglandin E_2 receptor EP1 transactivates EGFR/MET receptor tyrosine kinases and enhances invasiveness in human hepatocellular carcinoma cells. J Cell Physiol 207(1):261–270
10. Pai R et al (2002) Prostaglandin E_2 transactivates EGF receptor: a novel mechanism for promoting colon cancer growth and gastrointestinal hypertrophy. Nat Med 8(3):289–293
11. Sales KJ, Maudsley S, Jabbour HN (2004) Elevated prostaglandin EP2 receptor in endometrial adenocarcinoma cells promotes vascular endothelial growth factor expression via cyclic 3′,5′-adenosine monophosphate-mediated transactivation of the epidermal growth factor receptor and extracellular signal-regulated kinase 1/2 signaling pathways. Mol Endocrinol 18(6):1533–1545
12. Sales KJ et al (2004) Expression, localization, and signaling of prostaglandin $F_{2\alpha}$ receptor in human endometrial adenocarcinoma: regulation of proliferation by activation of the epidermal growth factor receptor and mitogen-activated protein kinase signaling pathways. J Clin Endocrinol Metab 89(2):986–993
13. Ali FY et al (2006) Role of prostacyclin versus peroxisome proliferator-activated receptor beta receptors in prostacyclin sensing by lung fibroblasts. Am J Respir Cell Mol Biol 34(2):242–246
14. Forman BM et al (1995) 15-Deoxy-delta 12,14-prostaglandin J_2 is a ligand for the adipocyte determination factor PPAR gamma. Cell 83(5):803–812
15. Kliewer SA et al (1995) A prostaglandin J_2 metabolite binds peroxisome proliferator-activated receptor gamma and promotes adipocyte differentiation. Cell 83(5):813–819
16. Lim H et al (1999) Cyclo-oxygenase-2-derived prostacyclin mediates embryo implantation in the mouse via PPARdelta. Genes Dev 13(12):1561–1574
17. Bos CL et al (2004) Prostanoids and prostanoid receptors in signal transduction. Int J Biochem Cell Biol 36(7):1187–1205
18. Hirata M et al (1991) Cloning and expression of cDNA for a human thromboxane A_2 receptor. Nature 349(6310):617–620
19. Hirata T, Narumiya S (2011) Prostanoid receptors. Chem Rev 111(10):6209–6230
20. Speirs CK et al (2010) Prostaglandin Gbetagamma signaling stimulates gastrulation movements by limiting cell adhesion through Snai1a stabilization. Development 137(8):1327–1337
21. Blomquist GJ et al (1982) Biosynthesis of linoleic acid in a termite, cockroach and cricket. Insect Biochem 12(3):349–353

22. Keith AD (1967) Fatty acid metabolism in *Drosophila melanogaster*: interaction between dietary fatty acids and de novo synthesis. Comp Biochem Physiol 21(3):587–600
23. Rapport EW, Stanley-Samuelson D, Dadd RH (1983) Ten generations of *Drosophila melanogaster* reared axenically on a fatty acid-free holidic diet. Arch Insect Biochem Physiol 1(3):243–250
24. Pages M et al (1986) Cyclooxygenase and lipoxygenase-like activity in *Drosophila melanogaster*. Prostaglandins 32(5):729–740
25. Shen LR et al (2010) *Drosophila* lacks C20 and C22 PUFAs. J Lipid Res 51(10):2985–2992
26. Steinhauer J et al (2009) *Drosophila* lysophospholipid acyltransferases are specifically required for germ cell development. Mol Biol Cell 20(24):5224–5235
27. Carvalho M et al (2012) Effects of diet and development on the *Drosophila* lipidome. Mol Syst Biol 8:600
28. Parisi M, Li R, Oliver B (2011) Lipid profiles of female and male *Drosophila*. BMC Res Notes 4:198
29. Scheitz CJ et al (2013) Heritability and inter-population differences in lipid profiles of *Drosophila melanogaster*. PLoS One 8(8):e72726
30. Kobayashi T, Narumiya S (2002) Function of prostanoid receptors: studies on knockout mice. Prostaglandins Other Lipid Mediat 68-69:557–573
31. Stanley D, Kim Y (2011) Prostaglandins and their receptors in insect biology. Front Endocrinol (Lausanne) 2:105
32. Tootle TL, Spradling AC (2008) *Drosophila* Pxt: a cyclooxygenase-like facilitator of follicle maturation. Development 135(5):839–847
33. Subbaramaiah K et al (2002) Cyclooxygenase-2 is overexpressed in HER-2/neu-positive breast cancer: evidence for involvement of AP-1 and PEA3. J Biol Chem 277(21):18649–18657
34. Edgar RC (2004) MUSCLE: multiple sequence alignment with high accuracy and high throughput. Nucleic Acids Res 32(5):1792–1797
35. Edgar RC (2004) MUSCLE: a multiple sequence alignment method with reduced time and space complexity. BMC Bioinformatics 5:113
36. Garavito RM, Malkowski MG, DeWitt DL (2002) The structures of prostaglandin endoperoxide H synthases-1 and -2. Prostaglandins Other Lipid Mediat 68-69:129–152
37. Tootle TL et al (2011) *Drosophila* eggshell production: identification of new genes and coordination by Pxt. PLoS One 6(5):e19943
38. Spradling AC (1993) Developmental genetics of oogenesis. In: Martinez-Arias B (ed) The development of *Drosophila melanogaster*. Cold Spring Harbor Laboratory Press, Plainview, pp 1–70
39. Groen CM et al (2012) *Drosophila* Fascin is a novel downstream target of prostaglandin signaling during actin remodeling. Mol Biol Cell 23(23):4567–4578
40. Spracklen AJ et al (2014) Prostaglandins temporally regulate cytoplasmic actin bundle formation during *Drosophila* oogenesis. Mol Biol Cell 25(3):397–411
41. Cavaliere V et al (2008) Building up the *Drosophila* eggshell: first of all the eggshell genes must be transcribed. Dev Dyn 237(8):2061–2072
42. Claycomb JM, Orr-Weaver TL (2005) Developmental gene amplification: insights into DNA replication and gene expression. Trends Genet 21(3):149–162
43. Kim JC et al (2011) Integrative analysis of gene amplification in *Drosophila* follicle cells: parameters of origin activation and repression. Genes Dev 25(13):1384–1398
44. Claycomb JM et al (2004) Gene amplification as a developmental strategy: isolation of two developmental amplicons in *Drosophila*. Dev Cell 6(1):145–155
45. Hudson AM, Cooley L (2002) Understanding the function of actin-binding proteins through genetic analysis of *Drosophila* oogenesis. Annu Rev Genet 36:455–488
46. Bownes M, Hames BD (1978) Genetic analysis of vitellogenesis in *Drosophila melanogaster*: the identification of a temperature-sensitive mutation affecting one of the yolk proteins. J Embryol Exp Morphol 47:111–120

47. Bownes M, Hodson BA (1980) Mutant Fs(1) 1163 of *Drosophila melanogaster* alters yolk protein secretion from the fat-body. Mol Gen Genet 180(2):411–418
48. Gutzeit HO (1986) The role of microfilaments in cytoplasmic streaming in *Drosophila* follicles. J Cell Sci 80:159–169
49. Theurkauf WE et al (1992) Reorganization of the cytoskeleton during *Drosophila* oogenesis: implications for axis specification and intercellular transport. Development 115(4):923–936
50. Montell DJ, Yoon WH, Starz-Gaiano M (2012) Group choreography: mechanisms orchestrating the collective movement of border cells. Nat Rev Mol Cell Biol 13(10):631–645
51. Guild GM et al (1997) Actin filament cables in *Drosophila* nurse cells are composed of modules that slide passively past one another during dumping. J Cell Biol 138(4):783–797
52. Huelsmann S, Ylanne J, Brown NH (2013) Filopodia-like actin cables position nuclei in association with perinuclear actin in *Drosophila* nurse cells. Dev Cell 26(6):604–615
53. Wheatley S, Kulkarni S, Karess R (1995) *Drosophila* nonmuscle myosin II is required for rapid cytoplasmic transport during oogenesis and for axial nuclear migration in early embryos. Development 121(6):1937–1946
54. Mahajan-Miklos S, Cooley L (1994) Intercellular cytoplasm transport during *Drosophila* oogenesis. Dev Biol 165(2):336–351
55. Cooley L, Verheyen E, Ayers K (1992) chickadee encodes a profilin required for intercellular cytoplasm transport during *Drosophila* oogenesis. Cell 69(1):173–184
56. Banan A et al (2000) Role of actin cytoskeleton in prostaglandin-induced protection against ethanol in an intestinal epithelial cell line. J Surg Res 88(2):104–113
57. Birukova AA et al (2007) Prostaglandins PGE_2 and PGI_2 promote endothelial barrier enhancement via PKA- and Epac1/Rap1-dependent Rac activation. Exp Cell Res 313(11):2504–2520
58. Bulin C et al (2005) Differential effects of vasodilatory prostaglandins on focal adhesions, cytoskeletal architecture, and migration in human aortic smooth muscle cells. Arterioscler Thromb Vasc Biol 25(1):84–89
59. Dormond O et al (2002) Prostaglandin E_2 promotes integrin alpha Vbeta 3-dependent endothelial cell adhesion, rac-activation, and spreading through cAMP/PKA-dependent signaling. J Biol Chem 277(48):45838–45846
60. Kawada N, Klein H, Decker K (1992) Eicosanoid-mediated contractility of hepatic stellate cells. Biochem J 285(pt 2):367–371
61. Peppelenbosch MP et al (1993) Epidermal growth factor-induced actin remodeling is regulated by 5-lipoxygenase and cyclooxygenase products. Cell 74(3):565–575
62. Martineau LC et al (2004) p38 MAP kinase mediates mechanically induced COX-2 and PG EP4 receptor expression in podocytes: implications for the actin cytoskeleton. Am J Physiol Renal Physiol 286(4):F693–F701
63. Spracklen AJ, Tootle TL (2013) The utility of stage-specific mid-to-late *Drosophila* follicle isolation. J Vis Exp 82:50493
64. Kraft R et al (2013) A cell-based fascin bioassay identifies compounds with potential anti-metastasis or cognition-enhancing functions. Dis Model Mech 6(1):217–235
65. Fabian L, Brill JA (2012) *Drosophila spermiogenesis:* big things come from little packages. Spermatogenesis 2(3):197–212
66. Akil M, Amos RS, Stewart P (1996) Infertility may sometimes be associated with NSAID consumption. Br J Rheumatol 35(1):76–78
67. Pall M, Friden BE, Brannstrom M (2001) Induction of delayed follicular rupture in the human by the selective COX-2 inhibitor rofecoxib: a randomized double-blind study. Hum Reprod 16(7):1323–1328
68. Smith G et al (1996) Reversible ovulatory failure associated with the development of luteinized unruptured follicles in women with inflammatory arthritis taking non-steroidal anti-inflammatory drugs. Br J Rheumatol 35(5):458–462
69. Lim H et al (1997) Multiple female reproductive failures in cyclooxygenase 2-deficient mice. Cell 91(2):197–208

70. Takahashi T et al (2006) Cyclooxygenase-2-derived prostaglandin E_2 directs oocyte maturation by differentially influencing multiple signaling pathways. J Biol Chem 281(48):37117–37129
71. Downs SM, Longo FJ (1983) Prostaglandins and preovulatory follicular maturation in mice. J Exp Zool 228(1):99–108
72. Lister AL, Van Der Kraak G (2008) An investigation into the role of prostaglandins in zebrafish oocyte maturation and ovulation. Gen Comp Endocrinol 159(1):46–57
73. Machado E et al (2007) Prostaglandin signaling and ovarian follicle development in the silkmoth, *Bombyx mori*. Insect Biochem Mol Biol 37(8):876–885
74. Chen WS et al (2001) Tumor invasiveness and liver metastasis of colon cancer cells correlated with cyclooxygenase-2 (COX-2) expression and inhibited by a COX-2-selective inhibitor, etodolac. Int J Cancer 91(6):894–899
75. Denkert C et al (2003) Elevated expression of cyclooxygenase-2 is a negative prognostic factor for disease free survival and overall survival in patients with breast carcinoma. Cancer 97(12):2978–2987
76. Gallo O et al (2002) Prognostic significance of cyclooxygenase-2 pathway and angiogenesis in head and neck squamous cell carcinoma. Hum Pathol 33(7):708–714
77. Khuri FR et al (2001) Cyclooxygenase-2 overexpression is a marker of poor prognosis in stage I non-small cell lung cancer. Clin Cancer Res 7(4):861–867
78. Lyons TR et al (2011) Postpartum mammary gland involution drives progression of ductal carcinoma in situ through collagen and COX-2. Nat Med 17(9):1109–1115
79. Rolland PH et al (1980) Prostaglandin in human breast cancer: evidence suggesting that an elevated prostaglandin production is a marker of high metastatic potential for neoplastic cells. J Natl Cancer Inst 64(5):1061–1070
80. Tsujii M, Kawano S, DuBois RN (1997) Cyclooxygenase-2 expression in human colon cancer cells increases metastatic potential. Proc Natl Acad Sci U S A 94(7):3336–3340
81. Yamakita Y, Matsumura F, Yamashiro S (2009) Fascin1 is dispensable for mouse development but is favorable for neonatal survival. Cell Motil Cytoskeleton 66(8):524–534
82. Albertson DG (2006) Gene amplification in cancer. Trends Genet 22(8):447–455
83. Santarius T et al (2010) A census of amplified and overexpressed human cancer genes. Nat Rev Cancer 10(1):59–64
84. Slamon DJ et al (1987) Human breast cancer: correlation of relapse and survival with amplification of the HER-2/neu oncogene. Science 235(4785):177–182
85. Tamma G et al (2003) The prostaglandin E_2 analogue sulprostone antagonizes vasopressin-induced antidiuresis through activation of Rho. J Cell Sci 116(pt 16):3285–3294
86. Pierce KL et al (1999) Activation of FP prostanoid receptor isoforms leads to Rho-mediated changes in cell morphology and in the cell cytoskeleton. J Biol Chem 274(50):35944–35949
87. Bearer EL, Prakash JM, Li Z (2002) Actin dynamics in platelets. Int Rev Cytol 217:137–182
88. Hamberg M, Svensson J, Samuelsson B (1975) Thromboxanes: a new group of biologically active compounds derived from prostaglandin endoperoxides. Proc Natl Acad Sci U S A 72(8):2994–2998
89. Moncada S, Higgs EA, Vane JR (1977) Human arterial and venous tissues generate prostacyclin (prostaglandin x), a potent inhibitor of platelet aggregation. Lancet 1(8001):18–20
90. Kloeze J (1966) Influence of prostaglandins on platelet adhesiveness and platelet aggregation. In: Prostaglandins: proceedings of the second nobel symposium. Almqvist & Wiksell/Interscience, Stockholm/New York/London [etc.]
91. Smith JB et al (1974) Prostaglandin D_2 inhibits the aggregation of human platelets. Thromb Res 5(3):291–299
92. Willis AL (1974) An enzymatic mechanism for the antithrombotic and antihemostatic actions of aspirin. Science 183(4122):325–327
93. Petrucci G et al (2011) Prostaglandin E_2 differentially modulates human platelet function through the prostanoid EP2 and EP3 receptors. J Pharmacol Exp Ther 336(2):391–402

94. Smith JP et al (2010) PGE_2 decreases reactivity of human platelets by activating EP2 and EP4. Thromb Res 126(1):e23–e29
95. Aszodi A et al (1999) The vasodilator-stimulated phosphoprotein (VASP) is involved in cGMP- and cAMP-mediated inhibition of agonist-induced platelet aggregation, but is dispensable for smooth muscle function. EMBO J 18(1):37–48
96. Bearer EL et al (2000) VASP protects actin filaments from gelsolin: an in vitro study with implications for platelet actin reorganizations. Cell Motil Cytoskeleton 47(4):351–364
97. Troys M, Vandekerckhove J, Ampe C (2008) Actin and actin-binding proteins in cancer progression and metastasis. In: Remedios C, Chhabra D (eds) Actin-binding proteins and disease. Springer, New York, pp 229–277
98. Li A et al (2010) The actin-bundling protein fascin stabilizes actin in invadopodia and potentiates protrusive invasion. Curr Biol 20(4):339–345
99. Qualtrough D et al (2009) The actin-bundling protein fascin is overexpressed in colorectal adenomas and promotes motility in adenoma cells in vitro. Br J Cancer 101(7):1124–1129
100. Chen L et al (2010) Migrastatin analogues target fascin to block tumour metastasis. Nature 464(7291):1062–1066
101. Hall A (2009) The cytoskeleton and cancer. Cancer Metastasis Rev 28(1-2):5–14
102. Bae YH et al (2009) Loss of profilin-1 expression enhances breast cancer cell motility by Ena/VASP proteins. J Cell Physiol 219(2):354–364
103. Hu LD et al (2008) EVL (Ena/VASP-like) expression is up-regulated in human breast cancer and its relative expression level is correlated with clinical stages. Oncol Rep 19(4):1015–1020
104. Bear JE, Gertler FB (2009) Ena/VASP: towards resolving a pointed controversy at the barbed end. J Cell Sci 122(pt 12):1947–1953
105. Hashimoto Y, Parsons M, Adams JC (2007) Dual actin-bundling and protein kinase C-binding activities of fascin regulate carcinoma cell migration downstream of Rac and contribute to metastasis. Mol Biol Cell 18(11):4591–4602
106. Vignjevic D et al (2006) Role of fascin in filopodial protrusion. J Cell Biol 174(6):863–875
107. Schoumacher M et al (2010) Actin, microtubules, and vimentin intermediate filaments cooperate for elongation of invadopodia. J Cell Biol 189(3):541–556
108. Yoder BJ et al (2005) The expression of fascin, an actin-bundling motility protein, correlates with hormone receptor-negative breast cancer and a more aggressive clinical course. Clin Cancer Res 11(1):186–192
109. Hashimoto Y, Skacel M, Adams JC (2005) Roles of fascin in human carcinoma motility and signaling: prospects for a novel biomarker? Int J Biochem Cell Biol 37(9):1787–1804
110. Chan C et al (2010) Fascin expression predicts survival after potentially curative resection of node-positive colon cancer. Am J Surg Pathol 34(5):656–666
111. Hashimoto Y et al (2004) The prognostic relevance of fascin expression in human gastric carcinoma. Oncology 67(3-4):262–270
112. Lee TK et al (2007) Fascin over-expression is associated with aggressiveness of oral squamous cell carcinoma. Cancer Lett 254(2):308–315
113. Li X et al (2008) Aberrant expression of cortactin and fascin are effective markers for pathogenesis, invasion, metastasis and prognosis of gastric carcinomas. Int J Oncol 33(1):69–79
114. Okada K et al (2007) Fascin expression is correlated with tumor progression of extrahepatic bile duct cancer. Hepatogastroenterology 54(73):17–21
115. Gertler FB et al (1996) Mena, a relative of VASP and Drosophila enabled, is implicated in the control of microfilament dynamics. Cell 87(2):227–239
116. Gertler F, Condeelis J (2011) Metastasis: tumor cells becoming MENAcing. Trends Cell Biol 21(2):81–90
117. Gurzu S et al (2013) The possible role of Mena protein and its splicing-derived variants in embryogenesis, carcinogenesis, and tumor invasion: a systematic review of the literature. Biomed Res Int 2013:365192

Chapter 14
Zebrafish as a Model Animal for Studying Lysophosphatidic Acid Signaling

Junken Aoki and Hiroshi Yukiura

Abstract Lysophosphatidic acid (LPA) is a second-generation lysophospholipid mediator that exerts multiple biological functions through its own cognate receptors. LPA is produced by specific enzymatic reactions and activates receptors with similar structures (Edg receptors and P2Y receptors), which results in a variety of actions from embryonic blood vessel formation to development of cancer. LPA-related genes are highly conserved in vertebrates. In the zebrafish genome, three LPA-producing enzymes and nine LPA receptors are present. In vitro experiments have shown that LPA-related genes in zebrafish are conserved biochemically. LPA-related genes can be up- and downregulated by injecting morpholino antisense oligonucleotides (MOs) specific to LPA-related genes or mRNAs in zebrafish embryos. Such tools help to assess the biological function of these genes. For example, knockdown of the LPA-produced enzyme autotaxin (ATX) in zebrafish embryos resulted in malformation of embryonic blood vessel formation, which has also been observed in embryos from ATX-knockout mice. Simultaneous inactivation of multiple genes is possible by injecting more than one MO in zebrafish embryos, which makes it possible to identify the LPA receptors responsible for embryonic blood vessel formation. Gene functions can be also eliminated in zebrafish embryos by pharmacological tools such as enzyme inhibitors or receptor antagonists. Interestingly, overexpression of ATX in zebrafish embryos resulted in duplication of the heart (two-heart phenotype) and the phenotype was canceled by treating the embryos with LPA receptor antagonists. The zebrafish system is a powerful tool not only for identification of gene functions but also for development of drugs against enzymes and receptors.

Keywords Lysophosphatidic acid • G protein-coupled receptor • Autotaxin • Zebrafish • Blood vessel formation • Embryo

J. Aoki (✉) • H. Yukiura
Department of Molecular and Cellular Biochemistry, Graduate School of Pharmaceutical Sciences, Tohoku University, 6-3, Aoba, Aramaki, Aoba-ku, Sendai, Miyagi 980-8578, Japan
e-mail: jaoki@m.tohoku.ac.jp

T. Yokomizo, M. Murakami (eds.), *Bioactive Lipid Mediators*,
DOI 10.1007/978-4-431-55669-5_14

Abbreviations

ATX	Autotaxin
Edg	Endothelial differentiation gene
hpf	Hours post fertilization
LPA	Lysophosphatidic acid
LPC	Lysophosphatidylcholine
MO	Morpholino antisense oligonucleotide
S1P	Sphingosine-1-phosphate

14.1 Introduction

Lysophosphatidic acid (LPA; 1- or 2-acyl-*sn*-glycerol-3-phosphate) is a small glycerophospholipid that has a wide range of actions through its receptors. Most of the actions of LPA were mediated by six G protein-coupled receptors (GPCRs) named LPA_{1-6}. Through the six LPA receptors, LPA has been shown to be involved in several physiological processes including neuronal development (LPA_1) [1], implantation (LPA_3) [2], blood vessel formation (LPA_4) [3], and hair follicle formation (LPA_6) [4]. LPA also has pathological roles such as progression of lung fibrosis (LPA_1) [5], cancer development (LPA_2) [6, 7], and drug- or irradiation-induced cell death in the intestine (LPA_2) [8]. LPA is produced extracellularly by two enzymes, autotaxin and PA-PLA$_1\alpha$ [9]. Studies of LPA synthetic pathways have revealed that ATX is involved in embryonic blood vessel formation [10, 11] whereas PA-PLA$_1\alpha$ is involved in hair follicle formation [12, 13]. The level of extracellularly produced LPA has been suggested to be downregulated by LPA-degrading enzymes called lipid phosphate phosphatases (LPPs) [14]. LPPs are membrane-bound enzymes that efficiently remove phosphate by their phosphatase activity.

14.2 LPA-Related Genes in Zebrafish

14.2.1 Zebrafish as a Model for Elucidating the Role of Human Genes

Zebrafish are widely used for studies of vertebrate gene function. Approximately 70 % of human genes have at least one obvious zebrafish orthologue [15]. The virtually transparent embryos of this species, and the ability to accelerate genetic studies by gene knockdown or overexpression, have led to the widespread use of zebrafish in the detailed investigation of vertebrate gene function and, increasingly, the study of human genetic disease. Fluorescent markers can be used in vivo to tag specific cell types and visualize their location and migration during embryogenesis. Moreover, well-developed videomicroscopic techniques have been available for detailed analyses of developmental stages. Analyses of vascular formation using mutants and antisense morpholino oligos (MOs), for example, have identified a number of molecules involved in vasculature development, including growth factors, cell adhesion molecules, and transcription factors [16]. These analyses have shown that the basic mechanisms of embryonic blood vessel formation are conserved in vertebrates. In addition to the traditional forward genetics, injection of morpholino oligonucleotides allows us to target gene knockdown more rapidly [17]. Moreover, new genome-editing tools such as TALENs (transcription activator-like effector nucleases) [18] and CRISPR (clustered regularly interspaced short palindromic repeats)/Cas [19] have also been applied to the zebrafish model, providing exciting new opportunities for high-efficiency mutagenesis.

14.2.2 Structure, Sequence Homology, and Biochemical Properties of LPA-Related Genes in Zebrafish

LPA-related genes such as LPA receptors, LPA-producing enzymes, and LPA-degrading enzymes are highly conserved in zebrafish (Table14.1). The zebrafish genome has homologous genes for autotaxin (2 genes), PA-PLA1α (2 genes), LPA receptors (9 genes), and LPP (6 genes) (Table 14.1; Fig. 14.1). As is often the case with zebrafish genes, there are two close homologues for ATX, PA-PLA_1α, LPA_2, LPA_5, LPA_6, LPP1, LPP2, and LPP3, which might be generated by gene duplication (Table 14.1). Thus, nine genes for LPA receptors (*lpa1, lpa2a, lpa2b, lpa3, lpa4, lpa5a, lpa5b, lpa6a, lpa6b*), two genes for ATX (*atxa, atxb*), two genes for PA-PLA_1α (*pla1aa, pla1ab*), and six genes for LPPs (*lpp1a, lpp1b, lpp2a, lpp2b, lpp3a, lpp3b*) are present. The amino acid sequences of these zebrafish LPA-related genes are similar to their mammalian homologues (Table 14.1; Fig. 14.1). These LPA-related genes are highly conserved between zebrafish and mammals, usually with 50–70 % amino acid sequence homology. It is noted that among the LPA-related genes, zebrafish LPA_1 shows approximately 90 % amino acid identity to mammalian LPA_1, suggesting that its role is conserved in a wide range of vertebrates. Most of these LPA-related genes in zebrafish were shown to conserve their biochemical functions. Indeed, seven LPA receptors (except for *lpa5a* and *lpa5b*) were activated by LPA to induce downstream G-protein signaling [20]. Two ATXs (*atxa* and *atxb*) also hydrolyzed lysophoshatidylcholine to produce LPA in vitro [20].

Table 14.1 Amino acid sequence homology between zebrafish and mammalian (human and mouse) lysophosphatidic acid (LPA)-related genes

	Homo sapiens (human)		*Mus musculus* (mouse)	
Zebrafish	Genes	Amino acid sequence homology (%)	Genes	Amino acid sequence homology (%)
atxa	*ATX*	66.7	*Atx*	66.1
atxb		66.6		66.1
papla1aa	*PAPLA1a*	49.3	*Papla1a*	47.3
papla1ab		48.7		47.9
lpa1	*LPAR1*	89.3	*Lpar1*	90.1
lpa2a	*LPAR2*	49.5	*Lpar2*	50.8
lpa2b		55.3		55.6
lpa3	*LPAR3*	62.0	*Lpar3*	61.1
lpa4	*LPAR4*	63.7	*Lpar4*	64.2
lpa5a	*LPAR5*	32.0	*Lpar5*	33.5
lpa5b		28.5		29.2
lpa6a	*LPAR6*	63.0	*Lpar6*	63.6
lpa6b		55.5		56.4
lpp1a	*LPP1*	69.2	*Lpp1*	65.5
lpp1b		62.5		60.2
lpp2a	*LPP2*	63.0	*Lpp2*	67.0
lpp2b		69.0		68.2
lpp3a	*LPP3*	73.2	*Lpp3*	71.3
lpp3b		47.8		47.3

14.2.3 Functional Aspects of LPA-Related Gene in Zebrafish

14.2.3.1 Embryonic Blood Vessel Formation

Autotaxin (ATX)-null mice die around embryonic day 9.5–10.5 with profound vascular defects in both the yolk sac and embryo, and with aberrant neural tube formation [10, 11, 21]. A number of mutants and knockout mice have shown phenotypes similar to those in ATX-knockout mice [22, 23]. However, the precise phenotypes of these mice have not been determined because real-time observation of blood vessel formation is impossible for mice. Introduction of mutant ATX, in which Thr^{210}, an amino acid responsible for the catalytic activity of ATX, was replaced with alanine, could not rescue the phenotype, indicating that the product of ATX, that is, LPA, is involved in embryonic vascular formation [24]. In addition, none of the LPA receptor knockout mice has shown a similar phenotype [1–3, 25, 26], and thus, it remains to be solved which LPA receptors are involved and how ATX regulates embryonic vasculature in the early developmental stages.

As stated earlier, zebrafish has two ATX orthologues, both of which have been shown to have lysophospholipase D activity to produce LPA. To examine the development of embryonic blood vessels called intersegmental vessels (ISVs) in zebrafish

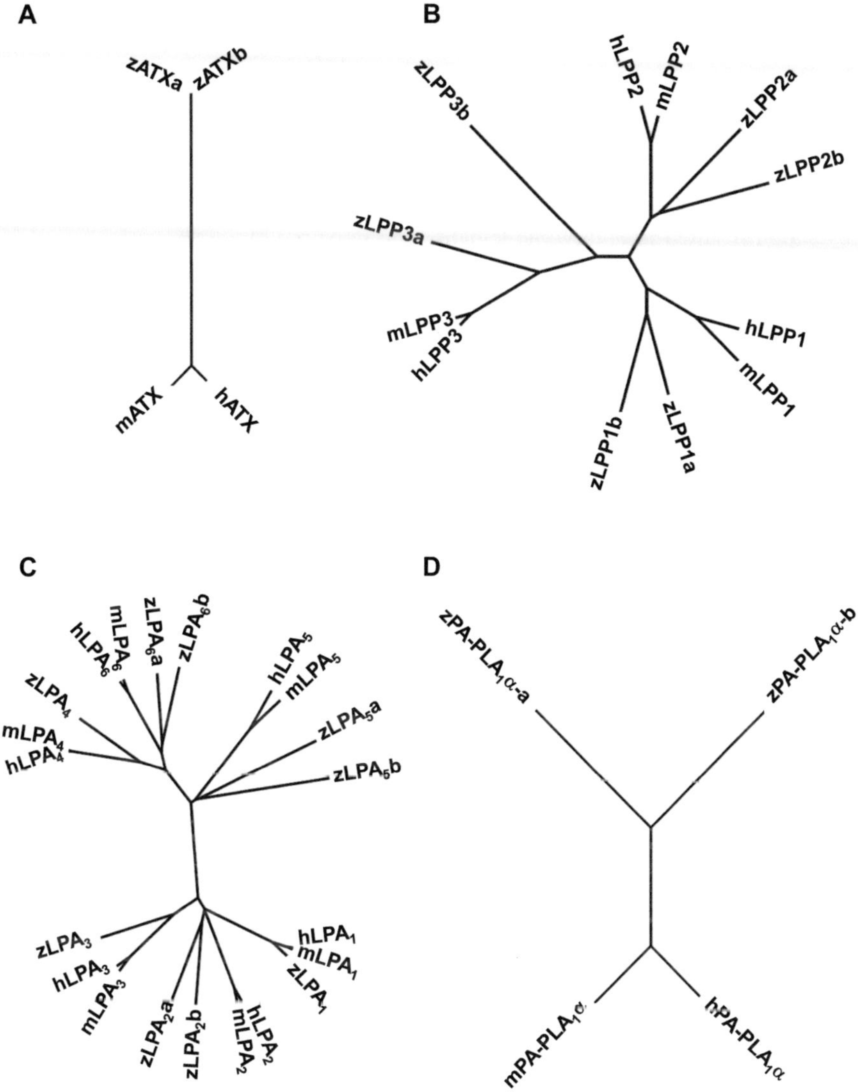

Fig. 14.1 Phylogenetic relationships of lysophosphatidic acid (LPA)-related genes [*ATX* (**a**), *LPP* (**b**), *LPAR* (**c**), and *PAPLA1A* (**d**)] based on analyses of individual genes from zebrafish, mouse, and human. Trees were all inferred using GENETYC-MAC Ver. 13.1.1

embryos, a transgenic line was used in which endothelial cells were labeled with EGFP [27]. Injection of embryos with ATX MO caused ISVs to stall in mid-course and to aberrantly connect to neighboring ISVs. The aberrant vascular network in ATX-downregulated embryos is not caused by abnormal proliferation of endothelial cells, because endothelial cells are differentiated and the number of the cells was normal. It should be stressed here that the zebrafish system makes it possible to precisely analyze blood vessel formation, which is very difficult in mice. Another important point was that the ISV phenotype has not been reported so far, indicating that the ATX–LPA system was a novel axis that regulates the embryonic blood vessel formation.

Knocking out LPA receptors in mice revealed the cellular processes specific to each of six LPA receptors, from brain development (LPA_1) to hair follicle formation (LPA_6). None of the individual knockouts was lethal. As already stated, ATX downregulation resulted in embryonic lethality and impaired blood vessel formation in both mice and zebrafish. Thus, it is possible that multiple LPA receptors redundantly regulate the embryonic blood vessel formation, or that novel LPA receptor(s) are involved. To suppress multiple LPA genes at a time in mice by crossing mice in which different LPA receptors are knocked out would require much time and labor. However, injecting zebrafish with MOs makes it possible to suppress multiple genes at a time. Simultaneous downregulation of multiple LPA receptors in zebrafish embryos revealed that LPA receptors have a redundant function in embryonic blood vessel formation. Downregulation of *lpa1* and *lpa4* caused abnormalities of blood vessel formation similar to those caused by *atx* downregulation. The phenotypic similarity strongly suggests that the LPA receptors and ATX act in the same axis governing embryonic blood vessel formation.

14.2.3.2 Neural Development and Regulation of Left–Right Asymmetry

Because zebrafish embryos with a partially established vascular system can develop for 7 days, other roles of ATX were uncovered by gene knockdown experiments using MOs. ATX is secreted by cells from the floor plate of the hindbrain and stimulates olig2-positive progenitor cells to differentiate into oligodendrocyte progenitors [28]. Dorsal forerunner cells (DFCs) regulate the formation of the central organ for establishing L-R asymmetry in zebrafish, called Kupffer's vesicle (KV). ATX–LPA_3 receptor signaling was found to induce calcium fluxes in DFCs, indicating that LPA is a regulator of L-R asymmetry in zebrafish embryos [29]. Our preliminary data suggest that ATX is also involved in the development of cartilage as ATX knockdown results in malformation of cartilage in zebrafish embryos.

14.2.3.3 Crosstalk Between LPA and S1P Signaling Revealed by Overexpression of Autotaxin in Zebrafish Embryos

Nakanaga et al. accidentally found that when ATX was overexpressed in zebrafish embryos by injecting atx mRNA, the embryos showed cardia bifida, a phenotype induced by downregulation of S1P signaling [30]. A similar cardiac phenotype was not induced when catalytically inactive ATX was introduced. The cardiac phenotype was synergistically enhanced when MOs against S1P receptor (s1pr2/mil) or S1P transporter (spns2) were introduced together with atx mRNA. The Atx-induced cardia bifida was prominently suppressed when embryos were treated with an MO against LPA_1. Thus, the study provided the first in vivo evidence of crosstalk between LPA and S1P signaling.

14.2.3.4 Zebrafish as a Tool for Drug Development

We have also tried to use the zebrafish system to evaluate small compounds for drug development. When zebrafish embryos injected with ATX mRNA were treated with an LPA receptor antagonist (Ki16425) (by just adding the compound to water in 96-well plates in which the embryos develop), it dramatically suppressed the cardia bifida phenotype [30]. The LPA antagonist was found to be active against zebrafish LPA receptors. However, our compounds that had ATX-inhibitory activity did not inhibit zebrafish ATX. Interestingly, overexpression of mammalian ATX instead of zebrafish ATX in zebrafish embryos induced a similar cardia bifida phenotype, and the phenotypes were efficiently suppressed by some of our ATX inhibitors specific for mammalian ATX. It should be noted that only a small fraction of such compounds suppressed the phenotype, even though all the compounds efficiently suppressed the ATX activity in a test tube. Such compounds were also found to be effective in vivo in mice. Thus our preliminary trial indicated that the zebrafish system is a powerful tool for in vivo evaluation of small compounds. Because the evaluation can be performed in 96-well plates, only small amounts of compounds are required, which makes it possible to evaluate the compounds in a chemical library in a first or second screening.

References

1. Contos JJ, Fukushima N, Weiner JA, Kaushal D, Chun J (2000) Requirement for the lpA1 lysophosphatidic acid receptor gene in normal suckling behavior. Proc Natl Acad Sci U S A 97:13384–13389
2. Ye X et al (2005) LPA3-mediated lysophosphatidic acid signalling in embryo implantation and spacing. Nature 435:104–108
3. Sumida H et al (2010) LPA4 regulates blood and lymphatic vessel formation during mouse embryogenesis. Blood 116:5060–5070
4. Pasternack SM et al (2008) G protein-coupled receptor P2Y5 and its ligand LPA are involved in maintenance of human hair growth. Nat Genet 40:329–334
5. Tager AM et al (2008) The lysophosphatidic acid receptor LPA1 links pulmonary fibrosis to lung injury by mediating fibroblast recruitment and vascular leak. Nat Med 14:45–54
6. Lin S et al (2009) The absence of LPA2 attenuates tumor formation in an experimental model of colitis-associated cancer. Gastroenterology 136:1711–1720
7. Lin S, Lee SJ, Shim H, Chun J, Yun CC (2010) The absence of LPA receptor 2 reduces the tumorigenesis by ApcMin mutation in the intestine. Am J Physiol Gastrointest Liver Physiol 299:G1128–G1138
8. Deng W et al (2002) Lysophosphatidic acid protects and rescues intestinal epithelial cells from radiation- and chemotherapy-induced apoptosis. Gastroenterology 123:206–216
9. Aoki J, Inoue A, Okudaira S (2008) Two pathways for lysophosphatidic acid production. Biochim Biophys Acta 1781:513–518

10. Tanaka M et al (2006) Autotaxin stabilizes blood vessels and is required for embryonic vasculature by producing lysophosphatidic acid. J Biol Chem 281:25822–25830
11. van Meeteren LA et al (2006) Autotaxin, a secreted lysophospholipase D, is essential for blood vessel formation during development. Mol Cell Biol 26:5015–5022
12. Kazantseva A et al (2006) Human hair growth deficiency is linked to a genetic defect in the phospholipase gene LIPH. Science 314:982–985
13. Inoue A et al (2011) LPA-producing enzyme PA-PLA(1)alpha regulates hair follicle development by modulating EGFR signalling. EMBO J 30:4248–4260
14. Brindley DN, Pilquil C (2009) Lipid phosphate phosphatases and signaling. J Lipid Res 50(suppl):S225–S230
15. Howe K et al (2013) The zebrafish reference genome sequence and its relationship to the human genome. Nature 496:498–503
16. Ellertsdottir E et al (2010) Vascular morphogenesis in the zebrafish embryo. Dev Biol 341:56–65
17. Corey DR, Abrams JM (2001) Morpholino antisense oligonucleotides: tools for investigating vertebrate development. Genome Biol 2: REVIEWS1015
18. Bedell VM et al (2012) In vivo genome editing using a high-efficiency TALEN system. Nature 491:114–118
19. Hwang WY et al (2013) Efficient genome editing in zebrafish using a CRISPR-Cas system. Nat Biotechnol 31:227–229
20. Yukiura H et al (2011) Autotaxin regulates vascular development via multiple lysophosphatidic acid (LPA) receptors in zebrafish. J Biol Chem 286:43972–43983
21. Fotopoulou S et al (2010) ATX expression and LPA signalling are vital for the development of the nervous system. Dev Biol 339:451–464
22. Ruppel KM et al (2005) Essential role for Galpha13 in endothelial cells during embryonic development. Proc Natl Acad Sci U S A 102:8281–8286
23. Kamijo H et al (2011) Impaired vascular remodeling in the yolk sac of embryos deficient in ROCK-I and ROCK-II. Genes Cells 16:1012–1021
24. Ferry G et al (2007) Functional invalidation of the autotaxin gene by a single amino acid mutation in mouse is lethal. FEBS Lett 581:3572–3578
25. Yang AH, Ishii I, Chun J (2002) In vivo roles of lysophospholipid receptors revealed by gene targeting studies in mice. Biochim Biophys Acta 1582:197–203
26. Lin ME, Rivera RR, Chun J (2012) Targeted deletion of LPA5 identifies novel roles for lysophosphatidic acid signaling in development of neuropathic pain. J Biol Chem 287:17608–17617
27. Lawson ND, Weinstein BM (2002) In vivo imaging of embryonic vascular development using transgenic zebrafish. Dev Biol 248:307–318
28. Yuelling LW, Waggener CT, Afshari FS, Lister JA, Fuss B (2012) Autotaxin/ENPP2 regulates oligodendrocyte differentiation in vivo in the developing zebrafish hindbrain. Glia 60:1605–1618
29. Lai SL et al (2012) Autotaxin/Lpar3 signaling regulates Kupffer's vesicle formation and left-right asymmetry in zebrafish. Development 139:4439–4448
30. Nakanaga K et al (2014) Overexpression of autotaxin, a lysophosphatidic acid-producing enzyme, enhances cardia bifida induced by hypo-sphingosine-1-phosphate signaling in zebrafish embryo. J Biochem 155:235–241

Chapter 15
Sphingosine 1-Phosphate Signaling via Transporters in Zebrafish and Mice

Yu Hisano, Tsuyoshi Nishi, and Atsuo Kawahara

Abstract The bioactive lipid mediator sphingosine 1-phosphate (S1P) plays a pivotal role in various cellular functions, such as proliferation, migration, and differentiation. S1P is intracellularly produced by sphingosine kinases and is released from the cells. Subsequently, the secreted S1P associates with S1P receptors (S1PRs) on a target cell surface, causing activation of downstream signaling pathways. The zebrafish (*Danio rerio*) is widely used as a vertebrate model organism to study the processes of organogenesis and morphogenesis. Spns2 was originally identified as an S1P transporter in zebrafish; Spns2 regulates the migration of cardiac progenitors via the S1PR2 receptor. Murine and human SPNS2 can also transport S1P from the cells. In mice, SPNS2 enables transport of S1P from vascular endothelial cells into the plasma and regulates lymphocyte egress from lymphoid organs. Recent remarkable developments in genome-editing technologies, such as transcription activator-like effector nucleases (TALENs) and the clustered regularly interspaced short palindromic repeats (CRISPR)/CRISPR-associated (Cas)9 system, allow researchers to introduce genomic modifications in various model animals. In this chapter, we review not only the physiological roles of S1P transporters in mammals and zebrafish but also the strategy for generating S1PR-knockout zebrafish using genome-editing technologies.

Keywords Sphingosine 1-phosphate (S1P) • Transporter • SPNS2 • Gene editing • TALEN • CRISPR/Cas9 system • Zebrafish

---

Y. Hisano (✉)
Laboratory for Developmental Gene Regulation, Brain Science Institute,
Riken, Wako, Saitama 351-0198, Japan
e-mail: y.hisano@riken.jp

T. Nishi
Department of Cell Membrane Biology, Institute of Scientific and Industrial Research, Osaka University, Ibaraki, Osaka 567-0047, Japan

Faculty of Pharmaceutical Science, Osaka University, Suita, Osaka 565-0871, Japan

A. Kawahara (✉)
Laboratory for Developmental Biology, Center for Medical Education and Sciences, Graduate School of Medical Science, University of Yamanashi,
Chimogatou 1110, Chuo, Yamanashi 409-3898, Japan
e-mail: akawahara@yamanashi.ac.jp

T. Yokomizo, M. Murakami (eds.), *Bioactive Lipid Mediators*,
DOI 10.1007/978-4-431-55669-5_15

15.1 Introduction

The bioactive lipid mediator sphingosine 1-phosphate (S1P) is involved in various physiological and pathological processes in immunity, allergy, and cancer [1–3]. Sphingosine kinases, SphK1 and SphK2, synthesize S1P from sphingosine via its phosphorylation. S1P is dephosphorylated to regenerate sphingosine by S1P phosphatases and extracellular lipid phosphate phosphatases, whereas S1P lyase irreversibly degrades S1P to ethanolamine phosphate and 2-hexadecenal [4]. S1P concentration may increase within a cellular compartment after various stimuli because SphK1 is located in the plasma membrane and SphK2 in the mitochondria, endoplasmic reticulum (ER), and nucleus. S1P functions not only as an intracellular second messenger [5] but also as an extracellular lipid mediator via G protein-coupled S1P receptors (S1PR1–S1PR5) on a target cell surface [6]. Thus, the system of S1P transport across the plasma membrane is necessary for intracellularly produced S1P to be recognized by S1P receptors.

Some *S1prs* (*S1pr1, S1pr2*, and *S1pr3*) are expressed in a wide variety of tissues in mice [7, 8], whereas *S1pr4* expression is detectable in the lymphoid organs and lung and *S1pr5* is expressed exclusively in the brain and skin [8]. Several types of cells are known to be S1P-releasing cells. Platelets were originally reported to release S1P in response to various stimuli, such as thrombin, collagen, and ADP [9]. It was reported that S1P released from platelets has an important role in maintaining vascular integrity during an immune response [10]. S1P release from platelets is inhibited by staurosporine (a protein kinase C inhibitor), glyburide (an ABCA1 transporter inhibitor), and R5421 (a phospholipid scramblase inhibitor) [11]. Furthermore, erythrocytes, endothelial cells, and mast cells were also reported to have the ability to release S1P [12–18].

S1P abundantly exists in blood plasma at concentrations ranging from nanomolar to micromolar. Because S1P is actively degraded in blood plasma, a high plasma S1P level should be maintained via continuous S1P supply from the S1P-producing cells. Both erythrocytes and endothelial cells are sources of plasma S1P [19], and S1P release from endothelial cells is mediated by the S1P transporter SPNS2 [20]. Spns2 was identified as the first physiological S1P transporter by the analysis of zebrafish mutant as described next [21]. The zebrafish is widely used as a vertebrate model organism for studies on molecular mechanisms of organogenesis because the embryo is transparent and develops rapidly. In this chapter, we review the physiological roles of S1P transporters in mammals and zebrafish, and describe the methods for analyzing S1P transporters and S1P signaling-related molecules in zebrafish using genome-editing technologies.

15.2 Functions of S1P Transporters in Mammals

The ATP-binding cassette (ABC) transporter superfamily, which is structurally conserved from prokaryotes to eukaryotes, is one of the active transport systems in the cell. Various hydrophobic and amphiphilic compounds can serve as substrates for

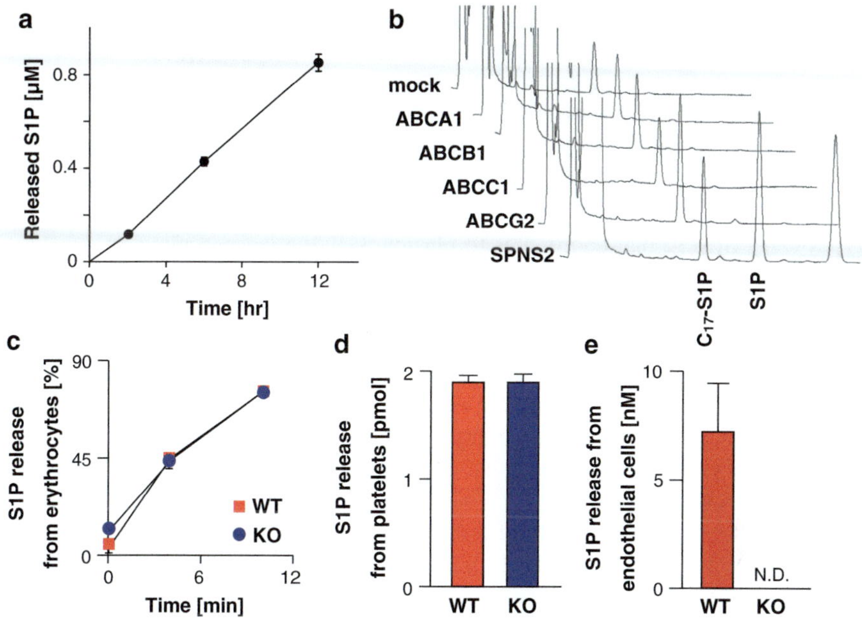

Fig. 15.1 (**a**) Chinese hamster ovary (CHO) cells expressing mouse SphK1 and human SPNS2 were incubated for 2, 6, or 12 h. The amounts of released S1P were measured using high-performance liquid chromatography (HPLC). (**b**) CHO/SphK1 cells (mock) or CHO/SphK1 cells expressing mouse ABCA1, human ABCB1, mouse ABCC1, human ABCG2, or human SPNS2 were incubated for 2 h after sphingosine stimulation (5 μM). Amounts of released S1P were measured using HPLC. C_{17}-S1P served as an internal standard [27]. (**c**) Erythrocytes from wild-type (WT) (*squares*) or *Spns2*-KO mice (*circles*) were incubated with [^{3}H]sphingosine for 0, 4, or 10 min and the [^{3}H]S1P released into the medium was quantified. (**d**) Platelets from WT or *Spns2*-KO mice were incubated for 10 min in the presence of thrombin, and the S1P released into the medium was measured using ultraperformance liquid chromatography with tandem mass spectrometry (UPLC–MS/MS). (**e**) Vascular endothelial cells from the aorta of WT or *Spns2*-KO mice were incubated for 4 h and the S1P released into the medium was measured using UPLC–MS/MS. In *Spns2*-KO endothelial cells, S1P in the medium was not detected (*N.D.*) [20]

several types of ABC transporters that utilize ATP hydrolysis as an energy source. ABCA1, ABCB1, ABCC1, and ABCG2 were independently reported to be involved in S1P release from several cell lines [22–25]. Furthermore, S1P release from platelets and erythrocytes is ATP dependent and is inhibited by glyburide (an ABCA1 transporter inhibitor) [11, 26]. As already described, SPNS2 functions physiologically as an S1P transporter [21] and belongs to the major facilitator superfamily (MFS) but not the ABC transporter superfamily. In Chinese hamster ovary (CHO) cells expressing SphK1, S1P is effectively synthesized and accumulated because of the lack of an S1P export system. The introduction of SPNS2 into these cells induces S1P release (Fig. 15.1). In contrast, introduction of ABCA1, ABCB1, ABCC1, or ABCG2 does not stimulate the release of S1P, indicating that SPNS2 is an exclusive S1P transporter capable of releasing S1P from the cells by itself (Fig. 15.1) [27]. If these ABC transporters are indeed involved in S1P release, they may require some type of modification or cofactor(s) for successful transport of S1P.

Consistent with the results from the cell culture system, *Spns2*-knockout (KO) mice exhibit a decreased plasma S1P level, at approximately 60 % of the level in wild-type (WT) mice, whereas the plasma S1P level of *Abca1*-KO or *Abcc1*-KO mice is not altered [14, 20, 28]. These data indicate that SPNS2 enables transport of S1P from S1P-producing cells into the plasma. One important question is which cell types carry SPNS2 that contributes to plasma S1P. We isolated platelets, erythrocytes, and endothelial cells from *Spns2*-KO mice and separately measured the release of S1P by these cells. S1P release from *Spns2*-KO endothelial cells was completely abolished, whereas this activity was not altered in *Spns2*-KO platelets and erythrocytes (Fig. 15.1) [20]. In line with this observation, a knockdown of *SPNS2* by siRNA suppressed the S1P secretion in human umbilical vein endothelial cells (HUVECs) and human pulmonary artery endothelial cells (HPAECs) [20]. In fact, SPNS2 expression is detectable in the heart and lung endothelial cells by means of an antibody [29].

The most remarkable feature of *Spns2*-KO mice is the depletion of peripheral blood of circulating lymphocytes [20, 28–31]. S1P signaling via S1PR1 regulates the lymphocyte egress from lymphoid organs, such as the thymus, bone marrow, and lymph nodes, into peripheral blood and lymph. Precursors from the bone marrow migrate to the thymus and mature into T cells in the medullary portion. S1PR1 expression is upregulated in the mature T cells, enabling them to exit from the thymus [32]. S1P is believed to be a key regulator of the egress of mature T cells via the activation of S1PR1. In the thymus of *Spns2*-KO mice, the population of mature T cells is increased and that of immature T cells is decreased [20]. Mature T cells isolated from *Spns2*-KO mice exhibit higher *S1pr1* mRNA expression and stronger migration mediated by S1P compared with WT cells [20]. These results suggest that thymocytes of *Spns2*-KO mice can normally mature and migrate toward S1P, but they cannot move from the thymus into the blood because of the lack of an SPNS2-mediated S1P transport system (Fig. 15.2). Consequently, the number of circulating T cells in the peripheral blood of *Spns2*-KO mice is significantly reduced. Thus, SPNS2 could be a novel target of immunosuppressive agents because SPNS2 participates in the regulation of the number of circulating T cells.

15.3 S1P Signaling in Zebrafish

Zebrafish are known as small tropical fish (4–5 cm long) native to India. A zebrafish embryo is transparent and its development progresses rapidly. Most of the organs start to function within several days after fertilization. One important point is that organogenesis and morphogenesis are well conserved between zebrafish and mammals; therefore, zebrafish is a convenient vertebrate model organism. A pair of zebrafish produces approximately 100–200 eggs, allowing researchers to perform genetic analysis such as screening of zebrafish mutants produced by random mutagenesis. In the forward genetic analysis, *s1pr2* was identified as a gene responsible

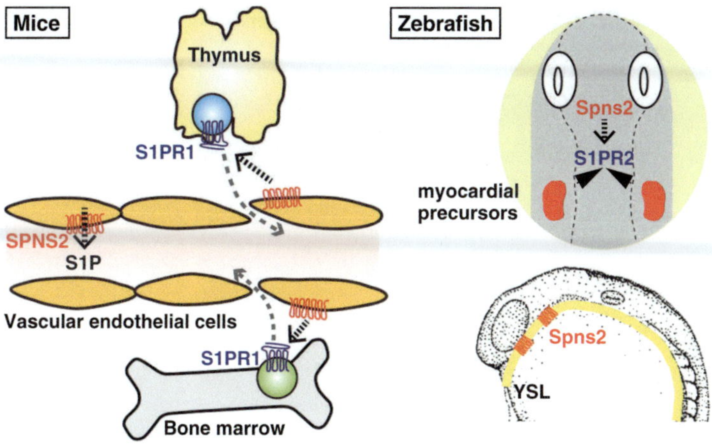

Fig. 15.2 Physiological roles of SPNS2. In mice, SPNS2 supplies S1P from vascular endothelial cells to plasma, and regulates lymphocyte egress from lymphoid organs, such as the thymus and bone marrow. S1P released by SPNS2 is believed to be recognized by S1PR1 on the surface of lymphocytes. In zebrafish, S1P released by Spns2 regulates cardiac progenitor migration by S1PR2-mediated signaling. *spns2* is strongly expressed in the yolk syncytial layer (YSL) [21]

for the *miles apart* mutation resulting in two hearts (*cardia bifida*) and tail blisters [33]. In vertebrates, cardiac progenitors are bilaterally located in the lateral plate mesoderm and migrate toward the midline where they merge to form a single primitive heart tube. Although *s1pr2* is expressed in the endoderm and cardiomyocytes, *s1pr2* mutants exhibit morphological defects in the anterior endoderm, which is necessary for cardiac progenitor migration [33]. S1PR2 is believed to be a partner of $G_{12/13}$, G_i, and G_q heterotrimeric G protein that activates their downstream signaling. In zebrafish, G_{13} was shown to be downstream of S1PR2 and to regulate migration of cardiac progenitors via a RhoGEF-dependent pathway [34]. Transplantation of endoderm deficient in either S1PR2 or G_{13} into WT embryos results in the development of two hearts; conversely, WT donor cells reverse the defects of a G_{13} knockdown [34]. These results suggest that activation of S1PR2 signaling in the endoderm is indispensable for the migration of cardiac progenitors.

Recently, Stainier's group and ours independently identified other *cardia bifida* mutants (*two of hearts* and *ko157*), which have two hearts and tail blisters identical to those of the *s1pr2* mutant (Fig. 15.3) [21, 35]. Both groups reported that the gene responsible for the mutations encodes a novel membrane protein Spns2. We demonstrated that Spns2 functions as an S1P transporter, as already mentioned. *spns2* transcripts are easily detectable in an extraembryonic tissue called the yolk syncytial layer (YSL), which is located immediately underneath the migrating cardiac progenitors [21]. A knockdown of *spns2* by antisense morpholino oligonucleotides (MOs) in the YSL at the shield stage induced *cardia bifida*; the defects of the *spns2* mutant were reversed by injection of *spns2* mRNA into the YSL at the shield stage [21]. These data suggest that Spns2 in the YSL is essential for the migration of cardiac progenitors (Fig. 15.2). Nevertheless, it is still unclear how S1P supplied from YSL is detected by S1PR2 in the endoderm or mesoderm.

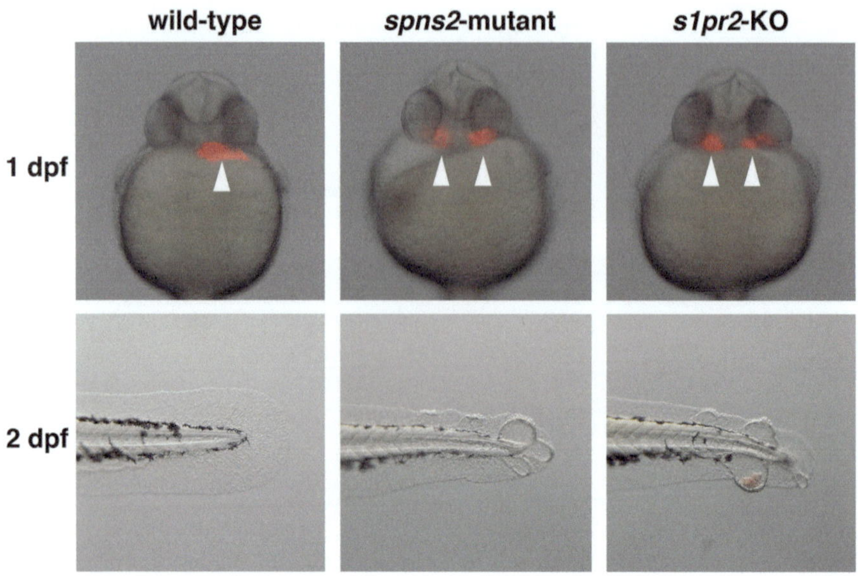

Fig. 15.3 Each *spns2*-mutant or *s1pr2*-KO embryo had two hearts (*cardia bifida*) and tail blisters. The positions of the hearts in wild-type, *spns2*-mutant, and *s1pr2*-KO embryo at 1 day post fertilization (*1 dpf*) are indicated by *white arrowheads*. Cardiac morphology (*ventral* view) was visualized using monomeric red fluorescent protein (*mRFP*) expression driven by the cardiac-specific promoter of *cmlc2* (gene of cardiac myosin light chain 2). Each *spns2*-mutant or *s1pr2*-KO embryo exhibited tail blisters at 2 dpf (*lateral* view)

S1P signaling via Spns2 and S1PR2 was reported to be involved in lower-jaw development [36]. S1PR2 in the endoderm is essential for lower-jaw development just as is S1PR2 for the migration of cardiac progenitors; however, the origin of S1P as supplied by Spns2 is unclear. We found that the Spns2–S1PR2 signaling pathway interacts with the cell adhesion molecule fibronectin during both cardiac and lower-jaw development [37]. Furthermore, expression of *endothelin receptor A* is down-regulated in the *spns2* mutant [37], suggesting that Spns2–S1PR2 signaling affects endothelin signaling involved in lower-jaw development. *In situ* hybridization analysis revealed that *spns2* mRNA is expressed in somites, myocardial precursors, and at the tip of the tail in addition to the YSL; these results indicate that Spns2 may participate in regulation of other types of organogenesis [21].

15.4 Reverse Genetics in Zebrafish

During the past few years, tremendous innovations took place in the field of genome-editing technologies, such as zinc-finger nucleases (ZFNs), transcription activator-like effector nucleases (TALENs), and the clustered regularly interspaced short palindromic repeats (CRISPR)/CRISPR-associated (Cas)9 system. These genome-editing technologies enable researchers to perform reverse genetic analysis in a variety of model organisms including zebrafish [38, 39]. TALENs are chimeric proteins consisting of a TALE domain for DNA recognition and a *Fok*I nuclease catalytic domain. Because *Fok*I functions as a dimer, a pair of TALENs causes a double-strand break (DSB) in the region between forward and reverse recognition sites of TALEN. However, the CRISPR/Cas9 system consists of a guide RNA (gRNA) and the nuclease Cas9. The gRNA contains a complementary sequence (20 nucleotides) targeting a genomic region next to a protospacer-adjacent motif (PAM). RNA-guided Cas9 recognizes the complex of gRNA and the target genomic sequence, resulting in DSB within the target site.

Targeted DSB induced by genome-editing technologies are repaired by nonhomologous end-joining (NHEJ) or homology-directed repair (HDR). In NHEJ, small insertion or deletion (indel) mutations are frequently introduced at the breakage locus, resulting in frameshift-mediated gene inactivation. HDR involves a donor template that contains a sequence homologous to the target locus and allows for precise genomic modification via replacement with the homologous fragment. We have developed a simple assay for measuring the genome-editing activity [40, 41]; thus, generation of KO zebrafish has become feasible and relatively easy.

Before the advent of genome-editing technologies, knockdown analysis using MO was a popular approach in studies of the function of a gene of interest. There are several disadvantages of MOs; these molecules are effective transiently because of their degradation and MOs have nonspecific effects such as activation of p53-mediated apoptosis [42]. Although it was reported that an *s1pr1* knockdown with MOs caused a loss of blood circulation and vascular defects [43–45], we found that *s1pr1*-KO zebrafish exhibit normal blood circulation without any obvious vascular defects [37]. This finding shows that analysis of KO zebrafish generated using the genome-editing technologies is a more accurate method for loss-of-function analysis. We also found that the phenotype of *s1pr2*-KO zebrafish generated using TALEN technology is identical to the phenotypes of previously described mutants, two hearts and tail blisters [37]. Thus, genome editing is a straightforward approach to studies of physiological functions of genes during development.

15.5 Conclusion

Physiologically important roles of S1P signaling have been elucidated in *in vitro* biochemical and *in vivo* biological studies, and the identification of the S1P transporter SPNS2 leads to a better understanding of lipid mediator signaling. An intercellular signaling pathway is typically composed of (1) production of signaling molecules, (2) secretion from the cells, and (3) recognition by receptors on the surface of target cells. Compared with the production and recognition of S1P by receptors, its secretion has been poorly understood, in spite of its importance in determining the spatiotemporal distribution of S1P. Analysis of *Spns2*-KO mice and *spns2*-mutant zebrafish revealed the physiological roles of S1P signaling mediated by the S1P transporter. Development of genome-editing technologies is essential for the discovery of new powerful tools for analysis of complex signaling processes. There are still some unidentified S1P transporters in erythrocytes and platelets; therefore, identification of such S1P transporters should help researchers to comprehend the whole set of events in the S1P signaling pathway.

15.6 Methods

15.6.1 Isolation of Mouse Platelets, Erythrocytes, and Endothelial Cells

Platelets and erythrocytes were isolated from whole blood, which was collected from mouse hearts using an acid citrate–dextrose solution as an anticoagulant. Platelet-rich plasma (PRP) was separated by centrifugation at 100 *g* for 15 min at room temperature. Platelets were prepared by centrifugation of PRP at 1000 *g* for 15 min at room temperature and were washed with buffer A (20 mM HEPES-NaOH pH 7.4, 3.3 mM NaH_2PO_4, 2.9 mM KCl, 1 mM $MgCl_2$, 138 mM NaCl, and 1 mg/ml glucose) containing 1 % bovine serum albumin (BSA). In the case of erythrocytes, whole blood was centrifuged at 100 *g* for 15 min at room temperature and the pelleted erythrocytes were washed twice with buffer A containing 1 % BSA.

Endothelial cells were isolated from the mouse aorta, according to the previously reported method with some modifications [46]. The mouse aorta was cut at the lower level of the abdominal aorta to release the blood and was perfused with phosphate-buffered saline (PBS) containing heparin (1000 units/ml). Subsequently, the aorta was excised between the aortic arch and the abdominal aorta and immersed in 20 % fetal bovine serum (FBS)–Dulbecco's modified Eagle's medium (DMEM) containing 100 units/ml heparin, 100 units/ml penicillin G, and 100 μg/ml streptomycin. A 24-gauge cannula was inserted into the proximal portion and the distal end was closed with a silk thread. Subsequently, the aorta was briefly washed with serum-free DMEM and filled with a solution of type II collagenase (2 mg/ml). After incubation for 45 min at 37 °C, endothelial cells were removed from the aorta by flushing it with 2 ml 20 % FBS–DMEM. The collected endothelial cells were incubated with 2 ml 20 % FBS–DMEM and cultured in a plate coated with type I collagen. After incubation for 2 h at 37 °C, the cells were washed with warmed 20 % FBS–DMEM to remove smooth muscle cells, and cultured in the G medium [20 % FBS, 100 units/ml penicillin G, 100 μg/ml streptomycin, 2 mM L-glutamine, 1× nonessential amino acids, 1× sodium pyruvate, 25 mM HEPES pH 7.0–7.6, 100 μg/ml heparin, 100 μg/μl endothelial cell growth supplement (ECGS), and DMEM] until confluent state.

15.6.2 Measurement of S1P Secretion

To measure the S1P-transporting activity, we established a CHO cell line stably expressing SphK1 (CHO/SphK1), in which S1P was effectively produced and accumulated inside the cells. Although CHO/SphK1 cells did not have the ability to release S1P, introduction of an exogenous S1P transporter caused S1P release into the medium. CHO/SphK1 cells expressing an S1P transporter candidate were washed with serum-free F-12 medium twice to remove S1P present in FBS and were incubated at 37 °C with the F-12 medium supplemented with 1 % BSA, 10 mM sodium glycerophosphate, 5 mM sodium fluoride, and 1 mM semicarbazide. After incubation for the indicated time periods, aliquots of the medium (200 μl) were collected and the detached cells were removed by centrifugation at 12,000 *g* for 5 min at 4 °C. The amount of released S1P was measured using high-performance liquid chromatography (HPLC).

A suspension of platelets (190 μl of 10^8 cells/ml) in buffer A containing 1 % BSA was incubated for 10 min at 37 °C. Thereafter, thrombin (10 μl; final concentration, 5 units/ml) was added to the mixture, followed by incubation for 10 min. Subsequently, the platelets and the medium were separated by centrifugation at 12,000 *g* for 5 s at 4 °C. An equal volume of methanol was added to the supernatant, which was centrifuged at 12,000 *g* for 5 min at 4 °C, and the resulting supernatant was applied to a Cosmospin filter G and analyzed using ultraperformance liquid chromatography with tandem mass spectrometry (UPLC–MS/MS).

S1P release from erythrocytes was measured using tritium-labeled sphingosine. A suspension of erythrocytes (190 μl of 10^7 cells/ml) in buffer A containing 1 % BSA was incubated for 5 min at 37 °C. Subsequently, [^{3}H]sphingosine (0.2 μM; 40 nCi per 10 μl) was added (final concentration of sphingosine, 10 nM). After incubation for the indicated time periods, the erythrocytes and the medium were separated by centrifugation at 12,000 *g* for 5 s at 4 °C. Lipids were extracted from the erythrocytes and from the supernatant by the Bligh–Dyer method [47] and then dried and resuspended in chloroform:methanol (2:1 v/v), followed by HPTLC in butanol/acetic acid/water (3:1:1 v/v). Radioactivity of the bands was quantified using the FLA-3000G Bioimaging Analyzer.

The isolated endothelial cells were cultured until confluent state. The cells were washed with a serum-free medium, and the culture medium was replaced with a release medium (endothelial cell serum-free defined medium consisting of 1 % BSA, 10 mM sodium glycerophosphate, 5 mM sodium fluoride, 1 mM semicarbazide, and 20 mM HEPES-KOH pH 7.4). After incubation for 4 h at 37 °C, aliquots (200 μl) of the release medium were taken, and the detached cells were removed by centrifugation at 12,000 *g* for 5 min at 4 °C. The amount of S1P in the release medium was measured using UPLC–MS/MS.

15.6.3 Establishment of KO Zebrafish Using TALENs or gRNA Plus Cas9

The strategy for generating KO zebrafish is shown in Fig. 15.4. First, genomic DNA was isolated from 1 dpf (day post fertilization) embryos injected with TALENs or gRNA plus Cas9. The genome-editing activity was evaluated via generation of heteroduplex bands in addition to homoduplex bands during polyacrylamide gel electrophoresis (heteroduplex mobility assay, HMA). Next, potential F_0 founders were grown to adulthood and mated with WT zebrafish. The germline transmission of indels was confirmed by the formation of heteroduplex bands. F_1 zebrafish containing a mutant allele were identified by genotyping of genomic DNA prepared from

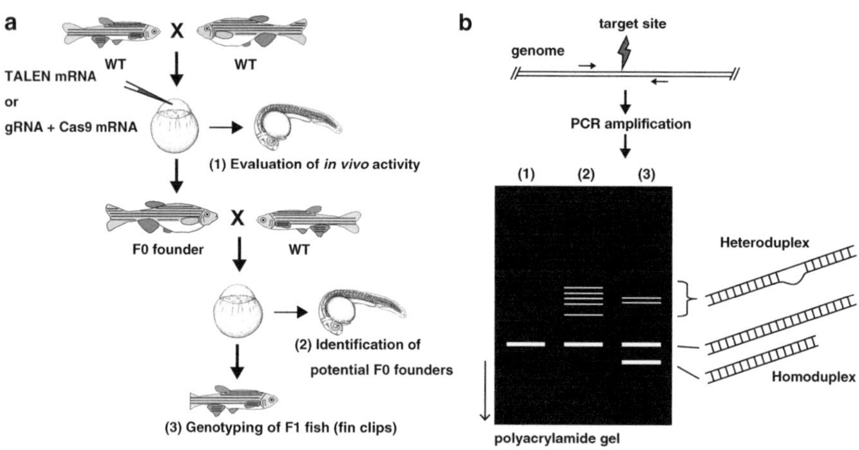

Fig. 15.4 (**a**) Strategy for preparation of KO zebrafish using genome-editing technologies. (*1*) Zebrafish genomic DNA samples were prepared from 1 day post fertilization (dpf) embryos injected with TALEN mRNA or gRNA plus Cas9 mRNA. Genome-editing activity was evaluated as generation of insertion and/or deletion (indel) mutations in a heteroduplex mobility assay (HMA). (*2*) Potential F_0 founders that were capable of producing indel mutations were grown to adulthood. We confirmed the germline transmission of indel mutations in F_1 embryos by mating the F0 founders with wild-type (*WT*) fish. (*3*) F_1 zebrafish containing mutant alleles were identified by genotyping of genomic DNA samples prepared from fin clips. At these three steps, we used HMA to detect indel mutations induced by TALENs or gRNA plus Cas9. (**b**) The principle of HMA. A genomic target region was amplified by means of PCR with locus-specific primers. During polyacrylamide gel electrophoresis, homoduplexes are separated by molecular weight, whereas heteroduplexes containing a mismatched region migrate more slowly than the homoduplexes because of the opened single-strand configuration. (*1*) PCR products from a WT genome contain one homoduplex band of the expected size. (*2*) Among the PCR products from F_0 embryos injected with TALENs or gRNA plus Cas9, ladder bands (heteroduplex bands) appeared above the homoduplex band because of various types of indel mutations in somatic cells. (*3*) When F_0 founder containing a deletion allele is mated with WT zebrafish, two homoduplex bands and two heteroduplex bands among PCR products should appear in F_1 embryos

fin clips. By in-crossing of F_1 zebrafish containing the mutant allele heterogeneously, we were able to analyze the morphological phenotypes of KO zebrafish, in which a target gene was zygotically inactivated.

As described earlier, we utilized HMA to detect genome modifications induced by TALENs or gRNA plus Cas9 [41]. TALEN- or gRNA/Cas9-injected embryos (1 dpf) were collected and incubated in lysis buffer (10 mM Tris-HCl pH 8.0, 0.1 mM EDTA, 0.2 % Triton X-100, and 0.2 mg/ml proteinase K) for more than 2 h at 55 °C, followed by the incubation for 10 min at 99 °C. After centrifugation (12,000 *g*, 5 min), the supernatant was used as a source of genomic DNA. Subsequently, 100- to 200-bp fragments containing the target genomic locus were amplified by polymerase chain reaction (PCR) with locus-specific primers. Homoduplexes from the WT genome were separated by molecular weight using polyacrylamide gel electrophoresis; heteroduplexes containing a mismatched region from the heterozygous mutant genome migrated more slowly than the homoduplexes because of an opened single-strand configuration (Fig. 15.4).

Acknowledgements We thank Drs. S. Ota, N. Kobayashi, and A. Yamaguchi for valuable discussion. This work was supported by the Funding Program for Next Generation World-Leading Researchers (NEXT Program) and by the Japan Society for the Promotion of Science (JSPS).

References

1. Wymann MP, Schneiter R (2008) Lipid signalling in disease. Nat Rev Mol Cell Biol 9(2):162–176. doi:10.1038/nrm2335
2. Brinkmann V (2007) Sphingosine 1-phosphate receptors in health and disease: mechanistic insights from gene deletion studies and reverse pharmacology. Pharmacol Ther 115(1):84–105. doi:10.1016/j.pharmthera.2007.04.006
3. Hisano Y, Nishi T, Kawahara A (2012) The functional roles of S1P in immunity. J Biochem 152(4):305–311. doi:10.1093/jb/mvs090
4. Strub GM, Maceyka M, Hait NC, Milstien S, Spiegel S (2010) Extracellular and intracellular actions of sphingosine-1-phosphate. Adv Exp Med Biol 688:141–155. doi:10.1007/978-1-4419-6741-1_10
5. Olivera A, Spiegel S (1993) Sphingosine-1-phosphate as second messenger in cell proliferation induced by PDGF and FCS mitogens. Nature 365(6446):557–560. doi:10.1038/365557a0
6. Kihara A, Mitsutake S, Mizutani Y, Igarashi Y (2007) Metabolism and biological functions of two phosphorylated sphingolipids, sphingosine 1-phosphate and ceramide 1-phosphate. Prog Lipid Res 46(2):126–144. doi:10.1016/j.plipres.2007.03.001
7. Chae SS, Proia RL, Hla T (2004) Constitutive expression of the S1P1 receptor in adult tissues. Prostaglandins Other Lipid Mediat 73(1-2):141–150. doi:10.1016/j.prostaglandins.2004.01.006
8. Ishii I, Friedman B, Ye X, Kawamura S, McGiffert C, Contos JJ, Kingsbury MA, Zhang G, Brown JH, Chun J (2001) Selective loss of sphingosine 1-phosphate signaling with no obvious phenotypic abnormality in mice lacking its G protein-coupled receptor, LP(B3)/EDG-3. J Biol Chem 276(36):33697–33704. doi:10.1074/jbc.M10441200
9. Yatomi Y, Ruan F, Hakomori S, Igarashi Y (1995) Sphingosine-1-phosphate: a platelet-activating sphingolipid released from agonist-stimulated human platelets. Blood 86(1):193–202

10. Herzog BH, Fu J, Wilson SJ, Hess PR, Sen A, McDaniel JM, Pan Y, Sheng M, Yago T, Silasi-Mansat R, McGee S, May F, Nieswandt B, Morris AJ, Lupu F, Coughlin SR, McEver RP, Chen H, Kahn ML, Xia L (2013) Podoplanin maintains high endothelial venule integrity by interacting with platelet CLEC-2. Nature 502(7469):105–109. doi:10.1038/nature12501
11. Kobayashi N, Nishi T, Hirata T, Kihara A, Sano T, Igarashi Y, Yamaguchi A (2006) Sphingosine 1-phosphate is released from the cytosol of rat platelets in a carrier-mediated manner. J Lipid Res 47(3):614–621. doi:10.1194/jlr.M500468-JLR200
12. Hanel P, Andreani P, Graler MH (2007) Erythrocytes store and release sphingosine 1-phosphate in blood. FASEB J 21(4):1202–1209. doi:10.1096/fj.06-7433com
13. Prieschl EE, Csonga R, Novotny V, Kikuchi GE, Baumruker T (1999) The balance between sphingosine and sphingosine-1-phosphate is decisive for mast cell activation after Fc epsilon receptor I triggering. J Exp Med 190(1):1–8. doi:10.1084/jem.190.1.1
14. Lee Y-M, Venkataraman K, Hwang S-I, Han DK, Hla T (2007) A novel method to quantify sphingosine 1-phosphate by immobilized metal affinity chromatography (IMAC). Prostaglandins Other Lipid Mediat 84(3-4):154–162. doi:10.1016/j.prostaglandins.2007.08.001
15. Bassi R, Anelli V, Giussani P, Tettamanti G, Viani P, Riboni L (2006) Sphingosine-1-phosphate is released by cerebellar astrocytes in response to bFGF and induces astrocyte proliferation through Gi-protein-coupled receptors. Glia 53(6):621–630. doi:10.1002/glia.20324
16. Anelli V, Bassi R, Tettamanti G, Viani P, Riboni L (2005) Extracellular release of newly synthesized sphingosine-1-phosphate by cerebellar granule cells and astrocytes. J Neurochem 92(5):1204–1215. doi:10.1111/j.1471-4159.2004.02955.x
17. Nieuwenhuis B, Luth A, Chun J, Huwiler A, Pfeilschifter J, Schafer-Korting M, Kleuser B (2009) Involvement of the ABC-transporter ABCC1 and the sphingosine 1-phosphate receptor subtype S1P(3) in the cytoprotection of human fibroblasts by the glucocorticoid dexamethasone. J Mol Med 87(6):645–657. doi:10.1007/s00109-009-0468-x
18. Tanfin Z, Serrano-Sanchez M, Leiber D (2011) ATP-binding cassette ABCC1 is involved in the release of sphingosine 1-phosphate from rat uterine leiomyoma ELT3 cells and late pregnant rat myometrium. Cell Signal 23(12):1997–2004. doi:10.1016/j.cellsig.2011.07.010
19. Pappu R, Schwab SR, Cornelissen I, Pereira JP, Regard JB, Xu Y, Camerer E, Zheng Y-W, Huang Y, Cyster JG, Coughlin SR (2007) Promotion of lymphocyte egress into blood and lymph by distinct sources of sphingosine-1-phosphate. Science 316(5822):295–298. doi:10.1126/science.1139221
20. Hisano Y, Kobayashi N, Yamaguchi A, Nishi T (2012) Mouse SPNS2 functions as a sphingosine-1-phosphate transporter in vascular endothelial cells. PLoS One 7(6), e38941. doi:10.1371/journal.pone.0038941
21. Kawahara A, Nishi T, Hisano Y, Fukui H, Yamaguchi A, Mochizuki N (2009) The sphingolipid transporter spns2 functions in migration of zebrafish myocardial precursors. Science 323(5913):524–527. doi:10.1126/science.1167449
22. Sato K, Malchinkhuu E, Horiuchi Y, Mogi C, Tomura H, Tosaka M, Yoshimoto Y, Kuwabara A, Okajima F (2007) Critical role of ABCA1 transporter in sphingosine 1-phosphate release from astrocytes. J Neurochem 103(6):2610–2619. doi:10.1111/j.1471-4159.2007.04958.x
23. Honig SM, Fu S, Mao X, Yopp A, Gunn MD, Randolph GJ, Bromberg JS (2003) FTY720 stimulates multidrug transporter- and cysteinyl leukotriene-dependent T cell chemotaxis to lymph nodes. J Clin Invest 111(5):627–637. doi:10.1172/JCI16200
24. Mitra P, Oskeritzian CA, Payne SG, Beaven MA, Milstien S, Spiegel S (2006) Role of ABCC1 in export of sphingosine-1-phosphate from mast cells. Proc Natl Acad Sci U S A 103(44):16394–16399. doi:10.1073/pnas.0603734103
25. Takabe K, Kim RH, Allegood JC, Mitra P, Ramachandran S, Nagahashi M, Harikumar KB, Hait NC, Milstien S, Spiegel S (2010) Estradiol induces export of sphingosine-1-phosphate from breast cancer cells via ABCC1 and ABCG2. J Biol Chem 285(14):10477–10486. doi:10.1074/jbc.M109.064162

26. Kobayashi N, Kobayashi N, Yamaguchi A, Nishi T (2009) Characterization of the ATP-dependent sphingosine 1-phosphate transporter in rat erythrocytes. J Biol Chem 284(32):21192–21200. doi:10.1074/jbc.M109.006163
27. Hisano Y, Kobayashi N, Kawahara A, Yamaguchi A, Nishi T (2011) The sphingosine 1-phosphate transporter, SPNS2, functions as a transporter of the phosphorylated form of the immunomodulating agent FTY720. J Biol Chem 286(3):1758–1766. doi:10.1074/jbc.M110.171116
28. Fukuhara S, Simmons S, Kawamura S, Inoue A, Orba Y, Tokudome T, Sunden Y, Arai Y, Moriwaki K, Ishida J, Uemura A, Kiyonari H, Abe T, Fukamizu A, Hirashima M, Sawa H, Aoki J, Ishii M, Mochizuki N (2012) The sphingosine-1-phosphate transporter Spns2 expressed on endothelial cells regulates lymphocyte trafficking in mice. J Clin Invest 122(4):1416–1426. doi:10.1172/JCI60746
29. Mendoza A, Bréart B, Ramos-Perez WD, Pitt LA, Gobert M, Sunkara M, Lafaille JJ, Morris AJ, Schwab SR (2012) The transporter spns2 is required for secretion of lymph but not plasma sphingosine-1-phosphate. Cell Rep 2(5):1104–1110. doi:10.1016/j.celrep.2012.09.021
30. Nagahashi M, Kim EY, Yamada A, Ramachandran S, Allegood JC, Hait NC, Maceyka M, Milstien S, Takabe K, Spiegel S (2013) Spns2, a transporter of phosphorylated sphingoid bases, regulates their blood and lymph levels, and the lymphatic network. FASEB J 27(3):1001–1011. doi:10.1096/fj.12-219618
31. Nijnik A, Clare S, Hale C, Chen J, Raisen C, Mottram L, Lucas M, Estabel J, Ryder E, Adissu H, Adams NC, Ramirez-Solis R, White JK, Steel KP, Dougan G, Hancock RE (2012) The role of sphingosine-1-phosphate transporter spns2 in immune system function. J Immunol 189(1):102–111. doi:10.4049/jimmunol.1200282
32. Zachariah MA, Cyster JG (2010) Neural crest-derived pericytes promote egress of mature thymocytes at the corticomedullary junction. Science 328(5982):1129–1135. doi:10.1126/science.1188222
33. Kupperman E, An S, Osborne N, Waldron S, Stainier DY (2000) A sphingosine-1-phosphate receptor regulates cell migration during vertebrate heart development. Nature 406(6792):192–195. doi:10.1038/35018092
34. Ye D, Lin F (2013) S1pr2/Galpha13 signaling controls myocardial migration by regulating endoderm convergence. Development 140(4):789–799. doi:10.1242/dev.085340
35. Osborne N, Brand-Arzamendi K, Ober EA, Jin S-W, Verkade H, Holtzman NG, Yelon D, Stainier DYR (2008) The spinster homolog, two of hearts, is required for sphingosine 1-phosphate signaling in zebrafish. Curr Biol 18(23):1882–1888. doi:10.1016/j.cub.2008.10.061
36. Balczerski B, Matsutani M, Castillo P, Osborne N, Stainier DYR, Crump JG (2012) Analysis of sphingosine-1-phosphate signaling mutants reveals endodermal requirements for the growth but not dorsoventral patterning of jaw skeletal precursors. Dev Biol 362(2):230–241. doi:10.1016/j.ydbio.2011.12.010
37. Hisano Y, Ota S, Takada S, Kawahara A (2013) Functional cooperation of spns2 and fibronectin in cardiac and lower jaw development. Biol Open 2(8):789–794. doi:10.1242/bio.20134994
38. Hisano Y, Ota S, Kawahara A (2014) Genome editing using artificial site-specific nucleases in zebrafish. Dev Growth Differ 56(1):26–33. doi:10.1111/dgd.12094
39. Gaj T, Gersbach CA, Barbas CF 3rd (2013) ZFN, TALEN, and CRISPR/Cas-based methods for genome engineering. Trends Biotechnol 31(7):397–405. doi:10.1016/j.tibtech.2013.04.004
40. Hisano Y, Ota S, Arakawa K, Muraki M, Kono N, Oshita K, Sakuma T, Tomita M, Yamamoto T, Okada Y, Kawahara A (2013) Quantitative assay for TALEN activity at endogenous genomic loci. Biol Open 2(4):363–367. doi:10.1242/bio.20133871
41. Ota S, Hisano Y, Muraki M, Hoshijima K, Dahlem TJ, Grunwald DJ, Okada Y, Kawahara A (2013) Efficient identification of TALEN-mediated genome modifications using heteroduplex mobility assays. Genes Cells 18(6):450–458. doi:10.1111/gtc.12050

42. Robu ME, Larson JD, Nasevicius A, Beiraghi S, Brenner C, Farber SA, Ekker SC (2007) p53 activation by knockdown technologies. PLoS Genet 3(5), e78. doi:10.1371/journal.pgen.0030078
43. Ben Shoham A, Malkinson G, Krief S, Shwartz Y, Ely Y, Ferrara N, Yaniv K, Zelzer E (2012) S1P1 inhibits sprouting angiogenesis during vascular development. Development. doi:10.1242/dev.078550
44. Gaengel K, Niaudet C, Hagikura K, Lavina B, Muhl L, Hofmann JJ, Ebarasi L, Nystrom S, Rymo S, Chen LL, Pang MF, Jin Y, Raschperger E, Roswall P, Schulte D, Benedito R, Larsson J, Hellstrom M, Fuxe J, Uhlen P, Adams R, Jakobsson L, Majumdar A, Vestweber D, Uv A, Betsholtz C (2012) The sphingosine-1-phosphate receptor S1PR1 restricts sprouting angiogenesis by regulating the interplay between VE-cadherin and VEGFR2. Dev Cell 23(3):587–599. doi:10.1016/j.devcel.2012.08.005
45. Mendelson K, Zygmunt T, Torres-Vazquez J, Evans T, Hla T (2013) Sphingosine 1-phosphate receptor signaling regulates proper embryonic vascular patterning. J Biol Chem 288(4):2143–2156. doi:10.1074/jbc.M112.427344
46. Kobayashi M, Inoue K, Warabi E, Minami T, Kodama T (2005) A simple method of isolating mouse aortic endothelial cells. J Atheroscler Thromb 12(3):138–142. doi:10.5551/jat.12.138
47. Bligh EG, Dyer WJ (1959) A rapid method of total lipid extraction and purification. Can J Physiol Pharmacol 37(8):911–917. doi:10.1139/o59-099

Part III
Lipid Mediators and Diseases

Chapter 16
Lipid Mediator LPA-Induced Demyelination and Self-Amplification of LPA Biosynthesis in Chronic Pain Memory Mechanisms

Hiroshi Ueda and Hitoshi Uchida

Abstract Chronic pain is considered to have a memory process because of its long-lasting nature even after the original cause such as nerve injury is resolved. This type contrasts to the cases with acute pain, nociceptive or inflammatory pain, which vanishes without delay after the cessation of stimulation or inhibition of the original inflammation. Lysophosphatidic acid (LPA) was identified to be a key initiator of neuropathic pain, one of the representative types of chronic pain, via activation of multiple machineries. Recent studies revealed that LPA induces LPA biosynthesis through actions of microglia and interleukin-1β. LPA_1 and LPA_3 receptor-mediated mechanisms are involved in this self-amplification of LPA production. Neuropathic pain is characterized as unique abnormal pain allodynia, in which gentle touch causes intense pain. The functional switch in allodynia is reasonably explained by demyelination, whose underlying mechanisms are also explained as downstream machineries of LPA and its LPA_1 receptor signaling. The conversion of tactile to intense pain caused by demyelination may be involved in the long-lasting feed-forward machineries in neuropathic pain. Recent reports describe the importance of endocannabinoids and new arachidonic acid metabolites in the regulation of chronic pain. This chapter also describes the possible relationships of LPA to these additional regulatory mechanisms.

Keywords Lysophosphatidic acid • Neuropathic pain • Demyelination • Axonal sprouting • Calpain • Epigenetic • Endocannabinoids • Soluble epoxide hydrolase

Abbreviations

ATX	autotaxin
BDNF	brain-derived neurotrophic factor
CFA	complete Freund's adjuvant

H. Ueda (✉) • H. Uchida
Department of Pharmacology and Therapeutic Innovation, Nagasaki University Graduate School of Biomedical Sciences, 1-14 Bunkyo-machi, Nagasaki 852-8521, Japan
e-mail: ueda@nagasaki-u.ac.jp

T. Yokomizo, M. Murakami (eds.), *Bioactive Lipid Mediators*,
DOI 10.1007/978-4-431-55669-5_16

$cPLA_2$	cytosolic phospholipase A_2
DRG	dorsal root ganglia
ERK	extracellular signal-regulated kinase
GABA	gamma-aminobutyric acid
IL-1β	interleukin-1β
$iPLA_2$	calcium-independent PLA_2
JNK	c-Jun N-terminal kinase
LPA	lysophosphatidic acid
LPC	lysophosphatidyl choline
MAG	myelin-associated glycoprotein
MAPK	mitogen-activated protein kinase
NSAID	nonsteroidal antiinflammatory drug
ROCK	Rho-kinase
sEH	soluble epoxide hydrolase
SP	substance, P

16.1 Introduction

Pain is classified into two groups, acute pain and chronic pain. Acute pain is further classified into nociceptive pain and inflammatory pain. Nociceptive pain occurs through an activation of unmyelinated C-fibers or myelinated Aδ-fibers upon thermal, mechanical, or chemical stimuli, whereas inflammatory pain occurs mainly through C-fibers, which are activated by inflammatory mediators in response to local inflammation. Both types of pain are caused by generation of action potentials initiated through an activation of specific ion channels or receptors on nociceptive endings [1]. The action potential later elicits a release of pain transmitters, such as substance P (SP) or glutamate, and then causes excitatory postsynaptic potentials in the superficial (lamina I and II) dorsal horn neurons of the spinal cord, whose fibers ultimately cross the spinal cord and relay the signal to the contralateral side of thalamus and cerebral cortex through secondary synapse. Usually, this type of pain sensation is transient and is abolished by the removal of the noxious signal with nonsteroidal anti-inflammatory drugs (NSAIDs) or the opioid-induced suppression of primary pain signals by driving the descending pain-inhibitory system through noradrenergic or serotonergic neurons [2]. Innocuous or tactile perception occurs through an activation of sensory organs (Merkel cells, Pacinian corpuscles, and hair follicles) and associated myelinated Aβ-fibers, which mostly innervate to the ipsilateral side of medulla oblongata neurons, although some Aβ-fibers innervate to the deeper (lamina IV) dorsal horn cells, as do noxious fibers.

In chronic pain such as the neuropathic pain paradigm, on the other hand, the damage to peripheral or central neurons in the pain pathway from the nociceptor endings in the skin or internal organs to somatosensory cortex produces a contingent of near-irreversible abnormal pain [3–8]. The abnormal pain is characterized as allodynia, hyperalgesia, or hypoesthesia, which are insensitive to opioids and NSAIDs [3–8]. As allodynia means the conversion of innocuous or tactile signals to painful ones, the nature of neuropathic pain in pain transmission is quite different from acute pain. This difference may be explained by functional crosstalk between noxious and innocuous fibers or respective central fibers. One possible hypothesis is that the demyelination of sensory fibers causes physical or electrical crosstalk from the loss of insulation and fiber sprouting and generates abnormal pain transmission by innocuous fibers [9]. Based on this speculation, we successfully discovered lysophosphatidic acid (LPA) to be a key molecule that causes neuropathic pain mechanisms, including demyelination [5, 6, 10].

16.2 Demyelination in Neuropathic Pain

One of the anatomical and histochemical events in mice subjected to different types of neuropathic pain model is demyelination [9], which occurs specifically in dorsal root fibers following sciatic nerve injury [11]. Pharmacological and biochemical analyses suggested that most positive actions, such as calcium channel $Ca_v\alpha_2\delta_1$ upregulation and sensory fiber-specific hyperalgesia, are closely related to myelinated A-fibers [5]. Based on these findings, we successfully revealed that LPA is the chemical signature in initiation of neuropathic pain including demyelination [4–6, 10, 12, 13]. Further confirmation has been obtained through neuropathic-like behavior and focal demyelination in animals after intrathecal and intratrigeminal injections of LPA [13–15].

16.2.1 LPA$_1$ Receptor-Mediated Mechanisms

Accumulating findings have suggested that central and peripheral demyelinating diseases, such as diabetes, Guillain–Barré syndrome, trigeminal neuralgia, and multiple sclerosis frequently accompany neuropathic pain [16–19]. However, little is known about the molecular relationship between demyelination and neuropathic pain. Partial injury of sciatic nerves causes significant demyelination in vivo in the sciatic nerve and dorsal root, except the spinal nerves [11]. As demyelination in the dorsal root, but not in the sciatic nerve, was abolished in *Lpar1*$^{-/-}$ mice [11], although application of LPA causes demyelination in all these sensory nerve regions [11], the dorsal root specificity following nerve injury seems to be attributed to LPA production in the spinal cord. Details of LPA biosynthesis are described later.

16.2.2 Calpain-Induced Degradation of Myelin Proteins

Significant rapid demyelination of dorsal root fibers by partial sciatic nerve ligation or intrathecal injection of LPA was abolished in *Lpar1*$^{-/-}$ mice [11, 13]. LPA induced downregulation of myelin-associated glycoprotein (MAG), and it was selectively inhibited by calpain inhibitors [14]. MAG is restricted to the innermost membrane facing the neuronal axon, and one of the MAG receptors is Nogo-66 receptor-1, which is mostly found in a tripartite complex composed of transmembrane co-receptors LINGO-1 and either p75^{NTR} or TROY [20, 21]. As this postreceptor signal uses RhoA-Rho-kinase (ROCK) signaling that inhibits tubulin and actin assembly [22], the loss of MAG is supposed to lead to a sprouting of sensory A-fibers, thereby possibly resulting in a formation of abnormal pain synapses in the spinal cord. Interestingly, nerve injury-induced calpain activation in the dorsal root was abolished in *Lpar1*$^{-/-}$ mice, although inflammatogenic complete Freund's adjuvant (CFA) did not induce calpain activation [14]. These findings were supported by pharmacological studies in which pretreatment with the calpain inhibitor E-64d [(2*S*,3*S*)-*trans*-epoxysuccinyl-L-leucylamido-3-methylbutane ethyl ester] or calpain inhibitor X (Z-Leu-Abu-CONH-ethyl) abolished nerve injury-induced neuropathic pain, but not CFA-induced pain [14]. Thus, it is suggested that the rapid downregulation of myelin protein is caused by activated calpain following an elevation of $[Ca^{2+}]_i$, possibly through an activation of LPA_1 receptor-$G_{q/11}$-phospholipase C and its subsequent production of inositol 1,4,5-trisphosphate.

16.2.3 Multiple Mechanisms Underlying RhoA-ROCK-Mediated Myelin Protein Gene Silencing

Downregulation of myelin proteins and gene expression is closely related to demyelination. Nerve injury- or intrathecal LPA-induced downregulation of myelin proteins was abolished in *Lpar1*$^{-/-}$ mice, or by pretreatment with RhoA inhibitor BoNT/C3 or ROCK inhibitor Y-27632 [13]. Recent studies revealed that RhoA-ROCK signaling is closely related to multiple mechanisms underlying downregulation of myelin genes. One of these is the mechanism through c-Jun N-terminal kinase (JNK) and Krox-20/Egr2, as follows: (1) c-Jun is a negative regulator to downregulate the transcription of pro-myelinating transcription factor Krox-20/Egr2 [23]; (2) c-Jun forms a nucleosome remodeling and histone deacetylation repressor complex with Mbd3, and represses the transcription of AP-1 target genes including c-Jun [24], but c-Jun phosphorylated by JNK loses this repression system; and (3) ROCK activation results in a sequential activation of MKK4 and JNK [25], which in turn phosphorylates the N-terminus of c-Jun [26, 27]. Therefore, we now hypothesize that the LPA_1 receptor could mediate a sequential activation of ROCK inhibitor-reversible JNK phosphorylation, followed by upregulation of c-Jun and downregulation of Krox-20/Egr2 (Fig. 16.1).

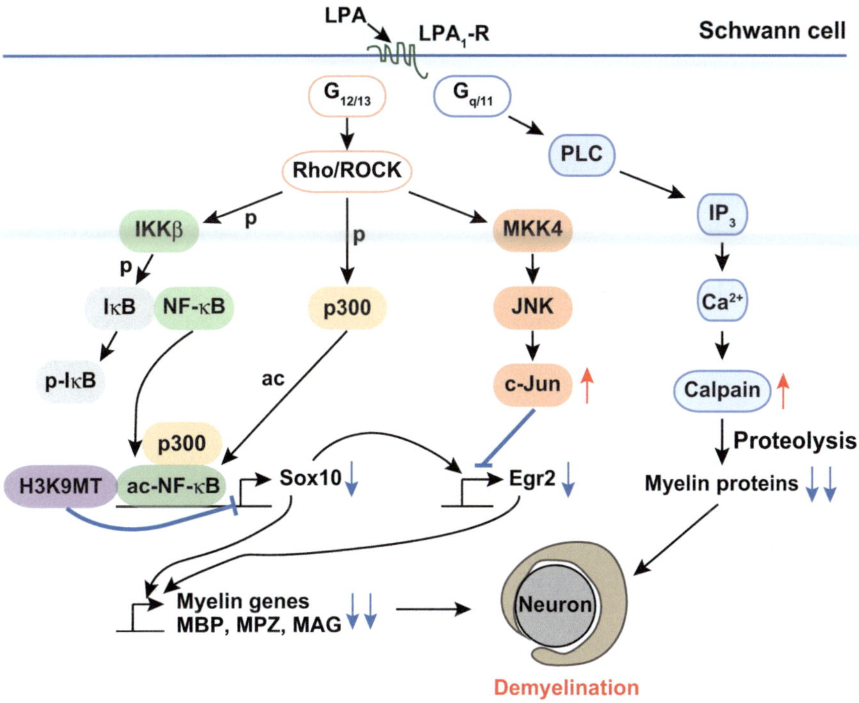

Fig. 16.1 Possible mechanisms for lysophosphatidic acid (LPA)-induced demyelination of Schwann cells. LPA-LPA$_1$ receptor signaling causes calpain-mediated protelolysis of myelin proteins, possibly via $G_{q/11}$-phospholipase C-inositol 1,4,5-triphosphate-Ca^{2+} pathway. In addition to posttranscriptional regulation, LPA might drive LPA$_1$ receptor-$G_{12/13}$-Rho-ROCK signaling cascade to cause IKKβ-mediated NF-κB activation in parallel with activation of p300, which acetylates NF-κB. Acetylated NF-κB could recruit H3K9 methyltransferase to downregulate transcription of Sox10, which activates transcription of Egr2. Simultaneously, ROCK might induce MKK4-JNK-c-Jun signaling to repress transcription of Egr2, which activates myelin protein genes in concert with Sox10

Another possible pathway might include epigenetic mechanisms. ROCK activates p300 histone acetyltransferase [28], which acetylates the NF-κB [29]. Then, acetylated NF-κB represses the transcription of Sox10 [30], which is a positive regulator of the transcription of Krox-20/Egr2 and myelin protein genes [31]. In addition to this, the RhoA-ROCK system mediates a phosphorylation and activation of IKKβ, plausibly followed by degradation of IκB and activation of NF-κB [32]. Therefore, we now hypothesize that LPA$_1$ receptor signaling could cause p300-mediated NF-κB acetylation via activation of $G_{12/13}$-RhoA-ROCK-IKKβ and following p300-mediated NF-κB acetylation to downregulate Sox10 transcription (Fig. 16.1).

16.2.4 Aberrant Sprouting and Synaptic Reorganization

Demyelination, a loss of insulation, may cause physical or electrical crosstalk among sensory fibers, which may lead to abnormal pain perception, as seen in allodynia. Physical contacts among A-fibers may also occur after demyelination and subsequent axonal sprouting, as seen in clinical observations, in which more axonal sprouting occurs as the neuropathy from mutation of myelin proteins advances [33]. This fact was also evidenced by co-culture experiments using myelinated fibers with isolated DRG neurons and Schwann cells, in which the addition of LPA caused sequential morphological changes, including collapse of the growth cone at 1 h, sprouting at nerve endings at 8 h, and axons at 18 h, and complete spreading of myelinated Schwann cells at 36 h [4]. We had the chance to observe the direct attachment of large-diameter fibers (possibly A-fibers) to small-diameter fibers (C-fibers) in the Remak bundle of mouse dorsal root after sciatic nerve injury [5].

Activation of the mitogen-activated protein kinase (MAPK) family, extracellular signal-regulated kinase ($ERK_{1/2}$) following membrane depolarization and Ca^{2+} influx, gives an insight into biochemical alterations during this synaptic reorganization [34, 35]. It is well documented that $ERK_{1/2}$ is rapidly phosphorylated following noxious stimulations in dorsal root ganglia (DRG) and spinal dorsal horn. The intensity-dependent manner of the stimulation allows visualization of the pain-signaling neuronal circuitry using phospho-$ERK_{1/2}$ as the functional marker [34]. A significant number of neurons at the superficial layer of the spinal dorsal horn show phospho-$ERK_{1/2}$ activity by Neurometer stimulation of Aδ- and C-fibers, but not by innocuous Aβ-fiber stimulation [35, 36]. Following sciatic nerve injury, on the other hand, Aβ-fiber stimulation resulted in localization of phospho-$ERK_{1/2}$ at the superficial layer of the dorsal horn. This mechanism can only be explained by massive structural reorganization of the synapse. Sprouting and crosstalk may be biophysical events, but the electrophysiological consequences remain to be clarified.

16.3 LPA Production

Although LPA is able to cause global demyelination in the dorsal root, spinal nerve, and sciatic nerve in ex vivo experiments, nerve injury-induced and LPA_1 receptor-dependent demyelination occurs specifically in the dorsal root remote from the injured sciatic nerve [11]. This fact suggests that the amount of LPA required for causing dorsal root demyelination after nerve injury may come from the spinal cord [11, 37].

16.3.1 Machineries of Nerve Injury-Specific LPA Production

The combined addition of SP and *N*-methyl-D-aspartate (NMDA) to spinal cord slices causes LPA production in the presence of autotaxin (ATX), which converted lysophosphatidylcholine (LPC) to LPA, although the single addition of SP or

NMDA alone did not [38]. This fact suggests that an intense pain signal is required for LPA production. Recent studies have revealed that phosphatidylcholine is converted to LPC by cytosolic phospholipase A_2 ($cPLA_2$) or calcium-independent PLA_2 ($iPLA_2$), both of which are regulated by Ca^{2+}-related mechanisms. $cPLA_2$ is activated by Ca^{2+} or phosphorylation by MAPK or protein kinase Cs [39–41], whereas $iPLA_2$ is activated by the calcium influx factor, which is produced following Ca^{2+} depletion in the endoplasmic reticulum [42, 43]. Therefore, the activation of both $cPLA_2$ and $iPLA_2$, which are predominantly expressed in neurons [44–46], may be induced by intense pain signals (SP and glutamate) after nerve injury. Using pharmacological studies with $cPLA_2$ or $iPLA_2$ inhibitor, this view was confirmed by recent studies that abolished neuropathic pain and LPA production [44, 47].

16.3.2 LPA Receptor-Mediated Amplification of LPA Production

As already mentioned, a single intrathecal administration of LPA causes week-long abnormal pain behavior that is similar to sciatic nerve injury-induced neuropathic pain [13]. When LPA levels in the spinal cord were measured by biological assay, LPA-induced amplification of LPA production was observed, and this amplification was abolished in *Lpar3*$^{-/-}$ but not *Lpar1*$^{-/-}$ mice. LPA measurement was then performed by use of MALDI-TOF-MS/MS: after the phostag modification of extracted LPA, sciatic nerve injury increased 18:1, 16:0, and 18:0 LPA [44]. This LPA-induced amplification of LPA production was substantially abolished in *Lpar3*$^{-/-}$ and *Lpar1*$^{-/-}$ mice. Furthermore, as LPA production was abolished by pretreatment with minocycline, a microglia inhibitor, it is evident that microglia have important functions in this mechanism.

Thus, the involvement of LPA_1 receptor in the amplification of LPA differs between the LPA (intrathecal, i.t.)-induced pain model and a nerve injury-induced one. Although detailed mechanisms remain elusive, we now speculate as follows: LPA produced by intense pain transmission may stimulate presynaptic LPA_1 receptors and locally release SP or glutamate, which in turn has an amplifying role in LPA production as well as microglial LPA_3 receptor-mediated mechanisms. However, exogenously added LPA may predominantly access microglia but not the presynaptic terminals. Recent studies have demonstrated that microglia have important functions in chronic pain after nerve injury [48, 49]. Once microglia are activated, they cause temporal enzyme activation and pro-inflammatory cytokine release, which in turn activates nearby astrocytes, microglia, and neurons [48, 49]. Activated astrocytes also release different types of cytokines and activate nearby cell types [48, 49]. We have reported that interleukin-1β (IL-1β) is produced by LPA (i.t.) or nerve injury, and neutralizing IL-1β antibody abolished neuropathic pain and LPA production [50] (Fig. 16.2).

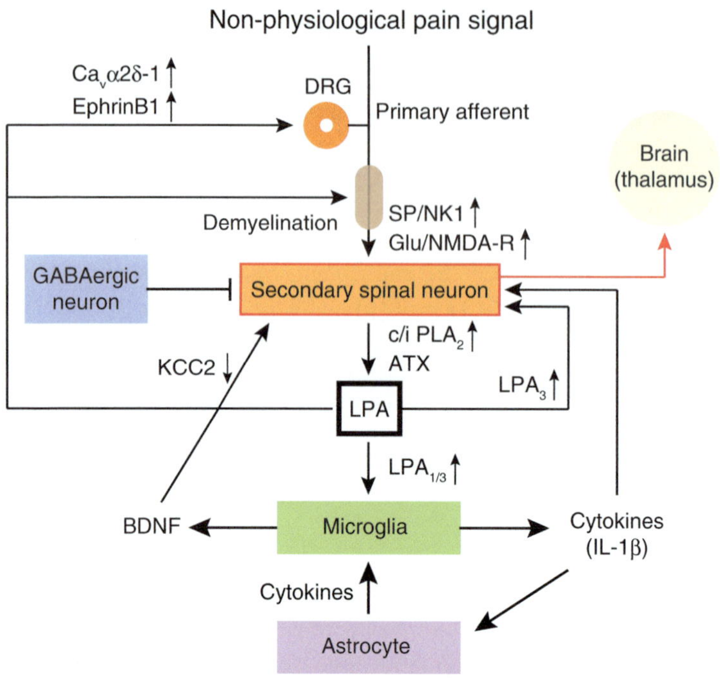

Fig. 16.2 Feed-forward mechanisms underlying LPA-induced neuropathic pain. Nonphysiological stimuli, including peripheral nerve injury, trigger presynaptic release of SP and glutamate, thereby resulting in activation of NK1 receptor and NMDA receptor, respectively, in the spinal dorsal horn. Subsequent activation of $cPLA_2$ and $iPLA_2$ in the secondary spinal neuron leads to production of LPC, which is converted to LPA by ATX. LPA_1 and LPA_3 receptor-mediated microglial activation is involved in biosynthesis of LPA, possibly via releasing IL-1β. On the other hand, LPA acts as a reverse signal to cause upregulation of pain-related genes, such as $Ca_v\alpha 2\delta 1$ and ephrinB1, in the dorsal root ganglion (DRG), and demyelination in the DRG via LPA_1 receptor. Also, LPA-induced brain-derived neurotrophic factor (BDNF) release from microglia might cause GABA switch via KCC2 downregulation. Finally, feed-forward mechanisms via LPA transmit enhance pain signal to the brain regions, including the thalamus

16.3.3 γ-Aminobutyric Acid (GABA) Switch Underlying LPA-Mediated Hyperalgesia

Our next question is what is the major machinery to cause exogenous LPA- or endogenously amplified LPA-induced hyperalgesia. Regarding this issue, we are also speculating on the involvement of LPA-stimulated microglia. In our previous study, LPA caused an increase in the expression of brain-derived neurotrophic factor (BDNF) in primary cultures of rat microglia, which express LPA_3, but not LPA_1 or LPA_2, receptors [51]. A previous study demonstrated that BDNF downregulates the neuron-specific K^+-Cl^- co-transporter KCC2, resulting in lower $[Cl^-]_i$ levels in neurons [52, 53]. Under this condition, GABA (or Glycine) receptor activation causes depolarization from the efflux of Cl^- [52], and the phenomenon is called the GABA (or Glycine) switch (Fig. 16.2). The current study using the microglia inhibitor minocycline revealed that LPA-induced microglia activation is involved in early-stage development, but not in the late-stage maintenance, of neuropathic pain [54]. In this study, the early treatment with minocycline abolished LPA-induced and nerve injury-induced neuropathic pain, LPA synthesis and its underlying activation of synthetic enzymes, and $cPLA_2$ and $iPLA_2$. As later treatment with minocycline failed to attenuate established neuropathic pain, as previously reported [55], microglia activation following LPA receptor signaling appears to be important for the initiation of neuropathic pain.

16.4 Other Lipid Mediators Involved in Chronic Pain Mechanisms

Endocannabinoids, such as anandamide and 2-arachidonoylglycerol (2-AG), are produced upon tissue damage or inflammation in neural and nonneural cells. They suppress sensitization and neurogenic inflammation by activating $G_{i/o}$-coupled CB_1- and CB_2-receptors [56, 57]. In the acute (nociceptive or inflammatory) pain state, CB_1 receptor is more abundantly expressed in sensory fibers (DRG neurons) than CB_2 receptor and have inhibitory functions in inflammatory pain [58]. In contrast, the numbers of CB_2 receptor in sensory fibers are markedly increased after peripheral nerve injury. As LPA production requires intense pain signals, it is abolished by pretreatment with morphine [59]. Therefore, it is plausible that endocannabinoids, which are produced in the early stages after tissue or nerve injury, may have similar preemptive functions to inhibit LPA-mediated neuropathic pain (Fig. 16.3a). Although endocannabinoids also interact with transient receptor potential vanilloid 1 (TRPV1) and G protein-coupled GPR55 receptor [56], their roles remain elusive. In addition to endocannabinoids, fatty acid ethanolamides, such as oleoylethanolmide and palmitoylethanolmide, have emerged as potential homeostatic regulators of pain [56]: they act as peroxisome proliferator-activated receptor-α agonists and exert profound anti-inflammatory effects in rats and mice [56].

Epoxygenated fatty acids, i.e., arachidonate-derived cytochrome P450 metabolites, also attract many concerns in terms of pain control [60]. Most recently, there are increasing reports that arachidonic acid-derived mediators, epoxyeicosatrienoic acids and dihydroxyeicosatrienoic acids, are analgesic/anti-inflammatory and algesic/pro-inflammatory, respectively [60]. The key enzyme to determine pain regulation is soluble epoxide hydrolase (sEH) [60]. The selective inhibitors of sEH showed not only anti-inflammatory actions but also mechanical allodynia observed in the streptozotocin-treated diabetic neuropathic pain model [61, 62]. Therefore, sEH may have unique roles in inflammatory and neuropathic pain (Fig. 16.3b).

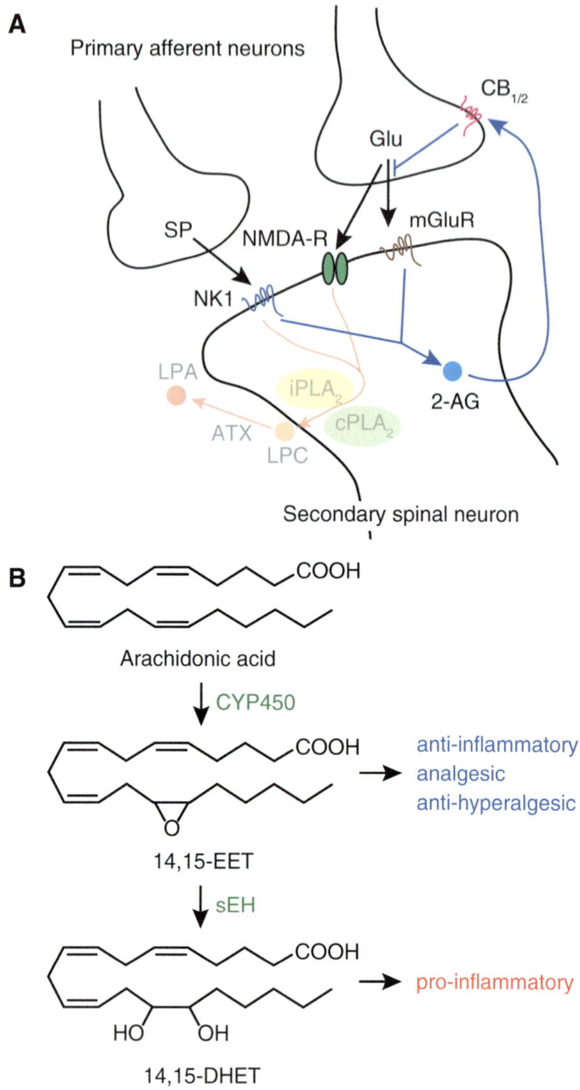

Fig. 16.3 Other lipid mediators and pain regulation. (**a**) The endocannabinoid 2-arachidonoylglycerol (2-AG) can be synthesized by an increase in $[Ca^{2+}]_i$ in secondary spinal neurons, possibly via activation of NK1 receptor and metabotropic glutamate receptors (mGluRs). 2-AG behaves as a reverse signal to activate presynaptic $CB_{1/2}$ receptors, thereby inhibiting glutamate release from nerve endings of primary afferents. Such mechanisms might block LPA production after nerve injury. (**b**) Arachidonic acid can be converted to epoxyeicosatrienoic acids (EETs) by cytochrome P450 (CYP450), which in turn are metabolized to dihydroxyeicosatrienoic acids (DHETs) by sEH. EETs exert anti-inflammatory, analgesic, and anti-hyperalgesic actions, whereas DHETs produce pro-inflammatory and algesic actions

16.5 Conclusion

Here, I propose a feed-forward system in which LPA-induced signals are involved in the mechanisms of neuropathic pain. In this hypothesis, the initial step is intense pain signal-induced LPA production in the dorsal horn. Thus, produced LPA in the spinal cord is then transported to the dorsal root, where LPA induces demyelination as a reverse signal. Demyelination may cause physical crosstalk between fibers and sprouting, both of which may lead to synaptic reorganization of primary afferent neurons underlying allodynia or hyperalgesia. The LPA also causes LPA biosynthesis through activation of LPA_3 receptor and microglia. Thus, LPA acts as a reverse signal, and LPA-induced amplification of LPA biosynthesis is a key mechanism for the feed-forward system underlying sustained neuropathic pain. However, LPA_1 receptor-mediated signaling does not function in inflammatory pain. It is plausible that endocannabinoids, which are rapidly and abundantly produced upon inflammatory stimulation, may inhibit the presynaptic pain transmitter release and repress the LPA production. As LPA not only stimulates the transcription of the endocannabinoid-degrading enzyme [63] but also inhibits ATX enzyme activity [64], the amplification of LPA production will be reactivated, although it will not proceed further without limitation. Recent studies reveal that the LPA-LPA_1 receptor signaling is also observed in mechanisms underlying various chronic pain models. Discovery of the best inhibitors for LPA receptors or biosynthetic enzymes is now needed.

References

1. Woolf CJ, Ma Q (2007) Nociceptors—noxious stimulus detectors. Neuron 55(3):353–364. doi:S0896-6273(07)00537-5
2. Fields HL, Basbaum AI, Heinricher MM (2006) Central nervous system mechanisms of pain modulation. In: McMahon SB, Koltzenburg M (eds) Wall and Melzack's textbook of pain. Churchill Livingstone, Oxford
3. Apkarian AV, Baliki MN, Geha PY (2009) Towards a theory of chronic pain. Prog Neurobiol 87(2):81–97. doi:S0301-0082(08)00113-5
4. Ueda H, Ueda M (2011) Lysophosphatidic acid as initiator of neuropathic pain: biosynthesis and demyelination. J Clin Lipid 6:147–158
5. Ueda H (2008) Peripheral mechanisms of neuropathic pain: involvement of lysophosphatidic acid receptor-mediated demyelination. Mol Pain 4:11. doi:1744-8069-4-11
6. Ueda H (2006) Molecular mechanisms of neuropathic pain–phenotypic switch and initiation mechanisms. Pharmacol Ther 109(1-2):57–77. doi:S0163-7258(05)00134-8
7. Saade NE, Jabbur SJ (2008) Nociceptive behavior in animal models for peripheral neuropathy: spinal and supraspinal mechanisms. Prog Neurobiol 86(1):22–47. doi:S0301-0082(08)00059-2
8. Costigan M, Scholz J, Woolf CJ (2009) Neuropathic pain: a maladaptive response of the nervous system to damage. Annu Rev Neurosci 32:1–32. doi:10.1146/annurev.neuro.051508.135531
9. Gillespie CS, Sherman DL, Fleetwood-Walker SM, Cottrell DF, Tait S, Garry EM, Wallace VC, Ure J, Griffiths IR, Smith A, Brophy PJ (2000) Peripheral demyelination and neuropathic pain behavior in periaxin-deficient mice. Neuron 26(2):523–531. doi:S0896-6273(00)81184-8
10. Ueda H, Matsunaga H, Olaposi OI, Nagai J (2013) Lysophosphatidic acid: chemical signature of neuropathic pain. Biochim Biophys Acta 1831(1):61–73. doi:S1388-1981(12)00185-0
11. Nagai J, Uchida H, Matsushita Y, Yano R, Ueda M, Niwa M, Aoki J, Chun J, Ueda H (2010) Autotaxin and lysophosphatidic acid 1 receptor-mediated demyelination of dorsal root fibers by sciatic nerve injury and intrathecal lysophosphatidylcholine. Mol Pain 6:78. doi:1744-8069-6-78
12. Ueda H (2013) Lysophosphatidic acid and neuropathic pain: demyelination and LPA biosynthesis. In: Lysophospholipid receptors: signaling and biochemistry. Wiley, Hoboken, pp 433–449

13. Inoue M, Rashid MH, Fujita R, Contos JJ, Chun J, Ueda H (2004) Initiation of neuropathic pain requires lysophosphatidic acid receptor signaling. Nat Med 10(7):712–718. doi:10.1038/nm1060 nm1060
14. Xie W, Uchida H, Nagai J, Ueda M, Chun J, Ueda H (2010) Calpain-mediated down-regulation of myelin-associated glycoprotein in lysophosphatidic acid-induced neuropathic pain. J Neurochem 113(4):1002–1011. doi:JNC6664
15. Ahn DK, Lee SY, Han SR, Ju JS, Yang GY, Lee MK, Youn DH, Bae YC (2009) Intratrigeminal ganglionic injection of LPA causes neuropathic pain-like behavior and demyelination in rats. Pain 146(1–2):114–120. doi:S0304-3959(09)00395-9
16. O'Connor AB, Schwid SR, Herrmann DN, Markman JD, Dworkin RH (2008) Pain associated with multiple sclerosis: systematic review and proposed classification. Pain 137(1):96–111. doi:S0304-3959(07)00460-5
17. Scrivani SJ, Mathews ES, Maciewicz RJ (2005) Trigeminal neuralgia. Oral Surg Oral Med Oral Pathol Oral Radiol Endod 100(5):527–538. doi:S1079-2104(05)00487-7
18. Said G (2007) Diabetic neuropathy: a review. Nat Clin Pract Neurol 3(6):331–340. doi:ncpneuro0504
19. Pentland B, Donald SM (1994) Pain in the Guillain-Barre syndrome: a clinical review. Pain 59(2):159–164. doi:0304-3959(94)90068-X
20. McDonald CL, Bandtlow C, Reindl M (2011) Targeting the Nogo receptor complex in diseases of the central nervous system. Curr Med Chem 18(2):234–244. doi:BSP/CMC/E-Pub/2011/016
21. Llorens F, Gil V, del Rio JA (2011) Emerging functions of myelin-associated proteins during development, neuronal plasticity, and neurodegeneration. FASEB J 25(2):463–475. doi:fj.10-162792
22. Ueda H (2011) Lysophosphatidic acid as the initiator of neuropathic pain. Biol Pharm Bull 34(8):1154–1158. doi:JST.JSTAGE/bpb/34.1154
23. Parkinson DB, Bhaskaran A, Arthur-Farraj P, Noon LA, Woodhoo A, Lloyd AC, Feltri ML, Wrabetz L, Behrens A, Mirsky R, Jessen KR (2008) c-Jun is a negative regulator of myelination. J Cell Biol 181(4): 625–637. doi:jcb.200803013
24. Aguilera C, Nakagawa K, Sancho R, Chakraborty A, Hendrich B, Behrens A (2011) c-Jun N-terminal phosphorylation antagonises recruitment of the Mbd3/NuRD repressor complex. Nature (Lond) 469(7329):231–235. doi:nature09607
25. Marinissen MJ, Chiariello M, Tanos T, Bernard O, Narumiya S, Gutkind JS (2004) The small GTP-binding protein RhoA regulates c-jun by a ROCK-JNK signaling axis. Mol Cell 14(1):29–41. doi:S1097276504001534
26. Das KC, Muniyappa H (2010) c-Jun-NH2 terminal kinase (JNK)-mediates AP-1 activation by thioredoxin: phosphorylation of cJun, JunB, and Fra-1. Mol Cell Biochem 337(1-2):53–63. doi:10.1007/s11010-009-0285-0
27. Muniyappa H, Das KC (2008) Activation of c-Jun N-terminal kinase (JNK) by widely used specific p38 MAPK inhibitors SB202190 and SB203580: a MLK-3-MKK7-dependent mechanism. Cell Signal 20(4):675–683. doi:S0898-6568(07)00374-9
28. Tanaka T, Nishimura D, Wu RC, Amano M, Iso T, Kedes L, Nishida H, Kaibuchi K, Hamamori Y (2006) Nuclear Rho kinase, ROCK2, targets p300 acetyltransferase. J Biol Chem 281(22):15320–15329. doi:M510954200
29. Chen LF, Greene WC (2004) Shaping the nuclear action of NF-kappaB. Nat Rev Mol Cell Biol 5(5):392–401 doi:10.1038/nrm1368 nrm1368
30. Chen Y, Wang H, Yoon SO, Xu X, Hottiger MO, Svaren J, Nave KA, Kim HA, Olson EN, Lu QR (2011) HDAC-mediated deacetylation of NF-kappaB is critical for Schwann cell myelination. Nat Neurosci 14(4):437–441. doi:nn.2780
31. Svaren J, Meijer D (2008) The molecular machinery of myelin gene transcription in Schwann cells. Glia 56(14):1541–1551. doi:10.1002/glia.20767

32. Kim HJ, Kim JG, Moon MY, Park SH, Park JB (2014) IkappaB kinase gamma/nuclear factor-kappaB-essential modulator (IKKgamma/NEMO) facilitates RhoA GTPase activation, which, in turn, activates Rho-associated KINASE (ROCK) to phosphorylate IKKbeta in response to transforming growth factor (TGF)-beta 1. J Biol Chem 289(3):1429–1440. doi:M113.520130
33. Misu K, Yoshihara T, Shikama Y, Awaki E, Yamamoto M, Hattori N, Hirayama M, Takegami T, Nakashima K, Sobue G (2000) An axonal form of Charcot-Marie-Tooth disease showing distinctive features in association with mutations in the peripheral myelin protein zero gene (Thr124Met or Asp75Val). J Neurol Neurosurg Psychiatry 69(6):806–811
34. Ji RR, Baba H, Brenner GJ, Woolf CJ (1999) Nociceptive-specific activation of ERK in spinal neurons contributes to pain hypersensitivity. Nat Neurosci 2(12):1114–1119. doi:10.1038/16040
35. Rosen LB, Ginty DD, Weber MJ, Greenberg ME (1994) Membrane depolarization and calcium influx stimulate MEK and MAP kinase via activation of Ras. Neuron 12(6):1207–1221. doi:0896-6273(94)90438-3
36. Matsumoto M, Xie W, Ma L, Ueda H (2008) Pharmacological switch in A beta-fiber stimulation-induced spinal transmission in mice with partial sciatic nerve injury. Mol Pain 4:25. doi:1744-8069-4-25
37. Fujita R, Kiguchi N, Ueda H (2007) LPA-mediated demyelination in *ex vivo* culture of dorsal root. Neurochem Int 50(2):351–355. doi:S0197-0186(06)00283-X
38. Inoue M, Ma L, Aoki J, Ueda H (2008) Simultaneous stimulation of spinal NK1 and NMDA receptors produces LPC which undergoes ATX-mediated conversion to LPA, an initiator of neuropathic pain. J Neurochem 107(6):1556–1565. doi:JNC5725
39. Durstin M, Durstin S, Molski TF, Becker EL, Sha'afi RI (1994) Cytoplasmic phospholipase A_2 translocates to membrane fraction in human neutrophils activated by stimuli that phosphorylate mitogen-activated protein kinase. Proc Natl Acad Sci U S A 91(8):3142–3146
40. Hirabayashi T, Murayama T, Shimizu T (2004) Regulatory mechanism and physiological role of cytosolic phospholipase A_2. Biol Pharm Bull 27(8):1168–1173. doi:JST.JSTAGE/bpb/27.1168
41. Xing M, Tao L, Insel PA (1997) Role of extracellular signal-regulated kinase and PKC alpha in cytosolic PLA_2 activation by bradykinin in MDCK-D1 cells. Am J Physiol 272(4 pt 1):C1380–C1387
42. Csutora P, Zarayskiy V, Peter K, Monje F, Smani T, Zakharov SI, Litvinov D, Bolotina VM (2006) Activation mechanism for CRAC current and store-operated Ca^{2+} entry: calcium influx factor and Ca^{2+}-independent phospholipase A_2beta mediated pathway. J Biol Chem 281(46):34926–34935. doi:M606504200
43. Smani T, Zakharov SI, Csutora P, Leno E, Trepakova ES, Bolotina VM (2004) A novel mechanism for the store-operated calcium influx pathway. Nat Cell Biol 6(2):113–120. doi:10.1038/ncb1089 ncb1089
44. Ma L, Nagai J, Chun J, Ueda H (2013) An LPA species (18:1 LPA) plays key roles in the self-amplification of spinal LPA production in the peripheral neuropathic pain model. Mol Pain 9(1):29. doi:1744-8069-9-29
45. Kurusu S, Matsui K, Watanabe T, Tsunou T, Kawaminami M (2008) The cytotoxic effect of bromoenol lactone, a calcium-independent phospholipase A_2 inhibitor, on rat cortical neurons in culture. Cell Mol Neurobiol 28(8):1109–1118. doi:10.1007/s10571-008-9287-9
46. Kishimoto K, Matsumura K, Kataoka Y, Morii H, Watanabe Y (1999) Localization of cytosolic phospholipase A_2 messenger RNA mainly in neurons in the rat brain. Neuroscience 92(3):1061–1077. doi:S0306-4522(99)00051-2
47. Ma L, Uchida H, Nagai J, Inoue M, Aoki J, Ueda H (2010) Evidence for *de novo* synthesis of lysophosphatidic acid in the spinal cord through phospholipase A_2 and autotaxin in nerve injury-induced neuropathic pain. J Pharmacol Exp Ther 333(2):540–546. doi:jpet.109.164830

48. Milligan ED, Watkins LR (2009) Pathological and protective roles of glia in chronic pain. Nat Rev Neurosci 10(1):23–36. doi:nrn2533
49. Scholz J, Woolf CJ (2007) The neuropathic pain triad: neurons, immune cells and glia. Nat Neurosci 10(11):1361–1368. doi:nn1992
50. Yano R, Ma L, Nagai J, Ueda H (2013) Interleukin-1beta plays key roles in LPA-induced amplification of LPA production in neuropathic pain model. Cell Mol Neurobiol 33(8):1033–1041. doi:10.1007/s10571-013-9970-3
51. Fujita R, Ma Y, Ueda H (2008) Lysophosphatidic acid-induced membrane ruffling and brain-derived neurotrophic factor gene expression are mediated by ATP release in primary microglia. J Neurochem 107(1):152–160. doi:JNC5599
52. Rivera C, Voipio J, Thomas-Crusells J, Li H, Emri Z, Sipila S, Payne JA, Minichiello L, Saarma M, Kaila K (2004) Mechanism of activity-dependent downregulation of the neuron-specific K-Cl cotransporter KCC2. J Neurosci 24(19):4683–4691. doi:10.1523/JNEUROSCI.5265-03.2004 24/19/4683
53. Rivera C, Li H, Thomas-Crusells J, Lahtinen H, Viitanen T, Nanobashvili A, Kokaia Z, Airaksinen MS, Voipio J, Kaila K, Saarma M (2002) BDNF-induced TrkB activation down-regulates the K^{+}–Cl^{-} cotransporter KCC2 and impairs neuronal Cl^{-} extrusion. J Cell Biol 159(5):747–752. doi:10.1083/jcb.200209011 jcb.200209011
54. Ma L, Nagai J, Ueda H (2010) Microglial activation mediates de novo lysophosphatidic acid production in a model of neuropathic pain. J Neurochem 115(3):643–653. doi:JNC6955
55. Raghavendra V, Tanga F, DeLeo JA (2003) Inhibition of microglial activation attenuates the development but not existing hypersensitivity in a rat model of neuropathy. J Pharmacol Exp Ther 306(2):624–630. doi:10.1124/jpet.103.052407 jpet.103.052407
56. Piomelli D, Sasso O (2014) Peripheral gating of pain signals by endogenous lipid mediators. Nat Neurosci 17(2):164–174. doi:nn.3612
57. Yao B, Mackie K (2009) Endocannabinoid receptor pharmacology. Curr Top Behav Neurosci 1:37–63. doi:10.1007/978-3-540-88955-7_2
58. Agarwal N, Pacher P, Tegeder I, Amaya F, Constantin CE, Brenner GJ, Rubino T, Michalski CW, Marsicano G, Monory K, Mackie K, Marian C, Batkai S, Parolaro D, Fischer MJ, Reeh P, Kunos G, Kress M, Lutz B, Woolf CJ, Kuner R (2007) Cannabinoids mediate analgesia largely via peripheral type 1 cannabinoid receptors in nociceptors. Nat Neurosci 10(7):870–879. doi:nn1916
59. Nagai J, Ueda H (2011) Pre-emptive morphine treatment abolishes nerve injury-induced lysophospholipid synthesis in mass spectrometrical analysis. J Neurochem 118(2):256–265. doi:10.1111/j.1471-4159.2011.07297.x
60. Morisseau C, Hammock BD (2013) Impact of soluble epoxide hydrolase and epoxyeicosanoids on human health. Annu Rev Pharmacol Toxicol 53:37–58. doi:10.1146/annurev-pharmtox-011112-140244
61. Wagner K, Inceoglu B, Dong H, Yang J, Hwang SH, Jones P, Morisseau C, Hammock BD (2013) Comparative efficacy of 3 soluble epoxide hydrolase inhibitors in rat neuropathic and inflammatory pain models. Eur J Pharmacol 700(1-3):93–101. doi:S0014-2999(12)01032-1
62. Inceoglu B, Wagner KM, Yang J, Bettaieb A, Schebb NH, Hwang SH, Morisseau C, Haj FG, Hammock BD (2012) Acute augmentation of epoxygenated fatty acid levels rapidly reduces pain-related behavior in a rat model of type I diabetes. Proc Natl Acad Sci U S A 109(28):11390–11395. doi:1208708109
63. Sordelli MS, Beltrame JS, Cella M, Gervasi MG, Perez Martinez S, Burdet J, Zotta E, Franchi AM, Ribeiro ML (2012) Interaction between lysophosphatidic acid, prostaglandins and the endocannabinoid system during the window of implantation in the rat uterus. PLoS One 7(9):e46059. doi:10.1371/journal.pone.0046059 PONE-D-12-19475
64. van Meeteren LA, Ruurs P, Christodoulou E, Goding JW, Takakusa H, Kikuchi K, Perrakis A, Nagano T, Moolenaar WH (2005) Inhibition of autotaxin by lysophosphatidic acid and sphingosine 1-phosphate. J Biol Chem 280(22):21155–21161. doi:M413183200

Chapter 17
Vascular Endothelial $S1P_2$ Receptor Limits Tumor Angiogenesis and Hyperpermeability

Noriko Takuwa, Yasuo Okamoto, Kazuaki Yoshioka, and Yoh Takuwa

Abstract The lipid mediator sphingosine-1-phosphate (S1P), which is constitutively produced by sphingosine kinase 1 (SphK1) in circulating erythrocytes and vascular endothelial cells, is exported into plasma to regulate vascular formation, vascular barrier function, vascular tonus, and lymphocyte trafficking through G protein-coupled S1P receptors. $S1P_1$, a principal endothelial S1P receptor, has crucial functions in developmental vascular formation and the maintenance of barrier function. The stabilizing action of $S1P_1$ on endothelial intercellular junctions leads to inhibition of sprouting angiogenesis. $S1P_1$ maintains barrier integrity through the activation of the small G protein Rac. Engagement of $S1P_1$ triggers internalization of $S1P_1$ into early endosomes, in which Rac activation occurs. The class IIα isoform of phosphatidylinositol 3-kinase (PI3K-C2α) is required for the internalization of activated $S1P_1$ and Rac activation on early endosomes. Endothelial cells also express $S1P_2$, which is a prototypic $G_{12/13}$-coupled chemorepellant receptor to activate RhoA with downstream inhibition of Rac and Akt. Endothelial $S1P_2$, together with $S1P_2$ in myeloid cells, mediates inhibition of tumor angiogenesis. This action, in concert with a direct inhibitory effect of $S1P_2$ on proliferation of tumor cells, suppresses tumor growth. Endothelial $S1P_2$ also suppresses disruption of intercellular junctions and resultant fatal vascular hyperpermeability in anaphylaxis, by limiting mediator-induced activation of eNOS through Rho-Rho kinase PTEN-dependent suppression of Akt. Thus, endothelial $S1P_1$ and $S1P_2$ have specialized, distinct roles in the regulation of angiogenesis and vascular barrier integrity.

Keywords Sphingosine-1-phosphate • $S1P_1$ • $S1P_2$ • Rac • Rho • PI3K-C2α • Endothelial cells • Angiogenesis • Anaphylaxis • Barrier function

---

N. Takuwa, M.D., Ph.D.
Department of Physiology, Kanazawa University Graduate School of Medicine, 13-1 Takaramachi, Kanazawa 920-8640, Japan

Department of Health and Medical Sciences, Ishikawa Prefectural Nursing University, Kahoku 929-1210, Japan

Y. Okamoto, M.D., Ph.D. • K. Yoshioka, Ph.D. • Y. Takuwa, M.D., Ph.D. (✉)
Department of Physiology, Kanazawa University Graduate School of Medicine, 13-1 Takaramachi, Kanazawa 920-8640, Japan
e-mail: ytakuwa@med.kanazawa-u.ac.jp

T. Yokomizo, M. Murakami (eds.), *Bioactive Lipid Mediators*,
DOI 10.1007/978-4-431-55669-5_17

Abbreviations

S1P	Sphingosine-1-phosphate
SphK	Sphingosine kinase
GPCR	G protein-coupled receptor
PLC	Phospholipase C
PI3K	Phosphatidylinositol 3-kinase
PI3K-C2α	Class IIα isoform of phosphatidylinositol 3-kinase
PAF	Platelet-activating factor
eNOS	Endothelial NO synthase
ERK	Extracellular signal-regulated kinase
JNK	Jun N-terminal kinase
MAPK	Mitogen-activated protein kinase
PTEN	Phosphatase and Tensin Homolog Deleted from Chromosome 10
HUVECs	Human umbilical vein endothelial cells
VEGF	Vascular endothelial growth factor

Lysophospholipid sphingosine-1-phosphate (S1P) is an extracellular signaling molecule abundantly present in the plasma. S1P signaling is pivotal for embryonic development of vascular and nervous systems and postnatal regulation of cardiovascular and immune cell physiology [1–7]. S1P signaling is also implicated in diverse pathological conditions, which include tumor angiogenesis, atherosclerosis, inflammation, and fibrosis [2, 3, 5, 8–10].

S1P was first recognized in the early 1990s as a bioactive lipid that exhibits mitogenic activity for cultured fibroblasts [11] and a unique bimodal activity to either stimulate or inhibit cell migration [12]. In the late 1990s the two related orphan G protein-coupled receptors (GPCR) that had been independently cloned from the vasculature in the early 1990s [13–15] were identified as the S1P-specific receptors $S1P_1$/endothelial differentiation gene-1 (Edg-1) and $S1P_2$/Edg-5/AGR16/H218 [16–18]. Three additional S1P-specific receptors, $S1P_3$, $S1P_4$, and $S1P_5$, have been thereafter identified so that the S1P receptor family contains five members, $S1P_1$–$S1P_5$, in mammals [19–21], among which $S1P_1$, $S1P_2$, and $S1P_3$ are the major ones, expressed in a wide variety of tissues including the vasculature and blood cells.

In this chapter, we briefly overview the S1P signaling mechanisms and then focus on distinctive yet concerted regulation of vascular formation and endothelial barrier function by $S1P_1$ and $S1P_2$ receptors, which are the most and the least abundantly expressed subtypes in endothelial cells, respectively.

17.1 Synthesis, Export, and Delivery of S1P to Endothelial Cells

S1P is synthesized through phosphorylation of membrane sphingolipid-derived sphingosine by the sphingosine kinases (SphKs) SphK1 and SphK2 [22]. S1P is either dephosphorylated back to sphingosine by lipid phosphate phosphatases (LPP1–LPP3) [23] or S1P phosphatases (SPP1, SPP2) [24], or cleaved to hexadecenal and ethanolamine-phosphate by S1P lyase (SPL) [25]. The products generated by S1P lyase are utilized to synthesize the glycerophospholipid phosphatidylethanolamine. Thus, intracellular S1P is an important intermediate of membrane lipid metabolism located at the intersection of sphingolipid and glycerolphospholipid metabolic pathways. Mice deficient for either SphK1 or SphK2 are viable and fertile, whereas double-knockout (KO) mice are embryonic lethal with an undetectable level of S1P, indicating that SphK1 and SphK2 are functionally redundant and solely responsible for S1P production in vivo [26].

The most distinct feature of the S1P signaling system is that the ligand S1P is persistently and abundantly present in plasma at a concentration about 1–3 μM [27–29]. S1P is produced mainly by the action of SphK1 in a constitutive manner, mostly by erythrocytes [28] and vascular endothelial cells [30], which are two major cell types in direct contact with plasma. SphK1KO, but not SphK2KO, mice show a 50 % reduction in plasma S1P level [27]. Blood flow shear stress on endothelial cells stimulates release of S1P, through downregulation of S1P lyase and S1P phosphatase [30]. In addition, in the case of vascular damage activated platelets release S1P [31]. S1P concentration in serum is approximately twofold greater than that of plasma [27], most likely because of S1P release from aggregated platelets [31]. S1P is exported from erythrocytes and endothelial cells through S1P transporters into plasma. Spns2 is responsible for S1P export from endothelial cells [32]. Although ABC family transporters were suggested to mediate S1P export from erythrocytes, the molecular entity of an S1P transporter in erythrocytes remains to be elucidated [33].

The majority (~98.5 %) of plasma S1P is bound to HDL (60 %), albumin (30 %), and other proteins, and free plasma S1P is estimated to be 15–45 nM. Apolipoprotein M in HDL particles provides a specific high-affinity binding site for S1P [34]. Many of the endothelium- and cardioprotective effects of HDL are attributable to HDL-associated S1P [34–36]. It is known that plasma S1P turns over rapidly [37]. Binding of a major portion of plasma S1P to the proteins enables stable delivery of S1P to target cells including endothelial cells and leukocytes. The confinement of S1P to the inside of blood vessels brings about a steep S1P concentration gradient, which is maintained between plasma and other extracellular compartments with the order of plasma>lymph>tissue interstitial fluid. The S1P concentration gradient allows $S1P_1$-dependent regulation of lymphocyte trafficking. The lymphatic endothelium is a source of S1P in lymph, which induces lymphatic patterning and drives chemotactic egress of lymphocytes from secondary lymphoid organs into lymph [38].

17.2 Signaling Mechanisms of $S1P_1$, $S1P_2$, and $S1P_3$ Receptors

$S1P_1$, $S1P_2$, and $S1P_3$ are widely expressed in most organs, mediating diverse actions of S1P [1–3, 5–10, 21]. Differing from $S1P_1$–$S1P_3$, the expression of $S1P_4$ is restricted to lymphoid tissues and the lung, and that of $S1P_5$ to the brain (especially oligodendrocytes), leukocytes, and spleen [1, 21].

The signaling mechanisms of $S1P_1$–$S1P_3$ are better characterized compared with $S1P_4$ and $S1P_5$. $S1P_1$, $S1P_2$, and $S1P_3$ activate overlapping yet distinctive intracellular signaling pathways (Fig. 17.1) [1–3, 7, 9, 16–21, 39–44]. $S1P_1$ couples exclusively to Gi to induce activation of Ras-ERK, class I PI3K-Akt, and small GTPase Rac signaling pathways. $S1P_1$ also moderately activates phospholipase C (PLC) and mobilizes Ca^{2+} (Fig. 17.1) [16, 17, 39–42]. In contrast to $S1P_1$, $S1P_2$ and $S1P_3$ couple to multiple G proteins, such as G_q, G_i, and $G_{12/13}$ [18–20]. $S1P_2$ stimulates Rho via $G_{12/13}$, PLC mainly via G_q, extracellular signal-regulated kinase (ERK) via G_i, and Jun N-terminal kinase (JNK) and p38 mitogen-activated protein kinase (MAPK) via pertussis toxin (PTX)-insensitive G protein (Fig. 17.1) [18]. Similar to $S1P_2$, $S1P_3$ also couples to G_q-mediated PLC stimulation, $G_{12/13}$-mediated Rho stimulation, and G_i-mediated ERK and Rac stimulation (Fig. 17.1) [19, 20]. Although $S1P_2$ and $S1P_3$ can similarly couple to G_q, G_i, and $G_{12/13}$ when overexpressed, an obvious difference in the two receptor subtypes exists in primary cells including mouse embryonic fibroblasts (MEFs): $S1P_2$ preferentially couples to the $G_{12/13}$-Rho pathway whereas $S1P_3$ preferentially couples to the G_q-PLC pathway [43, 44].

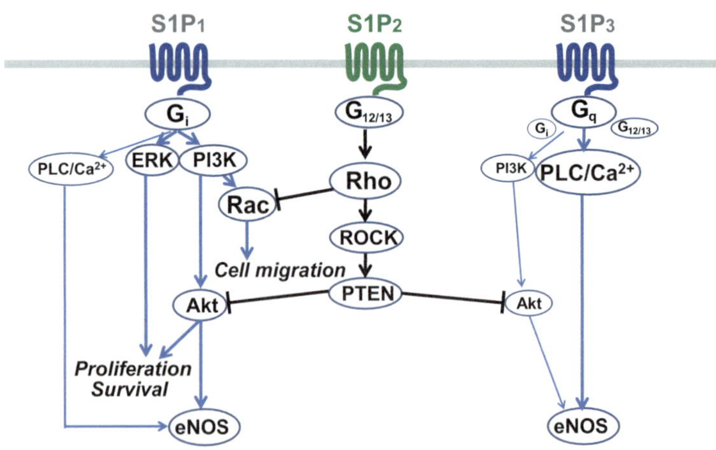

Fig. 17.1 Basic signaling mechanisms of $S1P_1$, $S1P_2$, and $S1P_3$. $S1P_1$ couples exclusively to G_i to activate Ras-ERK and PI3K(class I p110β)-Akt and -Rac, and also PLC-Ca^{2+} mobilization. $S1P_2$ and $S1P_3$ couple to multiple heterotrimeric G proteins. $S1P_2$ couples most prominently to $G_{12/13}$ to stimulate the Rho-ROCK(Rho kinase)-PTEN pathway. $S1P_3$ couples to G_q, stimulating PI3K(class I p110β)-Akt and PLC-Ca^{2+} mobilization

Because $S1P_1$, $S1P_2$, and $S1P_3$ are widely expressed, an integrated outcome of S1P signaling in a given cell type largely depends upon the relative expression levels of the S1P receptor subtypes. In addition, crosstalk between S1P receptor signaling and growth factor or cytokine receptor signaling has been reported.

17.3 Plasma S1P and Endothelial $S1P_1$ Contribute to Vascular Barrier Integrity

Anaphylaxis is elicited by antigen engagement to IgE on the surface of mast cells and basophils, which through FcεR activation leads to exocytosis of their granule contents such as histamine, and enzymatic generation of lipid mediators including platelet-activating factor (PAF) and leukotrienes in mast cells and macrophages. Such mediators of anaphylaxis activate the specific $G_{q/11}$-coupled receptors in vascular endothelial cells and airway smooth muscle cells to induce barrier disruption and bronchospasm, respectively, culminating in shock with cardiovascular collapse caused by plasma leakage, reduced vascular tone, and respiratory distress.

Vascular barrier integrity depends upon S1P in the plasma compartment. Camerer et al. [45] created "S1Pless" mice with a very low level (~30 nM) of plasma S1P by conditional knockout of one Sphk1 allele with poly(I:C)-treatment in mice of Mx1-Cre transgenic:Sphk1$^{flox/-}$:Sphk2$^{-/-}$ genotype. "S1Pless" mice showed increased vascular permeability at baseline and augmented extravasation of plasma either upon injection of the anaphylactic mediators histamine and PAF or upon passive systemic anaphylaxis. Resultantly, the lethality was much higher (50–80 %) in "S1Pless" mice than in non-lethal wild-type mice. PAF-induced lethality in "S1Pless" mice was completely prevented by prior transfusion of wild-type erythrocytes, a major source of plasma S1P. They also reported that Sphk1$^{-/-}$ mice, which had a 50 % reduction in plasma S1P level, showed increased lethality after PAF challenge. Olivera et al. [46] also reported that SphK1$^{-/-}$ mice showed more pronounced and sustained hypotension compared with wild-type mice after anaphylaxis or histamine injection, with progressive and irreversible hypothermia. Intravenous injection of S1P to SphK1$^{-/-}$ mice shortly after histamine administration led to rapid recovery from hypotension [46]. On the other hand, SphK2$^{-/-}$ mice, which had approximately twice as much plasma S1P as wild-type mice, showed facilitated recovery from shock compared with wild-type mice after anaphylaxis or histamine administration [46]. Thus, plasma S1P levels are inversely correlated with susceptibility to anaphylaxis. These observations indicate that plasma S1P plays a maintaining role in vascular barrier integrity at baseline and confers resistance to acute barrier-disruptive insult as in the case of anaphylactic shock.

The constitutive barrier-maintaining action of plasma S1P is mediated by $S1P_1$ in the endothelium: pretreatment with an $S1P_1$-specific agonist rescued "S1Pless" mice from lethality after PAF injection [45]. On the other hand, administration of an $S1P_1$-specific antagonist or pharmacological downregulation of $S1P_1$ rendered wild-type mice more susceptible to vascular leakage and lethality after PAF injection [46–49].

These observations indicate that $S1P_1$ mediates the barrier-protective effect of S1P (Fig. 17.2). Unexpectedly, however, inducible deletion of $S1P_1$ alleles in adult mice (because of embryonic lethality in $S1P_1$KO mice) did not compromise recovery from histamine injection [46]. One possibility to explain the discrepancy between the results obtained with different experimental models would be that relatively gradual deletion of $S1P_1$ function in inducible system allowed a compensatory rescue system to develop, whereas acute antagonism or downregulation of $S1P_1$, which rapidly became maximal, did not.

The maintenance of barrier integrity requires G_i protein [45]. $S1P_1$ is exclusively coupled to G_i to mediate multiple signaling pathways including Ras-ERK, PI3K-Akt, and Rac (Fig. 17.1). Previous in vitro studies showed that Rac and another Rho family GTPase Rho are both required for enhancing effect of S1P on VE–cadherin assembly at the adherens junction in HUVECs [50]. Because $S1P_1$ does not activate Rho, S1P receptor(s) that are capable of Rho activation, which include $S1P_2$ and $S1P_3$, also seem to be involved in the maintenance of barrier integrity. However, it is reported that genetic deletion of $S1P_3$ does not compromise vascular barrier integrity [46].

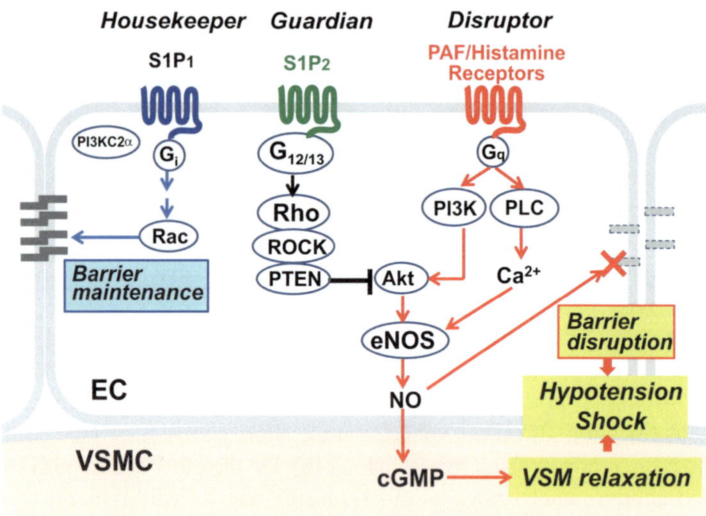

Fig. 17.2 $S1P_1$ and $S1P_2$ protect vascular barrier function through distinctive mechanisms. $S1P_1$ has a housekeeping function in endothelial barrier maintenance through mechanisms involving Rac activation, and in a manner dependent upon the class IIα isoform of PI3K (PI3K-C2α), which is required for internalization of activated $S1P_1$ receptor, endosomal Rac activation, and VE–cadherin assembly at the adherens junction. Endothelial barrier disruptors such as PAF and histamine act on G_q-coupled receptors to induce Ca^{2+} mobilization and Akt activation, which together lead to excess activation of eNOS, resulting in *S*-nitrosylation of β-catenin and barrier disruption. $S1P_2$ suppresses Akt and eNOS through the $G_{12/13}$-Rho-ROCK-PTEN pathway to prevent barrier disruption, acting as a guardian of endothelial barrier integrity. EC, endothelial cell; VSMC, vascular smooth muscle cell

17.4 Endothelial $S1P_2$ is Required for Barrier Protection Independently of $S1P_1$

At basal state, $S1P_2^{-/-}$ mice showed a similar extent of extravasation of injected Evans blue as wild-type mice, indicating a normal vascular barrier function at baseline in $S1P_2^{-/-}$ mice. However, $S1P_2^{-/-}$ mice were much more susceptible to anaphylaxis: they showed more exaggerated vascular leakage, hypotension, hypothermia, and lethality compared with wild-type mice on either antigen challenge or injections of anaphylactic mediators; nearly all $S1P_2$KO mice died 30 min after PAF injection, whereas none of the wild-type mice died [51]. Despite marked differences in the severity of anaphylaxis between $S1P_2^{-/-}$ and wild-type mice, plasma concentrations of histamine, a major mediator released from mast cells, were similar regardless of the genotypes [51], implying that $S1P_2$ deficiency did not affect mast cell activation. Olivera et al. [46] demonstrated delayed recovery from hypothermia after antigen challenge in $S1P_2^{-/-}$ mice compared with wild-type mice. Heterozygous $S1P_2^{+/-}$ mice also showed a similar extent of exaggerated hypothermia as homozygous $S1P_2^{-/-}$ mice [46]. Administration of S1P after histamine injection rapidly rescued $S1P_2^{+/-}$ but not $S1P_2^{-/-}$ mice from sustained hypothermia [46]. These findings indicate that reduced expression of $S1P_2$ in $S1P_2^{+/-}$ mice compromises barrier protection despite a normal level of $S1P_1$ expression, underscoring a unique and independent role for $S1P_2$ in barrier protection. In contrast to the studies by Olivera's group and ours, Oskeritzian et al. [52], reported totally different results showing that $S1P_2$ aggravated vascular leakage and lethality in anaphylaxis, which were attenuated in either $S1P_2$KO mice or mice receiving $S1P_2$-selective antagonist JTE013. Although the experimental anaphylactic models were different (passive anaphylaxis model in the study by Oskeritzian et al. vs. active anaphylaxis model in our study), the reason for the discrepancy is not known.

In endothelial cells, the anaphylactic mediators PAF and histamine stimulate both PLC and class I PI3Kβ to mobilize Ca^{2+} and activate Akt, respectively (Fig. 17.2). A rise in intracellular Ca^{2+} and stimulated Akt together fully activate eNOS to generate a large amount of NO. Therefore, the generation of NO is elevated in the vasculature in anaphylaxis. Either inhibition of PI3K or Akt [53], endothelial cell-specific genetic deletion of $G_{q/11}$ [54], or genetic deletion or pharmacological inhibition of eNOS [53] completely rescues otherwise fatal anaphylactic shock. NO leads to S-nitrosylation of β-catenin, which induces the disassembly of the VE–cadherin complex at adherens junctions and consequent barrier disruption [55, 56]. Thus, barrier disruption in anaphylactic shock totally depends on stimulation of NO production through $G_{q/11}$-PLC/PI3K-eNOS pathway [53, 54]. In $S1P_2$KO mice, activation of Akt is enhanced [57]. Resultantly, activation of eNOS and generation of NO in endothelial cells are exaggerated with augmented S-nitrosylation of β-catenin and disrupted VE–cadherin assembly. Genetic deletion of eNOS or pharmacological blockade of NOS almost completely rescues $S1P_2$KO mice from lethal shock [51]. These results reveal a novel barrier-protective role of endothelial $S1P_2$, that is, the prevention of exaggerated activation of endothelial

Akt-eNOS and thereby downstream *S*-nitrosylation of β-catenin and consequent barrier disruption (Fig. 17.2). $S1P_2$ also limits profound vasorelaxation and hypotension in anaphylaxis through suppressing NO-stimulated cGMP production in medial smooth muscle [51].

17.5 $S1P_2$ Inhibits Tumor Angiogenesis

$S1P_1$ mediates stimulatory effects of S1P on cell migration toward S1P, morphogenesis to form tube-like structures in vitro, and assembly of VE–cadherin at adherens junctions and consequent maintenance of endothelial barrier integrity. Global $S1P_1$KO mice and endothelial cell-specific conditional $S1P_1$KO mice are embryonic lethal because of defective vascular formation [58, 59]. Recent studies have disclosed that $S1P_1$ in endothelial cells inhibits, rather than stimulates, sprouting angiogenesis, which is otherwise excessive by the action of the angiogenic factor VEGF-A. Endothelial cell $S1P_1$ mediates promotion of VE–cadherin localization and assembly at adherens junctions, which suppresses tip cell formation and leads to the formation of a stabilized vascular network [60, 61].

$S1P_2$ is the first chemorepellant GPCR to be identified that mediates inhibition of cell migration through Rho and Rho-dependent inhibition of Rac [39]. In tumor cells such as B16 melanoma cells, which express $S1P_2$, S1P inhibits their migration, invasion of extracellular matrix, and hematogenous metastasis in a tail vein injection model [12, 62, 63]. The expression level of endogenous $S1P_2$ in cultured endothelial cells is substantially low or barely detectable [64]. In $S1P_2^{LacZ/+}$ mice in which LacZ gene was knocked in to $S1P_2$ locus, the expression of $S1P_2$ was identified in both of vascular endothelial cells and smooth muscle cells in the vasculature of various organs [57]. When overexpressed in endothelial cells, $S1P_2$ mediated S1P-induced inhibition of cell migration and tube formation in vitro and inhibition of angiogenesis in Matrigel plug assay in vivo [65]. Overexpression of $S1P_2$ in endothelial cells resulted in decreased barrier function when challenged with thrombin in vitro, for which Rho-ROCK-PTEN-dependent inhibition of Rac led to disassembly of VE–cadherin [66].

In the in vivo model of subcutaneous tumor inoculation, host $S1P_2$ that is expressed in both endothelial cells and angiogenic myeloid cells, the latter types of cells being recruited from bone marrow to tumor stroma, inhibited tumor angiogenesis and tumor progression [57]. Lung microvascular endothelial cells derived from $S1P_2$KO mice showed enhanced activities of Rac and Akt but not ERK, and stimulated cell proliferation, cell migration, and tube formation compared with those derived from wild-type mice. $S1P_2$-mediated inhibition of Akt is likely involved in inhibition of endothelial cell proliferation. $S1P_2$-mediated inhibition of Akt is the consequence of Rho kinase-dependent activation of PTEN (3′-specific phosphoinositide phosphatase). Rho kinase is dispensable, however, for $S1P_2$-mediated inhibition of Rac and cell migration, which are still observed in PTEN-deficient tumor cells [67]. $S1P_2$KO mice develop deafness and ataxia from aberrant vascular formation in the inner ear and loss of hair cells, and epileptic seizures and sudden death in the weaning period in C57BL/6 background [68–71].

$S1P_1$/$S1P_2$-double null and $S1P_1$/$S1P_2$/$S1P_3$-triple null embryos show increasingly more severe defects in vascular formation compared with single $S1P_1$-null embryos [72]. $S1P_3$KO mice show the normal phenotype whereas $S1P_2$/$S1P_3$-double null mice show partial embryonic and perinatal lethality [43, 44, 72]. Therefore, it is possible that $S1P_2$ and $S1P_3$ have some functional redundancy in the regulation of vascular formation.

17.6 $S1P_2$ and $S1P_3$ Are Involved in the Regulation of Vascular Tone

$S1P_2$ and $S1P_3$ are involved also in the regulation of vascular tone [73–75]. In vascular smooth muscle cells, excitatory agonists induce Rho activation and potentiate contraction through Rho kinase-dependent phosphorylation of the myosin-targeting subunit MYPT1 of myosin phosphatase and consequent inhibition of the myosin phosphatase activity [76–81]. The Rho-Rho kinase-dependent mechanism also involves phosphorylation of the myosin phosphatase inhibitor protein, CPI-17 [82]. Myosin phosphatase inhibition mediated by $G_{12/13}$-dependent activation of Rho-Rho kinase, together with activation of myosin light chain kinase by $G_{q/11}$-PLC-dependent Ca^{2+} mobilization and Ca^{2+} influx, efficiently increases sustained phosphorylation of myosin light chain and consequent vascular contraction [80]. In addition, S1P2-mediated contraction mechanism involves inhibition of eNOS as already described.

$S1P_3$ couples moderately to G_i besides $G_{q/11}$ and $G_{12/13}$ (Fig. 17.1) [19, 39, 43]. Differing from $S1P_2$, $S1P_3$ in endothelial cells rather stimulates eNOS via the Gi pathway as does $S1P_1$, leading to endothelial cell dependent vasorelaxation.

17.7 Class II PI 3-Kinase-C2α Is Required for $S1P_1$ Signaling and Delivery of VE–Cadherin to Adherens Junction

We have recently reported that the class IIα isoform of PI 3-kinase (PI3K-C2α), which specifically phosphorylates PI-4-P to produce PI-3,4-P_2 [83], has essential roles in endothelial cell functions, including embryonic and pathological angiogenesis and vascular barrier maintenance [84]. Global and endothelial cell-specific C2α-null mice were both embryonic lethal because of defective angiogenesis [84]. C2α-deficient heterozygous mice showed defective vascular barrier function, as evidenced by increased Evans blue dye leakage in the skin after subcutaneous injection of VEGF, and in the aorta and heart after chronic injection of angiotensin II, as well as reduced localization of VE–cadherin at the cell–cell junction in *en face* confocal observation of aortic intima. Consistently, PI3K-C2α-deficient heterozygous mice showed markedly increased lethality after anaphylaxis and formation of dissecting aneurysm of the aorta after chronic administration of angiotensin II [84].

We have shown in HUVECs that PI3K-C2α is localized in clathrin-coated vesicles, the *trans*-Golgi network, and early endosomes [84]. PI3K-C2α is required for $S1P_1$-mediated activation of Rac1, in which S1P-stimulated $S1P_1$ receptor is internalized to endosomal vesicular membrane where it co-localizes with active Rac1 [85]. PI3K-C2α is also required for delivery of VE–cadherin to the adherens junction, which depends upon endosomal Rac1 activation [84, 85]. S1P-induced activation of cell migration, lamellipodium formation, and tube formation, all of which depend upon Rac activation, were markedly inhibited by knockdown of either PI3K-C2α or class I p110β, but not p110α or Vps34 [85]. By contrast, only p110β was necessary for S1P-iduced activation of Akt. Fluorescent resonance energy transfer (FRET) imaging showed that S1P induced Rac1 activation in both the plasma membrane and endosomes. Knockdown of PI3K-C2α, but not p110β, markedly reduced S1P-induced $S1P_1$ internalization into endosomes and Rac1 activation. Pharmacological inhibition of endocytosis by Dynasore disrupted VE–cadherin assembly at the cell-to-cell junction and suppressed S1P-induced $S1P_1$ internalization and Rac1 activation, but not activation of Akt or ERK [85].

PI3K-C2α is required not only for $S1P_1$-mediated activation of Rac and Rac-dependent cellular processes, which depends upon endocytic internalization of the $S1P_1$ receptor and its signaling that emanates from endosomes, but also for VEGFR2-mediated RhoA activation and RhoA-dependent assembly of VE–cadherin, which also depends upon endocytic internalization of activated VEGFR2 [84].

17.8 Concluding Remarks

Endothelial barrier integrity requires the persistent presence of high concentrations of S1P in the plasma, which stimulates both $S1P_1$ and $S1P_2$ to exert barrier-protective effects. $S1P_1$, a principal S1P receptor in endothelial cells, has a housekeeping function in maintaining barrier function through mechanisms involving Rac activation, for which PI3K-C2α-dependent internalization of activated $S1P_1$ receptor is required. $S1P_2$ is much less abundant in endothelial cells than $S1P_1$. However, $S1P_2$ is crucial in mediating the barrier-protective role of S1P: $S1P_2$ counteracts anaphylactic mediator-induced barrier disruption through suppressing exaggerated activation of Akt and its target eNOS. $S1P_2$ also inhibits tumor angiogenesis, tumor invasion, and metastasis. These observations suggest that the $S1P_2$-selective agonist is a promising therapeutic tactic for cancer as well as anaphylaxis.

Acknowledgments This work was supported by grants from the Ministry of Education, Science, Sports and Culture of Japan, the Japan Society for the Promotion of Science, and grants for Core Research for Evolutional Science and Technology from JST, and IPNU Research Promotion Fund.

References

1. Blaho VA, Hla T (2014) An update on the biology of sphingosine 1-phosphate receptors. J Lipid Res 55:1596–1608
2. Takuwa Y, Ikeda H, Okamoto Y, Takuwa N, Yoshioka K (2013) Sphingosine-1-phosphate as a mediator involved in development of fibrotic diseases. Biochim Biophys Acta 1831:185–192. doi:10.1016/j.bbalip.2012.06.008
3. Takuwa Y, Okamoto Y, Yoshioka K, Takuwa N (2012) Sphingosine-1-phosphate signaling in physiology and diseases. Biofactors 38:329–337. doi:10.1002/biof.1030
4. Cyster JG, Schwab SR (2012) Sphingosine-1-phosphate and lymphocyte egress from lymphoid organs. Annu Rev Immunol 30:69–94. doi:10.1146/annurev-immunol-020711-075011
5. Skoura A, Hla T (2009) Lysophospholipid receptors in vertebrate development, physiology, and pathology. J Lipid Res 50(suppl):S293–S298. doi:10.1194/jlr.R800047-JLR200
6. Kono M, Allende ML, Proia RL (2008) Sphingosine-1-phosphate regulation of mammalian development. Biochim Biophys Acta 1781:435–441. doi:10.1016/j.bbalip.2008.07.001
7. Takuwa Y, Okamoto Y, Yoshioka K, Takuwa N (2008) Sphingosine-1-phosphate signaling and biological activities in the cardiovascular system. Biochim Biophys Acta 1781:483–488
8. Obinata H, Hla T (2012) Sphingosine 1-phosphate in coagulation and inflammation. Semin Immunopathol 34:73–91. doi:10.1007/s00281-011-0287-3
9. Takuwa N, Du W, Kaneko E, Okamoto Y, Yoshioka K, Takuwa Y (2011) Tumor-suppressive sphingosine-1-phosphate receptor-2 counteracting tumor-promoting sphingosine-1-phosphate receptor-1 and sphingosine kinase 1: Jekyll Hidden behind Hyde. Am J Cancer Res 1:460–481
10. Takuwa Y, Du W, Qi X, Okamoto Y, Takuwa N, Yoshioka K (2010) Roles of sphingosine-1-phosphate signaling in angiogenesis. World J Biol Chem 1:298–306. doi:10.4331/wjbc.v1.i10.298
11. Zhang H, Desai NN, Olivera A, Seki T, Brooker G, Spiegel S (1991) Sphingosine-1-phosphate, a novel lipid, involved in cellular proliferation. J Cell Biol 114:155–167
12. Sadahira Y, Ruan F, Hakomori S, Igarashi Y (1992) Sphingosine 1-phosphate, a specific endogenous signaling molecule controlling cell motility and tumor cell invasiveness. Proc Natl Acad Sci U S A 89:9686–9690
13. Hla T, Maciag T (1990) An abundant transcript induced in differentiating human endothelial cells encodes a polypeptide with structural similarities to G-protein-coupled receptors. J Biol Chem 265:9308–9313
14. Okazaki H, Ishizaka N, Sakurai T, Kurokawa K, Goto K, Kumada M, Takuwa Y (1993) Molecular cloning of a novel putative G protein-coupled receptor expressed in the cardiovascular system. Biochem Biophys Res Commun 190:1104–1109
15. MacLennan AJ, Browe CS, Gaskin AA, Lado DC, Shaw G (1994) Cloning and characterization of a putative G-protein coupled receptor potentially involved in development. Mol Cell Neurosci 5:201–209
16. Lee MJ, Van Brocklyn JR, Thangada S, Liu CH, Hand AR, Menzeleev R, Spiegel S, Hla T (1998) Sphingosine-1-phosphate as a ligand for the G protein-coupled receptor EDG-1. Science 279:1552–1555
17. Okamoto H, Takuwa N, Gonda K, Okazaki H, Chang K, Yatomi Y, Shigematsu H, Takuwa Y (1998) EDG1 is a functional sphingosine-1-phosphate receptor that is linked via a $G_{i/o}$ to multiple signaling pathways, including phospholipase C activation, Ca^{2+} mobilization, Ras-mitogen-activated protein kinase activation, and adenylate cyclase inhibition. J Biol Chem 273:27104–27110
18. Gonda K, Okamoto H, Takuwa N, Yatomi Y, Okazaki H, Sakurai T, Kimura S, Sillard R, Harii K, Takuwa Y (1999) The novel sphingosine 1-phosphate receptor AGR16 is coupled via pertussis toxin-sensitive and -insensitive G-proteins to multiple signalling pathways. Biochem J 337:67–75

19. Okamoto H, Takuwa N, Yatomi Y, Gonda K, Shigematsu H, Takuwa Y (1999) EDG3 is a functional receptor specific for sphingosine 1-phosphate and sphingosylphosphorylcholine with signaling characteristics distinct from EDG1 and AGR16. Biochem Biophys Res Commun 260:203–208
20. Sato K, Kon J, Tomura H, Osada M, Murata N, Kuwabara A, Watanabe T, Ohta H, Ui M, Okajima F (1999) Activation of phospholipase C-Ca^{2+} system by sphingosine 1-phosphate in CHO cells transfected with Edg-3, a putative lipid receptor. FEBS Lett 443:25–30
21. Chun J, Hla T, Lynch KR, Spiegel S, Moolenaar WH (2010) International Union of Basic and Clinical Pharmacology. LXXVIII. Lysophospholipid receptor nomenclature. Pharmacol Rev 62:579–587. doi:10.1124/pr.110.003111
22. Spiegel S, Milstien S (2007) Functions of the multifaceted family of sphingosine kinases and some close relatives. J Biol Chem 282:2125–2129
23. Waggoner DW, Xu J, Singh I, Jasinska R, Zhang QX, Brindley DN (1999) Structural organization of mammalian lipid phosphate phosphatases: implications for signal transduction. Biochim Biophys Acta 1439:299–316
24. Mandala SM, Thornton R, Galve-Roperh I, Poulton S, Peterson C, Olivera A, Bergstrom J, Kurtz MB, Spiegel S (2000) Molecular cloning and characterization of a lipid phosphohydrolase that degrades sphingosine-1-phosphate and induces cell death. Proc Natl Acad Sci U S A 97:7859–7864
25. Fyrst H, Saba J (2008) Sphingosine-1-phosphate lyase in developmental and disease: sphingolipid metabolism takes flight. Biochim Biophys Acta 1781:448–458
26. Mizugishi K, Yamashita T, Olivera A, Miller GF, Spiegel S, Proia RL (2005) Essential role for sphingosine kinases in neural and vascular development. Mol Cell Biol 25:11113–11121
27. Allende ML, Sasaki T, Kawai H, Olivera A, Mi Y, van Echten-Deckert G, Hajdu R, Rosenbach M, Keohane CA, Mandala S, Spiegel S, Proia RL (2004) Mice deficient in sphingosine kinase 1 are rendered lymphopenic by FTY720. J Biol Chem 279:52487–52492
28. Pappu R, Schwab SR, Cornelissen I, Pereira JP, Regard JB, Xu Y, Camerer E, Zheng YW, Huang Y, Cyster JG, Coughlin SR (2007) Promotion of lymphocyte egress into blood and lymph by distinct sources of sphingosine-1-phosphate. Science 316:295–298
29. Camerer E, Regard JB, Cornelissen I, Srinivasan Y, Duong DN, Palmer D, Pham TH, Wong JS, Pappu R, Coughlin SR (2009) Sphingosine-1-phosphate in the plasma compartment regulates basal and inflammation-induced vascular leak in mice. J Clin Invest 119:1871–1879
30. Venkataraman K, Lee YM, Michaud J, Thangada S, Ai Y, Bonkovsky HL, Parikh NS, Habrukowich C, Hla T (2008) Vascular endothelium as a contributor of plasma sphingosine 1-phosphate. Circ Res 102:669–676
31. Yatomi Y, Igarashi Y, Yang L, Hisano N, Qi R, Asazuma N, Satoh K, Ozaki Y, Kume S (1997) Sphingosine 1-phosphate, a bioactive sphingolipid abundantly stored in platelets, is a normal constituent of human plasma and serum. J Biochem 121:969–973
32. Kawahara A, Nishi T, Hisano Y, Fukui H, Yamaguchi A, Mochizuki N (2009) The sphingolipid transporter spns2 functions in migration of zebrafish myocardial precursors. Science 323:524–527. doi:10.1126/science.1167449
33. Kim RH, Takabe K, Milstien S, Spiegel S (2009) Export and functions of sphingosine-1-phosphate. Biochim Biophys Acta 1791:692–696. doi:10.1016/j.bbalip.2009.02.011
34. Christoffersen C, Obinata H, Kumaraswamy SB, Galvani S, Ahnström J, Sevvana M, Egerer-Sieber C, Muller YA, Hla T, Nielsen LB, Dahlbäck B (2011) Endothelium-protective sphingosine-1-phosphate provided by HDL-associated apolipoprotein M. Proc Natl Acad Sci U S A 108:9613–9618. doi:10.1073/pnas.1103187108
35. Okajima F (2002) Plasma lipoproteins behave as carriers of extracellular sphingosine 1-phosphate: is this an atherogenic mediator or an anti-atherogenic mediator? Biochim Biophys Acta 1582:132–137
36. Karliner JS (2013) Sphingosine kinase and sphingosine 1-phosphate in the heart: a decade of progress. Biochim Biophys Acta 1831:203–212. doi:10.1016/j.bbalip.2012.06.006

37. Kharel Y, Mathews TP, Gellett AM, Tomsig JL, Kennedy PC, Moyer ML, Macdonald TL, Lynch KR (2011) Sphingosine kinase type 1 inhibition reveals rapid turnover of circulating sphingosine 1-phosphate. Biochem J 440:345–353. doi:10.1042/BJ20110817
38. Pham TH, Baluk P, Xu Y, Grigorova I, Bankovich AJ, Pappu R, Coughlin SR, McDonald DM, Schwab SR, Cyster JG (2010) Lymphatic endothelial cell sphingosine kinase activity is required for lymphocyte egress and lymphatic patterning. J Exp Med 207:17–27. doi: 10.1084/jem.20091619
39. Okamoto H, Takuwa N, Yokomizo T, Sugimoto N, Sakurada S, Shigematsu H, Takuwa Y (2000) Inhibitory regulation of Rac activation, membrane ruffling, and cell migration by the G protein-coupled sphingosine-1-phosphate receptor EDG5 but not EDG1 or EDG3. Mol Cell Biol 20:9247–9261
40. Takuwa Y, Okamoto H, Takuwa N, Gonda K, Sugimoto N, Sakurada S (2001) Subtype-specific, differential activities of the EDG family receptors for sphingosine-1-phosphate, a novel lysophospholipid mediator. Mol Cell Endocrinol 177:3–11
41. Takuwa Y, Takuwa N, Sugimoto N (2002) The Edg family G protein-coupled receptors for lysophospholipids: their signaling properties and biological activities. J Biochem 131:767–771
42. Takuwa Y (2002) Subtype-specific differential regulation of Rho family G proteins and cell migration by the Edg family sphingosine-1-phosphate receptors. Biochim Biophys Acta 1582:112–120
43. Ishii I, Friedman B, Ye X, Kawamura S, McGiffert C, Contos JJ, Kingsbury MA, Zhang G, Brown JH, Chun J (2001) Selective loss of sphingosine 1-phosphate signaling with no obvious phenotypic abnormality in mice lacking its G protein-coupled receptor, LP(B3)/EDG-3. J Biol Chem 276:33697–33704
44. Ishii I, Ye X, Friedman B, Kawamura S, Contos JJ, Kingsbury MA, Yang AH, Zhang G, Brown JH, Chun J (2002) Marked perinatal lethality and cellular signaling deficits in mice null for the two sphingosine 1-phosphate (S1P) receptors, S1P(2)/LP(B2)/EDG-5 and S1P(3)/LP(B3)/EDG-3. J Biol Chem 277:25152–25159
45. Camerer E, Regard JB, Cornelissen I, Srinivasan Y, Duong DN, Palmer D, Pham TH, Wong JS, Pappu R, Coughlin SR (2009) Sphingosine-1-phosphate in the plasma compartment regulates basal and inflammation-induced vascular leak in mice. J Clin Invest 119:1871–1879
46. Olivera A, Eisner C, Kitamura Y, Dillahunt S, Allende L, Tuymetova G, Watford W, Meylan F, Diesner SC, Li L, Schnermann J, Proia RL, Rivera J (2010) Sphingosine kinase 1 and sphingosine-1-phosphate receptor 2 are vital to recovery from anaphylactic shock in mice. J Clin Invest 120:1429–1440. doi:10.1172/JCI40659
47. Sanna MG, Wang SK, Gonzalez-Cabrera PJ, Don A, Marsolais D, Matheu MP, Wei SH, Parker I, Jo E, Cheng WC, Cahalan MD, Wong CH, Rosen H (2006) Enhancement of capillary leakage and restoration of lymphocyte egress by a chiral S1P1 antagonist in vivo. Nat Chem Biol 2:434–441
48. Marsolais D, Rosen H (2009) Chemical modulators of sphingosine-1-phosphate receptors as barrier-oriented therapeutic molecules. Nat Rev Drug Discov 8:297–307. doi:10.1038/nrd2356
49. Brinkmann V, Cyster JG, Hla T (2004) FTY720: sphingosine 1-phosphate receptor-1 in the control of lymphocyte egress and endothelial barrier function. Am J Transplant 4:1019–1025
50. Lee MJ, Thangada S, Claffey KP, Ancellin N, Liu CH, Kluk M, Volpi M, Sha'afi RI, Hla T (1999) Vascular endothelial cell adherens junction assembly and morphogenesis induced by sphingosine-1-phosphate. Cell 99:301–312
51. Cui H, Okamoto Y, Yoshioka K, Du W, Takuwa N, Zhang W, Asano M, Shibamoto T, Takuwa Y (2013) Sphingosine-1-phosphate receptor 2 protects against anaphylactic shock through suppression of endothelial nitric oxide synthase in mice. J Allergy Clin Immunol 132:1205–1214.e9. doi:10.1016/j.jaci.2013.07.026
52. Oskeritzian CA, Price MM, Hait NC, Kapitonov D, Falanga YT, Morales JK, Ryan JJ, Milstien S, Spiegel S (2010) Essential roles of sphingosine-1-phosphate receptor 2 in human mast cell

activation, anaphylaxis, and pulmonary edema. J Exp Med 207:465–474. doi:10.1084/jem.20091513
53. Cauwels A, Janssen B, Buys E, Sips P, Brouckaert P (2006) Anaphylactic shock depends on PI3K and eNOS-derived NO. J Clin Invest 116:2244–2251
54. Korhonen H, Fisslthaler B, Moers A, Wirth A, Habermehl D, Wieland T, Schütz G, Wettschureck N, Fleming I, Offermanns S (2009) Anaphylactic shock depends on endothelial Gq/G11. J Exp Med 206:411–420. doi:10.1084/jem.20082150
55. Thibeault S, Rautureau Y, Oubaha M, Faubert D, Wilkes BC, Delisle C, Gratton JP (2010) S-nitrosylation of beta-catenin by eNOS-derived NO promotes VEGF-induced endothelial cell permeability. Mol Cell 39:468–476. doi:10.1016/j.molcel.2010.07.013
56. Marín N, Zamorano P, Carrasco R, Mujica P, González FG, Quezada C, Meininger CJ, Boric MP, Durán WN, Sánchez FA (2012) S-Nitrosation of β-catenin and p120 catenin: a novel regulatory mechanism in endothelial hyperpermeability. Circ Res 111:553–563. doi:10.1161/CIRCRESAHA.112.274548
57. Du W, Takuwa N, Yoshioka K, Okamoto Y, Gonda K, Sugihara K, Fukamizu A, Asano M, Takuwa Y (2010) $S1P_2$, the G protein-coupled receptor for sphingosine-1-phosphate, negatively regulates tumor angiogenesis and tumor growth in vivo in mice. Cancer Res 70:772–781. doi:10.1158/0008-5472.CAN-09-2722
58. Liu Y, Wada R, Yamashita T, Mi Y, Deng CX, Hobson JP, Rosenfeldt HM, Nava VE, Chae SS, Lee MJ, Liu CH, Hla T, Spiegel S, Proia RL (2000) Edg-1, the G protein-coupled receptor for sphingosine-1-phosphate, is essential for vascular maturation. J Clin Invest 106:951–961
59. Allende ML, Yamashita T, Proia RL (2003) G-protein-coupled receptor S1P1 acts within endothelial cells to regulate vascular maturation. Blood 102:3665–3667
60. Ben Shoham A, Malkinson G, Krief S, Shwartz Y, Ely Y, Ferrara N, Yaniv K, Zelzer E (2012) S1P1 inhibits sprouting angiogenesis during vascular development. Development 139:3859–3869
61. Gaengel K, Niaudet C, Hagikura K, Laviña B, Muhl L, Hofmann JJ, Ebarasi L, Nyström S, Rymo S, Chen LL, Pang MF, Jin Y, Raschperger E, Roswall P, Schulte D, Benedito R, Larsson J, Hellström M, Fuxe J, Uhlén P, Adams R, Jakobsson L, Majumdar A, Vestweber D, Uv A, Betsholtz C (2012) The sphingosine-1-phosphate receptor S1PR1 restricts sprouting angiogenesis by regulating the interplay between VE-cadherin and VEGFR2. Dev Cell 23:587–599. doi:10.1016/j.devcel.2012.08.005
62. Arikawa K, Takuwa N, Yamaguchi H, Sugimoto N, Kitayama J, Nagawa H, Takehara K, Takuwa Y (2003) Ligand-dependent inhibition of B16 melanoma cell migration and invasion via endogenous S1P2 G protein-coupled receptor. Ligand-dependent inhibition of B16 melanoma cell migration and invasion via endogenous S1P2 G protein-coupled receptor. J Biol Chem 278:32841–32851
63. Yamaguchi H, Kitayama J, Takuwa N, Arikawa K, Inoki I, Takehara K, Nagawa H, Takuwa Y (2003) Sphingosine-1-phosphate receptor subtype-specific positive and negative regulation of Rac and haematogenous metastasis of melanoma cells. Biochem J 374(pt 3):715–722
64. Ryu Y, Takuwa N, Sugimoto N, Sakurada S, Usui S, Okamoto H, Matsui O, Takuwa Y (2002) Sphingosine-1-phosphate, a platelet-derived lysophospholipid mediator, negatively regulates cellular Rac activity and cell migration in vascular smooth muscle cells. Circ Res 90:325–332
65. Inoki I, Takuwa N, Sugimoto N, Yoshioka K, Takata S, Kaneko S, Takuwa Y (2006) Negative regulation of endothelial morphogenesis and angiogenesis by S1P2 receptor. Biochem Biophys Res Commun 2006(346):293–300
66. Sanchez T, Skoura A, Wu MT, Casserly B, Harrington EO, Hla T (2007) Induction of vascular permeability by the sphingosine-1-phosphate receptor-2 (S1P2R) and its downstream effectors ROCK and PTEN. Arterioscler Thromb Vasc Biol 27:1312–1318
67. Malchinkhuu E, Sato K, Maehama T, Mogi C, Tomura H, Ishiuchi S, Yoshimoto Y, Kurose H, Okajima F (2008) S1P(2) receptors mediate inhibition of glioma cell migration through Rho signaling pathways independent of PTEN. Biochem Biophys Res Commun 366:963–968

68. MacLennan AJ, Benner SJ, Andringa A, Chaves AH, Rosing JL, Vesey R, Karpman AM, Cronier SA, Lee N, Erway LC, Miller ML (2006) The S1P2 sphingosine 1-phosphate receptor is essential for auditory and vestibular function. Hear Res 220:38–48
69. Kono M, Belyantseva IA, Skoura A, Frolenkov GI, Starost MF, Dreier JL, Lidington D, Bolz SS, Friedman TB, Hla T, Proia RL (2007) Deafness and stria vascularis defects in S1P2 receptor-null mice. J Biol Chem 282:10690–10696
70. Herr DR, Grillet N, Schwander M, Rivera R, Müller U, Chun J (2007) Sphingosine 1-phosphate (S1P) signaling is required for maintenance of hair cells mainly via activation of S1P2. J Neurosci 27:1474–1478
71. MacLennan AJ, Carney PR, Zhu WJ, Chaves AH, Garcia J, Grimes JR, Anderson KJ, Roper SN, Lee N (2001) An essential role for the H218/AGR16/Edg-5/LP(B2) sphingosine 1-phosphate receptor in neuronal excitability. Eur J Neurosci 14:203–209
72. Kono M, Mi Y, Liu Y, Sasaki T, Allende ML, Wu YP, Yamashita T, Proia RL (2004) The sphingosine-1-phosphate receptors S1P1, S1P2, and S1P3 function coordinately during embryonic angiogenesis. J Biol Chem 279:29367–29373
73. Ohmori T, Yatomi Y, Osada M, Kazama F, Takafuta T, Ikeda H, Ozaki Y (2003) Sphingosine 1-phosphate induces contraction of coronary artery smooth muscle cells via S1P2. Cardiovasc Res 58:170–177
74. Lorenz JN, Arend LJ, Robitz R, Paul RJ, MacLennan AJ (2007) Vascular dysfunction in S1P2 sphingosine 1-phosphate receptor knockout mice. Am J Physiol Regul Integr Comp Physiol 292:R440–R446
75. Murakami A, Takasugi H, Ohnuma S, Koide Y, Sakurai A, Takeda S, Hasegawa T, Sasamori J, Konno T, Hayashi K, Watanabe Y, Mori K, Sato Y, Takahashi A, Mochizuki N, Takakura N (2010) Sphingosine 1-phosphate (S1P) regulates vascular contraction via S1P3 receptor: investigation based on a new S1P3 receptor antagonist. Mol Pharmacol 77:704–713. doi:10.1124/mol.109.061481
76. Noda M, Yasuda-Fukazawa C, Moriishi K, Kato T, Okuda T, Kurokawa K, Takuwa Y (1995) Involvement of rho in GTP gamma S-induced enhancement of phosphorylation of 20 kDa myosin light chain in vascular smooth muscle cells: inhibition of phosphatase activity. FEBS Lett 367:246–250
77. Takuwa Y (1996) Regulation of vascular smooth muscle contraction. The roles of Ca^{2+}, protein kinase C and myosin light chain phosphatase. Jpn Heart J 37:793–813
78. Nagumo H, Sasaki Y, Ono Y, Okamoto H, Seto M, Takuwa Y (2000) Rho kinase inhibitor HA-1077 prevents Rho-mediated myosin phosphatase inhibition in smooth muscle cells. Am J Physiol Cell Physiol 278:C57–C65
79. Sakurada S, Okamoto H, Takuwa N, Sugimoto N, Takuwa Y (2001) Rho activation in excitatory agonist-stimulated vascular smooth muscle. Am J Physiol Cell Physiol 281:C571–C578
80. Takuwa Y (2003) Regulation of the Rho signaling pathway by excitatory agonists in vascular smooth muscle. Adv Exp Med Biol 538:67–75
81. Takagi T, Okamoto Y, Tomita S, Sato A, Yamaguchi S, Takuwa Y, Watanabe G (2011) Intraradial administration of fasudil inhibits augmented Rho kinase activity to effectively dilate the spastic radial artery during coronary artery bypass grafting surgery. J Thorac Cardiovasc Surg 142:e59–e65. doi:10.1016/j.jtcvs.2011.01.055
82. Koyama M, Ito M, Feng J, Seko T, Shiraki K, Takase K, Hartshorne DJ, Nakano T (2000) Phosphorylation of CPI-17, an inhibitory phosphoprotein of smooth muscle myosin phosphatase, by Rho-kinase. FEBS Lett 475:197–200
83. Posor Y, Eichhorn-Gruenig M, Puchkov D, Schöneberg J, Ullrich A, Lampe A, Müller R, Zarbakhsh S, Gulluni F, Hirsch E, Krauss M, Schultz C, Schmoranzer J, Noé F, Haucke V (2013) Spatiotemporal control of endocytosis by phosphatidylinositol-3,4-bisphosphate. Nature 499:233–237. doi:10.1038/nature12360

84. Yoshioka K, Yoshida K, Cui H, Wakayama T, Takuwa N, Okamoto Y, Du W, Qi X, Asanuma K, Sugihara K, Aki S, Miyazawa H, Biswas K, Nagakura C, Ueno M, Iseki S, Schwartz RJ, Okamoto H, Sasaki T, Matsui O, Asano M, Adams RH, Takakura N, Takuwa Y (2012) Endothelial PI3K-C2α, a class II PI3K, has an essential role in angiogenesis and vascular barrier function. Nat Med 18:1560–1569. doi:10.1038/nm.2928
85. Biswas K, Yoshioka K, Asanuma K, Okamoto Y, Takuwa N, Sasaki T, Takuwa Y (2013) Essential role of class II phosphatidylinositol-3-kinase-C2α in sphingosine 1-phosphate receptor-1-mediated signaling and migration in endothelial cells. J Biol Chem 2013(288):2325–2339. doi:10.1074/jbc.M112.409656

Chapter 18
Roles of Prostaglandins in Regulation of Pathological Angiogenesis and Lymphangiogenesis

Masataka Majima

Abstract Angiogenesis, the formation of new blood vessels from preexistent microvasculature, is an essential component of wound repair and tumor growth. Nonsteroidal anti-inflammatory drugs (NSAIDs) are known to suppress the incidence and progression of malignancies including colorectal cancers, and also to delay wound healing. However, the precise mechanisms are not fully elucidated. Recent results obtained from prostaglandin (PG) receptor knockout mice indicate that host stromal PGE type receptor signaling is crucial in tumor-associated angiogenesis. Implanted tumor growth and tumor-associated angiogenesis were markedly suppressed in EP3 receptor knockout mice ($EP3^{-/-}$), in comparison with their wild-type counterparts (WT). Tumor-associated angiogenesis in WT depends on vascular endothelial growth factor (VEGF). Major VEGF-expressing cells in stroma were CD3/Mac-1 double-negative fibroblasts, and stromal VEGF expression was markedly reduced in $EP3^{-/-}$. An EP3 receptor antagonist inhibited tumor growth and angiogenesis in WT. The wound healing process was significantly delayed in $EP3^{-/-}$. Transplantation of $EP3^{-/-}$ bone marrow cells revealed that recruitment of EP3-expressing bone marrow cells to wound granulation tissues was critical to the healing of wounds. Further, lymphangiogenesis was also enhanced by EP signaling in pathological conditions. These observations demonstrate the significance of EP receptor signaling to angiogenesis and lymphangiogenesis in vivo. Such signaling can be a good target for controlling angiogenesis and lymphangiogenesis in pathological conditions.

Keywords Prostaglandin • Thromboxane • Receptor • Angiogenesis • Lymphangiogenesis • Inflammation • Wound healing • Tumor • Vascular endothelial growth factor

M. Majima, M.D., Ph.D. (✉)
Department of Pharmacology, Kitasato University School of Medicine,
Kitasato 1-15-1, Minami-ku, Sagamihara, Kanagawa 252-0374, Japan
e-mail: mmajima@med.kitasato-u.ac.jp

T. Yokomizo, M. Murakami (eds.), *Bioactive Lipid Mediators*,
DOI 10.1007/978-4-431-55669-5_18

18.1 Angiogenesis

We are complex multicellular organisms, and all cells in the body require a finely controlled supply of oxygen and nutrients [1–3]. The diffusion of oxygen in the tissues is limited to 100–200 μm, and a highly developed vascular system has evolved to ensure that all cells are within this distance of a supply of oxygen. Angiogenesis, the process of new blood vessel development from preexisting vasculature, is indispensable to maintain the integrity of the body. This function involves endothelial cell division, selective degradation of the basement membrane and the surrounding extracellular matrix, endothelial cell migration, and the formation of a tubular structure. Once blood vessels have been established, the endothelial cells undergo tissue-specific changes to generate functionally active vessels. During embryogenesis, blood vessels are formed by the differentiation of endothelial cell precursors (angioblasts), which associate to form primitive vessels. This process is called vasculogenesis [3, 4].

Angiogenesis is subject to a complex control system with pro-angiogenic and antiangiogenic factors [3, 4]. In adults, angiogenesis is tightly controlled by this "angiogenic balance," that is, a physiological balance between the stimulatory and inhibitory signals for blood vessel formation. In normal status, the formation of new blood vessels occurs during wound healing, organ regeneration, and in the female reproductive system during ovulation, menstruation, and the formation of the placenta. It is also an important factor in several pathological processes such as tumor growth, rheumatoid arthritis, diabetic retinopathy, and psoriasis.

18.2 Angiogenic Switch

An angiogenic switch, which allows substantial blood supply, is activated in the quiescent vasculature around tumors, regulates tumor growth and metastasis, and decides ultimately the patient's prognosis [5]. Several growth factors, such as basic fibroblast growth factor, epidermal growth factor, and platelet-derived growth factor, that regulate tube formation of endothelial cells and pericyte structures are known. Besides these factors, the most important pro-angiogenic factor is vascular endothelial growth factor (VEGF). VEGF is a pro-angiogenic factor that is identified first as a factor acting only on the endothelial cells. VEGF also potentiates microvascular hyperpermeability, which can both precede and accompany angiogenesis. Inflammatory mediators such as prostaglandins (PGs) are known to enhance angiogenesis via upregulation of VEGF. In this review, we summarize the properties and functions of PGs in the process of angiogenesis in malignancy and other conditions. PGs may be one of the inducers of angiogenic switches.

18.3 Biosynthesis of Prostaglandins (PGs)

The arachidonic acid (AA) cascade is the biosynthetic pathway that involves the release of the n-6 polyunsaturated fatty acid AA from the sn-2 position of membrane phospholipids by a phospholipase A_2 (PLA_2) enzyme, and its subsequent metabolism to bioactive prostaglandins (PGs), thromboxanes, leukotrienes, and epoxy fatty acids, by cyclooxygenase (COX), lipoxygenase (LOX), and cytochrome P450 epoxygenase enzymes, coupled to specific terminal synthases. AA and its end products are involved in both physiological and pathological processes. Although pharmacological inhibitors can be used to investigate the role of key enzymes involved in AA release and metabolism of AA in physiological and pathological models, the lack, in some cases, of specific inhibitors or of a complete pharmacological inhibition, and standardization of dosing paradigms complicate the studies. Two COX isoforms have been identified: COX-1 is constitutively expressed in various tissues, whereas COX-2 is induced by mitogens, cytokines, and tumor promoters [6]. COX regulates the formation of an unstable endoperoxide intermediate, PGH_2, which, in turn, is metabolized to PGD_2, PGE_2, $PGF_{2\alpha}$, PGI_2, and thromboxane $(TX)A_2$ by cell-specific isomerases and synthases. Prostanoids formed are immediately released outside the cell; they are either chemically or metabolically unstable, and thus prostanoids are believed to work only locally, near their site of production. PGI_2 and TXA_2 spontaneously degrade into inactive compounds under physiological conditions, and other PGs are enzymatically inactivated during a single passage through the lung. PGD_2 and PGE_2 are slowly dehydrated in biological fluids containing serum albumin to be the cyclopentenone PGs.

18.4 Receptors for PGs

Prostanoids exert their actions via membrane receptors on the surface of target cells. Genes for each of these receptors have been disrupted and the corresponding knockout mice have been produced [7]. Furthermore, with the use of cloned receptors, agonists and antagonists highly selective for each of the four EP subtypes have been developed. Eight types and subtypes of membrane prostanoid receptors are conserved in mammals, from mouse to human [7]: the PGD receptor (DP), four subtypes of the PGE receptor (EP1, EP2, EP3, EP4), the PGF receptor (FP), the PGI receptor (IP), and the TXA receptor (TP). All receptors are G protein-coupled rhodopsin-type receptors with seven transmembrane domains, and each is encoded by different genes. There are several splice variants in the EP3, FP, and TP receptors, which differ only in their C-terminal tails. Among the eight types and subtypes, the IP, DP, EP2, and EP4 receptors mediate a cAMP rise, whereas the TP, FP, and

Table 18.1 Roles of prostaglandins (PGs) revealed by studies using prostaglandin receptor knockout mice

PGs	Receptors	Pathophysiological roles
PGD_2	DP	A mediator of allergic asthma
PGE_2	EP1	Inhibition of gastric motor activity
	EP2	Ovulation and fertilization, salt-sensitive hypertension
	EP3	A mediator of febrile response, acid-induced bicarbonate secretion in duodenum, urinary concentration, mast cell activation, angiogenesis, lymphangiogenesis
	EP4	Closure of ductus arteriosus, bone resorption, lymphangiogenesis
PGI_2	IP	Antithrombotic function, adaptive cytoprotection of stomach, acetic acid-induced writhing reaction
$PGF_2\alpha$	FP	An essential inducer of labor
TXA_2	TP	Hemostasis, endotoxicin-induced hepatic microcirculatory dysfunction, lymphangiogenesis

EP1 receptors induce calcium mobilization. EP3 has several splice variants, which increase or decrease cAMP levels and induce calcium mobilization. The effects of prostanoids on these G protein-coupled signaling pathways are reported to be changed in a ligand concentration-dependent manner. PGI_2 analogues bind to IP and activate adenylate cyclase via Gs in a dose-dependent manner, but higher concentrations of the ligand couple to IP and activate phospholipase C to mobilize calcium ions via Gq. There are four subtypes of the receptor for PGE_2, although the other prostanoids each have only a single receptor. The homology of amino acid sequences between different types of the receptors within each functional group is much higher than that found among the four PGE receptor subtypes. The phylogenetic tree derived from receptor homologies indicates that prostanoid receptors originated from the primitive PGE receptor. Other PGs and TX receptors subsequently evolved from functionally related PGE receptor subtypes by gene duplication. The roles of PGs in various physiological and pathophysiological processes have been clarified with mice deficient in each prostanoid receptor. The findings, including ours, with use of knockout mice are summarized in Table 18.1. In this review article, we discuss the significance of the findings related to angiogenesis together with lymphangiogenesis in vivo.

18.5 Roles of PGs in Angiogenesis Indispensable to Wound Healing and Recovery from Ischemia

Angiogenesis is believed to be an essential component of normal wound repair. Immediately following injury, it delivers oxygen, nutrients, and inflammatory cells to the site of injury. It also assists the development of granulation tissue formation and ultimately wound closure. Both angiogenic agonists and antagonists are identified at various stages of the wound repair process [8], suggesting a dynamic balance

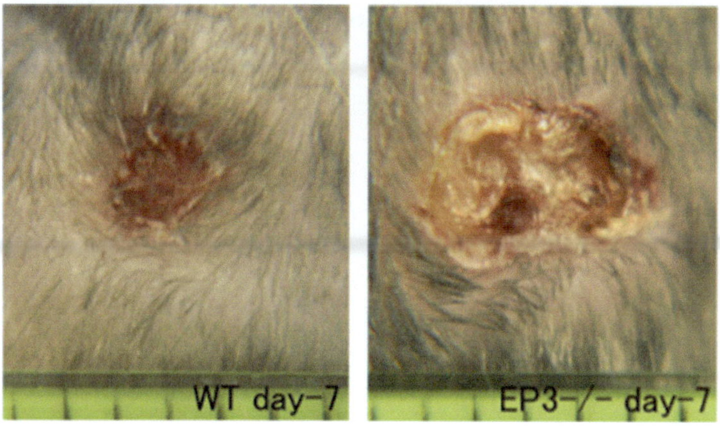

Fig. 18.1 Delay in wound healing in EP3 knockout mice. Typical appearance of wounds 7 days after wounding. Surgical wounds were made on the backs of EP3 receptor knockout mice ($EP3^{-/-}$) and of their wild-type counterparts (WT). Original diameter of the wounds was 8 mm. One division on the scale below the wound represents 1 mm (Cited from Kamoshita et al. [13] with permission)

of stimulators and inhibitors that favor either vascular growth or regression [9]. It has been reported that E-type PGs have a pro-angiogenic activity in corneal tests [10] and in the chorioallantoic membrane (CAM) technique [11]. Further, Form and Auerbach reported that PGE_2 strongly induced angiogenesis on the CAM of 8-day-old chicken embryos, but PGA_2, PGF_2, and a derivative of TXA_2 did not. A report [12] described that the endothelial migration was mediated by COX-2. This experiment was performed in vitro using confluent monolayer endothelial cells stimulated with PMA, and the authors also reported that corneal angiogenesis was suppressed by a COX-2 inhibitor, suggesting the involvement of COX-2 products in vivo. These results suggest that endogenous PGs regulate angiogenesis not only in physiological conditions but also in pathological states. The roles of COX-2-derived PGs in wound healing and the PG receptor signaling relevant to wound-induced angiogenesis were reported recently [13]. When full-thickness skin wounds were created in mice, and angiogenesis in wound granulation tissues was estimated, wound closure and reepithelization were delayed in mice treated with COX-2 inhibitors. The wound closure and reepithelization in EP3 receptor knockout mice ($EP3^{-/-}$) were significantly delayed compared with wild-type mice (WT) (Fig. 18.1), whereas those in $EP1^{-/-}$, $EP2^{-/-}$, and $EP4^{-/-}$ were not delayed. Wound-induced angiogenesis in $EP3^{-/-}$ was significantly inhibited compared with that in WT (Fig. 18.2). Reduced wound-induced angiogenesis in $EP3^{-/-}$ was accompanied by poor development of granulation tissues under the wound. VEGF expression in wound granulation tissues in $EP3^{-/-}$ was markedly less than that in WT. Wound closure in WT was delayed significantly by a VEGF-neutralizing antibody compared with control IgG. Wound-induced angiogenesis and wound closure were significantly suppressed in $EP3^{-/-}$ bone marrow transplantation mice, compared with those in WT bone marrow

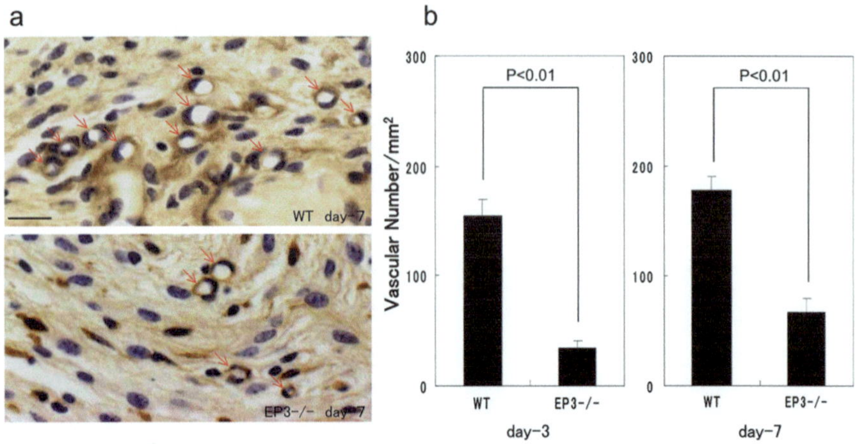

Fig. 18.2 Reduced angiogenesis in wound granulation tissues in EP3 knockout mice. (**a**) Immunohistochemical localization of CD31 in wound granulation tissues isolated from mice 7 days after wounding. (**b**) Vascular density in wound granulation tissues isolated from mice 3 and 7 days after wounding (Cited from Kamoshita et al. [13] with permission)

transplantation mice. These events were accompanied by reductions in accumulation of VEGF-expressing cells in wound granulation tissues and in recruitment of VEGF receptor 1-expressing leukocytes in peripheral circulation. These results indicate that the recruitment of EP3-expressing cells to wound granulation tissues is critical for surgical wound healing and angiogenesis via upregulation of VEGF [13].

Signaling through the thromboxane A_2 (TXA_2) receptor, TP, induces platelet adhesion. However, the cellular and molecular mechanisms of platelet-dependent angiogenesis, especially during recovery from ischemic conditions, and the involvement of TP signaling have not been fully elucidated. We recently identified that TP-dependent platelet adhesion contributed to angiogenesis in a mouse hindlimb ischemia model. Selective platelet adhesion to ischemic endothelial cells via P-selectin was identified in vivo microscopy. P-selectin glycoprotein ligand-1 (PSGL-1) was upregulated in ischemic muscles in a TP-dependent manner. These findings suggest that TP signaling facilitates angiogenesis via P-selectin-mediated platelet adhesion to the angiogenic regions and the supply of pro-angiogenic factors by the adherent platelets.

18.6 Roles of PGs in Tumor-Associated Angiogenesis

Several epidemiological studies revealed a 40 % to 50 % reduction in mortality from colorectal cancer in individuals taking nonsteroidal anti-inflammatory drugs (NSAIDs), and other evidence suggests that they also affect the incidence and progression of other types of cancer, pointing to a possible role of COX in tumor formation [14]. NSAIDs that inhibit COXs and suppress PG biosynthesis have been widely used as anti-inflammatory, anti-pyretic, and analgesic agents. Disruption of the COX-2 gene in mice reduced the number and size of intestinal polyps generated by a mutation in the adenomatous polyposis (APC) gene, thus verifying a role for COX-2 in the generation of colon tumors [15]. COX-2-selective inhibitors were expected to act as a "super aspirin" that would not exhibit the adverse effects typical of classical NSAIDs [6]. However, it has been reported that some organs, such as the kidney, expressed COX-2 constitutively [6], and that COX-2 was necessary for the kidney to mature after birth [16]. Thus, selective inhibition of PG receptor signaling may be a more effective form of treatment of patients than that of COX-2. A wide range of mechanisms of the anti-tumor actions of NSAIDs, some of which are unrelated to inhibition of COX activity, and of subsequent PG formation has been proposed [17–19].

Evidence that PG receptor signaling is relevant to tumor development has been accumulating through the use of PG receptor knockout mice (Table 18.2). We must

Table 18.2 PGE_2 and PGI_2 signaling in cancer development

Prostanoid	Receptor	G protein	Signaling effect	Relationship to cancer development	Refs
PGE_2	EP1	Gq	PtdIns response	EP1 receptor antagonist decreases the incidence of aberrant crypt foci in azoxymethane-treated mice	[20]
				EP1 receptor$^{-/-}$ mice are resistant to azoxymethane-induced aberrant crypt foci	[21]
PGE_2	EP2	Gs	cAMP increase	In EP2 receptor$^{-/-}$ mice, the number of ApcΔ716 intestinal polyps and the intensity of angiogenesis and VEGF expression are decreased	[22, 23]
				EP2 receptor$^{-/-}$ mice exhibit cancer-associated immunodeficiency and defective dendritic cell differentiation	[24]
				EP2 receptor$^{-/-}$ mice exhibit reduced tumor growth but normal tumor-associated angiogenesis and VEGF induction	[24]
PGE_2	EP3	Gs	cAMP increase	EP3 receptor$^{-/-}$ mice have reduced tumor-associated angiogenesis and tumor growth because induction of VEGF is reduced	[25]
		Gi	cAMP decrease	–	
		Gq	PtdIns response	–	
PGE_2	EP4	Gs	cAMP increase	EP4$^{-/-}$ mice decrease tumor cell motility	[26]
PGI_2	IP	Gs	cAMP increase	PGI_2 analogue inhibits metastasis	[27]
		Gq	PtdIns response	–	

Apc adenomatous polyposis, *PGE_2* prostaglandin E_2, *PGI_2* prostaglandin I_2, *PtdIns* phosphatidylinositol, *VEGF* vascular endothelial growth factor emphasize here that the models must be carefully selected according to the interests of researchers. Analysis of a tumor implantation model in some knockout mice is suitable for observing the host stromal responses that facilitate tumorigenesis, because the lack of a given receptor is observable in the host stroma, although tumor cells may, or may not, express the receptors in question, depending on the cell lines implanted. In tumor implantation experiments, differences in phenotype such as tumor growth and tumor-associated angiogenesis in mice are highly dependent on receptor signaling in the host. The models of other categories are developed to see the effect of PG receptor signaling on tumor cells themselves (tumor cell-autonomous effect), as induced by the mutation in tumor epithelial cells in addition to the host stromal effect. The most successful example of the latter is the marked reduction of polyp number in Apc^{716} mice (a model of FAP) against a $COX\text{-}2^{-/-}$ background, in comparison with control animals [15]. In $COX\text{-}2^{-/-}$ mice, COX-2 was deficient both in polyp epithelial cells and in stromal cells.

It was previously reported that PGE_2 can promote colorectal cancer growth, in part through the activation of the PGE_2 receptor subtypes EP1 and EP4. Experiments using knockout mice with a colon carcinogen, azoxymethane, revealed the significant suppression of aberrant crypt foci in EP1-deficient mice together with EP4-deficient mice. The suppression was limited in both cases, suggesting the possible involvement of other receptors or mechanisms. Moreover, aberrant crypt formation represents an initial step in carcinogenesis, and many events precede the development of colon cancer. It is likely that PGs are also involved in other steps and mechanisms. The expression of COX-2, but not COX-1, is elevated in many colorectal cancers, and the protein has been localized to both stromal and epithelial compartments. One mechanism by which elevated COX-2 promotes carcinogenesis is through stimulation of tumor-associated angiogenesis.

In tumor implantation models, the involvement of PGE_2-EP receptor signaling in tumor-induced angiogenesis was tested [28]. It was reported that among the four subtypes of EP receptor knockout mice, in $IP^{-/-}$, and in their WT counterparts, tumor-associated angiogenesis in $EP3^{-/-}$ mice was suppressed by nearly 80 %, although partial reduction of angiogenesis was observed in $EP2^{-/-}$ mice [28]. Histological examination of tumors formed in $EP3^{-/-}$ mice revealed a low level of vascularization and distinct boundaries with the surrounding normal tissue. In spite of the implantation of the same number of tumor cells, differences in tumor growth and tumor-associated angiogenesis were observed in these mice, strongly suggesting that the EP3 receptor in the host, not on the tumor cells, has a major, indeed, the critical role in tumor-induced angiogenesis and tumor growth. Staining for COX-2 was apparent in tumors together with the stromal cells and endothelial cells [28]. In WT mice, VEGF was abundant in the surrounding stromal cells, whose major components were Mac-1 and CD3-negative fibroblast-like cells. Expression of VEGF was markedly reduced in $EP3^{-/-}$ mice. Gel shift assays revealed that AP-1 may be important in VEGF expression and angiogenesis [28], although other factors, such as HIF-1, whose activation was related to EP1, EP2, and EP4 signaling [29, 30], were not ruled out.

It was reported that angiogenesis and growth of polyps were EP2-dependent when the APC gene was mutated [22, 23]. The reports of those authors stated that the major elements that express COX-2 are stromal cells around the intestinal polyps. The tumor stromal cells that produced VEGF so as to facilitate angiogenesis and tumor growth were CD3 and Mac-1 double-negative fibroblasts [28]. Fibroblasts exhibit heterogeneity in various biological parameters including PG-generating and receptor systems [31–33]. The authors of the foregoing report [23] did not show the microvessel density in $EP3^{-/-}$ with APC mutation, and the reduction percentage of angiogenesis in APC-mutated $EP2^{-/-}$ mice was approximately 30 % at best. The major EP receptor expressed in the subcutaneous tissues of WT mice was EP3 [28], which was not expressed in the intestine [22]. These findings suggest that tumor-associated angiogenesis may be regulated in a site-specific fashion, and that it may be related to the heterogeneity of the stromal fibroblasts. It was reported that EP2-null mice bearing subcutaneous tumor cells exhibited cancer-associated immunodeficiency and dendritic cell abnormalities, but surprisingly had no effects on tumor-associated angiogenesis and VEGF induction, in spite of a partial reduction in tumor growth. This finding indicates that intracellular signaling linked to EP2 receptor activation also may be heterogeneous.

The host microenvironment is believed to influence tumor progression [34, 35]. As already mentioned, PGs may be one of the important determinants of tumor–host communications. Examination of human colorectal cancer has revealed marked COX-2 expression not only in cancer cells but also in the stroma that surrounds them [36]. COX-2-deficient mice also exemplify the significance of stromal COX-2 in tumor-induced angiogenesis [37]. The study using $EP3^{-/-}$ mice revealed that COX-2-expressing stromal cells around the tumors, or the tumor cells themselves, or both, synthesize and release PGE_2 into the tumor microenvironment; and that PGE_2 then acts on the stromal cells that express EP3 receptor to induce pro-angiogenic factor production and consequent angiogenesis (Fig. 18.3). EP3 receptor signaling is important in angiogenesis promotion, but it cannot be ruled out that EP2 receptor signaling may facilitate angiogenesis synergistically.

18.7 Roles of PGs in Enhancement of Lymphangiogenesis In Vivo

The lymphatic vasculature forms a network of vessels that drain interstitial fluid from tissues and return this fluid to the blood. Lymphangiogenesis, the formation of lymphatic vessels from preexisting lymphatic vessels, is important in homeostasis of interstitial fluids, metabolism, and immunity. Recent evidence, however, indicates that lymphangiogenesis, similar to angiogenesis, also occurs during certain inflammatory and autoimmune conditions. Psoriasis, a chronic inflammatory skin disease, is characterized by extensive lymphangiogenesis [38], and lymphatic hyperplasia is frequently observed in rejected renal transplants from patients with this condition [25]. Furthermore, bacterial infection of the airway epithelia of mice induced intense lymphangiogenesis and upregulated VEGF-C and VEGF-D [39]. After inflammation has been resolved, new lymphatic networks form a drainage system for fluid and immune cells.

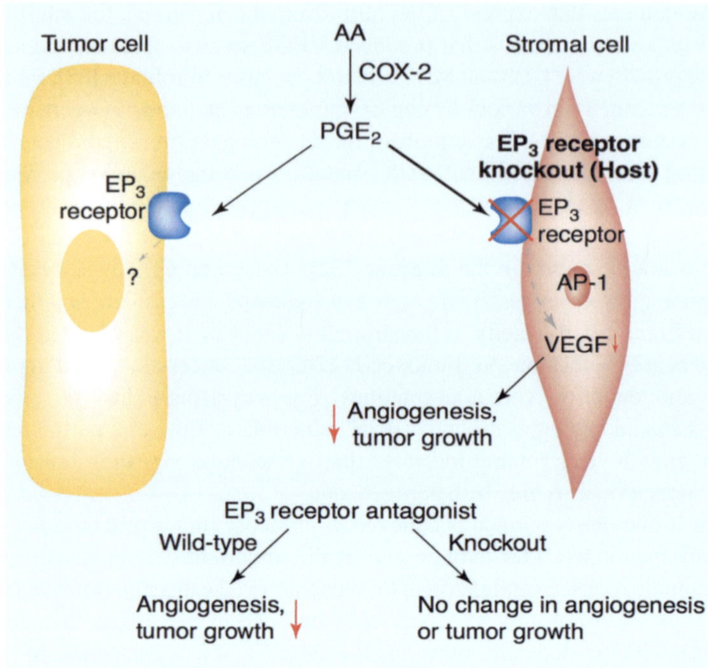

Fig. 18.3 Targets for controlling tumor-associated angiogenesis regulated by endogenous PGs. Stromal EP3 signaling is a key regulator of tumor angiogenesis and tumor growth. COX-2 inhibition and EP3 blockade are effective in preventing tumor growth and angiogenesis. Controlling host stromal function by modification of inflammatory signaling relevant to tumor angiogenesis may also become a useful strategy to treat solid tumors such as cancers

During chronic inflammation, angiogenesis is induced by a variety of inflammatory mediators, such as PGs. However, it remains unknown whether they enhance lymphangiogenesis. We recently clarified the roles of COX-2 and PGE_2 receptor signaling in enhancement of lymphangiogenesis during proliferative inflammation [40]. Lymphangiogenesis estimated by podoplanin/VEGF receptor-3/LYVE-1 expression was upregulated during proliferative inflammation seen around subcutaneous Matrigel plugs. A COX-2 inhibitor (celecoxib) significantly reduced lymphangiogenesis in a dose-dependent manner, whereas topical PGE_2 enhanced lymphangiogenesis. Topical injection of fluorescein isothiocyanate-dextran into the Matrigel revealed that lymphatic flow from the Matrigels was COX-2-dependent. Lymphangiogenesis was suppressed in the granulation tissues of mice lacking either EP3 or EP4, suggesting that these molecules are receptors in response to endogenous PGE_2. An EP3-selective agonist (ONO-AE-248) increased the expression of VEGF-C and VEGF-D in cultured macrophages, whereas an EP4-selective agonist (ONO-AE1-329) increased VEGF-C expression in cultured macrophages and increased VEGF-D expression in cultured fibroblasts. These results suggest that COX-2 and EP3/EP4 signaling contributes to lymphangiogenesis in proliferative inflammation and may become a therapeutic target for controlling lymphangiogenesis.

The same machinery was also active in the secondary lymphedema model, in which lymphedema was induced by a circumferential incision made in the tail of anesthetized mice to sever the dermal lymphatic vessels [41].

18.8 PGs as Targets for Controlling Angiogenesis In Vivo

As discussed, highly selective EP antagonists such as EP3 and EP2 receptor antagonists therefore exhibit beneficial actions on the stromal cells and may be a good choice as novel therapeutic tools against cancer. Administration of an EP3 antagonist to the tumor-bearing mice significantly suppressed tumor-associated angiogenesis and tumor growth in WT mice [28]. By contrast, administration of neither an EP1 nor an EP4 antagonist, both previously developed [21, 42], did so. Furthermore, such a preventive effect of an EP3 antagonist was not seen in EP3$^{-/-}$ mice [28], suggesting that EP3 receptors expressed on the tumor cells themselves did not exhibit any significant function in tumor-associated angiogenesis or tumor growth, because the EP3 antagonist administered may have effectively blocked the EP3 receptors on the tumor cells (Fig. 18.3). These facts confirmed the result obtained in EP3$^{-/-}$ mice, namely, that EP3 receptor signaling acts predominantly on the host stroma. This signaling on the stromal cells was relevant to the induction of the potent pro-angiogenic growth factor VEGF, and upregulated VEGF certainly has a pro-angiogenic action and facilitates tumor growth (Fig. 18.3) [28]. A highly selective EP3 antagonist therefore exhibits preventive action on the stromal cells and is expected to become a novel therapeutic tool against cancer.

Inflammation is a local protective response to harmful stimuli. Recent results have expanded the concept that inflammation is an important factor in facilitating tumor growth [43]. In fact, many cancers arise from the sites of infection, chronic irritation, and inflammation. PGs are the major pro-inflammatory mediators and increase inflammation induced by various chemical mediators. The tumor stromal reaction can be characterized by the proliferation of tissues including fibroblasts, which can facilitate angiogenesis and probably lymphangiogenesis. The results obtained from the sponge implantation model can support the significance of the proliferation or infiltration, or both, of inflammatory cells in the site where angiogenesis occurred [44–49]. Stromal fibroblasts may be derived from the bone marrow [43], and from tumor-associated macrophages as a paradigm for polarized M2 mononuclear phagocytes [50]. Targeting tumor angiogenesis with exogenous genes to tumor angiogenesis was performed by transplantation of genetically modified hematopoietic stem cells [51]. Thus, transplantation of EP receptor-null bone marrow cells may provide the means for targeted inhibition of tumor-associated angiogenesis. Control of the inflammatory responses in the tumor microenvironment where EP receptor-expressing cells are accumulating is also likely to be a novel therapeutic approach against cancer, which now annually causes some 550,000 deaths in the United States. [52].

References

1. Folkman J (2007) Angiogenesis: an organizing principle for drug discovery? Nat Rev Drug Discov 6(4):273–286
2. Carmeliet P (2005) Angiogenesis in life, disease and medicine. Nature 438(7070):932–936
3. Ferrara N, Kerbel RS (2005) Angiogenesis as a therapeutic target. Nature 438(7070):967–974
4. Jain RK (2003) Molecular regulation of vessel maturation. Nat Med 9(6):685–693
5. Hanahan D, Folkman J (1996) Patterns and emerging mechanisms of angiogenic switch during tumorigenesis. Cell 86:353–364
6. Katori M, Majima M (2000) Cyclooxygenase-2: its rich diversity of roles and possible application of its selective inhibitors. Inflamm Res 49:367–392
7. Narumiya S et al (1999) Prostanoid receptors: structures, properties, and function. Physiol Rev 79:1193–1226
8. Sivakumar B, Harry LE, Paleolog EM (2004) Modulating angiogenesis: more vs. less. JAMA 292:972–977
9. Neal MS (2001) Angiogenesis: is it the key to controlling the healing process? J Wound Care 10:281–287
10. Ziche M, Jones J, Gullino PM (1982) Role of prostaglandin E_1 and copper in angiogenesis. J Natl Cancer Inst 69:475–482
11. Spisni E, Manica F, Tomasi Y (1992) Involvement of prostanoids in the regulation of angiogenesis by polypeptide growth factors. Prostaglandins Leukot Essent Fatty Acids 47:111–115
12. Daniel TO et al (1999) Thromboxane A_2 is a mediator of cyclooxygenase-2-dependent endothelial migration and angiogenesis. Cancer Res 59:4574–4577
13. Kamoshita E et al (2007) Recruitment of a prostaglandin E receptor subtype, EP3-expressing bone marrow cells is crucial in wound-induced angiogenesis. Am J Pathol 169(4):1458–1472
14. Smalley W, Dubois RN (1997) Colorectal cancer and non steroidal anti-inflammatory drugs. Adv Pharmacol 39:1–20
15. Ohshima M et al (1996) Suppression of intestinal polyposis in $Apc^{\Delta 716}$ knockout mice by inhibition of cyclooxygenase 2 (COX-2). Cell 87:803–809
16. Norwood VF et al (2000) Postnatal development and progression of renal dysplasia in cyclooxygenase-2 null mice. Kidney Int 58:2291–2300
17. Marx J (2001) Anti-inflammatories inhibit cancer growth—but how? Science 291:581–582
18. Prescott SM, White RL (1996) Self-promotion? Intimate connections between APC and prostaglandin H synthase-2. Cell 87:783–786
19. Shiff SJ, Rigas B (1997) Nonsteroidal anti-inflammatory drugs and colorectal cancer: evolving concepts of their chemopreventive actions. Gastroenterology 113:1992–1998
20. Watanabe K et al (2000) Inhibitory effect of a prostaglandin E receptor subtype EP1 selective antagonist, ONO-8713, on development of azoxymethane-induced aberrant crypt foci in mice. Cancer Lett 156:57–61
21. Watanabe K (1999) Role of the prostaglandin E receptor subtype EP1 in colon carcinogenesis. Cancer Res 59:5093–5096
22. Sonoshita M et al (2001) Acceleration of intestinal polyposism through prostaglandin receptor EP2 in Apc(Delta 716) knockout mice. Nat Med 7:1048–1051
23. Seno H et al (2002) Cyclooxygenase 2- and prostaglandin E_2 receptor EP2-dependent angiogenesis in Apc(Delta716) mouse intestinal polyps. Cancer Res 62:506–511
24. Yang L et al (2003) Cancer-associated immunodeficiency and dendritic cell abnormalities mediated by the prostaglandin EP2 receptor. J Clin Invest 111:727–735

25. Kerjaschki D et al (2006) Lymphatic endothelial progenitor cells contribute to de novo lymphangiogenesis in human renal transplants. Nat Med 12:230–234
26. Yang L et al (2003) Cancer-associated immunodeficiency and dendritic cell abnormalities mediated by the prostaglandin EP2 receptor. J Clin Invest 111:727–735
27. Schirner M, Schneider MR (1997) Inhibition of metastasis by cicaprost in rats with established SMT2A mammary carcinoma growth. Cancer Detect Prev 21:44–50
28. Amano H et al (2003) Host prostaglandin E_2-EP3 signaling regulates tumor-associated angiogenesis and tumor growth. J Exp Med 197:221–232
29. Liu XH et al (2002) Prostaglandin E_2 induces hypoxia-inducible factor-1alpha stabilization and nuclear localization in a human prostate cancer cell line. J Biol Chem 277:50081–50086
30. Fukuda R et al (2003) Vascular endothelial growth factor gene expression in colon cancer cells exposed to prostaglandin E_2 is mediated by hypoxia-inducible factor 1. Cancer Res 63:2330–2334
31. Smith TJ et al (1995) Evidence for cellular heterogeneity in primary cultures of human orbital fibroblasts. J Clin Endocrinol Metab 80:2620–2625
32. Goldring SR et al (1990) Heterogeneity in hormone responses and patterns of collagen synthesis in cloned dermal fibroblasts. J Clin Invest 85:798–803
33. Ko SD et al (1977) Fibroblast heterogeneity and prostaglandin regulation of subpopulations. Proc Natl Acad Sci U S A 74:3429–3432
34. Fukumura D et al (1997) Effect of host microenvironment on the microcirculation of human colon adenocarcinoma. Am J Pathol 151:679–688
35. Gohongi T et al (1999) Tumor–host interactions in the gallbladder suppress distal angiogenesis and tumor growth: involvement of transforming growth factor beta. Nat Med 5:1203–1208
36. Sano H et al (1995) Expression of cyclooxygenase-1 and -2 in human colorectal cancer. Cancer Res 55:3785–3789
37. Williams CS et al (2000) Host cyclooxygenase-2 modulates carcinoma growth. J Clin Invest 105:1589–1594
38. Cliff S et al (1999) An in vivo study of the microlymphatics in psoriasis using fluorescence microlymphography. Br J Dermatol 140:61–66
39. Baluk P et al (2005) Pathogenesis of persistent lymphatic vessel hyperplasia in chronic airway inflammation. J Clin Invest 115:247–257
40. Hosono K et al (2011) Roles of prostaglandin E_2-EP3/EP4 receptor signaling in the enhancement of lymphangiogenesis during fibroblast growth factor-2-induced granulation formation. Arterioscler Thromb Vasc Biol 31:1049–1058
41. Kashiwagi S et al (2011) Role of COX-2 in lymphangiogenesis and restoration of lymphatic flow in secondary lymphedema. Lab Invest 91:1314–1325
42. Mutoh M et al (2002) Involvement of prostaglandin E receptor subtype EP4 in colon carcinogenesis. Cancer Res 62:28–32
43. Coussens LM, Werb Z (2002) Inflammation and cancer. Nature 420:860–867
44. Katada J et al (2002) Significance of vascular endothelial cell growth factor up-regulation mediated via a chymase-angiotensin-dependent pathway during angiogenesis in hamster sponge granulomas. J Pharmacol Exp Ther 302:949–956
45. Muramatsu M et al (2000) Chymase as a proangiogenic factor. A possible involvement of chymase-angiotensin-dependent pathway in the hamster sponge angiogenesis model. J Biol Chem 275:5545–5552
46. Hayashi I et al (2002) Suppressed angiogenesis in kininogen-deficiencies. Lab Invest 82:871–880
47. Majima M et al (2000) Cyclo-oxygenase-2 enhances basic fibroblast growth factor-induced angiogenesis through induction of vascular endothelial growth factor in rat sponge implants. Br J Pharmacol 130:641–649

48. Majima M et al (1997) Significant roles of inducible cyclooxygenase (COX)-2 in angiogenesis in rat sponge implants. Jpn J Pharmacol 75:105–114
49. Amano H et al (2001) Adenylate cyclase/protein kinase A signaling pathway enhances angiogenesis through induction of vascular endothelial growth factor in vivo. Jpn J Pharmacol 87:181–188
50. Mantovani A et al (2002) Macrophage polarization: tumor-associated macrophages as a paradigm for polarized M2 mononuclear phagocytes. Trends Immunol 23:549–555
51. De Palma M et al (2003) Targeting exogenous genes to tumor angiogenesis by transplantation of genetically modified hematopoietic stem cells. Nat Med 9:789–795
52. Greenlee RT et al (2000) Cancer statistics, 2000. CA Cancer J Clin 50:7–33

Chapter 19
Eicosanoids and Aortic Aneurysm

Utako Yokoyama, Ryo Ishiwata, and Yoshihiro Ishikawa

Abstract Some of the derivatives of arachidonic acid, namely, the 5-lipoxygenase (5-LO) metabolites leukotriene B_4 (LTB_4) and cysteinyl-leukotrienes (CysLTs) as well as the microsomal prostaglandin E_2 synthase-1 (mPGES-1) product prostaglandin E_2 (PGE_2), can act as potent pro-inflammatory mediators in vascular diseases. Abdominal aortic aneurysm (AAA) is a vascular pathology characterized by the infiltration of the media and adventitia by immune cells and the subsequent degradation of the medial elastic lamina layer. In human AAA, cyclooxygenase-2 (COX-2) and 5-LO are abundantly expressed, and the roles of PGE_2 and LTs in AAA have recently been a subject of intense investigation. In particular, the PGE_2 receptor EP4 has been suggested to promote cytokine production and proteolytic activation, and the inhibition of EP4 signaling attenuates the progression of murine models of AAA. The LTs LTB_4 and LTD_4 have been shown to stimulate the release of a variety of cytokines and other mediators that can enhance degradation of the extracellular matrix, such as macrophage inflammatory protein-1α and monocyte chemoattractant protein-1. Pharmacological inhibition of these eicosanoids attenuate murine models of AAA. Because no pharmacological therapies are currently available for inhibiting the progression of AAA, regulation of the PGE_2 and 5-LO pathways may serve as a new therapeutic strategy for AAA.

Keywords Prostaglandin • Leukotriene • EP4 • BLT1 • CysLT1 • Aortic aneurysm • Atherosclerosis • Elastic fibers • Matrix metalloprotease

19.1 Introduction

Eicosanoids, which include prostanoids and leukotrienes (LTs), have an important role in vascular physiology and pathophysiology. Eicosanoids are involved in a variety of disease conditions such as inflammation and abnormal immune responses. Prostanoid synthesis is initiated with the conversion of arachidonic acid (AA) into

U. Yokoyama (✉) • R. Ishiwata • Y. Ishikawa
Cardiovascular Research Institute, Yokohama City University School of Medicine,
3-9 Fukuura Kanazawaku, Yokohama 236-0004, Japan
e-mail: utako@yokohama-cu.ac.jp

T. Yokomizo, M. Murakami (eds.), *Bioactive Lipid Mediators*,
DOI 10.1007/978-4-431-55669-5_19

Table 19.1 Reported roles of eicosanoids in abdominal aortic aneurysm (AAA)

Reported functions	Proposed effector pathway(s)	Pathophysiological process	References
Inhibition of AAA	PGE_2-EP4	Inhibition of MCP-1 in macrophages	Tang et al. [1]
Progression of AAA	PGE_2-EP4	IL-6 production and MMP2 activation in SMCs	Yokoyama et al. [2]
Progression of AAA	PGE_2-EP4	MMP2 and MMP9 activation	Cao et al. [3]
Progression of AAA	LTD_4-CysLT1	MIP-1α production in macrophages	Zhao et al. [4]
No effect on AAA	5-LO	N.D.	Cao et al. [5]
Progression of AAA	LTD_4-CysLT1	MMP2 activation	Di Gennaro et al. [6]
Progression of AAA	LTB_4-BLT1	MCP-1, MIP-2, and IL-8 production;	Ahluwalia et al. [7]
		MMP2 and MMP9 activation in macrophages	
Progression of AAA	LTB_4-BLT1	Neutrophil chemotaxis	Houard et al. [8]
Progression of AAA	LTB_4-BLT1	MMP2 activation	Kristo et al. [9]

N.D. not determined

prostaglandin H_2 (PGH_2) through cyclooxygenase (COX). After the synthesis of PGH_2, it is converted into other prostanoids, including prostaglandin D_2 (PGD_2), prostaglandin E_2 (PGE_2), prostaglandin $F_2\alpha$ ($PGF_2\alpha$), prostaglandin I_2 (PGI_2), and thromboxane A_2 (TxA_2). In addition to this COX pathway, AA can be converted into 5-hydroperoxyeicosatetraenoic acid (5-HPETE) via the 5-lipoxygenase (5-LO) pathway. The subsequent conversion of 5-HPETE into LTA_4 is also catalyzed by 5-LO. LTA_4 then gives rise to LTB_4 or cysteinyl-LTs (CysLTs) such as LTC_4, LTD_4, and LTE_4. Among these eicosanoids, PGE_2 and LTs have been extensively examined in the field of vascular disease (Table 19.1). The present review is focused on the role of PGE_2 and LTs in the progression of aortic aneurysm and the potential for a therapeutic strategy based on regulating these signaling pathways.

19.2 Pathogenesis of Abdominal Aortic Aneurysm (AAA)

19.2.1 Clinical Features of AAA

Aortic aneurysm is the 13th leading cause of death in the United States, with approximately 15,000 deaths per year [10]. Abdominal aortic aneurysm (AAA) is the most common type, comprising more than 80 % of all aortic aneurysms [10].

Ultrasonographic screening studies have shown that a small AAA, 3–5 cm in diameter, is present in 4–5 % of men of more than 60 years old [11, 12]. A study that followed AAA patients for up to 6 years showed that 55 % of them exhibited increased aortic diameter, and 9 % exhibited a diameter greater than 6 cm, at which point the risk of rupture is significantly elevated [12]. After rupture occurs, the mortality rate exceeds 60 % [10]. Although an AAA typically continues to expand, increasing the risks of rupture and mortality, no effective pharmacological therapy to prevent the progression of an AAA is currently available. Patients with large AAAs with high risk of rupture are treated with open or endovascular repair.

It is widely accepted that the progression of AAA involves the complex interaction of diverse factors such as chronic inflammation, in contrast to thoracic aortic aneurysm (TAA), the cause of which is often attributed to genetic defects such as Marfan syndrome, Loeys–Dietz syndrome, and Ehlers–Danlos syndrome [13]. The risk of AAA is associated with age, gender, cigarette smoking, hypertension, and atherosclerosis [14]. Notably, AAA is primarily associated with atherosclerosis [15]. AAA was present in 10 % of patients with peripheral or cerebral arterial diseases [16], suggesting that AAA may share many of the risk factors of atherosclerotic stenosis. Nevertheless, the distinct mechanisms of AAA progression remain elusive.

19.2.2 Molecular Mechanisms of AAA Progression

AAA is characterized by chronic inflammation of the aortic wall initiated by infiltration of immune cells and excessive proteolytic degradation of elastic fibers [17]. Elastic fiber is a major extracellular matrix (ECM) in the tunica media of the aortic wall. The lamellae of elastic fibers are responsible for the elasticity of the large arteries, which allows the arteries to maintain sufficient blood pressure even during variations in hemodynamic stress. Mature elastic fibers appear to have a half-life of 70 years and are believed to be synthesized rarely in the adult abdominal aortae [14]. Indeed, elastic fibers comprise 35 % of the dry weight of the normal aortic media, although this is reduced to 8 % in AAA [18]. Hence, it is generally accepted that promotion of the degradation of elastic fibers results in arterial aneurysmal formation.

Activation of proteolytic enzymes, particularly matrix metalloproteinase-2 (MMP2) and MMP9, is considered to be a critical step for elastic fiber degradation. These MMPs exacerbate aortic dilatation, as demonstrated in studies using human patients or genetically engineered mice [15, 19]. MMP2 is mainly secreted by smooth muscle cells (SMCs) and macrophages, whereas MMP9 is predominantly secreted by macrophages that have infiltrated into the AAA wall [20].

In addition to MMPs, macrophages and other immune cells secrete various pro-inflammatory cytokines that play important roles, particularly in the initiation of aneurysms [10]. The immune cells that infiltrate into AAA tissues include T cells, B cells, neutrophils, and macrophages. In human AAA tissues, most inflammatory infiltrates are observed in the tunica adventitia [21]. Interleukin-6 (IL-6), tumor necrosis factor (TNF)-α, interleukin-1β (IL-1β), and monocyte chemoattractant protein-1 (MCP-1) are released in the AAA wall [22, 23]. In an experimental AAA model of $ApoE^{-/-}$ mice infused with angiotensin II, production of IL-6, IL-8, and MCP-1 was increased [24].

In the progression of AAA, increased neovascularization is thought to contribute to the introduction of immune cells into the tunica media. The tunica adventitia, the outermost tunic of the aortic wall, consists of fibroblasts and collagenous extracellular matrices (ECMs), in which small nutrient vessels called the vasa vasorum are dispersed. Under physiological conditions, the tunica media is poorly vascularized, and the inner smooth muscle layers depend largely on direct diffusion from the luminal side for their nutritional needs. In human AAA tissues, however, increased neovascularization into the tunica media is observed [25]. In a study of aneurysm rupture edges, medial neovascularization and the levels of pro-angiogenic cytokines such as vascular endothelial growth factor (VEGF) and MCP-1 were increased compared with nonruptured AAA [26]. Consistent with the human data, in the mouse model of AAA, in which periaortic $CaCl_2$ treatment induces chronic vascular inflammation, the expression levels of VEGF and MCP-1 were upregulated and the number of vasa vasorum was increased [27].

19.3 AAA and PGE_2 Signaling

19.3.1 PGE_2 and AAA

Among the prostanoids, PGE_2 is one of the major products generated by the actions of COX on AA and is well known to be an important mediator of inflammation, fever, and pain [28, 29]. In human AAA tissues, COX-2-dependent PGE_2 synthesis is induced during the development of the disease [22, 30]. In aneurysm walls, COX-2 is widely expressed in macrophages and SMCs along with locally synthesized PGE_2 [22]. PGE_2 synthesized by macrophages and SMCs increases the production of MMPs [31, 32] and stimulates the production of cytokines [22]. In addition to the presence of PGE_2 in AAA, it has been demonstrated that selective COX-2 inhibition, as induced by celecoxib or genetic disruption of COX-2, decreased angiotensin II-induced AAA formation in $ApoE^{-/-}$ mice [33, 34]. Microsomal PGE_2 synthase-1 (mPGES-1) is an inducible enzyme that catalyzes the isomerization of the COX product PGH_2 into PGE_2 and is expressed in response to inflammatory cytokines [35]. A study using deletion of mPGES-1 has also demonstrated that the inhibition of COX-2-PGE_2 decreased the rate of angiotensin II-induced AAA in $ApoE^{-/-}$ mice [36]. Progressed AAA is frequently associated with atherosclerosis. The biosynthesis of PGE_2 is increased in human atherosclerotic plaques [37] and has been implicated in atherosclerotic plaque rupture as well [38]. This accumulating evidence from both human and murine models of AAA suggests that COX-2-PGE_2 signaling is greatly involved in the pathogenesis of AAA.

19.3.2 Expression of EP Receptors in AAA

The biological effects of PGE_2 depend on the prostanoid EP receptor subtypes EP1 through EP4 [28]. Although it has been well recognized that EP2, EP3, and EP4 are present in vascular SMCs, EP4 is thought to play a pivotal role in AAA progression [39]. Bayston et al. examined the expression of EPs in human AAA tissues and found that EP2, EP3, and EP4 were expressed in AAA explants and that positive immunoreaction against EP4 was detected primarily in macrophages. Because EP4 is a predominant PGE_2 receptor expressed in macrophages [40], EP4 expression appears to be increased during the development of AAA, contemporaneous with the infiltration of macrophages.

In addition to macrophages, we have demonstrated through immunohistochemical examination that EP4 is abundantly expressed in the SMCs of human AAA explants [2]. Quantitative reverse transcriptase-polymerase chain reaction (RT-PCR) has revealed that the expression levels of EP4 mRNA are greater in primary cultures of SMCs from human AAA tissues than in those from normal aortae [2]. Interestingly, EP4 expression and elastic fiber degradation were both enhanced in aneurysmal areas compared to normal areas. Statistical analysis revealed that a significant correlation existed between EP4 expression level and the degree of elastic fiber degradation in eight human AAA tissue samples [2]. These studies of expression of PGE_2 receptors suggest that EP4 is highly expressed in both macrophages and SMCs in AAA.

19.3.3 The Roles of EP4 in Murine AAA Models

Based on the expression profiles of prostanoid receptors in AAA, two groups independently examined the role of EP4 signaling in AAA in murine models. They obtained similar results, indicating that pharmacological inhibition of the EP4 receptor with ONO-AE3-208 significantly inhibited progression and rupture rate of angiotensin II-induced AAA [2, 3]. Cao et al. demonstrated that the expression levels of macrophage inflammatory protein (MIP)-1α, interleukin-17 (IL-17), and the pan-T-cell marker CD90.2 were lower in mouse aortae treated with ONO-AE3-208 than in untreated aortae. Furthermore, in ONO-AE3-208-treated aortae, activation of MMP2 and MMP9, both of which degrade elastic fibers, was attenuated [3]. Not only pharmacological inhibition of EP4 but also global gene deletion of the EP4 receptor significantly inhibited the progression of angiotensin II- or calcium chloride-induced mouse AAA in which MMP2 and MMP9 activation was attenuated [2]. These data showing EP4-mediated proteolytic activation suggest that EP4 signaling promotes elastic fiber degradation.

We further examined the contribution of EP4 signaling to elastogenesis using fetal arteries and found that activation of the EP4 receptor in SMCs inhibited elastic fiber formation by inducing the degradation of lysyl oxidase, which crosslinks elastin [41]. Previous reports have demonstrated that aortic aneurysms and coronary dissections were related to a disturbance in lysyl oxidase expression in animal models and humans [42, 43]. Therefore, EP4 signaling may inhibit newly synthesized elastic fiber formation in AAAs as well.

Expression of EP4 is not restricted to macrophages and SMCs, but is also found in vascular endothelial cells [44]. In vascular endothelial cells, PGE_2 may have a role in angiogenesis. Activation of the EP4 receptor promoted in vitro tube formation of microvascular endothelial cells through protein kinase A (PKA) catalytic subunit-γ-mediated upregulation of endothelial nitric oxide synthase (eNOS) or VEGF production [45, 46]. In vivo experiments have indicated EP4-mediated angiogenesis as well [44, 45, 47]. Although the role of EP4 in the endothelial cells of AAA remains unclear, available evidence suggests that EP4 signaling may contribute to angiogenesis in atherosclerotic intima or adventitia.

The foregoing data, in particular those obtained in experiments using global deletion or pharmacological inhibition of EP4, suggest that PGE_2-EP4 signaling promotes AAA progression. However, EP4 seems to have a contrasting function in macrophages. To understand the cell type-specific effects of EP4, Tang et al. used hypercholesterolemic low density lipoprotein receptor knockout mice transplanted with either wild-type or EP4-deficient bone marrow and treated them with angiotensin II. When EP4 signaling was inhibited only in bone marrow-derived cells, inflammation and angiotensin II-induced AAA formation were enhanced [1]. This change most likely occurred because PGE_2 in blood cells had an anti-inflammatory effect, especially through reducing MCP-1 production [1].

19.3.4 Potential Effects of EP4 on Human AAA

Despite these conflicting results obtained with murine AAA models, some studies using human AAA explants have suggested that PGE_2-EP4 contributes to AAA progression through pro-inflammatory and proteolytic actions. Bayston et al. demonstrated that human AAA explants and macrophages isolated from aneurysm biopsy specimens secreted large amounts of IL-6 [48]. This IL-6 secretion was reduced in the presence of indomethacin, but the simultaneous addition of exogenous PGE_2 or 11-deoxy PGE_1 partially reversed the indomethacin inhibition, whereas the EP2 agonist butaprost or the EP1/3 agonist sulprostone had no effect [48]. We have reported on EP4-mediated IL-6 production in human AAA explants and SMCs isolated from human aneurysmal specimens [2] as well. Several reports have demonstrated that IL-6 is produced by EP4 stimulation in various cell types including macrophages, neutrophils, and fibroblasts [39]. Although downstream signaling of EP4 has not been reported to contribute to IL-6 production in AAA, Chen et al. have suggested that intracellular signaling events involving protein kinase A (PKA), protein kinase C (PKC), p38 mitogen-activated protein kinase (MAPK), and nuclear factor-kappa B (NF-κB) contribute to IL-6 induction by EP4 stimulation [49].

To examine the role of EP4 on proteolytic activation, we used human AAA explants and SMCs isolated from human aneurysmal specimens, and found that EP4 stimulation by ONO-AE1-329 increased MMP2 activation in both SMCs and vascular walls of human AAA [2]. Furthermore, inhibition of EP4 signaling with the EP4 antagonist ONO-AE3-208 reduced MMP2 activation and IL-6 production in cultured human AAA explants [2]. Based on previous reports, the inhibition of EP4 is predicted to rather enhance MCP-1 production in human AAA. In practice, however, inhibition of EP4 did not enhance MCP-1 production in cultured human AAA explants treated with EP4 antagonist ONO-AE3-208 at doses ranging from 1 to 100 nM [2].

Some clinical trials related to AAA have been conducted during the past decade. Because early reports strongly suggested that the COX-2-PGE_2 pathway was responsible for AAA progression, COX-2 inhibitors were expected to inhibit AAA. Contrary to these expectations, the administration of selective COX-2 inhibitors actually increased the frequency of adverse cardiovascular events [50, 51]. These results may indicate the non-selective inhibition of prostanoid production. Nonetheless, the inhibition of pathophysiological COX-2-dependent PGE_2 signaling would still remain an attractive therapeutic strategy. Selective inhibition of EP4 signaling may be a means of preventing aortic aneurysm formation. Further studies will be required to clarify the possibility of systemic administration of an EP4 antagonist as a pharmacological therapeutic strategy in AAA.

19.4 AAA and the 5-LO Pathway

19.4.1 Expression of the 5-LO Pathway in AAA

LTs are pro-inflammatory lipid mediators derived from the 5-LO pathway of AA metabolism. They are known to recruit and activate leukocytes at sites of inflammation [52, 53]. 5-LO, which acts in concert with 5-LO-activating protein (FLAP), is a rate-limiting enzyme in the production of LTs. During the past decade, the roles of the 5-LO-LT pathway in AAA have been extensively examined.

It is well recognized that 5-LO are abundantly expressed and downstream LTs are robustly produced in human atherosclerotic plaques [54] and associated with plaque instability [55]. In addition, Zhao et al. have demonstrated that 5-LO is highly expressed in the macrophages of the adventitia in a cholate-containing atherogenic diet-induced AAA model in $ApoE^{-/-}$ mice [4]. Similar findings have been reported using human AAA tissues. Biochemical analysis revealed an overexpression of 5-LO, FLAP, and LTC_4 synthase, but not of LTA_4 hydrolase, in the human AAA wall [6]. These three proteins, required for CysLTs biosynthesis, are expressed in the medial and adventitial layers of the AAA wall and are co-localized with immune cells including macrophages and neutrophils [6]. The specific receptor for LTB_4 is BLT1, a high-affinity receptor expressed in leukocytes, vascular SMCs, and endothelial cells that also mediates chemotaxis [56]. It has been demonstrated that production of LTB_4 and mRNA expression of BLT1 were increased in angiotensin II-induced AAA in $ApoE^{-/-}$ mice [7].

19.4.2 LTs in AAA

In accordance with these data on expression, other evidence has also demonstrated that the 5-LO pathways including LTB_4 and CysLTs are involved in the progression of AAA. Funk and coworkers found a protective effect of 5-LO gene deletion against cholate-containing atherogenic diet-induced AAA in $ApoE^{-/-}$ mice [4]. Using a zymogram assay, the authors demonstrated that MMP2 activity in aneurysmal aortae and plasma levels of MIP-1α were attenuated by 5-LO gene deletion. An in vitro experiment suggested the involvement of LTD_4 in macrophage-derived MIP-1α and endothelial cell-derived MIP-2 [4]. Another line of study also emphasized the role of CysLTs. Di Gennaro et al. demonstrated using HPLC that human AAA explants converted AA into significant quantities of CysLTs and, to a lesser extent, LTB_4 [6]. In organ culture of human AAA explants, administration of LTD_4 increased MMP2 activation, which was inhibited by the selective CysLT1 antagonist montelukast [6]. These data from mice and humans suggest that CysLTs such as LTD_4 play an important role in AAA. In keeping with the data on expression in AAA, these data indicate that the 5-LO pathway contribute to adventitial inflammation in the vascular wall.

Although some papers have emphasized the role of CysLTs rather than LTB_4, it has been demonstrated that genetic deletion or pharmacological inhibition of BLT1 receptor has a protective effect on angiotensin II-induced AAA [7, 9]. In these studies, diminished AAA formation in BLT1-deficient mice was associated with significant reductions in MMP2 and MMP9 and with the infiltration of macrophages and $CD4^+$ T cells [7]. Stimulation of LTB_4, but not LTD_4, increased MCP-1, MIP-2, and IL-8 proteins in freshly isolated human monocytes [7]. Pharmacological inhibition of BLT1, which can be achieved by CP-105696, decreased macrophage infiltration and MMP2 activation [9]. Using human AAA explants, Houard et al. recently proposed a role for LTB_4, derived from neutrophils within the intraluminal thrombus, as a chemotactic factor in AAA [8].

19.4.3 Therapeutic Strategies Against AAA Using 5-LO Pathway Inhibitors

As recent research has identified the importance of inflammation via the 5-LO pathway as a critical step in the initiation and perpetuation of atherosclerosis [57], drugs that inhibit the 5-LO pathway are the subject of current vascular research. In a phase II trial of the 5-LO inhibitor atreleuton, new coronary plaques were observed in 27.8 % of the placebo group but only 4.8 % of the treatment group [58]. Together with two other phase II trials, these phase II findings on atreleuton demonstrated inhibition of LTs, lowering of hsCRP, and potential for plaque stabilization. Investigation of the FLAP inhibitor celiflapon was examined in a phase III trial in 2006 [58]. Despite the relatively positive results that have emerged from these clinical trials on atherosclerosis, no clinical trial using the 5-LO pathway to inhibit AAA has been reported.

The effects of a compound that inhibits both the PGE_2 and 5-LO pathways in murine models of AAA have been recently reported [59]. A pirinixic acid derivative called LP105 potently inhibited 5-LO [60], COX-1, and mPGES-1 [61]. The researchers further demonstrated that LP105 interfered with the development of AAA in an angiotensin II-induced model in $ApoE^{-/-}$ mice [59]. In this model, MMP9 and the expression levels of several inflammatory cytokines including IL-6, IL-1β, and TNF-α were decreased by the administration of LP105.

19.5 Conclusions

The inflammatory process of AAA is associated with several pathological reactions characterized by the infiltration of immune cells in the adventitia and subsequent proteolytic activation in the medial layer. PGE_2 and LTs, which are primarily produced in macrophages that have infiltrated into the adventitia, participate in chronic inflammation of the aortic wall (Fig. 19.1). Recent studies have revealed that EP4 receptor signaling in SMCs appears to enhance cytokine production and MMP

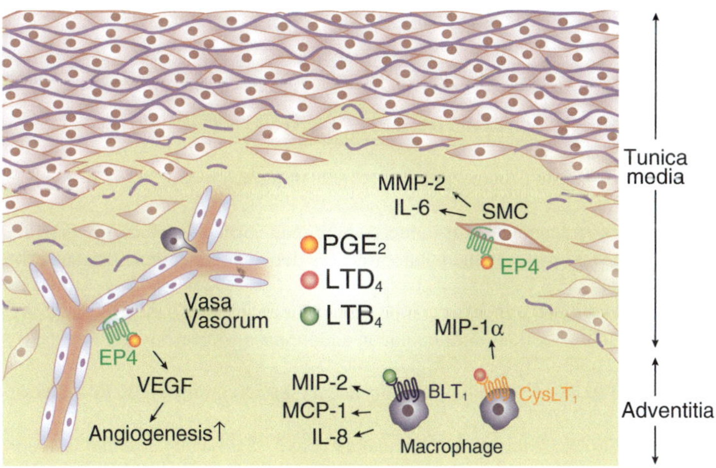

Fig. 19.1 Schematic diagram of the roles of eicosanoids in the progression of abdominal aortic aneurysm (AAA). Inflammatory infiltrates such as macrophages are enriched in the tunica adventitia of human AAA tissue. The expression of the PGE_2 receptor EP4 is upregulated in smooth muscle cells (SMCs) in AAA. EP4 signaling upregulates secretion of IL-6 in SMCs and induces activation of the elastic fiber-degrading enzyme MMP2. Stimulation of microvascular endothelial cells with EP4 induces VEGF expression and may contribute to the increased rate of neovascularization into the tunica media. Adventitial macrophages express the LT receptors $CysLT_1$ and BLT_1. LTD_4 via $CysLT_1$ upregulates MIP-1α expression, which may promote T-cell recruitment. LTB_4 via BLT_1 induces secretion of several cytokines by macrophages, e.g., IL-8, MCP-1, and MIP2

activation and to promote degradation of elastic fibers in the medial layer. CysLTs such as LTD_4 and LTB_4 are suggested to contribute to inflammatory response and proteolytic activation in the adventitia. Chronic inflammation of both medial and adventitial layers might synergistically exacerbate AAA.

Based on these findings, the regulation of PGE_2 and 5-LO-LT signaling may be useful for exploring pharmacological therapeutic strategies to treat or prevent AAA, a condition for which no pharmacological treatment is currently available. Further experimental and clinical studies are needed to determine the potential therapeutic strategies targeting these drugs in AAA.

Acknowledgments This work was supported by grants from the Ministry of Education, Culture, Sports, Science and Technology of Japan (U.Y. and Y.I.), a Grant-in-Aid for Scientific Research on Innovative Areas (U.Y.: 1123116514; YI, 22136009), the fund for Creation of Innovation Centers for Advanced Interdisciplinary Research Areas Program in the Project for Developing Innovation Systems from the Ministry of Education, Culture, Sports, Science and Technology (U.Y.), the Yokohama Foundation for Advanced Medical Science (U.Y.), the Vehicle Racing Commemorative Foundation (U.Y.), and the Takeda Science Foundation (U.Y. and Y.I.).

References

1. Tang EH et al (2011) Deletion of EP4 on bone marrow-derived cells enhances inflammation and angiotensin II-induced abdominal aortic aneurysm formation. Arterioscler Thromb Vasc Biol 31(2):261–269
2. Yokoyama U et al (2012) Inhibition of EP4 signaling attenuates aortic aneurysm formation. PLoS One 7(5):e36724
3. Cao RY et al (2012) Prostaglandin receptor EP4 in abdominal aortic aneurysms. Am J Pathol 181(1):313–321
4. Zhao L et al (2004) The 5-lipoxygenase pathway promotes pathogenesis of hyperlipidemia-dependent aortic aneurysm. Nat Med 10(9):966–973
5. Cao RY et al (2007) Angiotensin II-induced abdominal aortic aneurysm occurs independently of the 5-lipoxygenase pathway in apolipoprotein E-deficient mice. Prostaglandins Other Lipid Mediat 84(1–2):34–42
6. Di Gennaro A et al (2010) Increased expression of leukotriene C_4 synthase and predominant formation of cysteinyl-leukotrienes in human abdominal aortic aneurysm. Proc Natl Acad Sci U S A 107(49):21093–21097
7. Ahluwalia N et al (2007) Inhibited aortic aneurysm formation in BLT1-deficient mice. J Immunol 179(1):691–697
8. Houard X et al (2009) Differential inflammatory activity across human abdominal aortic aneurysms reveals neutrophil-derived leukotriene B_4 as a major chemotactic factor released from the intraluminal thrombus. FASEB J 23(5):1376–1383
9. Kristo F et al (2010) Pharmacological inhibition of BLT1 diminishes early abdominal aneurysm formation. Atherosclerosis 210(1):107–113
10. Annambhotla S et al (2008) Recent advances in molecular mechanisms of abdominal aortic aneurysm formation. World J Surg 32(6):976–986
11. Collin J et al (1988) Oxford screening programme for abdominal aortic aneurysm in men aged 65 to 74 years. Lancet 2(8611):613–615
12. Scott RA, Ashton HA, Kay DN (1991) Abdominal aortic aneurysm in 4237 screened patients: prevalence, development and management over 6 years. Br J Surg 78(9):1122–1125

13. Ruddy JM, Jones JA, Ikonomidis JS (2013) Pathophysiology of thoracic aortic aneurysm (TAA): is it not one uniform aorta? Role of embryologic origin. Prog Cardiovasc Dis 56(1):68–73
14. MacSweeney ST, Powell JT, Greenhalgh RM (1994) Pathogenesis of abdominal aortic aneurysm. Br J Surg 81(7):935–941
15. Guo DC et al (2006) Pathogenesis of thoracic and abdominal aortic aneurysms. Ann N Y Acad Sci 1085:339–352
16. MacSweeney ST et al (1993) High prevalence of unsuspected abdominal aortic aneurysm in patients with confirmed symptomatic peripheral or cerebral arterial disease. Br J Surg 80(5):582–584
17. Freestone T et al (1995) Inflammation and matrix metalloproteinases in the enlarging abdominal aortic aneurysm. Arterioscler Thromb Vasc Biol 15(8):1145–1151
18. Campa JS, Greenhalgh RM, Powell JT (1987) Elastin degradation in abdominal aortic aneurysms. Atherosclerosis 65(1-2):13–21
19. Longo GM et al (2002) Matrix metalloproteinases 2 and 9 work in concert to produce aortic aneurysms. J Clin Invest 110(5):625–632
20. Thompson RW, Parks WC (1996) Role of matrix metalloproteinases in abdominal aortic aneurysms. Ann N Y Acad Sci 800:157–174
21. Bobryshev YV, Lord RS (2001) Vascular-associated lymphoid tissue (VALT) involvement in aortic aneurysm. Atherosclerosis 154(1):15–21
22. Walton LJ et al (1999) Inhibition of prostaglandin E_2 synthesis in abdominal aortic aneurysms: implications for smooth muscle cell viability, inflammatory processes, and the expansion of abdominal aortic aneurysms. Circulation 100(1):48–54
23. Treska V et al (2002) Inflammation in the wall of abdominal aortic aneurysm and its role in the symptomatology of aneurysm. Cytokines Cell Mol Ther 7(3):91–97
24. Tieu BC et al (2009) An adventitial IL-6/MCP1 amplification loop accelerates macrophage-mediated vascular inflammation leading to aortic dissection in mice. J Clin Invest 119(12):3637–3651
25. Holmes DR et al (1995) Medial neovascularization in abdominal aortic aneurysms: a histopathologic marker of aneurysmal degeneration with pathophysiologic implications. J Vasc Surg 21(5):761–771; discussion 771–772
26. Choke E et al (2006) Abdominal aortic aneurysm rupture is associated with increased medial neovascularization and overexpression of proangiogenic cytokines. Arterioscler Thromb Vasc Biol 26(9):2077–2082
27. Kaneko H et al (2011) Role of vascular endothelial growth factor-A in development of abdominal aortic aneurysm. Cardiovasc Res 91(2):358–367
28. Woodward DF, Jones RL, Narumiya S (2011) International Union of Basic and Clinical Pharmacology. LXXXIII: classification of prostanoid receptors, updating 15 years of progress. Pharmacol Rev 63(3):471–538
29. Sugimoto Y, Narumiya S (2007) Prostaglandin E receptors. J Biol Chem 282(16):11613–11617
30. Holmes DR et al (1997) Prostaglandin E_2 synthesis and cyclooxygenase expression in abdominal aortic aneurysms. J Vasc Surg 25(5):810–815
31. Khan KM, Howe LR, Falcone DJ (2004) Extracellular matrix-induced cyclooxygenase-2 regulates macrophage proteinase expression. J Biol Chem 279(21):22039–22046
32. Corcoran ML et al (1992) Interleukin 4 inhibition of prostaglandin E_2 synthesis blocks interstitial collagenase and 92-kDa type IV collagenase/gelatinase production by human monocytes. J Biol Chem 267(1):515–519
33. King VL et al (2006) Selective cyclooxygenase-2 inhibition with celecoxib decreases angiotensin II-induced abdominal aortic aneurysm formation in mice. Arterioscler Thromb Vasc Biol 26(5):1137–1143
34. Gitlin JM et al (2007) Genetic deficiency of cyclooxygenase-2 attenuates abdominal aortic aneurysm formation in mice. Cardiovasc Res 73(1):227–236
35. Samuelsson B, Morgenstern R, Jakobsson PJ (2007) Membrane prostaglandin E synthase-1: a novel therapeutic target. Pharmacol Rev 59(3):207–224

36. Wang M et al (2008) Microsomal prostaglandin E synthase-1 deletion suppresses oxidative stress and angiotensin II-induced abdominal aortic aneurysm formation. Circulation 117(10):1302–1309
37. Cipollone F et al (2005) Association between prostaglandin E receptor subtype EP4 overexpression and unstable phenotype in atherosclerotic plaques in human. Arterioscler Thromb Vasc Biol 25(9):1925–1931
38. Linton MF, Fazio S (2008) Cyclooxygenase products and atherosclerosis. Drug Discov Today Ther Strateg 5(1):25–36
39. Yokoyama U et al (2013) The prostanoid EP4 receptor and its signaling pathway. Pharmacol Rev 65(3):1010–1052
40. Takayama K et al (2002) Prostaglandin E_2 suppresses chemokine production in human macrophages through the EP4 receptor. J Biol Chem 277(46):44147–44154
41. Yokoyama U et al (2014) Prostaglandin E_2 inhibits elastogenesis in the ductus arteriosus via EP4 signaling. Circulation 129(4):487–496
42. Nakashima Y, Sueishi K (1992) Alteration of elastic architecture in the lathyritic rat aorta implies the pathogenesis of aortic dissecting aneurysm. Am J Pathol 140(4):959–969
43. Sibon I et al (2005) Lysyl oxidase deficiency: a new cause of human arterial dissection. Heart 91(5):e33
44. Rao R et al (2007) Prostaglandin E_2–EP4 receptor promotes endothelial cell migration via ERK activation and angiogenesis in vivo. J Biol Chem 282(23):16959–16968
45. Zhang Y, Daaka Y (2011) PGE_2 promotes angiogenesis through EP4 and PKA Cgamma pathway. Blood 118(19):5355–5364
46. Yanni SE et al (2009) The role of PGE_2 receptor EP4 in pathologic ocular angiogenesis. Invest Ophthalmol Vis Sci 50(11):5479–5486
47. Kuwano T et al (2004) Cyclooxygenase 2 is a key enzyme for inflammatory cytokine-induced angiogenesis. FASEB J 18(2):300–310
48. Bayston T et al (2003) Prostaglandin E_2 receptors in abdominal aortic aneurysm and human aortic smooth muscle cells. J Vasc Surg 38(2):354–359
49. Chen BC et al (2006) Peptidoglycan-induced IL-6 production in RAW 264.7 macrophages is mediated by cyclooxygenase-2, PGE_2/PGE4 receptors, protein kinase A, I kappa B kinase, and NF-kappa B. J Immunol 177(1):681–693
50. Ray WA et al (2002) COX-2 selective non-steroidal anti-inflammatory drugs and risk of serious coronary heart disease. Lancet 360(9339):1071–1073
51. McGettigan P, Henry D (2006) Cardiovascular risk and inhibition of cyclooxygenase: a systematic review of the observational studies of selective and nonselective inhibitors of cyclooxygenase 2. JAMA 296(13):1633–1644
52. Peters-Golden M, Henderson WR Jr (2007) Leukotrienes. N Engl J Med 357(18):1841–1854
53. Funk CD (2005) Leukotriene modifiers as potential therapeutics for cardiovascular disease. Nat Rev Drug Discov 4(8):664–672
54. Spanbroek R et al (2003) Expanding expression of the 5-lipoxygenase pathway within the arterial wall during human atherogenesis. Proc Natl Acad Sci U S A 100(3):1238–1243
55. Cipollone F et al (2005) Association between 5-lipoxygenase expression and plaque instability in humans. Arterioscler Thromb Vasc Biol 25(8):1665–1670
56. Funk CD (2001) Prostaglandins and leukotrienes: advances in eicosanoid biology. Science 294(5548):1871–1875
57. Vila L (2004) Cyclooxygenase and 5-lipoxygenase pathways in the vessel wall: role in atherosclerosis. Med Res Rev 24(4):399–424
58. Berman JP, Farkouh ME, Rosenson RS (2013) Emerging anti-inflammatory drugs for atherosclerosis. Expert Opin Emerg Drugs 18(2):193–205
59. Revermann M et al (2011) A pirinixic acid derivative (LP105) inhibits murine 5-lipoxygenase activity and attenuates vascular remodelling in a murine model of aortic aneurysm. Br J Pharmacol 163(8):1721–1732
60. Werz O et al (2008) Novel and potent inhibitors of 5-lipoxygenase product synthesis based on the structure of pirinixic acid. J Med Chem 51(17):5449–5453
61. Koeberle A et al (2008) Pirinixic acid derivatives as novel dual inhibitors of microsomal prostaglandin E_2 synthase-1 and 5-lipoxygenase. J Med Chem 51(24):8068–8076

Chapter 20
Cysteinyl Leukotrienes and Disease

Laura B. Fanning and Joshua A. Boyce

Abstract Cysteinyl leukotrienes (cys-LTs) are peptide-conjugated lipid inflammatory mediators generated predominantly by hematopoietic effector cells. Although they were originally recognized based on their capacity to potently induce vascular leakage and smooth muscle contraction, they are now known to elicit a wide range of effects on hematopoietic and structural cells and to have a key role in the initiation and amplification of type 2 immunity and eosinophilic inflammation. The three ligands comprising the cys-LTs, that is, leukotriene (LT)C_4, LTD_4, and LTE_4, mediate their effects through at least three G protein-coupled receptors (GPCRs), termed the type 1 and type 2 cys-LT receptors ($CysLT_1R$ and $CysLT_2R$, respectively), and the recently identified third receptor, GPR99, the only receptor with a binding preference for LTE_4. This chapter is concerned primarily with the expanding role of the cys-LTs and their receptors in asthma and aspirin-exacerbated respiratory disease, and touches on the potential implications for other pathobiological processes.

Keywords Cysteinyl leukotriene • Asthma • Allergic inflammation • Atopy • Prostaglandin • Lipid mediator

20.1 Introduction

Cysteinyl leukotrienes (cys-LTs) are metabolites of arachidonic acid generated by myeloid effector cells in response to activation. Cys-LTs were initially recognized as potent smooth muscle constrictors [1] and mediators of vascular leak [2], fueling interest in their potential role in asthma and allergic disease. More than three

---

L.B. Fanning, M.D. • J.A. Boyce, M.D. (✉)
Department of Medicine, Harvard Medical School, Boston, MA, USA

Division of Rheumatology, Immunology, and Allergy, Jeff and Penny Vinik Center for Allergic Disease Research, Brigham and Women's Hospital,
1 Jimmy Fund Way, Smith Building, Room 638, Boston, MA 02115, USA
e-mail: jboyce@rics.bwh.harvard.edu

T. Yokomizo, M. Murakami (eds.), *Bioactive Lipid Mediators*,
DOI 10.1007/978-4-431-55669-5_20

decades after the derivation of their chemical structures, they have fulfilled "Koch's postulates" in that regard. Cys-LT biosynthesis is a characteristic feature of asthma and allergic rhinitis [3, 4] and increases concomitantly with disease activity [5]. Pharmacological inhibitors of 5-LO and the type 1 receptor for cys-LTs ($CysLT_1R$), respectively, are among the first successful treatments for asthma that were developed based on specific molecular targets. Their efficacy validates the role of the cys-LTs in asthma pathogenesis.

This chapter focuses primarily on the role of the cys-LTs in asthma in humans because of the wealth of pharmacological, functional, immunohistochemical, and physiological data in the area. Nonetheless, an emerging body of evidence also implicates their function in cardiovascular disease [6–10], respiratory viral infections [11], obstructive sleep apnea [12–15], chronic urticaria [16, 17], and Henoch–Schoenlein vasculitis [18]. Confirmatory therapeutic intervention studies are necessary to firmly implicate cys-LTs in the pathobiology of these conditions. Additional functions of cys-LTs identified in mouse model systems are covered elsewhere in this volume, and mouse studies are discussed only in instances that directly inform observations made in humans.

20.2 Biosynthesis and Cellular Sources of the cys-LTs

20.2.1 Biosynthetic Pathway

Following the release of arachidonic acid mediated by a cytosolic phospholipase A_2 ($cPLA_2$) [19], 5-lipoxygenase (5-LO) translocates to the nuclear envelope [20] and acts with a 5-LO-activating protein (FLAP) [21] to catalyze the formation of LTA_4, the unstable precursor of both LTC_4, the parent cys-LT, and the dihydroxy LT, LTB_4. LTC_4 results from conjugation of LTA_4 to reduced glutathione catalyzed by the membrane-anchored enzyme LTC_4 synthase (LTC_4S) [22]. LTC_4 is then exported from the cells of origin by specific energy-dependent transporters [23]. In the extracellular compartment, LTC_4 is converted to LTD_4, a powerful but short-lived agonist of smooth muscle contraction, by a gamma glutamyl leukotrienease [24]. LTD_4 is then metabolized to LTE_4, a stable (and therefore abundant) metabolite, by dipeptidases [25]. LTE_4 is excreted in the urine, and its measurement is commonly used in clinical studies as a "time-weighted" reflection of cys-LT generation [26]. Thus, three separate ligands (LTC_4 and its extracellular metabolites LTD_4 and LTE_4) result from a single intracellular synthetic reaction that depends on the sequential functions of 5-LO and LTC_4S (Fig. 20.1) [27]. These three ligands possess a common lipid backbone and conjugation to cysteine. They differ not only in the amino acids retained from the glutathione adduct, but also in half-life, receptor specificity, and function, as detailed following.

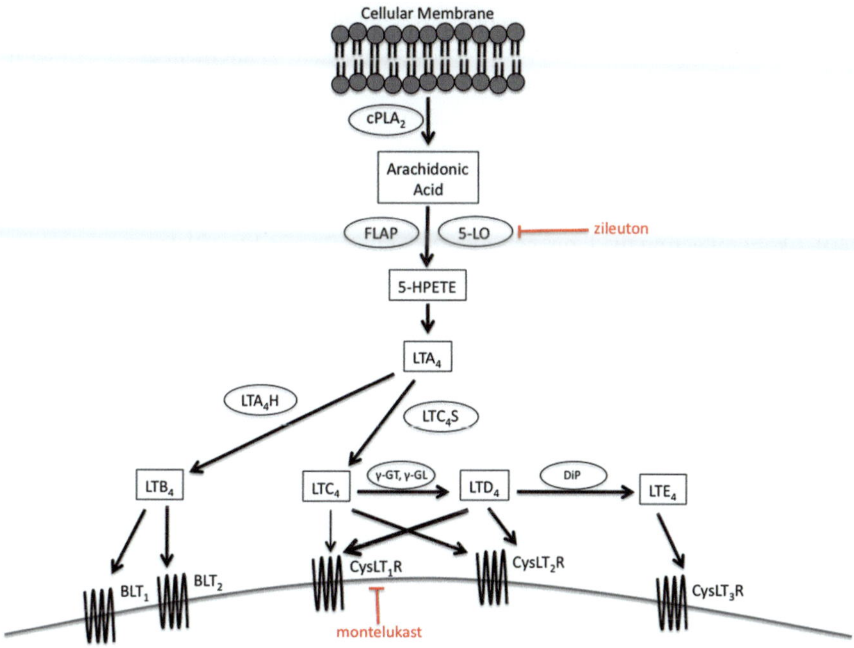

Fig. 20.1 Leukotriene biosynthesis. 5-Lipoxygenase (5-LO) and 5-LO-activating protein (FLAP) metabolize arachidonic acid to leukotriene (LT)A_4, which can be converted to LTB_4 (by LTA_4H) or the cysteinyl leukotrienes (cys-LTs), by leukotriene C_4 synthase (LTC_4S). LTs then bind to the indicated G protein-coupled receptors. *cPLA₂* cytosolic phospholipase A_2, *BLT* leukotriene B receptor, *DiP* dipeptidase, *g-GL* g-glutamyl leukotrienase, *g-GT* g-glutamyl transpeptidase, *5-HPETE* 5-hydroxyperoxyeicosatetraenoic acid. (Reprinted from *Annals of Allergy, Asthma, & Immunology*, Volume 111/No. 3. Fanning LB and Boyce JA, Lipid mediators and allergic diseases, pp. 155–162. Copyright 2013, with permission from Elsevier.)

20.2.2 Cellular Sources

Mast cells [28], basophils [29], eosinophils [30], monocyte/macrophages [31], and myeloid dendritic cells [32] all possess the complete repertoire of enzymes and proteins needed to convert endogenous arachidonic acid to LTC_4. In allergic diseases, cys-LTs are thought to originate from eosinophils and basophils (which accumulate at sites of inflammation) and activated resident tissue mast cells (which respond to cross-linkage of their high-affinity receptor for IgE, FcεRI, in response to specific allergens). In bronchial biopsies from atopic asthmatic individuals, mast cells are the primary site of localization of LTC_4S [33], potentially reflecting the strongly inducible nature of the enzyme in this cell type when exposed to the Th2 cytokine IL-4 [34]. Cys-LT concentrations in the bronchoalveolar lavage fluid obtained from atopic asthmatic subjects increase sharply following challenge by inhalation with a specific allergen [3], concomitantly with the release of other mast cell-derived mediators (histamine, tryptase, and prostaglandin D_2 (PGD_2)) [35–37] that are released or generated de novo following allergen-induced mast cell activation. Basophils, which also express FcεRI and respond to allergen, are recruited to the bronchial mucosa in the hours after challenge and may contribute to the production of cys-LTs during the allergen-induced late-phase response [38]. Increases in urinary LTE_4 occurring following anaphylaxis likely reflect contributions from both mast cells and basophils [39]. Patients with eosinophilic pneumonia exhibit markedly elevated levels of urinary LTE_4 relative to asthmatic and nonasthmatic controls [40], likely reflecting the large burden of eosinophils capable of LTC_4 generation.

Similar to the foregoing myeloid cells, platelets express LTC_4S [41]. However, platelets lack 5-LO and FLAP and therefore cannot generate LTA_4 from endogenous arachidonic acid. However, platelets can convert LTA_4 to LTC_4 if the unstable precursor is provided from an adjacent cell source [42]. Activated platelets display P-selectin on their surfaces, which permits their binding to neutrophils, monocytes, and eosinophils, all of which have an active 5-LO pathway [43]. Activated neutrophils, in particular, generate LTA_4 in molar excess of their capacity to convert it to LTB_4, their sole terminal 5-LO pathway product. The excess LTA_4, once released, can be converted to LTC_4 by the adherent platelets. Thus, platelet-adherent neutrophils comprise a de novo unit potentially capable of amplifying the production of cys-LTs in circumstances in which platelets become activated by tissue injury or inflammation. In addition, the presence of adherent platelets primes neutrophils for augmented 5-LO pathway activity by an unknown mechanism [44]. The bidirectional cooperation between neutrophils and platelets may markedly amplify the production of cys-LTs, particularly in aspirin-exacerbated respiratory disease (AERD) [41], in which platelet–leukocyte complexes are detected with aberrantly high frequencies.

20.3 Functions of cys-LTs in Human Airways

The chemical characterization of the cys-LTs, followed by the availability of the synthetic compounds, permitted analyses of their pharmacological properties in humans and preclinical models. These studies accurately predicted the existence of individual receptors with respective preferences for each of the three cys-LTs [45] decades before the cloning and characterization of these receptors [46–48].

20.3.1 Exogenous cys-LTs as Bronchoconstrictors

The effects of inhaled cys-LTs on airway physiology were studied in healthy and asthmatic volunteers subjected to challenges with LTC_4, LTD_4, or LTE_4. The doses of both LTC_4 [49] and LTD_4 [50] required to induce bronchoconstriction in healthy subjects were 1000 to 5000 fold lower than the doses of histamine necessary to induce an equivalent fall in airflow. The cys-LT-mediated changes in lung function were substantially more prolonged than those in response to histamine and noteworthy for the lack of associated cough or hoarseness. The fall in lung function induced by inhalation of LTD_4 occurred more rapidly than that induced by LTC_4 [51]. This difference, in retrospect, might reflect the time required for LTC_4 to be converted to LTD_4, or the differences between the two ligands in their receptor-binding properties. Although LTD_4 was also a markedly potent bronchoconstrictor in subjects with asthma [51], asthmatic subjects were not substantially more responsive to LTD_4 than were nonasthmatic controls when corrected for histamine reactivity.

Similar to LTC_4 and LTD_4, LTE_4 also causes bronchoconstriction, but is less potent than its precursors, inducing reductions in airflow at doses about 1 log lower than histamine in normal and asthmatic subjects [52]. A separate study, however, noted that subjects with asthma exhibited selective hyperresponsiveness to LTE_4, but not to LTC_4, when compared with nonasthmatic individuals (16 fold more sensitive) [53]. Moreover, individuals with AERD demonstrated even greater sensitivity (~1 log fold) to LTE_4, but identical responsiveness to LTC_4, when compared with aspirin-tolerant asthmatics [54]. These observations fueled speculation that LTE_4 acted by receptor-dependent mechanisms distinct from those of LTC_4 and LTD_4, and that these mechanisms might be altered as a consequence (and potentially a pathogenetic factor) of asthma.

20.3.2 Nonbronchoconstricting Effects of cys-LTs in Human Airways

Preincubation of guinea pig tracheal smooth muscle with LTE_4 shifted the sensitivity of the organ to constriction in response to the subsequent administration of histamine [45]. This potentiation of histamine sensitivity was not observed in response to LTC_4 or LTD_4 and could be blocked by treatment of the tracheal rings with the cyclooxygenase (COX) inhibitor indomethacin. Pretreatment with LTE_4 also potentiated histamine contractility of surgically excised human bronchi. This potentiation was inhibited by pretreatment with either indomethacin or with a T-prostanoid (TP) receptor antagonist [55]. In subjects with asthma, inhalation of LTE_4 shifted the dose–response curve to inhaled histamine challenge at 4 and 7 h [56]. As predicted from the in vitro studies, this effect was sensitive to inhibition by the oral administration of indomethacin. Thus, LTE_4 enhances end-organ reactivity to histamine through induction of COX product(s), potentially thromboxane A_2 (TXA_2) or PGD_2, both of which mediate their effects on the airway by TP receptors [57, 58].

When inhaled by subjects with mild asthma, LTE_4 caused an increase in the numbers of eosinophils, and to a lesser extent, neutrophils, in bronchial biopsies obtained 4 h after the inhalation challenge [59]. In a second study, subjects with asthma underwent sputum induction (as well as bronchial biopsies in a subset) after inhalations of LTD_4, LTE_4, and diluent. LTE_4, but not LTD_4, caused sharp increases an increase in sputum eosinophils at both 7 and 24 h, and sputum basophils at 7 h when the dosages of the two cys-LTs were titrated to achieve a comparable degree of bronchoconstriction [60]. Bronchial biopsies revealed an eosinophil-rich submucosal infiltrate 4 h after the inhalation of LTE_4, but not LTD_4 or diluent. Thus, although LTE_4 is weaker than its precursors as a constrictor, it is unique for its ability to induce the accumulation of immune effector cells. The fact that LTE_4 has the longest biological half-life of the three cys-LTs suggests that it may contribute to the perpetuation of tissue eosinophilia in circumstances in which cys-LT synthesis is increased.

20.3.3 Potential Cell Targets and Mechanisms of Action of LTE_4 in the Airway

Although the mechanisms and cell targets that account for LTE_4-induced COX activation and eosinophilic pulmonary inflammation in humans remain to be determined, in vitro and in vivo models support very different mechanisms of action for LTE_4 than for LTC_4 or LTD_4. Although the two classical G protein-coupled receptors (GPCRs) for cys-LTs ($CysLT_1R$ and the type 2 receptor for cys-LTs, $CysLT_2R$) are expressed by both hematopoietic and nonhematopoietic cells in human airways [47, 61], neither of these receptors binds LTE_4 with high affinity [46, 47]. One study demonstrated that LTE_4, but not LTC_4 or LTD_4, could induce the production of PGD_2 by a human mast cell line, LAD2 [62]. Importantly, this induction was not altered by knockdown of either $CysLT_1R$ or $CysLT_2R$, and required LTE_4-mediated upregulation of COX-2 and signaling through peroxisome proliferator-activated receptor-γ (PPAR-γ). Because PGD_2 is both a bronchoconstrictor (acting at TP receptors) [57] and is chemotactic for eosinophils and basophils [acting at D-prostanoid 2 (DP_2) receptors] [63], the unique properties of LTE_4 could reflect, in part, the activation of pulmonary mast cells and secondary generation of PGD_2.

In another study, intrapulmonary administration of LTE_4, but not LTD_4, potentiated the recruitment of eosinophils to the airways of ovalbumin (OVA)-sensitized mice if administered with low-dose OVA [64]. This mechanism was independent of $CysLT_1R$ and $CysLT_2R$, but depended exquisitely on platelets and $P2Y_{12}$ receptors, which recognize adenosine diphosphate (ADP). The requirement for platelets was not the result of direct platelet activation by LTE_4, suggesting that platelets in this model may be the target of a mediator released by another cell type that expresses a true LTE_4 receptor. Moreover, $P2Y_{12}$ receptors do not bind LTE_4 directly, but are involved in certain LTE_4-initiated signaling functions [65]. For example, knockdown of $P2Y_{12}$ receptors in LAD2 cells markedly suppressed the capacity of LTE_4 to induce PGD_2 generation and chemokine production [64]. It seems possible that $P2Y_{12}$ receptors, which amplify activation signals to other agonists through autocrine ADP-mediated circuits in platelets [66], could amplify signaling through a true LTE_4 receptor on mast cells or other cell types. To date, the recently identified GPR99 is the only GPCR with a preference for LTE_4 binding [48]. The distribution of GPR99, its expression in human airways, and its role in responses to LTE_4 in general and asthma in particular remain to be determined.

20.4 Functions of cys-LTs and Their Receptors on Human Hematopoietic Cells

As noted, a wide range of hematopoietic cells express $CysLT_1R$ and $CysLT_2R$ [67–70]. Several in vitro studies support potentially important functions of cys-LTs in activating immune effector cells relevant to asthma. Because primary cells express various combinations of two or more cys-LT receptors (including yet to be identified potential receptors), their functional responses to cys-LTs ex vivo often diverge from the pharmacology predicted from studies of transfected receptors overexpressed in isolation. These divergent responses reflect physical or functional interactions between known receptors [71], as well as the potential existence of previously unidentified receptors. Additionally, the expressions of cys-LT receptors on cells that also generate cys-LTs (e.g., mast cells, eosinophils) suggest the potential for autocrine/intracrine signaling functions.

20.4.1 Mast Cells

Human mast cells derived in vitro from both cord blood [68, 69] and peripheral blood [72] express both $CysLT_1R$ and $CysLT_2R$. LTC_4 and LTD_4 both elicit strong calcium fluxes [68] and phosphorylation of extracellular signal-related kinase (ERK) [73] in primary cord blood mast cells, as well as the LAD2 cell line [62]. Priming of primary human cord blood mast cells with the Th2 cytokine IL-4 amplifies both calcium flux (particularly in response to LTC_4) and ERK activation in response to the cys-LTs without changing the levels of $CysLT_1R$ or $CysLT_2R$ expression [68, 74]. This action differs from that in monocytes, in which IL-4 and IL-13 upregulate the expression of $CysLT_1R$ [75] and $CysLT_2R$ [76]. LTC_4 and LTD_4 both behave as potent agonists for the production of cytokines (IL-5, tumor necrosis factor, and IL-13) by IL-4-primed human mast cells by a $CysLT_1R$-dependent mechanism [73]. LTD_4 also induces proliferation of human mast cells derived from cord blood or peripheral blood by inducing $CysLT_1R$-dependent transactivation of the c-kit tyrosine kinase [72, 77]. Interestingly, knockdown of $CysLT_2$ receptors augments both proliferation and cytokine generation by mast cells in response to stimulation by LTD_4, indicating that $CysLT_2R$ inhibits $CysLT_1R$ functions on this cell type [71, 77]. $CysLT_2$ forms heterodimers with $CysLT_1R$ on mast cells, interfering with its presentation at the plasma membrane [71]. IL-4, an accessory mitogen for mast cells, requires both the induced expression of LTC_4S and endogenous function of $CysLT_1R$ to drive mast cell proliferation [72]. Of note, both $CysLT_1R$ and $CysLT_2R$ proteins localize to the nuclear envelope of human mast cells (the location of LTC_4S), and are thus potentially positioned to induce intracrine signaling in response to newly synthesized LTC_4.

20.4.2 Eosinophils

Eosinophils express both $CysLT_1R$ and $CysLT_2R$, with the latter receptor being expressed far more abundantly at the mRNA level [78]. Human peripheral blood eosinophils secrete IL-4 in response to stimulation with IL-16, eotaxin, or RANTES by a mechanism that is sensitive to the inhibition of 5-LO, suggesting a requirement for the synthesis of endogenous cys-LTs [70]. Interestingly, permeabilized eosinophils secrete IL-4 in response to LTC_4, but not LTD_4 or LTE_4, by a mechanism that is resistant to conventional $CysLT_1R$ antagonists but sensitive to pertussis toxin [70]. These findings argue that intracellular cys-LT receptors exhibit a preference for LTC_4, again supporting an intracrine function.

20.4.3 Lymphocytes

Human $CD4^+$ Th2 cells express $CysLT_1R$, as well as the DP_2 receptor for PGD_2. Stimulation of human Th2 cells with PGD_2 ex vivo induces the release of IL-5 and IL-13 [79]. Although cys-LTs are comparatively weak as agonists for cytokine generation by Th2 cells, they synergize markedly with PGD_2 for this function, with LTE_4 being more potent than either LTC_4 or LTD_4 [80]. The effect of LTE_4 on T cells is blocked by treatment of the cells with the $CysLT_1R$ antagonist montelukast. The sensitivity to montelukast is at odds with the rank-order potency of LTE_4 over LTD_4 for these T-cell-directed effects. Whether a montelukast-sensitive receptor for LTE_4 that is not $CysLT_1R$ exists is unknown but has been suggested by previous studies of LAD2 cells [62]. The recently identified LTE_4 receptor GPR99 is resistant to blockade by $CysLT_1R$ antagonists and therefore not likely to account for the effects of LTE_4 observed in T cells [48].

Type 2 innate lymphoid helper cells (ILC2 cells) are robust sources of Th2-type cytokines that reside constitutively in tissues and are activated directly by epithelially derived cytokines such as IL-33, IL-25 and thymic stromal lymphopoietin [81]. Similar to Th2 cells, ILC2 cells express DP_2 receptors [81], and respond to the combination of PGD_2 and cys-LTs in a synergistic manner by generating IL-13 and IL-5, as well as IL-9, GM-CSF, and IL-8 [82]. Interestingly, although ILC2 cells do not produce IL-4 when activated by IL-33 or IL-25, they generate significant amounts of this cytokine when activated by mast cell supernatants, an activity attributed to the combination of PGD_2 and cys-LTs [82]. The cys-LT receptor(s) expressed by human ILC2 cells are not yet known, although their counterparts in the mouse lung strongly express $CysLT_1R$, and generate IL-4, IL-5, and IL-13 when stimulated with LTD_4 or LTE_4 [83]. Interestingly, the effects of LTD_4 in the mouse lung ILC2 cells are blocked by montelukast, but the effects of LTE_4 are montelukast resistant, suggesting the potential contribution from GPR99 or a different LTE_4 receptor.

20.4.4 Platelets

Human platelets express both $CysLT_1R$ and $CysLT_2R$ proteins [84, 85]. Stimulation of human platelets with LTC_4, LTD_4, and LTE_4 elicits their release of the chemokine RANTES [84]. A recent mouse study indicated that LTC_4 can strongly potentiate allergen-induced eosinophilic pulmonary inflammation through the activation of platelets through $CysLT_2R$ [85]. Whether platelets have analogous CysLT2R-dependent functions in humans is presently unknown.

20.4.5 Dendritic Cells

Myeloid dendritic cells in human peripheral blood express $CysLT_1R$ [86]. The administration of montelukast to atopic asthmatic individuals blocked the reduction in blood dendritic cells elicited by allergen inhalation challenges. Several in vitro studies also suggest a role for cys-LTs in modulating dendritic cell function. Human monocyte-derived dendritic cells (moDCs) matured with LPS had a 50 % reduction in $CysLT_1R$ expression but an increase in $CysLT_2R$. In contrast, moDCs treated with polyI:C had no change in receptor expression. This downregulation of $CysLT_1R$ by LPS was prevented by COX inhibitors. Furthermore, DCs matured with polyI:C demonstrated chemotaxis in response to LTD_4, as well as increased migration in response to CCL19. This response to LTD_4 was seen only weakly in DCs matured with LPS [87]. Human moDCs cultured with LTC_4 release the eosinophil chemoattractant RANTES and induce T-cell proliferation. Both of these effects are blocked by montelukast [88]. Whether $CysLT_1R$ participates in priming for Th2 responses to dust mite and mold allergens in humans, as it does in mice [89], remains to be seen.

20.5 The Role of cys-LTs and Their Receptors in Asthma

The potency of the cys-LTs as contractile agonists for human airways, and the observation that urinary levels of LTE_4 were elevated in asthmatic subjects presenting to the emergency room for spontaneous disease exacerbations [5], suggested that drugs targeting the synthesis of cys-LTs or blocking their receptors had therapeutic potential in asthma. Indeed, the 5-LO inhibitor zileuton and the antagonists of $CysLT_1R$ were among the earliest drugs developed for the treatment of asthma that were based on a disease-related molecular target.

20.5.1 Clinical Efficacy of cys-LT Pathway-Targeted Drugs

In double-blind, placebo-controlled trials of patients with mild to moderate asthma who were not receiving inhaled glucocorticoids, treatment with zileuton significantly increased baseline FEV_1 while reducing exacerbations of asthma [90, 91]. The effect of zileuton on airway function occurred despite relatively modest effects on basal concentrations of urinary LTE_4 compared with placebo (~40 % decrease) [91]. Treatment of mild to moderate asthmatic subjects with antagonists of $CysLT_1R$ also improved FEV_1 compared with placebo while reducing the frequency of exacerbations [92]. Montelukast also reduces the frequency of exacerbations of wheezing in preschool children [93]. When added to standard bronchodilator therapy, intravenous administration of montelukast significantly increased peak expiratory flow rates in three studies of adult asthmatic subjects presenting to the emergency department when compared with placebo [94–96], although a pediatric study failed to replicate these results [97]. These observations demonstrate that the cys-LTs contribute substantially to airflow obstruction during asthma exacerbations, at least in adults, and that signaling through $CysLT_1R$ is at least partially responsible.

In addition to spontaneous exacerbations of asthma, blockade of $CysLT_1R$ attenuates the decrements in airflow that occur in response to a number of provocative challenges. Challenges of susceptible individuals with exercise [98, 99] or inhalation of specific allergens [99, 100] result in increases in urinary LTE_4 and accompanying airflow obstruction that is attenuated by the administration of $CysLT_1R$ antagonists. Inhaled adenosine [101] and mannitol [102] both induce bronchoconstriction in asthmatic subjects that is sensitive to inhibition by pretreatment with montelukast. Exercise, allergen, adenosine, and mannitol hold in common the feature of inducing mast cell activation. Thus, these studies support both the pathogenetic relevance of mast cell-derived cys-LTs in physiological provocations and the role of $CysLT_1R$ in these contexts.

20.5.2 Determinants of Responsiveness to Leukotriene Pathway-Active Drugs

Clinical responses to zileuton and $CysLT_1R$ antagonists are not uniform and are at least partly determined by environmental and genetic controls. Asthmatic children exposed to cigarette smoke exhibit significantly greater protection from albuterol usage when treated with montelukast than do children not exposed to smoke [103], and children with the highest ratios of urinary LTE_4 to exhaled nitric oxide are the most protected [104]. Promoter polymorphisms in the *ALOX5* gene (encoding 5-LO) that alter transcription of the enzyme significantly influence the extent to which treatment with zileuton changes baseline lung function [105–107]. The presence of a common polymorphic variant (A(−444)C) of *LTC4S*, the gene encoding LTC_4S, was associated with a larger increment in baseline FEV_1 with administration of pranlukast in a Japanese cohort than occurred in individuals without the variant allele. Another study showed that individuals bearing at least one copy of the same *LTC4S* variant exhibited a larger suppression of exhaled nitric oxide in response to the administration of montelukast when compared to those with two wild-type alleles [108]. Variants of *ALOX5* and *MRP1* (encoding an LTC_4 transporter protein) are significantly associated with montelukast-induced changes in FEV_1, and variants in *LTC4S* and in *LTA4H* (encoding the terminal enzyme involved in the production of LTB_4) are associated with changes in exacerbation rates [109].

20.6 Aspirin-Exacerbated Respiratory Disease

No disease is more clearly linked to dysregulation of the cys-LT system than AERD. It is characterized by adult-onset asthma, severe rhinosinusitis with nasal polyps, and idiosyncratic, cys-LT-dependent respiratory reactions to aspirin and other nonselective inhibitors of COX [110]. Individuals with AERD exhibit levels of urinary LTE_4 that exceed those found in aspirin tolerant asthmatics by about three- to fourfold [41, 111]. The administration of nonselective COX inhibitors results in a dramatic further increase in urinary LTE_4 levels, concomitantly with bronchoconstriction, incremental sinonasal congestion, rhinorrhea, and ocular itching and discharge. The administration of either zileuton or $CysLT_1R$ antagonists attenuates the severity of aspirin-induced bronchoconstriction in AERD [112]. Both classes of drugs are also superior to placebo for improving sinonasal function [113, 114]. Thus, cys-LTs are not only overproduced in AERD but are highly relevant to disease pathogenesis.

20.6.1 Mechanisms of Dysregulated cys-LT Pathway Activity in AERD

Cell-specific dysregulation of 5-LO/LTC_4S pathway components likely contributes to the overproduction of cys-LTs in AERD. Eosinophils in bronchial [115] and nasal biopsies [116] from patients with AERD markedly overexpress LTC_4S, but not 5-LO or FLAP, when compared with eosinophils in biopsies from aspirin-tolerant controls. The numbers of LTC_4S^+ eosinophils in the biopsies correlated strongly with the levels of cys-LTs in the bronchoalveolar lavage (BAL) fluid and inversely with the dose of aspirin required to induce a reduction in FEV_1. Because eosinophils are the dominant cell type in the respiratory tissue lesions in AERD, their selective overexpression of LTC_4S may ensure the efficient conversion of LTA_4 to LTC_4 in the inflammatory milieu, although the mechanism responsible for the activation of eosinophils, or for their selective overexpression of LTC_4S in AERD, remains unknown. A study conducted in Poland reported that the presence of the

A(−444)C variant allele, which enhances the production of LTC_4 by eosinophils [117], was a risk factor for AERD [118]. However, this association was not replicated on studies of other populations [119].

A recent study reported remarkably increased numbers of platelet-adherent eosinophils, monocytes, and neutrophils in the peripheral blood of patients with AERD compared to samples from aspirin-tolerant asthmatic and nonasthmatic controls [41]. This increase in adherent platelets in the blood was paralleled by large numbers of extravasated platelet-adherent leukocytes in nasal polyps from patients with AERD. As noted previously, platelets express LTC_4S, and adherent platelets accounted for as much as 60–70 % of the LTC_4S activity associated with peripheral blood granulocytes obtained from subjects with AERD [41]. Moreover, removal of adherent platelets decreased the generation of LTC_4 by ionophore-stimulated granulocytes from subjects with AERD by approximately 70 %. The percentages of platelet-adherent neutrophils, eosinophils, and monocytes in the blood of patients with AERD correlated strongly with the amount of urinary LTE_4 detected at baseline. These findings are consistent with a substantial contribution from platelet-dependent transcellular mechanism of LTC_4 generation to the high-level production of cys-LTs in AERD. The underlying basis accounting for the high frequencies of adherent platelets is presently unknown but is suggestive of a stimulus for ongoing platelet activation and P-selectin expression in vivo.

Although the numbers of mast cells in the bronchial and nasal mucosa of patients with AERD do not differ substantially from their numbers in the tissues of aspirin-tolerant asthmatic controls [115, 116, 120], mast cells likely contribute to cys-LT generation during reactions to COX-1 inhibitors. As is the case in allergen challenges of atopic individuals, provocative aspirin challenges of patients with AERD result in the release of mast cell-associated mediators such as tryptase and PGD_2 into the plasma or lavage fluids [121]. The administration of mast cell-stabilizing cromone drugs blocks not only the change in lung function but also the rise in urinary LTE_4 that accompanies reactions [122]. Interestingly, the administration of zileuton before intranasal lysine aspirin challenge blocks the release of tryptase and histamine as determined by their measurements in nasal lavage fluid [121], suggesting that mast cells may be both sources and targets of cys-LT-induced activation during aspirin-induced reactions in AERD. The existence of an autocrine/paracrine cys-LT-dependent mechanism of mast cell activation is supported by a recent model of AERD in mice [123], as well as earlier studies of cultured human mast cells [73].

20.6.2 End-Organ Reactivity to cys-LTs in AERD

In addition to dysregulated cys-LT generation, subjects with AERD show enhanced end-organ reactivity to cys-LTs. Compared with aspirin-tolerant asthmatic controls, individuals with AERD demonstrate bronchoconstriction at significantly lower doses of inhaled LTE_4 [54] and LTD_4 [124]. The numbers and percentages of

$CysLT_1R$-positive mast cells, eosinophils, and monocytes in nasal biopsies from patients with AERD exceed those observed in the tissues of aspirin-tolerant asthmatic controls [125, 126], whereas there are no differences in $CysLT_2R$ expression. Interestingly, bronchial reactivity to inhaled LTD_4 in AERD or aspirin-tolerant asthma does not correlate with the numbers of $CysLT_1R$- or $CysLT_2R$-expressing cells in bronchial biopsies [124]. The percentages of hematopoietic cells in nasal biopsies that express $CysLT_1R$ protein decreases following desensitization to aspirin [125], a procedure that attenuates bronchial reactivity to LTE_4 [127]. As already noted, the cellular distribution of GPR99 in the airway and the potential role of this receptor in mediating responsiveness to cys-LTs remain to be defined.

20.6.3 Control of LT Pathway Activity by Prostaglandin E_2

Although the dysregulated expression and function of LT pathway enzymes in AERD likely predispose to overproduction of cys-LTs, it cannot account alone for the pathognomonic cys-LT-dependent respiratory reactions to drugs that block COX-1. Prostaglandin E_2 (PGE_2), a ubiquitous COX product, is generated from arachidonic acid metabolized by either COX-1 or COX-2 and one or more PGE_2 synthases (PGES). Although most cell types express COX-1 constitutively, COX-2 is expressed inducibly or upregulated in many cell types in response to environmental danger signals such as lipopolysaccharide [128]. In cells that generate PGE_2 (including macrophages, fibroblasts, and epithelial cells), COX-2 induction is accompanied by upregulated expression of microsomal PGES-1 (mPGES-1), a membrane-anchored enzyme that preferentially converts the COX-2-derived precursor PGH_2 to PGE_2 [129]. The simultaneously induced expressions of COX-2 and mPGES-1 substantially increase the rate of PGE_2 synthesis during inflammatory responses. Four different GPCRs, termed the EP_1, EP_2, EP_3, and EP_4 receptors, mediate the effects of PGE_2. EP_2 and EP_4 receptors share the property of activating stimulatory G (Gs) proteins that in turn activate adenylate cyclase, increasing intracellular levels of cyclic adenosine monophosphate (cAMP) and activating protein kinase A (PKA) [130]. PKA, in turn, can phosphorylate 5-LO and suppress its catalytic function [131, 132]. As a result, PGE_2 may serve as an endogenous inhibitor of cys-LT production, because all 5-LO-expressing hematopoietic cell types express EP_2 and EP_4 receptors. Indeed, inhalation of PGE_2 by subjects with AERD completely blocks subsequent bronchoconstriction and increases in urinary LTE_4 occurring in response to aspirin challenge [133], verifying the anti-leukotriene effect of PGE_2.

There is considerable evidence that AERD involves dysregulated PGE_2 synthesis. Extracted lipid fractions from nasal polyps excised from patients with AERD contain markedly less PGE_2 than does control nasal tissue [134]. mRNA levels encoding COX-2 are low to absent in nasal polyps from subjects with AERD [135], and fibroblasts cultured from AERD nasal polyps display markedly diminished interleukin-1β-induced COX-2 expression and PGE_2 synthesis when compared with polyp fibroblasts from aspirin-tolerant controls [136]. An unbiased methylomic analysis of nasal polyps identified the gene encoding mPGES-1 (*PTGES*) as among the most hypermethylated genes in polyps from subjects with AERD compared to those from aspirin-tolerant controls. Thus, AERD may involve epigenetic silencing of the COX-2/mPGES-1 system. Because COX-2 is largely resistant to aspirin at doses that trigger reactions in AERD, its absence would have the predicted consequence of a failure to maintain tissue levels of PGE_2 in the setting of COX-1 inhibition, thus removing a tenuous "brake" on an upregulated system for cys-LT generation. Indeed, mice lacking mPGES-1 (and therefore lacking COX-2-derived PGE_2) are markedly impaired in their capacity to maintain PGE_2 generation in the lung when challenged with inhaled lysine aspirin, which also elicits mast cell activation, cys-LT release, and cys-LT-dependent bronchoconstriction in this strain [123].

The effects of putative lesions that impair the function of the COX-2/mPGES-1 system in AERD may be amplified by parallel abnormalities in the EP receptor system. An immunohistochemical study of sinonasal biopsy tissue revealed that the percentages of eosinophils, neutrophils, mast cells, and T cells expressing the EP_2 receptor protein were significantly lower than the corresponding subsets of cells in aspirin-tolerant controls [137]. In contrast, the percentages of cells expressing the EP_1, EP_3, and EP_4 receptors did not differ between AERD and control tissues. A similar trend was reported for leukocytes in bronchial biopsies from subjects with AERD and aspirin-tolerant controls [124]. In contrast to the findings of these immunohistochemical analyses, a recent study using flow cytometry detected no differences in EP_2 receptor protein expression by eosinophils, neutrophils, or platelets between subjects with AERD and aspirin-tolerant healthy and asthmatic controls [44]. Nevertheless, this study reported that granulocytes from individuals with AERD were markedly resistant to the effects of PGE_2 and other cyclic AMP-activating agonists in terms of their capacity to suppress the generation of both LTB_4 and LTC_4 in response to activation by the bacterial tripeptide fMLP. This PGE_2 resistance was attributed to aberrantly low levels of expression of the PKA catalytic γ-subunit, as well as diminished basal PKA activity, in granulocytes from subjects with AERD relative to controls [44]. These studies suggest that several steps in the EP receptor system and its downstream effectors can contribute to AERD by dysregulating 5-LO pathway activity.

20.7 Conclusions

Decades after their discovery, the cys-LTs remain a topic of intense interest and to date are the only successful pharmacotherapeutic mediator target for asthma. The application of molecular technology and experimental systems have verified the identity of at least three receptors, suggesting additional potential therapeutic applications. The astonishing degree of heterogeneity among human subjects in both the production of cys-LTs and the responsiveness to cys-LTs (as exemplified by studies of AERD), the range of responsiveness to the targeted therapies, and the numbers of mechanisms that up- or downregulate components of the system hold clues both to fundamental biology and to underlying genetic and epigenetic modifications that might serve as more precise guides to therapy in the future.

References

1. Weiss JW, Drazen JM, McFadden ER Jr, Weller PF, Corey EJ, Lewis RA, Austen KF (1982) Comparative bronchoconstrictor effects of histamine, leukotriene C, and leukotriene D in normal human volunteers. Trans Assoc Am Physicians 95:30–35
2. Soter NA, Lewis RA, Corey EJ, Austen KF (1983) Local effects of synthetic leukotrienes (LTC_4, LTD_4, LTE_4, and LTB_4) in human skin. J Invest Dermatol 80(2):115–119
3. Wenzel SE, Larsen GL, Johnston K, Voelkel NF, Westcott JY (1990) Elevated levels of leukotriene C_4 in bronchoalveolar lavage fluid from atopic asthmatics after endobronchial allergen challenge. Am Rev Respir Dis 142(1):112–119
4. Figueroa DJ, Borish L, Baramki D, Philip G, Austin CP, Evans JF (2003) Expression of cysteinyl leukotriene synthetic and signalling proteins in inflammatory cells in active seasonal allergic rhinitis. Clin Exp Allergy 33(10):1380–1388
5. Drazen JM, O'Brien J, Sparrow D, Weiss ST, Martins MA, Israel E, Fanta CH (1992) Recovery of leukotriene E_4 from the urine of patients with airway obstruction. Am Rev Respir Dis 146(1):104–108
6. Iovannisci DM, Lammer EJ, Steiner L, Cheng S, Mahoney LT, Davis PH, Lauer RM, Burns TL (2007) Association between a leukotriene C_4 synthase gene promoter polymorphism and coronary artery calcium in young women: the Muscatine Study. Arterioscler Thromb Vasc Biol 27(2):394–399
7. Freiberg JJ, Tybjaerg-Hansen A, Sillesen H, Jensen GB, Nordestgaard BG (2008) Promotor polymorphisms in leukotriene C_4 synthase and risk of ischemic cerebrovascular disease. Arterioscler Thromb Vasc Biol 28(5):990–996
8. Freiberg JJ, Tybjaerg-Hansen A, Nordestgaard BG (2010) Novel mutations in leukotriene C_4 synthase and risk of cardiovascular disease based on genotypes from 50,000 individuals. J Thromb Haemost 8(8):1694–1701
9. Maznyczka A, Braund P, Mangino M, Samani NJ (2008) Arachidonate 5-lipoxygenase (5-LO) promoter genotype and risk of myocardial infarction: a case-control study. Atherosclerosis 199(2):328–332
10. Ingelsson E, Yin L, Back M (2012) Nationwide cohort study of the leukotriene receptor antagonist montelukast and incident or recurrent cardiovascular disease. J Allergy Clin Immunol 129(3):702–707
11. Bisgaard H (2003) A randomized trial of montelukast in respiratory syncytial virus postbronchiolitis. Am J Respir Crit Care Med 167(3):379–383
12. Goldbart AD, Goldman JL, Veling MC, Gozal D (2005) Leukotriene modifier therapy for mild sleep-disordered breathing in children. Am J Respir Crit Care Med 172(3):364–370
13. Goldbart AD, Krishna J, Li RC, Serpero LD, Gozal D (2006) Inflammatory mediators in exhaled breath condensate of children with obstructive sleep apnea syndrome. Chest 130(1):143–148
14. Stanke-Labesque F, Back M, Lefebvre B, Tamisier R, Baguet JP, Arnol N, Levy P, Pepin JL (2009) Increased urinary leukotriene E_4 excretion in obstructive sleep apnea: effects of obesity and hypoxia. J Allergy Clin Immunol 124(2):364–370
15. Shen Y, Xu Z, Shen K (2011) Urinary leukotriene E_4, obesity, and adenotonsillar hypertrophy in Chinese children with sleep disordered breathing. Sleep 34(8):1135–041

16. Erbagci Z (2002) The leukotriene receptor antagonist montelukast in the treatment of chronic idiopathic urticaria: a single-blind, placebo-controlled, crossover clinical study. J Allergy Clin Immunol 110(3):484–488
17. Khan S, Lynch N (2012) Efficacy of montelukast as added therapy in patients with chronic idiopathic urticaria. Inflamm Allergy Drug Targets 11(3):235–243
18. Wu SH, Liao PY, Chen XQ, Yin PL, Dong L (2014) Add-on therapy with montelukast in treatment of Henoch-Schonlein purpura. Pediatr Int 56(3):315–322
19. Clark JD, Milona N, Knopf JL (1990) Purification of a 110-kilodalton cytosolic phospholipase A_2 from the human monocytic cell line U937. Proc Natl Acad Sci U S A 87(19):7708–7712
20. Malaviya R, Malaviya R, Jakschik BA (1993) Reversible translocation of 5-lipoxygenase in mast cells upon IgE/antigen stimulation. J Biol Chem 268(7):4939–4944
21. Reid GK, Kargman S, Vickers PJ, Mancini JA, Leveille C, Ethier D, Miller DK, Gillard JW, Dixon RA, Evans JF (1990) Correlation between expression of 5-lipoxygenase-activating protein, 5-lipoxygenase, and cellular leukotriene synthesis. J Biol Chem 265(32):19818–19823
22. Lam BK, Penrose JF, Freeman GJ, Austen KF (1994) Expression cloning of a cDNA for human leukotriene C_4 synthase, an integral membrane protein conjugating reduced glutathione to leukotriene A_4. Proc Natl Acad Sci U S A 91(16):7663–7667
23. Leier I, Jedlitschky G, Buchholz U, Cole SP, Deeley RG, Keppler D (1994) The MRP gene encodes an ATP-dependent export pump for leukotriene C_4 and structurally related conjugates. J Biol Chem 269(45):27807–27810
24. Shi ZZ, Han B, Habib GM, Matzuk MM, Lieberman MW (2001) Disruption of gamma-glutamyl leukotrienase results in disruption of leukotriene D_4 synthesis in vivo and attenuation of the acute inflammatory response. Mol Cell Biol 21(16):5389–5395
25. Lee CW, Lewis RA, Corey EJ, Austen KF (1983) Conversion of leukotriene D_4 to leukotriene E_4 by a dipeptidase released from the specific granule of human polymorphonuclear leucocytes. Immunology 48(1):27–35
26. Asano K, Lilly CM, O'Donnell WJ, Israel E, Fischer A, Ransil BJ, Drazen JM (1995) Diurnal variation of urinary leukotriene E_4 and histamine excretion rates in normal subjects and patients with mild-to-moderate asthma. J Allergy Clin Immunol 96(5 pt 1):643–651
27. Fanning LB, Boyce JA (2013) Lipid mediators and allergic diseases. Ann Allergy Asthma Immunol 111(3):155–162
28. Murphy RC, Hammarstrom S, Samuelsson B (1979) Leukotriene C: a slow-reacting substance from murine mastocytoma cells. Proc Natl Acad Sci U S A 76(9):4275–4279
29. Warner JA, Peters SP, Lichtenstein LM, Hubbard W, Yancey KB, Stevenson HC, Miller PJ, MacGlashan DW Jr (1989) Differential release of mediators from human basophils: differences in arachidonic acid metabolism following activation by unrelated stimuli. J Leukoc Biol 45(6):558–571
30. Weller PF, Lee CW, Foster DW, Corey EJ, Austen KF, Lewis RA (1983) Generation and metabolism of 5-lipoxygenase pathway leukotrienes by human eosinophils: predominant production of leukotriene C_4. Proc Natl Acad Sci U S A 80(24):7626–7630
31. Abe M, Hugli TE (1988) Characterization of leukotriene C_4 synthetase in mouse peritoneal exudate cells. Biochim Biophys Acta 959(3):386–398
32. Barrett NA, Maekawa A, Rahman OM, Austen KF, Kanaoka Y (2009) Dectin-2 recognition of house dust mite triggers cysteinyl leukotriene generation by dendritic cells. J Immunol 182(2):1119–1128
33. Seymour ML, Rak S, Aberg D, Riise GC, Penrose JF, Kanaoka Y, Austen KF, Holgate ST, Sampson AP (2001) Leukotriene and prostanoid pathway enzymes in bronchial biopsies of seasonal allergic asthmatics. Am J Respir Crit Care Med 164(11):2051–2056

34. Hsieh FH, Lam BK, Penrose JF, Austen KF, Boyce JA (2001) T helper cell type 2 cytokines coordinately regulate immunoglobulin E-dependent cysteinyl leukotriene production by human cord blood-derived mast cells: profound induction of leukotriene C_4 synthase expression by interleukin 4. J Exp Med 193(1):123–133
35. Wenzel SE, Westcott JY, Smith HR, Larsen GL (1989) Spectrum of prostanoid release after bronchoalveolar allergen challenge in atopic asthmatics and in control groups. An alteration in the ratio of bronchoconstrictive to bronchoprotective mediators. Am Rev Respir Dis 139(2):450–457
36. Wenzel SE, Westcott JY, Larsen GL (1991) Bronchoalveolar lavage fluid mediator levels 5 minutes after allergen challenge in atopic subjects with asthma: relationship to the development of late asthmatic responses. J Allergy Clin Immunol 87(2):540–548
37. Wenzel SE, Fowler AA III, Schwartz LB (1988) Activation of pulmonary mast cells by bronchoalveolar allergen challenge. In vivo release of histamine and tryptase in atopic subjects with and without asthma. Am Rev Respir Dis 137(5):1002–1008
38. Gauvreau GM, Watson RM, O'Byrne PM (1999) Protective effects of inhaled PGE_2 on allergen-induced airway responses and airway inflammation. Am J Respir Crit Care Med 159(1):31–36
39. Taniguchi M, Higashi N, Ono E, Mita H, Akiyama K (2008) Hyperleukotrieneuria in patients with allergic and inflammatory disease. Allergol Int 57(4):313–320
40. Ono E, Taniguchi M, Mita H, Higashi N, Fukutomi Y, Tanimoto H, Sekiya K, Oshikata C, Tsuburai T, Tsurikisawa N et al (2008) Increased urinary leukotriene E_4 concentration in patients with eosinophilic pneumonia. Eur Respir J 32(2):437–442
41. Laidlaw TM, Kidder MS, Bhattacharyya N, Xing W, Shen S, Milne GL, Castells MC, Chhay H, Boyce JA (2012) Cysteinyl leukotriene overproduction in aspirin-exacerbated respiratory disease is driven by platelet-adherent leukocytes. Blood 119(16):3790–3798
42. Maclouf JA, Murphy RC (1988) Transcellular metabolism of neutrophil-derived leukotriene A_4 by human platelets. A potential cellular source of leukotriene C_4. J Biol Chem 263(1):174–181
43. Maugeri N, Evangelista V, Celardo A, Dell'Elba G, Martelli N, Piccardoni P, de Gaetano G, Cerletti C (1994) Polymorphonuclear leukocyte–platelet interaction: role of P-selectin in thromboxane B_2 and leukotriene C_4 cooperative synthesis. Thromb Haemost 72(3):450–456
44. Laidlaw TM, Cutler AJ, Kidder MS, Liu T, Cardet JC, Chhay H, Feng C, Boyce JA (2014) Prostaglandin E resistance in granulocytes from patients with aspirin-exacerbated respiratory disease. J Allergy Clin Immunol 133:1692–1701
45. Lee TH, Austen KF, Corey EJ, Drazen JM (1984) Leukotriene E_4-induced airway hyperresponsiveness of guinea pig tracheal smooth muscle to histamine and evidence for three separate sulfidopeptide leukotriene receptors. Proc Natl Acad Sci U S A 81(15):4922–4925
46. Lynch KR, O'Neill GP, Liu Q, Im DS, Sawyer N, Metters KM, Coulombe N, Abramovitz M, Figueroa DJ, Zeng Z et al (1999) Characterization of the human cysteinyl leukotriene CysLT1 receptor. Nature 399(6738):789–793
47. Heise CE, O'Dowd BF, Figueroa DJ, Sawyer N, Nguyen T, Im DS, Stocco R, Bellefeuille JN, Abramovitz M, Cheng R et al (2000) Characterization of the human cysteinyl leukotriene 2 receptor. J Biol Chem 275(39):30531–30536
48. Kanaoka Y, Maekawa A, Austen KF (2013) Identification of GPR99 as a potential third cysteinyl leukotriene receptor with a preference for leukotriene E_4. J Biol Chem 288:10967–10972
49. Weiss JW, Drazen JM, Coles N, McFadden ER Jr, Weller PF, Corey EJ, Lewis RA, Austen KF (1982) Bronchoconstrictor effects of leukotriene C in humans. Science 216(4542):196–198
50. Weiss JW, Drazen JM, McFadden ER Jr, Weller P, Corey EJ, Lewis RA, Austen KF (1983) Airway constriction in normal humans produced by inhalation of leukotriene D. Potency, time course, and effect of aspirin therapy. JAMA 249(20):2814–2817

51. Griffin M, Weiss JW, Leitch AG, McFadden ER Jr, Corey EJ, Austen KF, Drazen JM (1983) Effects of leukotriene D on the airways in asthma. N Engl J Med 308(8):436–439
52. Davidson AB, Lee TH, Scanlon PD, Solway J, McFadden ER Jr, Ingram RH Jr, Corey EJ, Austen KF, Drazen JM (1987) Bronchoconstrictor effects of leukotriene E_4 in normal and asthmatic subjects. Am Rev Respir Dis 135(2):333–337
53. Arm JP, O'Hickey SP, Hawksworth RJ, Fong CY, Crea AE, Spur BW, Lee TH (1990) Asthmatic airways have a disproportionate hyperresponsiveness to LTE_4, as compared with normal airways, but not to LTC_4, LTD_4, methacholine, and histamine. Am Rev Respir Dis 142(5):1112–1118
54. Christie PE, Schmitz-Schumann M, Spur BW, Lee TH (1993) Airway responsiveness to leukotriene C_4 (LTC_4), leukotriene E_4 (LTE_4) and histamine in aspirin-sensitive asthmatic subjects. Eur Respir J 6(10):1468–1473
55. Jacques CA, Spur BW, Johnson M, Lee TH (1991) The mechanism of LTE_4-induced histamine hyperresponsiveness in guinea-pig tracheal and human bronchial smooth muscle, in vitro. Br J Pharmacol 104(4):859–866. PMCID:PMC1908836
56. Christie PE, Hawksworth R, Spur BW, Lee TH (1992) Effect of indomethacin on leukotriene-induced histamine hyperresponsiveness in asthmatic subjects. Am Rev Respir Dis 146(6):1506–1510
57. Beasley RC, Featherstone RL, Church MK, Rafferty P, Varley JG, Harris A, Robinson C, Holgate ST (1989) Effect of a thromboxane receptor antagonist on PGD_2- and allergen-induced bronchoconstriction. J Appl Physiol (1985) 66(4):1685–1693
58. Jones GL, Saroea HG, Watson RM, O'Byrne PM (1992) Effect of an inhaled thromboxane mimetic (U46619) on airway function in human subjects. Am Rev Respir Dis 145(6):1270–1274
59. Laitinen LA, Laitinen A, Haahtela T, Vilkka V, Spur BW, Lee TH (1993) Leukotriene E_4 and granulocytic infiltration into asthmatic airways. Lancet 341(8851):989–990
60. Gauvreau GM, Parameswaran KN, Watson RM, O'Byrne PM (2001) Inhaled leukotriene E_4, but not leukotriene D_4, increased airway inflammatory cells in subjects with atopic asthma. Am J Respir Crit Care Med 164(8 pt 1):1495–1500
61. Zhu J, Qiu YS, Figueroa DJ, Bandi V, Galczenski H, Hamada K, Guntupalli KK, Evans JF, Jeffery PK (2005) Localization and upregulation of cysteinyl leukotriene-1 receptor in asthmatic bronchial mucosa. Am J Respir Cell Mol Biol 33(6):531–540
62. Paruchuri S, Jiang Y, Feng C, Francis SA, Plutzky J, Boyce JA (2008) Leukotriene E_4 activates peroxisome proliferator-activated receptor gamma and induces prostaglandin D_2 generation by human mast cells. J Biol Chem 283(24):16477–16487
63. Hirai H, Tanaka K, Yoshie O, Ogawa K, Kenmotsu K, Takamori Y, Ichimasa M, Sugamura K, Nakamura M, Takano S et al (2001) Prostaglandin D_2 selectively induces chemotaxis in T helper type 2 cells, eosinophils, and basophils via seven-transmembrane receptor CRTH2. J Exp Med 193(2):255–261
64. Paruchuri S, Tashimo H, Feng C, Maekawa A, Xing W, Jiang Y, Kanaoka Y, Conley P, Boyce JA (2009) Leukotriene E_4-induced pulmonary inflammation is mediated by the P2Y12 receptor. J Exp Med 206(11):2543–2555
65. Nonaka Y, Hiramoto T, Fujita N (2005) Identification of endogenous surrogate ligands for human P2Y12 receptors by in silico and in vitro methods. Biochem Biophys Res Commun 337(1):281–288
66. Hollopeter G, Jantzen HM, Vincent D, Li G, England L, Ramakrishnan V, Yang RB, Nurden P, Nurden A, Julius D et al (2001) Identification of the platelet ADP receptor targeted by antithrombotic drugs. Nature 409(6817):202–207
67. Bautz F, Denzlinger C, Kanz L, Mohle R (2001) Chemotaxis and transendothelial migration of $CD34^+$ hematopoietic progenitor cells induced by the inflammatory mediator leukotriene D_4 are mediated by the 7-transmembrane receptor CysLT1. Blood 97(11):3433–3440

68. Mellor EA, Maekawa A, Austen KF, Boyce JA (2001) Cysteinyl leukotriene receptor 1 is also a pyrimidinergic receptor and is expressed by human mast cells. Proc Natl Acad Sci U S A 98(14):7964–7969
69. Mellor EA, Frank N, Soler D, Hodge MR, Lora JM, Austen KF, Boyce JA (2003) Expression of the type 2 receptor for cysteinyl leukotrienes (CysLT2R) by human mast cells: functional distinction from CysLT1R. Proc Natl Acad Sci U S A 100(20):11589–11593
70. Bandeira-Melo C, Woods LJ, Phoofolo M, Weller PF (2002) Intracrine cysteinyl leukotriene receptor-mediated signaling of eosinophil vesicular transport-mediated interleukin-4 secretion. J Exp Med 196(6):841–850
71. Jiang Y, Borrelli LA, Kanaoka Y, Bacskai BJ, Boyce JA (2007) CysLT2 receptors interact with CysLT1 receptors and down-modulate cysteinyl leukotriene dependent mitogenic responses of mast cells. Blood 110(9):3263–3270
72. Jiang Y, Kanaoka Y, Feng C, Nocka K, Rao S, Boyce JA (2006) Interleukin 4-dependent mast cell proliferation requires autocrine/intracrine cysteinyl leukotriene-induced signaling. J Immunol 177(5):2755–2759
73. Mellor EA, Austen KF, Boyce JA (2002) Cysteinyl leukotrienes and uridine diphosphate induce cytokine generation by human mast cells through an interleukin 4-regulated pathway that is inhibited by leukotriene receptor antagonists. J Exp Med 195(5):583–592
74. Lin DA, Boyce JA (2005) IL-4 regulates MEK expression required for lysophosphatidic acid-mediated chemokine generation by human mast cells. J Immunol 175(8):5430–5438
75. Thivierge M, Stankova J, Rola-Pleszczynski M (2001) IL-13 and IL-4 up-regulate cysteinyl leukotriene 1 receptor expression in human monocytes and macrophages. J Immunol 167(5):2855–2860
76. Early SB, Barekzi E, Negri J, Hise K, Borish L, Steinke JW (2007) Concordant modulation of cysteinyl leukotriene receptor expression by IL-4 and IFN-gamma on peripheral immune cells. Am J Respir Cell Mol Biol 36(6):715–720
77. Jiang Y, Borrelli L, Bacskai BJ, Kanaoka Y, Boyce JA (2009) P2Y6 receptors require an intact cysteinyl leukotriene synthetic and signaling system to induce survival and activation of mast cells. J Immunol 182(2):1129–1137
78. Mita H, Hasegawa M, Saito H, Akiyama K (2001) Levels of cysteinyl leukotriene receptor mRNA in human peripheral leucocytes: significantly higher expression of cysteinyl leukotriene receptor 2 mRNA in eosinophils. Clin Exp Allergy 31(11):1714–1723
79. Xue L, Gyles SL, Wettey FR, Gazi L, Townsend E, Hunter MG, Pettipher R (2005) Prostaglandin D_2 causes preferential induction of proinflammatory Th2 cytokine production through an action on chemoattractant receptor-like molecule expressed on Th2 cells. J Immunol 175(10):6531–6536
80. Xue L, Barrow A, Fleming VM, Hunter MG, Ogg G, Klenerman P, Pettipher R (2012) Leukotriene E_4 activates human Th2 cells for exaggerated proinflammatory cytokine production in response to prostaglandin D_2. J Immunol 188(2):694–702
81. Mjosberg JM, Trifari S, Crellin NK, Peters CP, van Drunen CM, Piet B, Fokkens WJ, Cupedo T, Spits H (2011) Human IL-25- and IL-33-responsive type 2 innate lymphoid cells are defined by expression of CRTH2 and CD161. Nat Immunol 12(11):1055–1062
82. Xue L, Salimi M, Panse I, Mjosberg JM, McKenzie AN, Spits H, Klenerman P, Ogg G (2013) Prostaglandin D activates group 2 innate lymphoid cells through chemoattractant receptor-homologous molecule expressed on T2 cells. J Allergy Clin Immunol 133:1184–1194
83. Doherty TA, Khorram N, Lund S, Mehta AK, Croft M, Broide DH (2013) Lung type 2 innate lymphoid cells express cysteinyl leukotriene receptor 1, which regulates TH2 cytokine production. J Allergy Clin Immunol 132(1):205–213
84. Hasegawa S, Ichiyama T, Hashimoto K, Suzuki Y, Hirano R, Fukano R, Furukawa S (2010) Functional expression of cysteinyl leukotriene receptors on human platelets. Platelets 21(4):253–259

85. Cummings HE, Liu T, Feng C, Laidlaw TM, Conley PB, Kanaoka Y, Boyce JA (2013) Cutting edge: leukotriene C_4 activates mouse platelets in plasma exclusively through the type 2 cysteinyl leukotriene receptor. J Immunol 191:5807–5810
86. Parameswaran K, Liang H, Fanat A, Watson R, Snider DP, O'Byrne PM (2004) Role for cysteinyl leukotrienes in allergen-induced change in circulating dendritic cell number in asthma. J Allergy Clin Immunol 114(1):73–79
87. Thivierge M, Stankova J, Rola-Pleszczynski M (2006) Toll-like receptor agonists differentially regulate cysteinyl-leukotriene receptor 1 expression and function in human dendritic cells. J Allergy Clin Immunol 117(5):1155–1162
88. Ilarraza R, Wu Y, Adamko DJ (2012) Montelukast inhibits leukotriene stimulation of human dendritic cells in vitro. Int Arch Allergy Immunol 159(4):422–427
89. Barrett NA, Rahman OM, Fernandez JM, Parsons MW, Xing W, Austen KF, Kanaoka Y (2011) Dectin-2 mediates Th2 immunity through the generation of cysteinyl leukotrienes. J Exp Med 208:593–604
90. Liu MC, Dube LM, Lancaster J (1996) Acute and chronic effects of a 5-lipoxygenase inhibitor in asthma: a 6-month randomized multicenter trial. Zileuton Study Group. J Allergy Clin Immunol 98(5 pt 1):859–871
91. Israel E, Rubin P, Kemp JP, Grossman J, Pierson W, Siegel SC, Tinkelman D, Murray JJ, Busse W, Segal AT et al (1993) The effect of inhibition of 5-lipoxygenase by zileuton in mild-to-moderate asthma. Ann Intern Med 119(11):1059–1066
92. Israel E, Chervinsky PS, Friedman B, Van BJ, Skalky CS, Ghannam AF, Bird SR, Edelman JM (2002) Effects of montelukast and beclomethasone on airway function and asthma control. J Allergy Clin Immunol 110(6):847–854
93. Fitzgerald DA, Mellis CM (2006) Leukotriene receptor antagonists in virus-induced wheezing: evidence to date. Treat Respir Med 5(6):407–417
94. Camargo CA Jr, Smithline HA, Malice MP, Green SA, Reiss TF (2003) A randomized controlled trial of intravenous montelukast in acute asthma. Am J Respir Crit Care Med 167(4):528–533
95. Adachi M, Taniguchi H, Tohda Y, Sano Y, Ishine T, Smugar SS, Hisada S (2012) The efficacy and tolerability of intravenous montelukast in acute asthma exacerbations in Japanese patients. J Asthma 49(6):649–656
96. Camargo CA Jr, Gurner DM, Smithline HA, Chapela R, Fabbri LM, Green SA, Malice MP, Legrand C, Dass SB, Knorr BA et al (2010) A randomized placebo-controlled study of intravenous montelukast for the treatment of acute asthma. J Allergy Clin Immunol 125(2):374–380
97. Morris CR, Becker AB, Pinieiro A, Massaad R, Green SA, Smugar SS, Gurner DM (2010) A randomized, placebo-controlled study of intravenous montelukast in children with acute asthma. Ann Allergy Asthma Immunol 104(2):161–171
98. Kikawa Y, Miyanomae T, Inoue Y, Saito M, Nakai A, Shigematsu Y, Hosoi S, Sudo M (1992) Urinary leukotriene E_4 after exercise challenge in children with asthma. J Allergy Clin Immunol 89(6):1111–1119
99. Reiss TF, Hill JB, Harman E, Zhang J, Tanaka WK, Bronsky E, Guerreiro D, Hendeles L (1997) Increased urinary excretion of LTE_4 after exercise and attenuation of exercise-induced bronchospasm by montelukast, a cysteinyl leukotriene receptor antagonist. Thorax 52(12):1030–1035
100. Bancalari L, Conti I, Giannessi D, Lazzerini G, Dente FL, De CR, Paggiaro PL (1999) Early increase in urinary leukotriene E_4 (LTE_4) is dependent on allergen dose inhaled during bronchial challenge in asthmatic subjects. Allergy 54(12):1278–1285
101. Rorke S, Jennison S, Jeffs JA, Sampson AP, Arshad H, Holgate ST (2002) Role of cysteinyl leukotrienes in adenosine 5′-monophosphate induced bronchoconstriction in asthma. Thorax 57(4):323–327
102. Brannan JD, Gulliksson M, Anderson SD, Chew N, Kumlin M (2003) Evidence of mast cell activation and leukotriene release after mannitol inhalation. Eur Respir J 22(3):491–496

103. Rabinovitch N, Strand M, Stuhlman K, Gelfand EW (2008) Exposure to tobacco smoke increases leukotriene E4-related albuterol usage and response to montelukast. J Allergy Clin Immunol 121(6):1365–1371
104. Rabinovitch N, Graber NJ, Chinchilli VM, Sorkness CA, Zeiger RS, Strunk RC, Bacharier LB, Martinez FD, Szefler SJ (2010) Urinary leukotriene E_4/exhaled nitric oxide ratio and montelukast response in childhood asthma. J Allergy Clin Immunol 126(3):545–551
105. In KH, Asano K, Beier D, Grobholz J, Finn PW, Silverman EK, Silverman ES, Collins T, Fischer AR, Keith TP et al (1997) Naturally occurring mutations in the human 5-lipoxygenase gene promoter that modify transcription factor binding and reporter gene transcription. J Clin Invest 99(5):1130–1137
106. Silverman ES, Drazen JM (1999) The biology of 5-lipoxygenase: function, structure, and regulatory mechanisms. Proc Assoc Am Physicians 111(6):525–536
107. Silverman E, In KH, Yandava C, Drazen JM (1998) Pharmacogenetics of the 5-lipoxygenase pathway in asthma. Clin Exp Allergy 28(suppl 5):164–170
108. Whelan GJ, Blake K, Kissoon N, Duckworth LJ, Wang J, Sylvester JE, Lima JJ (2003) Effect of montelukast on time-course of exhaled nitric oxide in asthma: influence of LTC_4 synthase A(-444)C polymorphism. Pediatr Pulmonol 36(5):413–420
109. Lima JJ, Zhang S, Grant A, Shao L, Tantisira KG, Allayee H, Wang J, Sylvester J, Holbrook J, Wise R et al (2006) Influence of leukotriene pathway polymorphisms on response to montelukast in asthma. Am J Respir Crit Care Med 173(4):379–385
110. Laidlaw TM, Boyce JA (2013) Pathogenesis of aspirin-exacerbated respiratory disease and reactions. Immunol Allergy Clin N Am 33(2):195–210
111. Christie PE, Tagari P, Ford-Hutchinson AW, Charlesson S, Chee P, Arm JP, Lee TH (1991) Urinary leukotriene E_4 concentrations increase after aspirin challenge in aspirin-sensitive asthmatic subjects. Am Rev Respir Dis 143(5 pt 1):1025–1029
112. White A, Ludington E, Mehra P, Stevenson DD, Simon RA (2006) Effect of leukotriene modifier drugs on the safety of oral aspirin challenges. Ann Allergy Asthma Immunol 97(5):688–693
113. Dahlen B, Nizankowska E, Szczeklik A, Zetterstrom O, Bochenek G, Kumlin M, Mastalerz L, Pinis G, Swanson LJ, Boodhoo TI et al (1998) Benefits from adding the 5-lipoxygenase inhibitor zileuton to conventional therapy in aspirin-intolerant asthmatics. Am J Respir Crit Care Med 157(4 pt 1):1187–1194
114. Dahlen SE, Malmstrom K, Nizankowska E, Dahlen B, Kuna P, Kowalski M, Lumry WR, Picado C, Stevenson DD, Bousquet J et al (2002) Improvement of aspirin-intolerant asthma by montelukast, a leukotriene antagonist: a randomized, double-blind, placebo-controlled trial. Am J Respir Crit Care Med 165(1):9–14
115. Cowburn AS, Sladek K, Soja J, Adamek L, Nizankowska E, Szczeklik A, Lam BK, Penrose JF, Austen FK, Holgate ST et al (1998) Overexpression of leukotriene C_4 synthase in bronchial biopsies from patients with aspirin-intolerant asthma. J Clin Invest 101(4):834–846
116. Adamjee J, Suh YJ, Park HS, Choi JH, Penrose JF, Lam BK, Austen KF, Cazaly AM, Wilson SJ, Sampson AP (2006) Expression of 5-lipoxygenase and cyclooxygenase pathway enzymes in nasal polyps of patients with aspirin-intolerant asthma. J Pathol 209(3):392–399
117. Sampson AP, Siddiqui S, Buchanan D, Howarth PH, Holgate ST, Holloway JW, Sayers I (2000) Variant LTC_4 synthase allele modifies cysteinyl leukotriene synthesis in eosinophils and predicts clinical response to zafirlukast. Thorax 55(suppl 2):S28–S31
118. Sanak M, Pierzchalska M, Bazan-Socha S, Szczeklik A (2000) Enhanced expression of the leukotriene C_4 synthase due to overactive transcription of an allelic variant associated with aspirin-intolerant asthma. Am J Respir Cell Mol Biol 23(3):290–296
119. Van SR, Stevenson DD, Baldasaro M, Lam BK, Zhao J, Yoshida S, Yandora C, Drazen JM, Penrose JF (2000) 5′-Flanking region polymorphism of the gene encoding leukotriene C_4 synthase does not correlate with the aspirin-intolerant asthma phenotype in the United States. J Allergy Clin Immunol 106(1 pt 1):72–76

120. Sousa A, Pfister R, Christie PE, Lane SJ, Nasser SM, Schmitz-Schumann M, Lee TH (1997) Enhanced expression of cyclo-oxygenase isoenzyme 2 (COX-2) in asthmatic airways and its cellular distribution in aspirin-sensitive asthma. Thorax 52(11):940–945
121. Fischer AR, Rosenberg MA, Lilly CM, Callery JC, Rubin P, Cohn J, White MV, Igarashi Y, Kaliner MA, Drazen JM et al (1994) Direct evidence for a role of the mast cell in the nasal response to aspirin in aspirin-sensitive asthma. J Allergy Clin Immunol 94(6 pt 1):1046–1056
122. Yoshida S, Amayasu H, Sakamoto H, Onuma K, Shoji T, Nakagawa H, Tajima T (1998) Cromolyn sodium prevents bronchoconstriction and urinary LTE_4 excretion in aspirin-induced asthma. Ann Allergy Asthma Immunol 80(2):171–176
123. Liu T, Laidlaw TM, Katz HR, Boyce JA (2013) Prostaglandin E_2 deficiency causes a phenotype of aspirin sensitivity that depends on platelets and cysteinyl leukotrienes. Proc Natl Acad Sci U S A 110:16987–16992
124. Corrigan CJ, Napoli RL, Meng Q, Fang C, Wu H, Tochiki K, Reay V, Lee TH, Ying S (2012) Reduced expression of the prostaglandin E_2 receptor E-prostanoid 2 on bronchial mucosal leukocytes in patients with aspirin-sensitive asthma. J Allergy Clin Immunol 129(6):1636–1646
125. Sousa AR, Parikh A, Scadding G, Corrigan CJ, Lee TH (2002) Leukotriene-receptor expression on nasal mucosal inflammatory cells in aspirin-sensitive rhinosinusitis. N Engl J Med 347(19):1493–1499
126. Corrigan C, Mallett K, Ying S, Roberts D, Parikh A, Scadding G, Lee T (2005) Expression of the cysteinyl leukotriene receptors $cysLT_1$ and $cysLT_2$ in aspirin-sensitive and aspirin-tolerant chronic rhinosinusitis. J Allergy Clin Immunol 115(2):316–322
127. Arm JP, O'Hickey SP, Spur BW, Lee TH (1989) Airway responsiveness to histamine and leukotriene E_4 in subjects with aspirin-induced asthma. Am Rev Respir Dis 140(1):148–153
128. Balzary RW, Cocks TM (2006) Lipopolysaccharide induces epithelium- and prostaglandin E_2-dependent relaxation of mouse isolated trachea through activation of cyclooxygenase (COX)-1 and COX-2. J Pharmacol Exp Ther 317(2):806–812
129. Uematsu S, Matsumoto M, Takeda K, Akira S (2002) Lipopolysaccharide-dependent prostaglandin E_2 production is regulated by the glutathione-dependent prostaglandin E_2 synthase gene induced by the Toll-like receptor 4/MyD88/NF-IL6 pathway. J Immunol 168(11):5811–5816
130. Sugimoto Y, Narumiya S (2007) Prostaglandin E receptors. J Biol Chem 282(16):11613–11617
131. Luo M, Jones SM, Phare SM, Coffey MJ, Peters-Golden M, Brock TG (2004) Protein kinase A inhibits leukotriene synthesis by phosphorylation of 5-lipoxygenase on serine 523. J Biol Chem 279(40):41512–41520
132. Feng C, Beller EM, Bagga S, Boyce JA (2006) Human mast cells express multiple EP receptors for prostaglandin E_2 that differentially modulate activation responses. Blood 107(8):3243–3250
133. Sestini P, Armetti L, Gambaro G, Pieroni MG, Refini RM, Sala A, Vaghi A, Folco GC, Bianco S, Robuschi M (1996) Inhaled PGE_2 prevents aspirin-induced bronchoconstriction and urinary LTE_4 excretion in aspirin-sensitive asthma. Am J Respir Crit Care Med 153(2):572–575
134. Yoshimura T, Yoshikawa M, Otori N, Haruna S, Moriyama H (2008) Correlation between the prostaglandin D_2/E_2 ratio in nasal polyps and the recalcitrant pathophysiology of chronic rhinosinusitis associated with bronchial asthma. Allergol Int 57(4):429–436
135. Picado C, Fernandez-Morata JC, Juan M, Roca-Ferrer J, Fuentes M, Xaubet A, Mullol J (1999) Cyclooxygenase-2 mRNA is downexpressed in nasal polyps from aspirin-sensitive asthmatics. Am J Respir Crit Care Med 160(1):291–296

136. Roca-Ferrer J, Garcia-Garcia FJ, Pereda J, Perez-Gonzalez M, Pujols L, Alobid I, Mullol J, Picado C (2011) Reduced expression of COXs and production of prostaglandin E_2 in patients with nasal polyps with or without aspirin-intolerant asthma. J Allergy Clin Immunol 128(1):66–72
137. Ying S, Meng Q, Scadding G, Parikh A, Corrigan CJ, Lee TH (2006) Aspirin-sensitive rhinosinusitis is associated with reduced E-prostanoid 2 receptor expression on nasal mucosal inflammatory cells. J Allergy Clin Immunol 117(2):312–318

Chapter 21
Lipid Mediators and Skin Diseases

Tetsuya Honda and Kenji Kabashima

Abstract Lipid mediators, such as prostanoids and leukotrienes, are metabolites of arachidonic acid released in various pathophysiological conditions that exert a range of actions mediated through their respective receptors expressed on target cells. Although it has been difficult to analyze the physiological role of prostanoids or leukotrienes, recent developments in the disruption of both the respective genes and receptor-selective compounds have enabled us to investigate the physiological roles for lipid mediators. It has been demonstrated that each receptor of lipid mediators has multiple functions in a context-dependent manner, which sometimes results in opposite—excitatory and inhibitory—outcomes. The balance of the production and the receptor expression of each lipid mediator has been proved to be important for maintaining the homeostasis of our body. Here, we review new findings on the functions of prostanoids and leukotrienes in skin inflammatory diseases, focusing on contact dermatitis, atopic dermatitis, and psoriasis.

Keywords Prostanoid • Leukotriene • Atopic dermatitis • Contact dermatitis • Psoriasis

21.1 Introduction

The skin is an organ that serves as an interface between the host and the environment. The skin provides not only mechanical barrier functions, to restrict water loss and prevent the entry of harmful environmental substances and microorganisms, but

T. Honda
Department of Dermatology, Kyoto University Graduate School of Medicine, 54 Shogoin-Kawara, Sakyo, Kyoto 606-8507, Japan

Center for Innovation in Immunoregulative Technology and Therapeutics, Kyoto University Graduate School of Medicine, Kyoto 606-8501, Japan

K. Kabashima (✉)
Department of Dermatology, Kyoto University Graduate School of Medicine, 54 Shogoin-Kawara, Sakyo, Kyoto 606-8507, Japan
e-mail: kaba@kuhp.kyoto-u.ac.jp

T. Yokomizo, M. Murakami (eds.), *Bioactive Lipid Mediators*,
DOI 10.1007/978-4-431-55669-5_21

also an active barrier that provides the first line of immunological defense against infections [1, 2]. The skin is composed of the epidermis and dermis, and each layer is composed of several cell types, such as keratinocytes and dendritic cells (DCs), which are important for maintenance of skin homeostasis and for induction of skin diseases, such as atopic dermatitis (AD) and psoriasis. Lipid mediators, such as prostanoids and leukotrienes (LTs), are the candidates for the regulation of its balance [3–5].

When tissues are exposed to diverse pathophysiological stimuli, arachidonic acid (AA) is released from membrane phospholipids and converted to lipid mediators, such as prostanoids, LTs, and hydroxy-eicosatetraenoic acids (HETEs). Prostanoids are formed by the cyclooxygenase (COX) pathway, whereas LTs and HETEs are formed by the 5-, 12-, and 15-lipoxygenase (LO) pathways. COX has two isoforms, COX-1 and COX-2: COX-1 is constitutively expressed in cells, while COX-2 requires specific stimulation by substances such as acetone and phorbol ester [6]. The COX reaction results in the formation of an unstable endoperoxide intermediate, prostaglandin (PG) H_2, which, in turn, is metabolized to PGD_2, PGE_2, $PGF_{2\alpha}$, PGI_2, and thromboxane (TX) A_2 by their specific synthases. LTs include LTB_4 and cystenyl (Cys) LTs: CysLTs further include LTC_4, LTD_4, and LTE_4.

Prostanoids are released from cells immediately after their formation. Because they are chemically and metabolically unstable, they usually function only locally through membrane receptors on target cells [6]. Nine types and subtypes of membrane prostanoid receptors are conserved in mammals from mouse to human: two subtypes of the PGD receptor (DP and chemoattractant receptor homologous molecule expressed on Th2 cells, CRTH2), four subtypes of the PGE receptor (EP1, EP2, EP3, and EP4), the PGF receptor (FP), the PGI receptor (IP), and the TXA receptor (TP). All are G protein-coupled rhodopsin-type receptors with seven transmembrane domains. LTs also exert their functions through their specific G protein-coupled receptors on the cell surface. LTB_4 binds to two kinds of receptors, BLT1 and BLT2, and LTC_4 binds to $CysLT_1$ and $CysLT_2$.

Recently, individual prostanoids and LT receptor gene-deficient mice have been used as models to dissect the respective roles of each receptor in combination with the use of compounds that selectively bind to the receptors as agonists or antagonists [3–5]. These genetic and pharmacological approaches have revealed new roles for lipid mediators and their receptors in inflammatory skin diseases. In this review, we describe the current investigative status of prostanoids and LTs in skin inflammatory diseases, focusing on contact dermatitis, atopic dermatitis, and psoriasis, and discuss the clinical potentials of receptor-selective drugs.

21.2 Prostanoids and LTs in Contact Dermatitis

Contact dermatitis, such as metal allergy or plant allergy, is one of the most frequent skin inflammatory diseases [7, 8]. Most of the chemicals that induce contact dermatitis are small compound called haptens. The development of contact dermatitis consists of two phases, the sensitization and elicitation phases. In the sensitization phase, hapten is captured by cutaneous DCs, which migrate to skin-draining lymph nodes and present the antigen to naïve T cells. Then, the naïve T cells differentiate to the antigen-specific effector T cells. In the elicitation phase, the effector T cells are recruited to the skin exposed to the haptens and are activated by the skin DCs to produce inflammatory cytokines. Contact hypersensitivity (CHS) is a frequently used mouse model of contact dermatitis. By using this model, prostanoids and LTs have been shown to be essential in each phase of contact dermatitis.

21.2.1 Prostanoids and LTs in the Sensitization Phase of Contact Dermatitis

Migration and maturation of cutaneous DCs, such as Langerhans cells (LCs) and dermal DCs, are the critical steps for sensitization, and several lipid mediators have been reported to regulate this process [9–11]. On hapten application to the skin, PGE_2 is produced by keratinocytes and acts at EP4 on LCs to facilitate initiation of cutaneous immune responses by promoting migration and maturation of cutaneous DCs [9].

In contrast, PGE_2-EP3 signaling suppresses DC migration and maturation after hapten application and is suggested to function suppressively to prevent excessive skin inflammation [10]. PGD_2 is also reported to inhibit cutaneous DC migration [12, 13]. PGD_2 induced by percutaneous infection with the helminth parasite *Schistosoma mansoni* specifically impedes the migration of LCs through the DP receptor [13]. Administration of a DP agonist, BW245C, inhibits migration of LCs and attenuates OVA-induced dermatitis [12]. Consistently, DP-deficient mice exhibit enhanced cutaneous DC migration and exacerbated inflammation in murine CHS [14]. These activities of lipid mediators are not only limited to prostanoids.

LC migration from the skin to the draining lymph nodes utilizes multidrug resistance-associated protein 1 as a LTC_4 transporter [15]. BLT1-deficient mice exhibit reduced numbers of migrating cutaneous DCs after hapten application, suggesting that LTB_4-BLT1 signaling also promotes DC migration [11].

When cutaneous DCs migrate to draining lymph nodes, DCs present antigens to naïve T cells to prime them. Subsequently, the engagement of the antigen complex by T-cell receptors triggers clonal expansion and differentiation of T cells. $CD4^+$ helper T (Th) cells are differentiated into at least three subsets: Th1, Th2, and Th17. Similarly, $CD8^+$ cytotoxic T (Tc) cells undergo differentiation into two subsets: Tc1 cells and Tc17 cells. CHS is mainly mediated by Tc1/Th1 cells and to some extent by Tc17/Th17 cells [7].

Although the suppressive activity of PGE_2 on Th1 differentiation in vitro has been known since the 1980s, the in vivo role of PGE_2 on Th differentiation has only recently been addressed. In the sensitization phase of CHS, PGE_2 produced by DCs stimulate EP1 receptors on naïve $CD4^+$ and $CD8^+$ T cells and promote Th1 and Tc1 differentiation [16]. Accordingly, EP1-deficient mice exhibit reduced Th1 and Tc1 differentiation and CHS responses [16]. Signaling from EP2 and EP4 receptors also facilitates Th1 differentiation [17]. In addition to EP receptor signaling, IP signaling also promotes Th1 and Tc1 differentiation in CHS [18]. Other than Th1/Tc1 differentiation, PGE_2 promotes Th17 differentiation and expansion through EP2 and EP4 receptors by increasing interleukin (IL)-23 production from DCs and upregulation of IL-23R on Th17 cells [17]. Consistently, administration of EP4 antagonist suppresses Th17 differentiation and expansion in CHS [17]. Prostanoids also regulate DC–T-cell interaction in the priming of naïve T cells. Cutaneous DCs produce abundant TXA_2, which acts on naïve T cells to impair the DC–T-cell interaction [19]. Predictably, TP-deficient mice or wild-type mice treated with a TP antagonist, S-145, during the sensitization period exhibit enhanced CHS responses, indicating that TP signaling negatively regulates the priming of T cells [19].

21.2.2 Prostanoids and LTs in the Elicitation Phase of Contact Dermatitis

After establishment of the sensitization phase, antigen re-challenge onto the skin stimulates keratinocytes (KCs) to produce memory T cell-attracting chemokines, such as CCL27, and neutrophil-attracting chemokines, such as CXCL1 and CXCL2, and to evoke inflammation, in a stage called the elicitation phase [7]. It has been demonstrated that these chemokines are induced by PGE_2 [20], and several prostanoid receptors are also involved in this phase. For example, PGD_2 promotes neutrophil infiltration through CRTH2 and contributes to the progression of inflammation [21]. Accordingly, administration of a CRTH2 antagonist attenuates the CHS response [22]. On the other hand, stimulation of the EP3 receptors on KCs inhibits the chemokine expression in KCs and suppresses the CHS response [23].

The role of LT receptors in the elicitation phase is less clear. As BLT1 is expressed on effector CD8 or CD4 T cells and mediates the infiltration of those cells into skin [24, 25] and LTB_4 has been reported to enhance CCL27 production from keratinocytes [26], LTB_4-BLT1-dependent mechanisms may also operate in the elicitation phase of CHS. The possible roles of individual prostanoid and LT receptors in contact dermatitis are summarized in Figs. 21.1 and 21.2.

21.3 Prostanoids and LTs in Atopic Dermatitis

Atopic dermatitis (AD) is a common pruritic and chronic inflammatory skin disease that is regarded as one of the Th2 diseases. In the dermis, a cellular infiltrate is present consisting of lymphocytes, monocytes, and mast cells. In biopsy specimens from patients with AD, PGE_2 has been determined in biologically active amounts in both lesional and perilesional skin [27]. In contrast, normal levels of eicosanoids

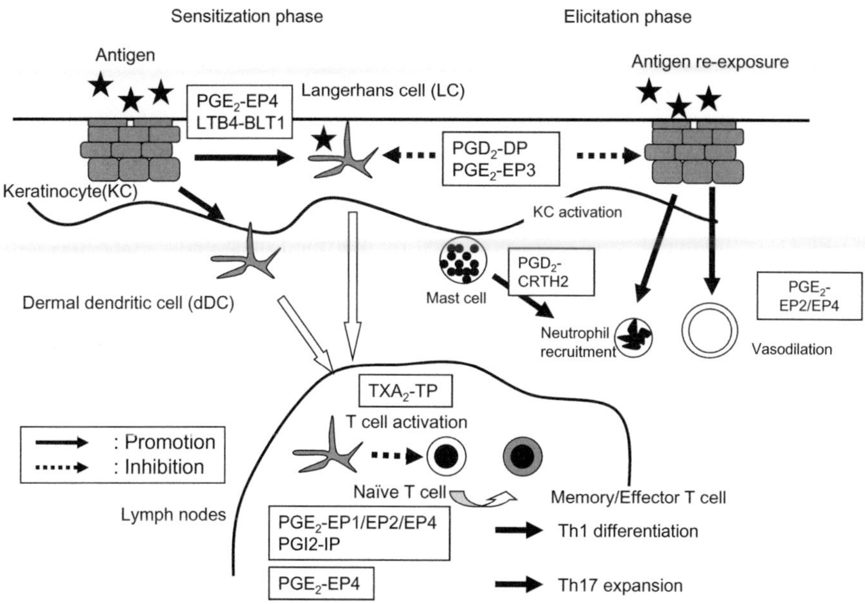

Fig. 21.1 Roles of prostanoids and leukotrienes (LTs) in the development of contact dermatitis. Schematic summary of possible roles of prostanoids and LTs in the sensitization phase (*left*) and the elicitation phase (*right*) of contact dermatitis

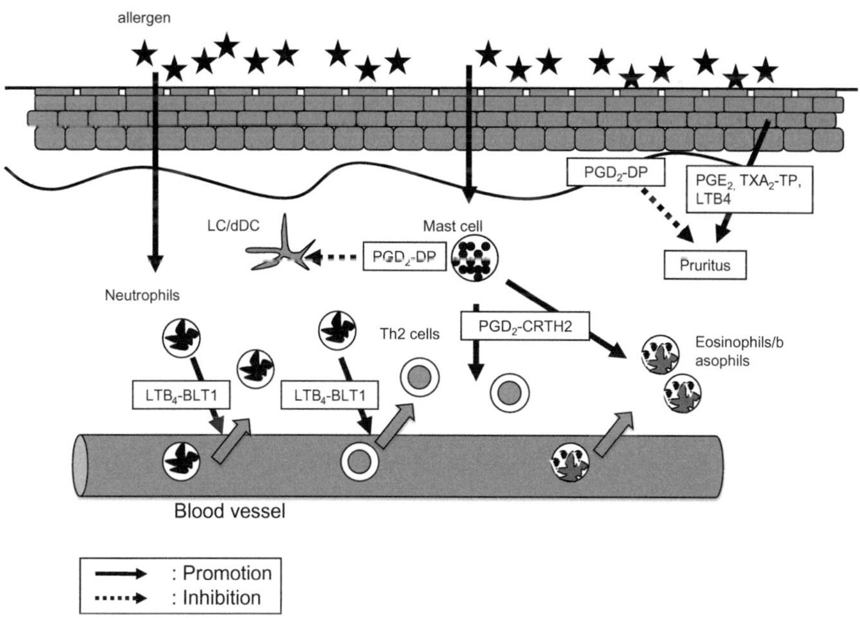

Fig. 21.2 Roles of prostanoids and LTs in the development of atopic dermatitis. Schematic summary of possible roles of prostanoids and LTs in the development of atopic dermatitis

were found in the uninvolved skin of these patients [27]. In an OVA-induced mouse AD model, COX-2-deficient mice exhibited both enhanced eosinophil infiltration and elevated IL-4 expression in the skin lesion with elevated serum IgE and IgG1 [28]. In in vitro studies, PGE_2 drives Ig class switching to IgE by acting at EP2 and EP4 on B cells under LPS and IL-4 stimulation [29]. These results suggest that COX-2-derived PGE_2 plays both protective and promoting roles in the development of AD.

PGD_2 is the major prostanoid produced by activated mast cells. PGD_2 has two types of receptors, DP and CRTH2. The effect of PGD_2 on skin inflammation is not so simple and is extremely context dependent, because DP and CRHT2 possess independent, sometimes opposite, functions even in the same pathological conditions [30–33]. However, CRTH2 generally seems to be pro-inflammatory and DP has both pro- and anti-inflammatory effects. CRTH2 induces chemotaxis in Th2 cells, eosinophils, and basophils with enhanced degranulation [34, 35]. In response to PGD_2, CRTH2 also induces Th2 cell and neutrophil migration into inflammatory skin sites [21]. Virtually all $CRTH2^+$ $CD4^+$ lymphocytes have a pure Th2 phenotype and occupy not all, but a large proportion, of circulating Th2 cells in both normal and AD subjects. In AD patients, a preferential increase of $CRTH2^+$ cells was noted within the disease-related cutaneous lymphocyte-associated antigen-positive $CD4^+$ T-cell compartment [36]. CRTH2-deficient mice exhibit reduced inflammation in the OVA-induced AD model [30]. These results suggest the importance of CRTH2 on Th2 cells in AD, although there remains a need to clarify the respective roles of DP and CRTH2 in AD.

As for the role of LTs in the pathogenesis of AD, it has recently been reported that the LTB_4-BLT1 axis of neutrophils is critical to recruit the Th2 cells into skin in the OVA-induced mice AD model [37].

Pruritus is also an important hallmark of AD. PGE_2 is known to evoke pruritus in AD patients [38]. PGD_2, but not a CRTH2 agonist, 13,14-dihydro-15-keto-PGD_2, reduced scratching behavior in NC/Nga AD model mice, suggesting that DP suppresses pruritic activity [39]. In addition, TXA_2 and LTB_4 are known to mediate the itch sensation [40, 41].

Taken together, these results suggest that appropriate control of lipid mediator production or signaling in the skin can become a novel drug target for AD (Fig. 21.3).

21.4 Prostanoids and LTs in Psoriasis

Psoriasis is a common chronic inflammatory skin disorder characterized by epidermal hyperplasia and widespread erythema. Although the etiology of the disease is still unknown, it has been revealed that cytokines such as IL-17, IL-22, IL-23, and tumor necrosis factor (TNF)-α are pivotal [42].

In the 1980s, several prostanoids and LTs were reported to be abundantly present in human psoriatic skin lesions, suggesting involvement of lipid mediators [43].

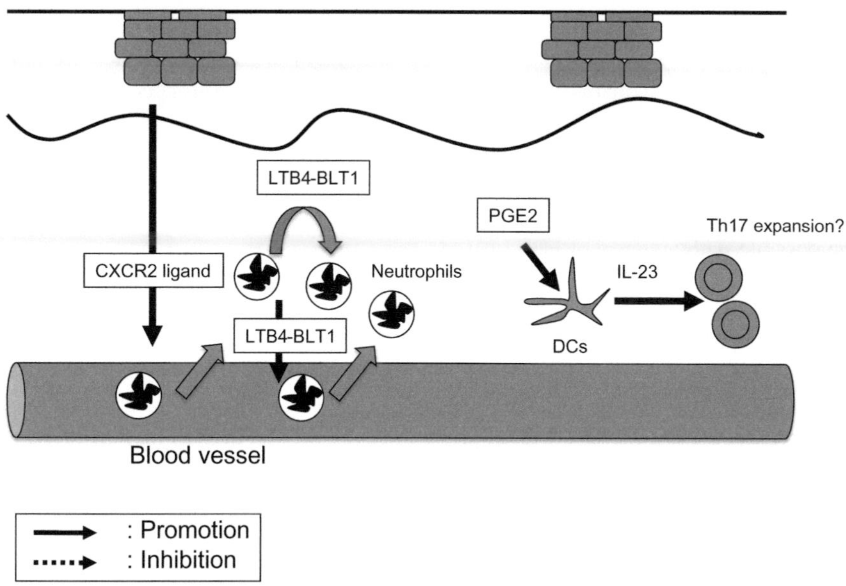

Fig. 21.3 Roles of prostanoids and LTs in the development of psoriasis. LTB_4 and CXCR2 ligand cooperatively promote neutrophil infiltration into skin. PGE_2 may promote IL-23 production from skin DCs and may contribute to the expansion of Th17 cells

However, the roles of prostanoids and LTs in the development of psoriasis remain mostly unclear because of the lack of an appropriate animal psoriasis model.

Recently, it has been reported that daily topical application of imiquimod, an agonist for Toll-like receptor 7 and 8, induced psoriasis-like dermatitis in mice [44]. Using this model, it has recently been shown that BLT1 and CXCR2, a chemokine receptor, work cooperatively for neutrophil infiltration in the psoriasis lesion [45]. This result is in line with previous reports that a 5-lipoxygenase inhibitor improved the clinical symptoms of psoriasis whereas indomethacin exacerbated the symptoms [43].

In in vitro studies, it has been reported that PGE_2 produced by fibroblasts promoted IL-23 production from DCs, which supported the expansion of Th17 cells, suggesting the possibility that this system may work in psoriasis [46, 47]. We are currently investigating the role of prostanoids in psoriasis by applying each prostanoid to receptor-deficient mice in this model (manuscript in preparation).

21.5 Prostanoids in Other Skin Inflammatory Diseases

Finally, we discuss a couple of skin inflammatory diseases in which lipid mediators have suggested to have close relationship with their pathogenesis.

Eosinophilic pustular folliculitis (EPF) is a chronic intractable pruritic dermatosis characterized by massive eosinophil infiltrates involving the pilosebaceous units with unknown mechanisms [48]. Indomethacin, a COX inhibitor, is successfully used to treat patients with EPF, suggesting that COX metabolites such as PGs are involved in the etiology of EPF. The recent report suggests that eotaxin-3 production from sebocytes, induced by PGD_2 and its immediate metabolite 15-deoxy-Δ-12,14-PGJ_2 via the peroxisome proliferator-activated receptor-γ pathway, may be the possible pathogenesis of EPF [49].

Pellagra is another example with a close relationship with lipid mediators. Pellagra is a photosensitivity syndrome characterized by three "D's"—diarrhea, dermatitis, and dementia—as a result of niacin deficiency. Although the molecular mechanisms of photosensitivity dermatitis have been remained unclear, a newly developed murine model of pellagra has revealed that the photosensitivity was mediated through PGE_2-EP4 signaling via ROS production in keratinocytes [50].

Ultraviolet B-induced dermatitis is also mediated by PGE_2 through EP2 and EP4 receptors [51]. Arachidonic acid-induced dermatitis is a murine model of acute skin inflammation. Application of arachidonic acid to the skin increases vascular permeability and causes edema and inflammatory infiltration. PGE_2 has been known as a potentiation of acute inflammation [52], and PGE_2-EP3 signaling mediates this process by inducing mast cell activation and subsequent histamine release and IL-6 production [53, 54].

21.6 Conclusions

In this review, we have summarized current findings on the actions of prostanoids and LTs and their receptors in various skin inflammatory diseases. Although the function of each lipid mediator is various and complex, selective manipulation of the actions mediated by each receptor may provide a novel therapeutic strategy for cutaneous inflammatory disorders.

Acknowledgments The authors declare no conflicts of interest.

References

1. Pasparakis M, Haase I, Nestle FO (2014) Mechanisms regulating skin immunity and inflammation. Nat Rev Immunol 14(5):289–301. doi:10.1038/nri3646
2. Heath WR, Carbone FR (2013) The skin-resident and migratory immune system in steady state and memory: innate lymphocytes, dendritic cells and T cells. Nat Immunol 14(10):978–985. doi:10.1038/ni.2680
3. Honda T, Tokura Y, Miyachi Y, Kabashima K (2010) Prostanoid receptors as possible targets for anti allergic drugs: recent advances in prostanoids on allergy and immunology. Curr Drug Targets 11(12):1605–1613

4. Hirata T, Narumiya S (2012) Prostanoids as regulators of innate and adaptive immunity. Adv Immunol 116:143–174. doi:10.1016/B978-0-12-394300-2.00005-3
5. Yokomizo T (2011) Leukotriene B_4 receptors: novel roles in immunological regulations. Adv Enzyme Regul 51(1):59–64. doi:10.1016/j.advenzreg.2010.08.002
6. Narumiya S, Sugimoto Y, Ushikubi F (1999) Prostanoid receptors: structures, properties, and functions. Physiol Rev 79(4):1193–1226
7. Honda T, Egawa G, Grabbe S, Kabashima K (2013) Update of immune events in the murine contact hypersensitivity model: toward the understanding of allergic contact dermatitis. J Invest Dermatol 133(2):303–315. doi:10.1038/jid.2012.284
8. Peiser M, Tralau T, Heidler J, Api AM, Arts JH, Basketter DA, English J, Diepgen TL, Fuhlbrigge RC, Gaspari AA, Johansen JD, Karlberg AT, Kimber I, Lepoittevin JP, Liebsch M, Maibach HI, Martin SF, Merk HF, Platzek T, Rustemeyer T, Schnuch A, Vandebriel RJ, White IR, Luch A (2012) Allergic contact dermatitis: epidemiology, molecular mechanisms, in vitro methods and regulatory aspects. Current knowledge assembled at an international workshop at BfR, Germany. Cell Mol Life Sci 69(5):763–781. doi:10.1007/s00018-011-0846-8
9. Kabashima K, Sakata D, Nagamachi M, Miyachi Y, Inaba K, Narumiya S (2003) Prostaglandin E_2-EP4 signaling initiates skin immune responses by promoting migration and maturation of Langerhans cells. Nat Med 9(6):744–749. doi:10.1038/nm872
10. Shiraishi N, Nomura T, Tanizaki H, Nakajima S, Narumiya S, Miyachi Y, Tokura Y, Kabashima K (2013) Prostaglandin E_2-EP3 axis in fine-tuning excessive skin inflammation by restricting dendritic cell functions. PLoS One 8(7):e69599. doi:10.1371/journal.pone.0069599
11. Del Prete A, Shao WH, Mitola S, Santoro G, Sozzani S, Haribabu B (2007) Regulation of dendritic cell migration and adaptive immune response by leukotriene B_4 receptors: a role for LTB_4 in up-regulation of CCR7 expression and function. Blood 109(2):626–631. doi:10.1182/blood-2006-02-003665
12. Angeli V, Staumont D, Charbonnier AS, Hammad H, Gosset P, Pichavant M, Lambrecht BN, Capron M, Dombrowicz D, Trottein F (2004) Activation of the D prostanoid receptor 1 regulates immune and skin allergic responses. J Immunol 172(6):3822–3829
13. Angeli V, Faveeuw C, Roye O, Fontaine J, Teissier E, Capron A, Wolowczuk I, Capron M, Trottein F (2001) Role of the parasite-derived prostaglandin D_2 in the inhibition of epidermal Langerhans cell migration during schistosomiasis infection. J Exp Med 193(10):1135–1147
14. Yamamoto Y, Otani S, Hirai H, Nagata K, Aritake K, Urade Y, Narumiya S, Yokozeki H, Nakamura M, Satoh T (2011) Dual functions of prostaglandin D_2 in murine contact hypersensitivity via DP and CRTH2. Am J Pathol 179(1):302–314. doi:10.1016/j.ajpath.2011.03.047
15. Robbiani DF, Finch RA, Jager D, Muller WA, Sartorelli AC, Randolph GJ (2000) The leukotriene C_4 transporter MRP1 regulates CCL19 (MIP-3beta, ELC)-dependent mobilization of dendritic cells to lymph nodes. Cell 103(5):757–768
16. Nagamachi M, Sakata D, Kabashima K, Furuyashiki T, Murata T, Segi-Nishida E, Soontrapa K, Matsuoka T, Miyachi Y, Narumiya S (2007) Facilitation of Th1-mediated immune response by prostaglandin E receptor EP1. J Exp Med 204(12):2865–2874. doi:10.1084/jem.20070773
17. Yao C, Sakata D, Esaki Y, Li Y, Matsuoka T, Kuroiwa K, Sugimoto Y, Narumiya S (2009) Prostaglandin E_2-EP4 signaling promotes immune inflammation through Th1 cell differentiation and Th17 cell expansion. Nat Med 15(6):633–640. doi:10.1038/nm.1968
18. Nakajima S, Honda T, Sakata D, Egawa G, Tanizaki H, Otsuka A, Moniaga CS, Watanabe T, Miyachi Y, Narumiya S, Kabashima K (2010) Prostaglandin I_2-IP signaling promotes Th1 differentiation in a mouse model of contact hypersensitivity. J Immunol 184(10):5595–5603. doi:10.4049/jimmunol.0903260
19. Kabashima K, Murata T, Tanaka H, Matsuoka T, Sakata D, Yoshida N, Katagiri K, Kinashi T, Tanaka T, Miyasaka M, Nagai H, Ushikubi F, Narumiya S (2003) Thromboxane A_2 modulates interaction of dendritic cells and T cells and regulates acquired immunity. Nat Immunol 4(7):694–701. doi:10.1038/ni943
20. Kanda N, Mitsui H, Watanabe S (2004) Prostaglandin E_2 suppresses CCL27 production through EP2 and EP3 receptors in human keratinocytes. J Allergy Clin Immunol 114(6):1403–1409. doi:10.1016/j.jaci.2004.08.041

21. Takeshita K, Yamasaki T, Nagao K, Sugimoto H, Shichijo M, Gantner F, Bacon KB (2004) CRTH2 is a prominent effector in contact hypersensitivity-induced neutrophil inflammation. Int Immunol 16(7):947–959. doi:10.1093/intimm/dxh096
22. Boehme SA, Chen EP, Franz-Bacon K, Sasik R, Sprague LJ, Ly TW, Hardiman G, Bacon KB (2009) Antagonism of CRTH2 ameliorates chronic epicutaneous sensitization-induced inflammation by multiple mechanisms. Int Immunol 21(1):1–17. doi:dxn118
23. Honda T, Matsuoka T, Ueta M, Kabashima K, Miyachi Y, Narumiya S (2009) Prostaglandin E_2-EP3 signaling suppresses skin inflammation in murine contact hypersensitivity. J Allergy Clin Immunol 124(4):809–818 e802. doi:10.1016/j.jaci.2009.04.029
24. Goodarzi K, Goodarzi M, Tager AM, Luster AD, von Andrian UH (2003) Leukotriene B_4 and BLT1 control cytotoxic effector T cell recruitment to inflamed tissues. Nat Immunol 4(10):965–973. doi:10.1038/ni972
25. Tager AM, Bromley SK, Medoff BD, Islam SA, Bercury SD, Friedrich EB, Carafone AD, Gerszten RE, Luster AD (2003) Leukotriene B_4 receptor BLT1 mediates early effector T cell recruitment. Nat Immunol 4(10):982–990. doi:10.1038/ni970
26. Kanda N, Watanabe S (2007) Leukotriene B_4 enhances tumour necrosis factor-alpha-induced CCL27 production in human keratinocytes. Clin Exp Allergy 37(7):1074–1082. doi:10.1111/j.1365-2222.2007.02743.x
27. Fogh K, Herlin T, Kragballe K (1989) Eicosanoids in skin of patients with atopic dermatitis: prostaglandin E_2 and leukotriene B_4 are present in biologically active concentrations. J Allergy Clin Immunol 83(2 pt 1):450–455
28. Laouini D, Elkhal A, Yalcindag A, Kawamoto S, Oettgen H, Geha RS (2005) COX-2 inhibition enhances the TH2 immune response to epicutaneous sensitization. J Allergy Clin Immunol 116(2):390–396. doi:10.1016/j.jaci.2005.03.042
29. Fedyk ER, Phipps RP (1996) Prostaglandin E_2 receptors of the EP2 and EP4 subtypes regulate activation and differentiation of mouse B lymphocytes to IgE-secreting cells. Proc Natl Acad Sci U S A 93(20):10978–10983
30. He R, Oyoshi MK, Wang JY, Hodge MR, Jin H, Geha RS (2010) The prostaglandin D_2 receptor CRTH2 is important for allergic skin inflammation after epicutaneous antigen challenge. J Allergy Clin Immunol 126(4):784–790. doi:10.1016/j.jaci.2010.07.006
31. Matsushima Y, Satoh T, Yamamoto Y, Nakamura M, Yokozeki H (2011) Distinct roles of prostaglandin D_2 receptors in chronic skin inflammation. Mol Immunol 49(1-2):304–310. doi:10.1016/j.molimm.2011.08.023
32. Sarashina H, Tsubosaka Y, Omori K, Aritake K, Nakagawa T, Hori M, Hirai H, Nakamura M, Narumiya S, Urade Y, Ozaki H, Murata T (2014) Opposing immunomodulatory roles of prostaglandin D_2 during the progression of skin inflammation. J Immunol 192(1):459–465. doi:10.4049/jimmunol.1302080
33. Satoh T, Moroi R, Aritake K, Urade Y, Kanai Y, Sumi K, Yokozeki H, Hirai H, Nagata K, Hara T, Utsuyama M, Hirokawa K, Sugamura K, Nishioka K, Nakamura M (2006) Prostaglandin D_2 plays an essential role in chronic allergic inflammation of the skin via CRTH2 receptor. J Immunol 177(4):2621–2629
34. Hirai H, Tanaka K, Yoshie O, Ogawa K, Kenmotsu K, Takamori Y, Ichimasa M, Sugamura K, Nakamura M, Takano S, Nagata K (2001) Prostaglandin D_2 selectively induces chemotaxis in T helper type 2 cells, eosinophils, and basophils via seven-transmembrane receptor CRTH2. J Exp Med 193(2):255–261
35. Yoshimura-Uchiyama C, Iikura M, Yamaguchi M, Nagase H, Ishii A, Matsushima K, Yamamoto K, Shichijo M, Bacon KB, Hirai K (2004) Differential modulation of human basophil functions through prostaglandin D_2 receptors DP and chemoattractant receptor-homologous molecule expressed on Th2 cells/DP2. Clin Exp Allergy 34(8):1283–1290. doi:10.1111/j.1365-2222.2004.02027.x

36. Iwasaki M, Nagata K, Takano S, Takahashi K, Ishii N, Ikezawa Z (2002) Association of a new-type prostaglandin D_2 receptor CRTH2 with circulating T helper 2 cells in patients with atopic dermatitis. J Invest Dermatol 119(3):609–616. doi:10.1046/j.1523-1747.2002.01862.x
37. Oyoshi MK, He R, Li Y, Mondal S, Yoon J, Afshar R, Chen M, Lee DM, Luo HR, Luster AD, Cho JS, Miller LS, Larson A, Murphy GF, Geha RS (2012) Leukotriene B_4-driven neutrophil recruitment to the skin is essential for allergic skin inflammation. Immunity 37(4):747–758. doi:10.1016/j.immuni.2012.06.018
38. Neisius U, Olsson R, Rukwied R, Lischetzki G, Schmelz M (2002) Prostaglandin E_2 induces vasodilation and pruritus, but no protein extravasation in atopic dermatitis and controls. J Am Acad Dermatol 47(1):28–32
39. Arai I, Takano N, Hashimoto Y, Futaki N, Sugimoto M, Takahashi N, Inoue T, Nakaike S (2004) Prostanoid DP1 receptor agonist inhibits the pruritic activity in NC/Nga mice with atopic dermatitis. Eur J Pharmacol 505(1-3):229–235. doi:10.1016/j.ejphar.2004.10.031
40. Andoh T, Haza S, Saito A, Kuraishi Y (2011) Involvement of leukotriene B_4 in spontaneous itch-related behaviour in NC mice with atopic dermatitis-like skin lesions. Exp Dermatol 20(11):894–898. doi:10.1111/j.1600-0625.2011.01346.x
41. Andoh T, Nishikawa Y, Yamaguchi-Miyamoto T, Nojima H, Narumiya S, Kuraishi Y (2007) Thromboxane A_2 induces itch-associated responses through TP receptors in the skin in mice. J Invest Dermatol 127(8):2042–2047. doi:10.1038/sj.jid.5700810
42. Wagner EF, Schonthaler HB, Guinea-Viniegra J, Tschachler E (2010) Psoriasis: what we have learned from mouse models. Nat Rev Rheumatol 6(12):704–714. doi:10.1038/nrrheum.2010.157
43. Ikai K (1999) Psoriasis and the arachidonic acid cascade. J Dermatol Sci 21(3):135–146
44. van der Fits L, Mourits S, Voerman JS, Kant M, Boon L, Laman JD, Cornelissen F, Mus AM, Florencia E, Prens EP, Lubberts E (2009) Imiquimod-induced psoriasis-like skin inflammation in mice is mediated via the IL-23/IL-17 axis. J Immunol 182(9):5836–5845. doi:10.4049/jimmunol.0802999
45. Sumida H, Yanagida K, Kita Y, Abe J, Matsushima K, Nakamura M, Ishii S, Sato S, Shimizu T (2014) Interplay between CXCR2 and BLT1 facilitates neutrophil infiltration and resultant keratinocyte activation in a murine model of imiquimod-induced psoriasis. J Immunol 192(9):4361–4369. doi:10.4049/jimmunol.1302959
46. Schirmer C, Klein C, von Bergen M, Simon JC, Saalbach A (2010) Human fibroblasts support the expansion of IL-17-producing T cells via up regulation of IL-23 production by dendritic cells. Blood 116(10):1715–1725. doi:10.1182/blood-2010-01-263509
47. Sheibanie AF, Tadmori I, Jing H, Vassiliou E, Ganea D (2004) Prostaglandin E_2 induces IL-23 production in bone marrow-derived dendritic cells. FASEB J 18(11):1318–1320. doi:10.1096/fj.03-1367fje
48. Ofuji S, Furukawa F, Miyachi Y, Ohno S (1984) Papuloerythroderma. Dermatologica 169(3):125–130
49. Nakahigashi K, Doi H, Otsuka A, Hirabayashi T, Murakami M, Urade Y, Zouboulis CC, Tanizaki H, Egawa G, Miyachi Y, Kabashima K (2012) PGD_2 induces eotaxin-3 via PPAR-gamma from sebocytes: a possible pathogenesis of eosinophilic pustular folliculitis. J Allergy Clin Immunol 129(2):536–543. doi:10.1016/j.jaci.2011.11.034
50. Sugita K, Ikenouchi-Sugita A, Nakayama Y, Yoshioka H, Nomura T, Sakabe J, Nakahigashi K, Kuroda E, Uematsu S, Nakamura J, Akira S, Nakamura M, Narumiya S, Miyachi Y, Tokura Y, Kabashima K (2013) Prostaglandin E_2 is critical for the development of niacin-deficiency-induced photosensitivity via ROS production. Sci Rep 3:2973. doi:10.1038/srep02973
51. Kabashima K, Nagamachi M, Honda T, Nishigori C, Miyachi Y, Tokura Y, Narumiya S (2007) Prostaglandin E_2 is required for ultraviolet B-induced skin inflammation via EP2 and EP4 receptors. Lab Invest J Tech Methods Pathol 87(1):49–55. doi:10.1038/labinvest.3700491

52. Williams TJ, Morley J (1973) Prostaglandins as potentiators of increased vascular permeability in inflammation. Nature 246(5430):215–217
53. Morimoto K, Shirata N, Taketomi Y, Tsuchiya S, Segi-Nishida E, Inazumi T, Kabashima K, Tanaka S, Murakami M, Narumiya S, Sugimoto Y (2014) Prostaglandin E_2-EP3 signaling induces inflammatory swelling by mast cell activation. J Immunol 192(3):1130–1137. doi:10.4049/jimmunol.1300290
54. Goulet JL, Pace AJ, Key ML, Byrum RS, Nguyen M, Tilley SL, Morham SG, Langenbach R, Stock JL, McNeish JD, Smithies O, Coffman TM, Koller BH (2004) E-prostanoid-3 receptors mediate the proinflammatory actions of prostaglandin E_2 in acute cutaneous inflammation. J Immunol 173(2):1321–1326

Chapter 22
Roles and Actions of Arachidonic Acid-Derived Bioactive Lipids in Stress-Related Behaviors

Tomoyuki Furuyashiki and Shiho Kitaoka

Abstract Excessive or prolonged stress causes cognitive and emotional changes and is thought to be a risk factor for psychiatric disorders. Recent studies in rodents showed roles and actions of arachidonic acid (AA)-derived bioactive lipids, namely, prostaglandin (PG) E_2 and endocannabinoids (eCB), and their receptors in emotional regulation under psychological stress induced by social and environmental stimuli. Stress exposure increases synthesis of PGE_2 in the brain, which suppresses emotional impulsivity under acute stress and facilitates depression and anxiety-like behaviors under repeated stress. This PGE_2 action is mediated, at least in part, through dopaminergic regulation by EP1, a PGE receptor subtype. Stress exposure also increases synthesis of 2-arachidonoylglycerol (2-AG), one eCB species, which suppresses depression and anxiety-like behaviors through multiple brain structures through its receptor CB1. Thus, stress activates both the PGE_2-EP1 pathway and the 2-AG-CB1 pathway, which have distinct, mostly opposing, roles in emotional regulation under stress. COX-1, a PG synthase enriched in microglia, is critical for stress-induced behavioral changes as well as PGE_2 synthesis in the brain. Given a recent report that PGE_2 synthesis in the brain mostly depends on 2-AG metabolism to AA, stress-induced 2-AG synthesis may underlie concomitant PGE_2 synthesis. Collectively, the PGE_2-EP1 and 2-AG-CB1 pathways as well as their crosstalk may be targets for pharmaceutical development for stress-related pathophysiology in psychiatric disorders.

Keywords Rodent • Stress • Depression • Anxiety • Arachidonic acid • Prostaglandin E_2 • 2-Arachidonoylglycerol • Dopamine

---

T. Furuyashiki, M.D., Ph.D. (✉) • S. Kitaoka, Ph.D.
Division of Pharmacology, Kobe University Graduate School of Medicine,
Kusunoki-cho 7-5-1, Chuo-ku, Kobe 650-0017, Japan
e-mail: tfuruya@med.kobe-u.ac.jp

T. Yokomizo, M. Murakami (eds.), *Bioactive Lipid Mediators*,
DOI 10.1007/978-4-431-55669-5_22

22.1 Introduction

Stress is defined as the state of perturbed homeostasis evoked by various aversive stimuli, such as social and environmental stimuli. Excessive or prolonged stress causes cognitive and emotional changes, including depression, anxiety, and attentional deficit, across species, and is thought to precipitate psychiatric disorders, such as mood disorders and schizophrenia. However, because the mechanism underlying these changes remains unknown, a pharmaceutical strategy targeting stress has not been established. Recent rodent studies showed that arachidonic acid (AA)-derived bioactive lipids, such as prostaglandin (PG) E_2 and endocannabinoids (eCBs), regulate stress-induced behavioral changes [11, 12, 23, 30]. Clinical studies have also suggested roles of these bioactive lipids in psychiatric disorders. For example, several clinical reports showed that add-on therapy with nonsteroidal anti-inflammatory drugs (NSAIDs) that block PG synthesis augments therapeutic effects of antidepressants in depressive patients [2, 31, 33]. On the other hand, rimonabant, an antagonist for one of the eCB receptors called CB1, increases risk for depression and anxiety [6]. This review introduces recent studies about the roles and actions of PGE_2 and eCB in rodent stress models, which have been often used as a preclinical model of depression [35].

22.2 Synthesis of PGE_2 and Its Receptor Subtypes

PGE_2 is derived from AA by sequential actions of cyclooxygenase (COX) and PGE synthase [32, 47]. There are two COX isoforms, COX-1 and COX-2. Although these isoforms are often regarded as constitutive and inducible, respectively, both these isoforms are constitutively expressed in the brain [11]. COX-1 is expressed in microglia, whereas COX-2 is primarily expressed in neurons in cortical structures including the cerebral cortex and the hippocampus. PGE_2 exerts its actions through binding to four cognate G protein-coupled receptors named EP1, EP2, EP3, and EP4 [34]. EP1 is primarily coupled to intracellular Ca^{2+} increase. Stimulation of EP2 and EP4 facilitates cAMP production through Gs. EP3 stimulation typically suppressed cAMP production through Gi, although this action may vary depending on EP3 isoforms (Fig. 22.1a). Each of these receptor subtypes is expressed in neurons in specific brain areas, although EP1 distribution remains less well characterized [11]. Cultured glial cells express all these receptors, but glial expression of these receptors in physiological conditions remains poorly characterized.

22.3 Roles of PGE_2 and Its Receptor Subtypes in Stress-Related Behaviors

Because NSAIDs block PG synthesis, PGE_2 synthesized in the brain has been studied as an inflammation-related molecule that mediates pain, febrile, and neuroendocrine responses to peripheral inflammation [11, 49]. In contrast, the function of

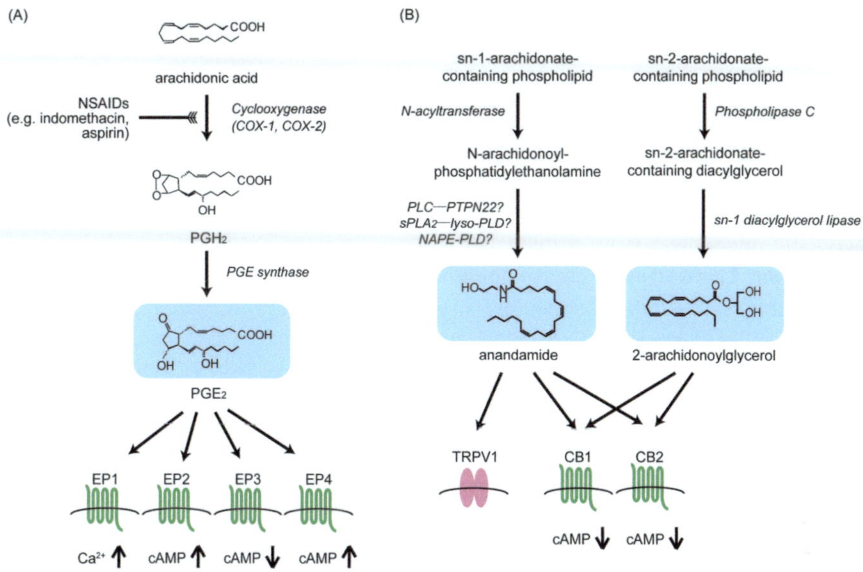

Fig. 22.1 Biosynthesis of arachidonic acid (AA)-derived bioactive lipids and their receptors. (**a**) Prostaglandin (PG)E_2 biosynthesis and its receptors. PGE_2 is derived from arachidonic acid (AA) by sequential actions of cyclooxygenase (COX) and PGE synthase. NSAIDs, such as indomethacin and aspirin, suppress PG production by inhibiting COX-1 and COX-2. PGE_2 exerts its functions through binding to four G protein-coupled receptors (GPCRs) named EP1, EP2, EP3, and EP4, each of which is coupled to a distinct signaling pathway. (**b**) eCB biosynthesis and its receptors. Anandamide (AEA) is synthesized from AA through three distinct pathways. 2-Arachidonoylglycerol (2-AG) is synthesized from AA by sequential actions of phospholipase C and sn-1 diacylglycerol lipase. AEA and 2-AG act as agonists for the GPCRs CB1 and CB2, whereas AEA can also bind to TRPV1

brain PGE_2 in physiological conditions remained poorly characterized. A decade ago, it was reported that mice lacking the PGE receptor subtype EP1 exhibit abnormal emotional behavior under acute psychological challenge [27]. For example, although wild-type mice showed social interaction with a juvenile intruder mouse, EP1-knockout mice rather showed aggressive behavior to the latter. These mice also failed to show cliff avoidance, thus jumping off the elevated platform within several minutes, whereas wild-type mice stayed on the platform during this observation period. In contrast, several other behavioral measures, such as locomotor activity and short-term spatial memory, appeared to be normal. Therefore, EP1 is critical for behavioral control under acute social and environmental challenges. Social interaction and cliff avoidance were also impaired in mice treated with ONO-8713, s specific EP1 antagonist. Because intracerebroventricular injection of an EP1 agonist suppresses electric shock-induced fighting, the EP1 action in behavioral regulation is likely to be located in the brain.

It was frequently reported that the levels of inflammation-related molecules including PGE_2 are increased in blood samples taken from depressive patients (e.g., [25]; for review, [50]). As described in the Introduction, several clinical studies showed that NSAIDs such as celecoxib and aspirin augmented the therapeutic effects of antidepressants in depressive patients [2, 31], and the effect of celecoxib was further confirmed by a meta-analysis based on the results from several clinical reports [33]. These clinical studies led to the hypothesis that PGE_2 may be involved in the pathophysiology of depression.

To test that hypothesis, a role of PGE_2 was examined in repeated social defeat stress in rodents [45], which has been considered to be a mouse model of depression [35]. In this stress model, a male mouse of C57BL/6 background is subjected to agonistic encounters from a male ICR mouse selected based on a high level of aggression. This social defeat is applied for 10 min daily for 10 consecutive days, and various behavioral changes are measured. Typically, repeated social defeat stress induces depression-like behaviors, such as social avoidance and reduced sucrose preference (anhedonia), and increases anxiety-like behaviors as often measured by the elevated plus maze test and the light–dark box test [35]. Repeated social defeat increases PGE_2 content in the brain, and EP1-knockout mice failed to show social avoidance and elevated anxiety [45]. However, EP1-knockout mice showed submissive posture, an immediate behavioral response to social defeat, to a normal level. These results showed that the PGE_2-EP1 pathway is not critical for perception of repeated social defeat, but for long-term behavioral changes after repeated social defeat (Fig. 22.2a).

22.4 EP1-Mediated Dopaminergic Regulation and Its Behavioral Effect Under Stress

The PGE_2-EP1 pathway regulates stress-related behaviors through the dopaminergic pathway, at least in part. Dopamine is a neuromodulator critical for various neural functions ranging from motor to emotional to cognitive functions. Dopamine neurons are mostly, but not exclusively, located in the midbrain, namely, the ventral tegmental area (VTA) and the substantia nigra pars compacta (SNpc). There are several groups of dopamine neurons. Dopamine neurons in the SNpc project to the dorsal striatum. In the VTA, there are at least two groups of VTA dopamine neurons, each of which projects to the ventral striatum (or the nucleus accumbens) and the medial prefrontal cortex, respectively.

EP1-deficient mice showed elevated dopamine turnover, a biochemical index of dopamine release, in frontal cortex and striatum, compared with wild-type mice [27]. Consistently, equilibrium in vivo microdialysis showed that the extracellular concentration of dopamine is elevated in the striatum of EP1-deficient mice, compared with wild-type mice [46]. Furthermore, systemic treatment with SCH23390, a dopamine D1 receptor antagonist, suppressed electric shock-induced fighting

Fig. 22.2 (continued) release appears to be involved in this 2-AG synthesis, at least in the prefrontal cortex. The 2-AG-CB1 pathway in respective brain areas attenuates stress-induced behavioral changes through distinct mechanisms. Note that red and gray lines indicate active and inactive processes in respective conditions

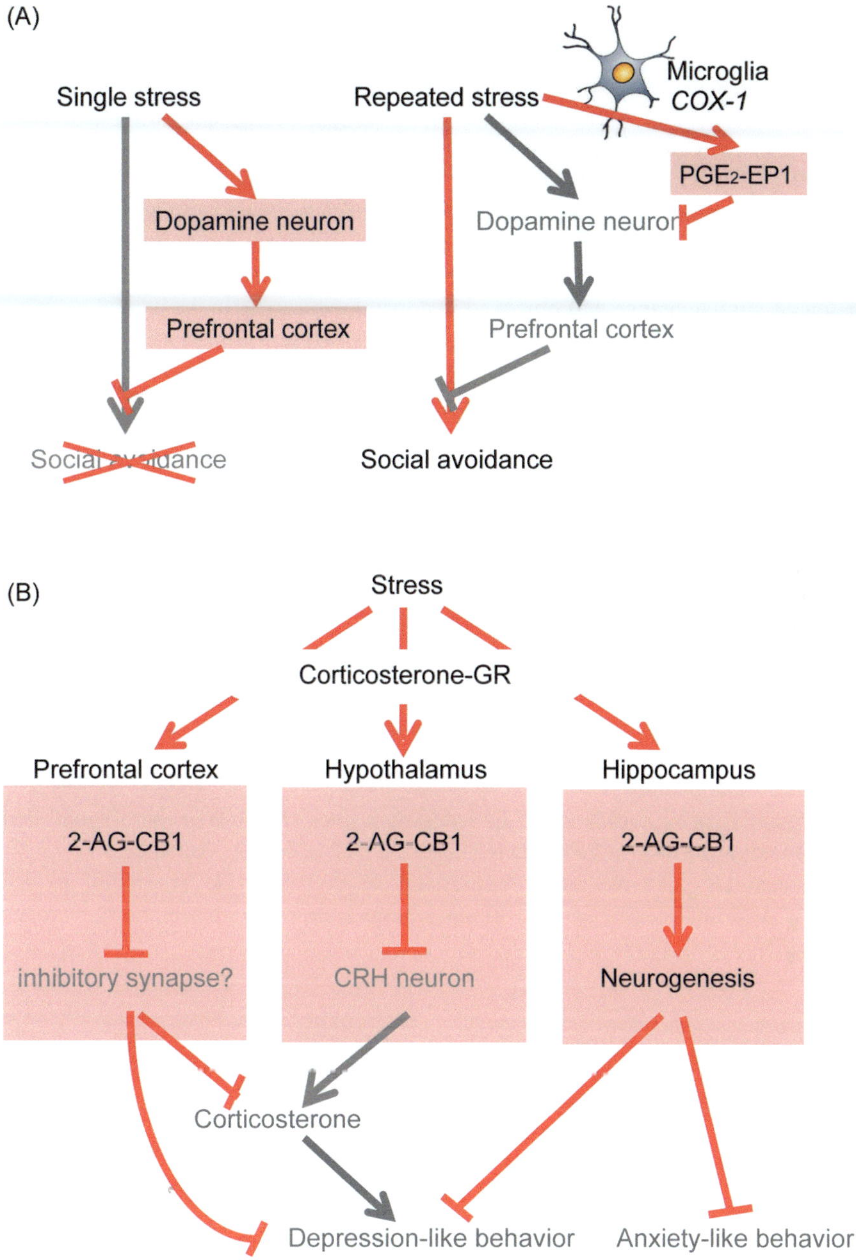

Fig. 22.2 Roles of PGE_2 and eCB in stress-related behaviors. (**a**) The PGE_2-EP1 pathway facilitates stress-related behaviors. A single stress exposure activates the dopaminergic pathway projecting to the prefrontal cortex, which suppresses stress-induced social avoidance. After repeated stress, PGE_2 attenuates this dopaminergic pathway through EP1, thereby leading to social avoidance. Because COX-1, a PG synthase enriched in microglia, is critical for PGE_2 synthesis and social avoidance upon stress exposure, microglia are likely to be the cellular source of PGE_2 synthesis that contributes to stress-induced social avoidance. (**b**) A role of eCB-CB1 pathway in attenuating stress-related behaviors. Stress exposure increases 2-AG synthesis in multiple brain areas, such as the prefrontal cortex, hypothalamus, and hippocampus. Stress-evoked corticosterone

[27]. These results suggested that EP1 suppresses dopaminergic activity in multiple brain structures, and that disinhibited dopaminergic activity caused by EP1 deficiency causes aggressive behavior.

EP1 is also critical for dopaminergic changes induced by repeated social defeat stress [45]. Immunostaining for c-Fos, a marker for neuronal activity, showed that single exposure to social defeat stress activates dopamine neurons, and that this stress response of VTA dopamine neurons is attenuated with repetition of social defeat. A similar change was observed in dopamine turnover in the medial prefrontal cortex: Single exposure to social defeat increased dopamine turnover, and this response was attenuated with stress repetition. In EP1-deficient mice, whereas a single exposure to defeat normally increased c-Fos expression in VTA dopamine neurons and dopamine turnover in the medial prefrontal cortex, repeated stress-induced attenuation of both these two indices was abolished. This finding showed that the PGE_2-EP1 pathway mediates suppression, but not facilitation, of dopaminergic pathway under stress condition.

In repeated social defeat stress, dopaminergic activity in the medial prefrontal cortex suppresses induction of social avoidance, thereby leading to stress resilience [5, 45]. This idea is supported by several findings. Thus, dopamine turnover in the medial prefrontal cortex was negatively correlated to the level of social avoidance induced by repeated social defeat, and pharmacological damage or optogenetic inhibition of dopaminergic projection to the medial prefrontal cortex facilitated induction of social avoidance by social defeat stress. Notably, pharmacological blockade of dopamine D1-like receptors by their antagonist, SCH23390, restored induction of social avoidance in EP1-deficient mice, suggesting that disinhibited dopaminergic activity in these mice causes the lack of social avoidance.

Collectively, EP1-mediated suppression of dopaminergic projection to the medial prefrontal cortex appears to underlie at least some of behavioral changes induced by repeated stress (Fig. 22.2a).

22.5 The Mechanism of Action of PGE_2-EP1 Pathway in the Dopaminergic System

Previous studies showed that PGE_2-EP1 can regulate the dopaminergic pathway at least through two mechanisms.

First, EP1 stimulation can inhibit dopamine neurons through augmenting GABAergic inhibitory synaptic inputs [46]. EP1 immunostaining showed EP1 localization at GABAergic synaptic terminals formed on SNpc dopamine neurons. In midbrain slices, pharmacological stimulation of EP1 by ONO-DI-004, an EP1 agonist, increased evoked inhibitory postsynaptic currents in SNpc dopamine neurons, and this EP1 action was abolished by pharmacological blockade or genetic deletion of EP1. This EP1 action, if extrapolated to VTA dopamine neurons, can explain the EP1-mediated suppression of dopaminergic activity under stress conditions as described, although this possibility remains to be proven using conditional EP1-knockout mice.

Second, EP1 stimulation can alter the intracellular signaling of dopamine receptors. This mechanism was suggested on the basis of the finding that EP1 deficiency attenuates Thr34 phosphorylation of DARPP-32 induced by dopamine D1 receptor stimulation [24]. Consistent with this finding, in HEK293T cells overexpressing both EP1 and D1 receptors, these receptors form a biochemical complex, and pharmacological stimulation of EP1 by its agonist ONO-DI-004 facilitates cAMP production induced by D1 receptor agonists [10]. Originally, this EP1 action appeared to be opposite to the EP1 action implicated in behavioral regulation under stress. However, further analysis revealed that dopamine D1 receptor signaling with or without EP1 stimulation utilizes a distinct molecular pathway [10]. Thus, EP1-mediated facilitation of D1-induced cAMP production is mediated by $G_{\beta\gamma}$ subunits and adenylyl cyclase 7, a $G_{\beta\gamma}$-sensitive adenylyl cyclase isoform, whereas, without EP1 stimulation, D1-induced cAMP production was mediated through adenylyl cyclase 5, a $G_{\beta\gamma}$-insensitive and Ca^{2+}-suppressed adenylyl cyclase isoform. Whether and how this EP1 action on dopamine receptor signaling could be involved in stress-induced behavioral changes warrants future investigation.

22.6 Synthesis and Metabolism of Endocannabinoids and Their Receptors

eCBs are endogenous ligands for receptors that mediate the psychotic and addictive actions of Δ^9-tetrahydrocannabinol (THC) in cannabis, and include 2-AG and anandamide (arachidonoylethanolamide; AEA) [8, 39]. Recent studies have revealed a critical role for eCBs and their receptors in emotional regulation without or with stress. Before describing these studies in detail, we briefly introduce the synthesis and metabolism of eCBs and their receptors. The reader can refer to comprehensive reviews elsewhere [8, 21, 39].

For the synthesis of 2-AG, phospholipase C (PLC) catalyzes the conversion from *sn-2*-arachidonate-containing phospholipids to *sn-2*-arachidonate-containing diacylglycerol (DAG), which is then metabolized to 2-AG by *sn-1*-diacylglycerol lipase (DGL). For the synthesis of AEA, multiple enzymatic cascades via phospholipases of different classes, namely, PLA_2, PLC, and PLD, have been proposed, although the contribution of respective cascades in physiological and pathophysiological contexts remains to be established (Fig. 22.1b). For inactivation of 2-AG, monoacylglycerol lipase (MGL) metabolizes 2-AG to AA and glycerol. AEA is metabolized to AA and ethanolamine by fatty acid amide hydrolase (FAAH) for its inactivation.

2-AG and AEA act as agonists for the G protein-coupled receptors CB1 and CB2, both of which are primarily coupled to G_i-mediated inhibition of cAMP production. CB1 is mainly, but not exclusively, localized at presynaptic terminals of neurons in the brain. CB1 expression was expressed in multiple brain areas, such as the striatum, thalamus, hypothalamus, and cerebellar cortex [21]. In addition, CB1 is strongly expressed in CCK-positive inhibitory neurons in the cerebral cortex and the hippocampus, whereas it is also weakly expressed in excitatory neurons in these brain structures. CB2 expression is well known in immune cells in peripheral tissues and in microglial cells in the brain, although its expression in neurons was also reported. In addition to CB1 and CB2, AEA is known to act as an agonist for a capsaicin receptor cation channel called TRPV1.

For neural functions, the 2-AG-CB1 pathway mediates synaptic retrograde signaling for synaptic plasticity in both excitatory and inhibitory synapses in many brain areas [21]. Depolarization-induced increase in intracellular Ca^{2+} at the postsynaptic site is thought to facilitate synthesis of 2-AG, which diffuses through the synaptic cleft and suppresses release of synaptic vesicles through presynaptic CB1. On the other hand, the role and action of AEA in synaptic plasticity remain elusive.

22.7 Roles of eCBs in Emotional Regulation

Rodent studies showed roles of CB1 in emotional regulation, although results are not entirely consistent across studies. For example, systemic injection with the CB1 agonists CP55940 and WIN55212-2 at relatively low doses suppresses anxiety-like behavior in the elevated plus maze test, whereas systemic injection with the CB1 antagonists AM251 and rimonabant (SR141716) enhances such anxiety-like behavior [38]. Consistently, CB1-deficient mice showed elevated anxiety-like behaviors in elevated or lighted compartments [16, 26]. Because serotonergic neuron-selective knockout of CB1 also augments the level of anxiety-like behaviors [9], it is plausible that CB1 reduces the level of anxiety through inhibiting synaptic transmission from serotonergic neurons. In contrast to these findings, it was also reported that systemic treatment with rimonabant rather suppressed anxiety-like behaviors in the elevated plus maze test [15].

Roles of CB1 in depression-like behaviors are also reported, although results are again not entirely consistent across studies. CB1-deficient mice showed typical depression-like behaviors, such as increased immobility time in the forced swim test and reduced motivation for sucrose reward [1, 42]. Consistent with this finding, Δ^9-THC reduced immobility time in the forced swim test, and this effect was blocked by rimonabant treatment, and thus is likely to be CB1 dependent [17]. However, it was also reported that systemic treatment with rimonabant alone reduced immobility time in the forced swim test [17], implying the presence of multiple CB1 actions in depression-like behaviors. Because conditional deletion of CB1 in GABAergic neurons abolishes rimonabant-induced suppression of depression-like behaviors [17], this antidepressant-like effect of remonabant involves disinhibition of GABAergic synaptic transmission, at least in part.

In contrast to CB1, roles of CB2 in emotional regulation have been much less well characterized, except in a couple of cases. Thus, CB2-deficient mice showed increased anxiety-like behaviors in elevated and lighted compartments as well as increased immobility time in the forced swim test [37]. In contrast, transgenic mice overexpressing CB2 in the brain showed a reduced level of anxiety [13].

Because pharmacological inhibition for eCB metabolism augments eCB functions at the site of endogenous eCB synthesis, several behavioral studies using inhibitors for eCB-metabolizing enzymes were also performed. For example, pharmacological inhibition of FAAH, an AEA-metabolizing enzyme, by URB597 reduces the level of anxiety measured by the elevated plus maze and depression-like behaviors in the tail suspension test and the forced swim test [14, 22, 38]. Similarly, systemic treatment with JZL184, an inhibitor for the 2-AG-metabolizing enzyme MGL, also suppresses anxiety-like behavior in the elevated plus maze test [43]. Although the anxiolytic effects of URB597 and JZL194 are apparently similar, the mechanisms of actions of these drugs may be mediated through CB1 and CB2, respectively. This notion is supported by the finding that the anxiolytic effect of URB597 is blocked by rimonabant, a CB1 antagonist, whereas CB2 antagonists, such as SR144528 and AM630, suppress the anxiolytic effect of JZL194 [4]. The anxiolytic effect of URB597 may be mediated through CB1 in the medial prefrontal cortex, as local injection of URB597 and AEA to the medial prefrontal cortex at low doses also suppresses anxiety-like behaviors [41]. In contrast, it was reported that local injection of the same drugs to the same brain area, but at high doses, increases anxiety-like behaviors [41]. This anxiogenic effect was blocked by capsazepine, a TRPV1 antagonist, and thus is likely to be mediated by TRPV1. Therefore, the level of AEA in the medial prefrontal cortex needs to be tightly regulated to achieve the optimal level of anxiety.

22.8 Roles of eCBs in Stress Responses

Most, if not all, studies in rodents showed that stress exposure of various conditions, such as single and repeated restraint stress and repeated social defeat stress, augments 2-AG synthesis in the medial prefrontal cortex [9, 18, 40]. This 2-AG synthesis is blocked by RU486, a GR antagonist, suggesting a role for glucocorticoid release in this process [18]. On the other hand, 2-AG synthesis in the medial prefrontal cortex appears to suppress stress-induced glucocorticoid release, as local injection of AM251, a CB1 antagonist, to the medial prefrontal cortex augments corticosterone release upon acute restraint stress [18]. Therefore, the 2-AG-CB1 pathway in the medial prefrontal cortex may provide a negative feedback loop for stress-induced glucocorticoid release. The eCB-CB1 pathway is also involved in stress-induced behavioral changes. Thus, local injection of AM251 to the medial prefrontal cortex facilitates the effect of chronic mild stress on the time of immobility in the forced swim test [29]. Similarly, systemic injection with URB597, a FAAH inhibitor, for several consecutive weeks ameliorates chronic mild stress-induced anhedonia as measured by decreased sucrose preference [3]. Local injection of URB597 to the medial prefrontal cortex also reduces immobility in the forced swim test [28]. These behavioral findings suggest that stress exposure activates the eCB-CB1 pathway in the medial prefrontal cortex, thereby counteracting neuroendocrine and behavioral responses to stress.

Besides 2-AG synthesis in the medial prefrontal cortex, repeated social defeat stress facilitates 2-AG synthesis in other structures, such as the hippocampus and the hypothalamus [9]. CB1 stimulation in the hypothalamus suppresses excitatory synaptic inputs to CRH-secreting neurons, thereby inhibiting glucocorticoid release [7]. CB1 in the hippocampus appears to regulate neurogenesis in the subgranular zone of the dentate gyrus of the hippocampus, which is critical for the behavioral effect of antidepressants at least in rodents. Thus, in CB1-deficient mice, hippocampal neurogenesis was reduced by half [20], and HU210, a synthetic cannabis, causes anxiolytic and antidepressant-like actions in a manner dependent on hippocampal neurogenesis [19].

Collectively, stress activates the eCB-CB1 pathway in multiple brain structures, including the prefrontal cortex, hippocampus, and hypothalamus, which counteracts stress responses, perhaps through distinct mechanisms (Fig. 22.2b).

22.9 Stress-Induced Synthesis of PGE_2 and eCBs in the Brain

As already described, stress augments synthesis of PGE_2 and eCBs in the brain. COX-1-dependent PGE_2 synthesis appears to be critical for stress-induced behavioral changes. Systemic injection with sc-560, a COX-1-selective inhibitor, or genetic deletion of COX-1 abolishes social avoidance induced by repeated social defeat, whereas neither pharmacological blockade nor genetic deletion of COX-2 affects the social avoidance [45]. Our preliminary results as well as those of others [36] showed that COX-1 is critical for PGE_2 synthesis in the brain without or with stress exposure. Combined with the finding that COX-1 is selectively expressed in microglia in the brain [45], these findings lead to the hypothesis that COX-1-expressing microglia may be the source of PGE_2 in the brain, especially under stress conditions. Consistent with this hypothesis, repeated social defeat stress induces histological changes of microglia, such as enlargement of the cell body and hypertrophy of microglial processes, which is reminiscent of microglial activation [45].

Because repeated social defeat stress did not change the expression level of COX-1 [45], repeated stress may increase the supply of AA for COX-1-mediated PG production. It has been established that cytosolic $PLA_{2\alpha}$ is critical for the release of AA from phospholipids of the cell membrane for various physiological and pathophysiological functions [44]. However, studies using an MGL inhibitor and MGL-knockout mice recently showed that PGE_2 synthesis in brain highly depends on a pool of AA supplied from MGL-mediated 2-AG metabolism [36]. Therefore, one can hypothesize that stress-induced increase in PGE_2 synthesis in the brain results from stress-induced 2-AG synthesis, although this hypothesis remains to be experimentally tested (Fig. 22.3).

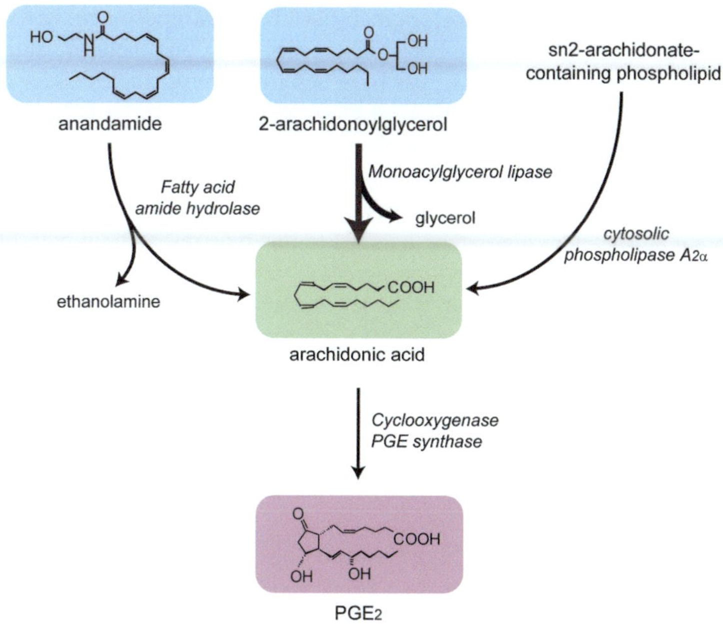

Fig. 22.3 Crosstalk between eCBs and PGE_2. Anandamide (AEA) is metabolized by fatty acid amide hydrolase (FAAH) into arachidonic acid (AA) and ethanolamine. 2-Arachidonoylglycerol (2-AG) is metabolized by monoacylglycerol lipase (MGL) into AA and glycerol. It is known that 2-AG is much more abundant than AEA in the brain. It is established that $cPLA_{2\alpha}$-mediated AA release from the cell membrane is coupled to PGE_2 synthesis for various physiological and pathophysiological functions. However, it was recently shown that PGE_2 synthesis in the brain primarily utilizes a pool of AA derived from 2-AG metabolism. Therefore, it can be hypothesized that stress-induced 2-AG synthesis leads to concomitant PGE_2 synthesis

The mechanism for stress-induced increase in 2-AG production remains poorly understood. It is well known that stimulation of multiple GPCR-type neurotransmitter receptors can evoke 2-AG synthesis through the PLC-DGL pathway at the postsynaptic site of neurons [21]. It was also reported that stimulation of P2X7 purinergic receptor stimulates 2-AG production from cultured microglia [48]. Given a link between 2-AG metabolism and PGE_2 production in the brain, identification of the cellular source for stress-induced 2-AG synthesis will also help our understanding of the mechanism for stress-induced PGE_2 synthesis.

22.10 Concluding Remarks

This review summarizes recent studies in rodents about the roles and actions of the AA-derived bioactive lipids, PGE_2 and eCBs, in regulating emotional behaviors, especially under stressful conditions. Given historical backgrounds, functions of

PGE_2 and eCBs in the brain have originally been studied in different contexts, in sickness behaviors and addictive behaviors, respectively, and therefore studied by different groups of researchers. As has been described, most studies so far suggest distinct, and mostly opposing, actions of PGE_2 and eCBs in stress-induced behavioral changes. Based on this idea, the balance between PGE_2 and eCBs pathways appears to be a critical determinant for stress susceptibility as well as the level of each pathway. An emerging coupling between eCB metabolism and PGE_2 synthesis in the brain all the more highlights the need to integrate research on both of these bioactive lipids in the same behavioral context. Understanding the roles of PGE_2 and eCBs as well as their crosstalk at the level of synthesis, metabolism, and receptor signaling will allow identifying novel targets for pharmaceutical intervention for stress-related pathophysiology in psychiatric disorders.

References

1. Aso E, Ozaita A, Valdizán EM et al (2008) BDNF impairment in the hippocampus is related to enhanced despair behavior in CB1 knockout mice. J Neurochem 105:565–572
2. Berk M, Dean O, Drexhage H et al (2013) Aspirin: a review of its neurobiological properties and therapeutic potential for mental illness. BMC Med 11:74
3. Bortolato M, Mangieri RA, Fu J et al (2007) Antidepressant-like activity of the fatty acid amide hydrolase inhibitor URB597 in a rat model of chronic mild stress. Biol Psychiatry 62:1103–1110
4. Busquets-Garcia A, Puighermanal E, Pastor A et al (2011) Differential role of anandamide and 2-arachidonoylglycerol in memory and anxiety-like responses. Biol Psychiatry 70:479–486
5. Chaudhury D, Walsh JJ, Friedman AK et al (2013) Rapid regulation of depression-related behaviours by control of midbrain dopamine neurons. Nature 493:532–536
6. Christensen R, Kristensen PK, Bartels EM et al (2007) Efficacy and safety of the weight-loss drug rimonabant: a meta-analysis of randomised trials. Lancet 370:1706–1713
7. Di S, Malcher-Lopes R, Halmos KC et al (2003) Nongenomic glucocorticoid inhibition via endocannabinoid release in the hypothalamus: a fast feedback mechanism. J Neurosci 23:4850–4857
8. Di Marzo V (2008) Targeting the endocannabinoid system: to enhance or reduce? Nat Rev Drug Discov 7:438–455
9. Dubreucq S, Matias I, Cardinal P et al (2012) Genetic dissection of the role of cannabinoid type-1 receptors in the emotional consequences of repeated social stress in mice. Neuropsychopharmacology 37:1885–1900
10. Ehrlich AT, Furuyashiki T, Kitaoka S et al (2013) Prostaglandin E receptor EP1 forms a complex with dopamine D1 receptor and directs D1-induced cAMP production to adenylyl cyclase 7 through mobilizing $G_{\beta\gamma}$ subunits in human embryonic kidney 293T cells. Mol Pharmacol 84:476–486
11. Furuyashiki T, Narumiya S (2011) Stress responses: the contribution of prostaglandin E_2 and its receptors. Nat Rev Endocrinol 7:163–175
12. Furuyashiki T (2012) Roles of dopamine and inflammation-related molecules in behavioral alterations caused by repeated stress. J Pharmacol Sci 120:63–69
13. García-Gutiérrez MS, Manzanares J (2011) Overexpression of CB2 cannabinoid receptors decreased vulnerability to anxiety and impaired anxiolytic action of alprazolam in mice. J Psychopharmacol 25:111–120

14. Gobbi G, Bambico FR, Mangieri R et al (2005) Antidepressant-like activity and modulation of brain monoaminergic transmission by blockade of anandamide hydrolysis. Proc Natl Acad Sci U S A 102:18620–18625
15. Griebel G, Stemmelin J, Scatton B (2005) Effects of the cannabinoid CB1 receptor antagonist rimonabant in models of emotional reactivity in rodents. Biol Psychiatry 57:261–267
16. Haller J, Bakos N, Szirmay M et al (2002) The effects of genetic and pharmacological blockade of the CB1 cannabinoid receptor on anxiety. Eur J Neurosci 16:1395–1398
17. Häring M, Grieb M, Monory K et al (2013) Cannabinoid CB_1 receptor in the modulation of stress coping behavior in mice: the role of serotonin and different forebrain neuronal subpopulations. Neuropharmacology 65:83–89
18. Hill MN, McLaughlin RJ, Pan B et al (2011) Recruitment of prefrontal cortical endocannabinoid signaling by glucocorticoids contributes to termination of the stress response. J Neurosci 31:10506–10515
19. Jiang W, Zhang Y, Xiao L et al (2005) Cannabinoids promote embryonic and adult hippocampus neurogenesis and produce anxiolytic- and antidepressant-like effects. J Clin Invest 115:3104–3116
20. Jin K, Xie L, Kim SH et al (2004) Defective adult neurogenesis in CB1 cannabinoid receptor knockout mice. Mol Pharmacol 66:204–208
21. Kano M, Ohno-Shosaku T, Hashimotodani Y et al (2009) Endocannabinoid-mediated control of synaptic transmission. Physiol Rev 89:309–380
22. Kathuria S, Gaetani S, Fegley D et al (2003) Modulation of anxiety through blockade of anandamide hydrolysis. Nat Med 9:76–81
23. Katona I, Freund TF (2012) Multiple functions of endocannabinoid signaling in the brain. Annu Rev Neurosci 35:529–558
24. Kitaoka S, Furuyashiki T, Nishi A et al (2007) Prostaglandin E_2 acts on EP1 receptor and amplifies both dopamine D1 and D2 receptor signaling in the striatum. J Neurosci 27:12900–12907
25. Lieb J, Karmali R, Horrobin D (1983) Elevated levels of prostaglandin E_2 and thromboxane B2 in depression. Prostaglandins Leukot Med 10:361–367
26. Martin M, Ledent C, Parmentier M et al (2002) Involvement of CB1 cannabinoid receptors in emotional behaviour. Psychopharmacology (Berl) 159:379–387
27. Matsuoka Y, Furuyashiki T, Yamada K et al (2005) Prostaglandin E receptor EP1 controls impulsive behavior under stress. Proc Natl Acad Sci U S A 102:16066–16071
28. McLaughlin RJ, Hill MN, Bambico FR et al (2012) Prefrontal cortical anandamide signaling coordinates coping responses to stress through a serotonergic pathway. Eur Neuropsychopharmacol 22:664–671
29. McLaughlin RJ, Hill MN, Dang SS et al (2013) Upregulation of CB_1 receptor binding in the ventromedial prefrontal cortex promotes proactive stress-coping strategies following chronic stress exposure. Behav Brain Res 237:333–337
30. Mechoulam R, Parker LA (2013) The endocannabinoid system and the brain. Annu Rev Psychol 64:21–47
31. Muller N, Schwarz MJ, Dehning S et al (2006) The cyclooxygenase-2 inhibitor celecoxib has therapeutic effects in major depression: results of a double-blind, randomized, placebo controlled, add-on pilot study to reboxetine. Mol Psychiatry 11:680–684
32. Murakami M, Nakatani Y, Tanioka T, Kudo I (2002) Prostaglandin E synthase. Prostaglandins Other Lipid Mediat 68-69:383–399
33. Na KS, Lee KJ, Lee JS et al (2014) Efficacy of adjunctive celecoxib treatment for patients with major depressive disorder: a meta-analysis. Prog Neuropsychopharmacol Biol Psychiatry 48:79–85
34. Narumiya S, Sugimoto Y, Ushikubi F (1999) Prostanoid receptors: structures, properties, and functions. Physiol Rev 79:1193–1226
35. Nestler EJ, Hyman SE (2010) Animal models of neuropsychiatric disorders. Nat Neurosci 13:1161–1169

36. Nomura DK, Morrison BE, Blankman JL et al (2011) Endocannabinoid hydrolysis generates brain prostaglandins that promote neuroinflammation. Science 334:809–813
37. Ortega-Alvaro A, Aracil-Fernández A, García-Gutiérrez MS et al (2011) Deletion of CB2 cannabinoid receptor induces schizophrenia-related behaviors in mice. Neuropsychopharmacology 36:1489–1504
38. Patel S, Hillard CJ (2006) Pharmacological evaluation of cannabinoid receptor ligands in a mouse model of anxiety: further evidence for an anxiolytic role for endogenous cannabinoid signaling. J Pharmacol Exp Ther 318:304–311
39. Piomelli D (2003) The molecular logic of endocannabinoid signaling. Nat Rev Neurosci 4:873–884
40. Rademacher DJ, Meier SE, Shi L et al (2008) Effects of acute and repeated restraint stress on endocannabinoid content in the amygdala, ventral striatum, and medial prefrontal cortex in mice. Neuropharmacology 54:108–116
41. Rubino T, Realini N, Castiglioni C et al (2008) Role in anxiety behavior of the endocannabinoid system in the prefrontal cortex. Cereb Cortex 18:1292–1301
42. Sanchis-Segura C, Cline BH, Marsicano G et al (2004) Reduced sensitivity to reward in CB1 knockout mice. Psychopharmacology (Berl) 176:223–232
43. Sciolino NR, Zhou W, Hohmann AG (2011) Enhancement of endocannabinoid signaling with JZL184, an inhibitor of the 2-arachidonoylglycerol hydrolyzing enzyme monoacylglycerol lipase, produces anxiolytic effects under conditions of high environmental aversiveness in rats. Pharmacol Res 64:226–234
44. Shimizu T, Ohto T, Kita Y (2006) Cytosolic phospholipase A_2: biochemical properties and physiological roles. IUBMB Life 58:328–333
45. Tanaka K, Furuyashiki T, Kitaoka S et al (2012) Prostaglandin E_2-mediated attenuation of mesocortical dopaminergic pathway is critical for susceptibility to repeated social defeat stress in mice. J Neurosci 32:4319–4329
46. Tanaka Y, Furuyashiki T, Momiyama T et al (2009) Prostaglandin E receptor EP1 enhances GABA-mediated inhibition of dopaminergic neurons in the substantia nigra pars compacta and regulates dopamine level in the dorsal striatum. Eur J Neurosci 30:2338–2346
47. Vane JR, Bakhle YS, Botting RM (1998) Cyclooxygenases 1 and 2. Annu Rev Pharmacol Toxicol 38:97–120
48. Witting A, Walter L, Wacker J et al (2004) P2X7 receptors control 2-arachidonoylglycerol production by microglial cells. Proc Natl Acad Sci U S A 101:3214–3219
49. Zeilhofer HU, Brune K (2006) Analgesic strategies beyond the inhibition of cyclooxygenases. Trends Pharmacol Sci 27:467–474
50. Zorrilla EP, Luborsky L, McKay JR et al (2001) The relationship of depression and stressors to immunological assays: a meta-analytic review. Brain Behav Immun 15:199–226

Part IV
Protocols for Analyzing Lipid Mediators

Chapter 23
Basic Techniques for Lipid Extraction from Tissues and Cells

Toshiaki Okuno and Takehiko Yokomizo

Abstract Lipids as well as proteins, nucleic acids, and carbohydrates are essential components of animals and plants, and their chemical structures and biological functions are highly variable. However, most lipids are insoluble in aqueous solutions, and careful manipulation using organic solvents is required for the extraction and purification of lipids. In this chapter, we describe basic techniques for the extraction and purification of lipids from animal tissues and cells using the Bligh and Dyer method, solvent fractionation, and column chromatography. General methods for the extraction and purification of lipids are mentioned here; refer to other chapters for detailed information on individual lipids.

Keywords Lipid extraction • Bligh and Dyer • Solvent fractionation • Ion-exchange chromatography • DEAE cellulose column • Reversed-phase column

23.1 Handling of Lipids

Lipids are usually defined as molecules that are soluble in organic solvents such as chloroform, methanol, hexane, and ether. However, some lipids [e.g., lysophospholipids, glycolipids, and phosphoinositides (PIPs)] are readily soluble in aqueous solutions but are only slightly soluble in organic solvents. Lipid compounds should be stored in appropriate organic solvents according to their polarity. For example, polar lipids should be stored in a chloroform:methanol solution with a ratio of 2:1, whereas nonpolar lipids should be stored in pure chloroform. Organic solvents containing lipids should be stored in glass vials, and plastic lids that have Teflon welded into their top are recommended. When it is important to completely avoid contamination, such as in mass spectrometric analyses, lipid solutions should be transferred using glass pipettes, glass syringes, or autopipettors that contain Teflon and have glass surfaces. Organic solvents should always be prepared in high-quality, fresh

T. Okuno, Ph.D. (✉) • T. Yokomizo
Department of Biochemistry, Juntendo University Graduate School of Medicine,
2-1-1 Hongo, Bunkyo-ku, Tokyo 113-8421, Japan
e-mail: tokuno@juntendo.ac.jp

T. Yokomizo, M. Murakami (eds.), *Bioactive Lipid Mediators*,
DOI 10.1007/978-4-431-55669-5_23

solvents that have a low water content [high performance liquid chromatography (HPLC) or liquid chromatography–mass spectrometry (LC-MS) grade]. In general, lipids are stored in a deep freezer (below −20 ° C) and in glass containers at relatively high concentrations. Lipid samples should be stored in containers that have a blanket of inert gas such as nitrogen or argon in their headspace. In optimal storage conditions, light and oxygen are absent to avoid isomerization and oxidation of the double bonds of lipids. When removed from the freezer, lipid samples should be allowed to reach room temperature before the cap is opened because moisture in the atmosphere may hydrolyze the ester or amide bonds of lipids.

23.2 Extraction of Lipids

Animal tissues should be prepared immediately before lipid extraction to prevent hydrolysis and oxidation during the extraction procedures. Some lipids are rapidly produced or degraded after the animal is killed. When animal tissues cannot be processed immediately, they should be frozen in liquid N_2 as rapidly as possible and stored at −80 ° C. Generally, many lipids are tightly bound with proteins in vivo via hydrogen bonds, ionic bonds, and hydrophobic forces. Polar solvents such as methanol are used to separate lipids from proteins, whereas nonpolar solvents such as chloroform are used to dissolve lipids. Lipids are usually extracted at low temperatures (4 ° C) as soon as possible after animal tissues are removed. Two structural features of phospholipids include the nonpolar hydrocarbon chains of the fatty acids and the polar phosphate-containing headgroups. The combination of polar and nonpolar groups within the molecule affects its solubility and extraction efficiency in organic solvents. Hence, the use of a single organic solvent is not suitable for the extraction of all lipid species. The mixture of chloroform and methanol in appropriate ratios is used for the efficient extraction of phospholipids. This method was developed by Folch et al. [1] and uses chloroform:methanol in a ratio of 2:1 and large volumes of water to wash out the nonlipid components [1]. Although this extraction procedure is extremely efficient, rapid, and conducted at room temperature (or below), the formation of emulsions is a major drawback. Despite a modification (including salt in the media) made by Folch et al. [2], this procedure has been replaced by the most widely adapted method for lipid extraction in use today. The method of Bligh and Dyer [3] was originally designed to extract lipids from fish muscle [3]. This method is advantageous for tissues containing a high percentage of water. The method is a variation of Folch's extraction and calculates the amount of water present in the sample so that the final composition of chloroform:methanol:water is 1:2:0.8, creating a single extraction phase. The entire procedure is extremely rapid and is the most efficient means to extract lipids. After the addition of equal volumes of water and chloroform, total lipids can be recovered in the chloroform-rich lower phase (Fig. 23.1). We describe an example of lipid extraction using the Bligh and Dyer method. Please also refer to the useful and detailed websites of 'Avanti Polar Lipids' and 'LIPID MAPS,' etc. [4–6].

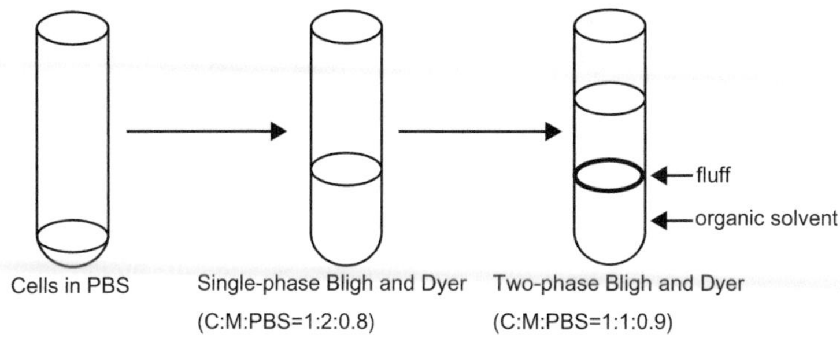

Fig. 23.1 Lipid extraction by the Bligh and Dyer method. *C* chloroform, *M* methanol, *PBS* phosphate-buffered saline

23.2.1 Bligh and Dyer Extraction of Total Lipids from Cells

1. Collect and suspend cells (~10^8 cells) in 0.8 ml phosphate-buffered saline (PBS) and transfer to a glass centrifuge tube with a Teflon cap.
2. Add 3 ml ice-cold chloroform/methanol solution (1:2) and mix to generate a single-phase Bligh and Dyer mixture (final ratio of chloroform:methanol:PBS = 1:2:0.8). Vortex vigorously and incubate for 5 min.
3. Add 1 ml chloroform and 1 ml PBS and mix to generate a two-phase solution (final ratio of chloroform:methanol:PBS = 1:1:0.9).
4. Centrifuge at 1000 *g* for 2 min. The lower phase is composed of chloroform:methanol:water in a ratio of 86:14:1 (by volume) and contains lipids; the upper phase consists of the same solvents in the ratio of 3:48:47 (by volume) and contains the majority of nonlipid contaminants.
5. Recover the lower phase carefully with a Pasteur pipette. Do not remove the white material at the interface, which contains denatured proteins.
6. (Optional) Add 1 ml chloroform, centrifuge, and recover the lower solution.
7. Concentrate the lower solution under a N_2 stream, resuspend the resulting pellets in chloroform:methanol solution (2:1), and store at −30° C. Generally, about 10 mg dried lipid is recovered from 0.1 g cells or tissues.

23.2.2 Bligh and Dyer Extraction of Total Lipids from Tissues

The Bligh and Dyer method is also suitable for lipid extraction from animal tissues (brain, liver, intestine, etc.). The volumes of extraction solvents should be adjusted to maintain the chloroform:methanol:water ratio at 1:2:0.8 for a single-phase Bligh and Dyer mixture and at 1:1:0.9 for the formation of a two-phase solution (*3*, Fig. 23.1). The exact volumes must be determined empirically for each sample type when extracting lipids from tissues. The water volume should be considered the aqueous component of the tissue. Different tissues have different water contents, and the endogenous water content contributes to the aqueous component in the single-phase mixture. Animal tissues often require homogenization in water or PBS using a Potter-Elvehjem device or a Polytron homogenizer before and after the addition of chloroform and methanol. An example protocol to extract lipids from rat liver using the Bligh and Dyer method is provided next.

1. Add 3 ml water, 20 ml methanol, and 10 ml chloroform (ice-cold) to rat liver (6–7 g) and homogenize on ice using a Polytron homogenizer to generate a single-phase Bligh and Dyer mixture (final ratio of chloroform:methanol:water = 1:2:0.8). Homogenize the mixture thoroughly, vortex vigorously, and incubate for 10–15 min.
2. Pour the solution into a 50-ml glass centrifuge tube with a Teflon cap and centrifuge at 2000 rpm for 5 min. Collect the solution by decantation.
3. Resuspend the resulting pellets in 10 ml chloroform, 20 ml methanol, and 8 ml water. Homogenize again on ice using a Polytron homogenizer.
4. Pour the solution into a 50-ml glass centrifuge tube and centrifuge at 2000 rpm for 5 min. Collect the solution by decantation and combine with the previous solution.
5. Add 20 ml chloroform and 20 ml water to the solution and mix to generate a two-phase solution (final ratio of chloroform:methanol:water = 1:1:0.9).
6. Pour the solution into two 50-ml glass centrifuge tubes and centrifuge at 2000 rpm for 5 min.
7. Recover the lower phase carefully with a Pasteur pipette. Do not remove the white material at the interface, which contains denatured proteins.
8. (Optional) Add 10 ml chloroform to two 50-ml glass centrifuge tubes, centrifuge, and recover the lower phase with a Pasteur pipette.
9. Concentrate the solution with a rotary evaporator or under a N_2 stream, resuspend the resulting pellets in chloroform/methanol solution (2:1), and store at −30 ° C.

23.3 Solvent Fractionation

Solvent fractionation is a simple method that is sometimes the most efficient means to crudely separate lipids (Fig. 23.2). This approach depends on the differential solubility of lipids in various organic solvents.

23.3.1 Separation of Polar and Nonpolar Lipids

Acetone precipitation is currently used for one-step separation of polar lipids (e.g., phospholipids and glycolipids) from neutral and nonpolar lipids (e.g., triglycerides, cholesterol, and the majority of fatty acids).

1. Evaporate ~100 mg crude lipid extract in a glass centrifuge tube under a N_2 stream and add 20–30 volumes (~5 ml) of acetone.
2. Vortex the glass centrifuge tube for 1 min and leave on ice for 1 h.

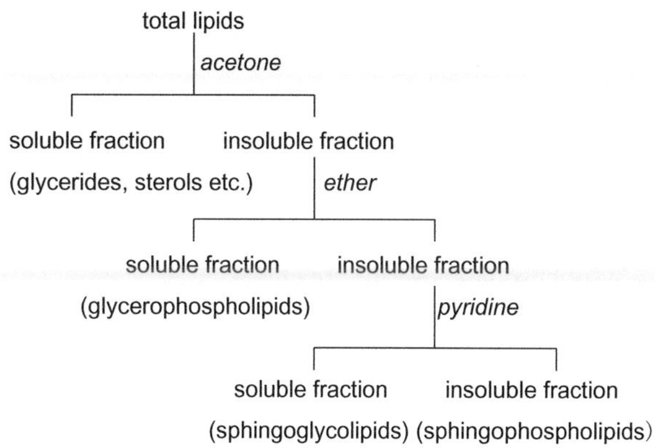

Fig. 23.2 Solvent fractionation of total lipids

3. Centrifuge the tube at 2000 rpm for 5 min and carefully collect the supernatant with a Pasteur pipette.
4. Repeat the procedure (steps 1–3) and combine the acetone extracts.
5. Dry the pellet, which is rich in phospholipids and glycolipids, under a N_2 stream, redissolve in chloroform/methanol mixture (2:1), and store at −30° C for further analysis.
6. The acetone extract contains glycerides, sterols, sterol esters, carotenoids, lipid-soluble vitamins, and fatty acids. Dry the acetone extract under a N_2 stream, redissolve in chloroform:methanol mixture (2:1), and store at −30° C.

23.4 Purification of Lipids by Chromatography

The separation of lipids by ion-exchange chromatography is based on the different ionic groups present in the molecules. Nonionic, acidic, and zwitterionic lipids are separated on several ion-exchange materials: diethylaminoethyl (DEAE) cellulose, triethylaminoethyl cellulose, or ion-exchange resins. DEAE is most frequently used to separate lipids of different classes. Carboxymethyl cellulose is also used for preparative separation of phospholipid classes.

23.4.1 Purification of Acidic Phospholipids by Butanol Extraction Using a DEAE Cellulose Column

1. Equilibrate the DEAE cellulose column (Wako or Sigma) with 1 ml methanol, 1 ml chloroform:methanol solution (1:1), and 1 ml chloroform three times each.
2. Homogenize ~10^6 cells or ~100 mg tissue in 2 ml 1-butanol.
3. Add 2 ml PBS to the homogenate and vortex for 3 min on ice.
4. Centrifuge at 2000 *g* for 5 min and recover the supernatant.
5. Add 1 ml methanol and 1 ml chloroform to the extract solution and apply to a pre-equilibrated DEAE cellulose column.

6. Wash the column with 1 ml chloroform:methanol solution (1:1) three times. Neutral lipids, phosphatidylcholine, and phosphatidylethanolamine are eluted.
7. Elute the column with 1 ml chloroform:methanol:28 % ammonia solution:glacial acetic acid solution (200:100:3:0.9) three times. Phosphatidylserine, phosphatidylglycerol, phosphatidylinositol, phosphatidic acid, and cardiolipin are eluted.
8. Elute the column with 1 ml chloroform:methanol:28 % ammonia solution:glacial acetic acid solution (200:100:10:6.7) three times. Lysophosphatidic acid and lysophosphatidylserine are eluted.
9. Elute the column with 1 ml chloroform:methanol/12N HCl:water solution (60:60:5:5) three times. Phosphoinositide (PIP) is eluted.

23.4.2 Purification of Acidic Phospholipids by Methanol Extraction Using a Reversed-Phase Column

1. Equilibrate the column (Oasis HLB, 3 ml/60 mg; Waters) with 3 ml methanol and 3 ml water containing formic acid (pH 2.0–3.0).
2. Homogenize ~10^6 cells or ~100 mg tissue in 1 ml methanol and incubate on ice for 1 h.
3. Centrifuge at 2000 *g* for 5 min and recover the supernatant (extract solution).
4. Add 9 ml water containing formic acid (pH 2.0–3.0) to the extract solution (final concentration of methanol is 10 %) and apply to a pre-equilibrated reversed-phase column.
5. Wash the column with 3 ml water.
6. Elute the column with 3 ml hexane. Neutral lipids are eluted.
7. Elute the column with 3 ml methylformate. Fatty acids and oxidized fatty acids are eluted.
8. Elute the column with 3 ml methanol. Phospholipids and oxidized phospholipids are eluted.

References

1. Folch J, Ascoli I, Lees M, Meath JA, Le BN (1951) Preparation of lipid extracts from brain tissue. J Biol Chem 191:833–841
2. Folch J, Lees M, Sloane Stanley GH (1957) A simple method for the isolation and purification of total lipids from animal tissues. J Biol Chem 226:497–509
3. Bligh EG, Dyer WJ (1959) A rapid method for total lipid extraction and purification. Can J Biochem Physiol 37:911–917
4. Avanti Polar Lipids. Inc. Alabama 'Technical Support' and 'Lipid Extraction.' http://avantilipids.com/
5. LIPID MAPS. University of California, San Diego 'Protocols' http://www.lipidmaps.org/protocols/index.html and 'Lipidmics methods' http://www.lipidmaps.org/resources/tutorials/lipidomicsmethods.html
6. Brown, HA (ed) (2007) Methods in enzymology, vol 432. Academic, San Diego

Chapter 24
Comprehensive Analysis of Eicosanoids

Yoshihiro Kita and Takao Shimizu

Abstract Arachidonic acid released from phospholipids is converted to a series of metabolites referred to as eicosanoids in vivo, many of which have indispensable roles in normo- and pathophysiology. As their contents in tissues and biological fluids are low in general, a sensitive and specific quantification method is necessary for basic as well as clinical studies. This chapter describes methods to quantify various eicosanoids at one time with high sensitivity using a liquid chromatography-tandem mass spectrometry. The principle of reversed-phase chromatography of eicosanoids and quantification using selected reaction monitoring mode of triple quadrupole mass spectrometers, as well as sample collection and pretreatment methods including extraction from tissues or biological fluids, are described.

Keywords Eicosanoid • Prostaglandin • Leukotriene • Arachidonic acid • Lipid mediator • Mass spectrometry • Quantification

24.1 Introduction

Eicosanoids are metabolites of 20-carbon polyunsaturated fatty acids (PUFAs), such as arachidonic acid (AA), eicosapentaenoic acid (EPA), or dihomo-γ-linolenic acid (DGLA) [1, 2]. Prostaglandins (PGs), thromboxane (TX), and leukotrienes (LTs) are major eicosanoid lipid mediators produced in vivo and function through

Y. Kita (✉)
Department of Lipidomics, Graduate School of Medicine, The University of Tokyo, 7-3-1 Hongo, Bunkyo-ku, Tokyo 113-0033, Japan

Life Sciences Core Facility, Graduate School of Medicine, The University of Tokyo, 7-3-1 Hongo, Bunkyo-ku, Tokyo 113-0033, Japan
e-mail: kita@m.u-tokyo.ac.jp

T. Shimizu
Department of Lipidomics, Graduate School of Medicine, The University of Tokyo, 7-3-1 Hongo, Bunkyo-ku, Tokyo 113-0033, Japan

Department of Lipid Signaling, Research Institute, National Center for Global Health and Medicine, 1-21-1 Toyama, Shinjuku-ku, Tokyo 162-8655, Japan

T. Yokomizo, M. Murakami (eds.), *Bioactive Lipid Mediators*,
DOI 10.1007/978-4-431-55669-5_24

binding to cognate G protein-coupled receptors (GPCR) that mediate various normo- and pathophysiological functions [2]. In addition to studying lipid-ligand GPCRs and lipid mediator-producing enzymes, quantitative analysis of eicosanoids is of great importance to understand their functions [3]. Although enzyme-linked immunosorbent assay (ELISA) kits are available for major eicosanoids, liquid chromatography–mass spectrometry (LC-MS)-based quantification is more suited for comprehensive analysis. In this chapter, we describe a methodology of eicosanoid quantification from biological samples.

24.2 Sample Collection and Storage

Biological samples should be immediately frozen to avoid potential changes after collection. Typically, samples are snap frozen with liquid nitrogen or dry ice and then stored at –80 °C until use. For aqueous samples such as plasma, urine, or culture media, approximately 10 μM indomethacin can be added immediately after collection to avoid ex vivo formation of prostanoids. Long-term storage is often inevitable, especially for samples of human origin; the preservability of eicosanoids has not been reported in a comprehensive manner, but there is a report describing possible degradation of eicosanoids by freeze–thaw cycles with ordinary laboratory freezers [4]. Sample amount depends on expected eicosanoid content in the sample versus sensitivity of the mass spectrometer, but it is typically less than 1 g or 1 ml, which fits subsequent extraction procedures.

24.3 Eicosanoid Extraction and Pretreatment

Eicosanoids in tissues and biological fluids are extracted and cleaned up before analysis. Tissue homogenization in aqueous buffer (e.g., saline or phosphate-buffered saline) can be performed in most biochemical or bioanalytical laboratories using ordinary homogenizers. Indomethacin (~10 μM) and/or butylated hydroxytoluene [BHT; 0.1 % (w/v)] can be included in the homogenization buffer to avoid any ex vivo eicosanoid formation or oxidation. The homogenates are mixed with approximately 10 volumes of methanol for 1 h at 4 °C for extraction and then centrifuged to remove debris and protein precipitates. A more preferable alternative is to use a cryomill. An SK-mill (Tokken, Chiba, Japan) or similar can be used to powder frozen tissues in a polypropylene tube without thawing. For example, up to 100 mg frozen tissue can be crushed by a SK-100 mill in a 2-ml Eppendorf Safe-Lock tube with a stainless steel crusher, to which 1 ml methanol is added directly for extraction. After extraction, debris and protein precipitates are removed using a high-speed centrifuge (10,000 *g*, 10 min, 4 °C). The use of a cryomill minimizes the risk of sample deterioration while providing better throughput. For both methods, deuterated internal standards for LC-MS quantification are included at this step. For

Fig. 24.1 Solid-phase extraction of eicosanoids

Oasis HLB cartridge (1cc/10 mg, Waters)
Conditioning
↓ Methanol
↓ 0.03% formic acid-water
Sample Loading
↓ Dilute 1 mL methanol solution with 3-4 mL of 0.03% formic acid-water and immidiately load onto the cartridge
Wash
↓ 0.03% formicacid-water
↓ 15% ethanol-0.03% formic acid-water
↓ petroleum ether
Elution
↓ 0.2% formic acid-methanol, 0.2 mL
to LC-MS analysis

biological fluids, samples are mixed with ten (or more) volumes of methanol and internal standards, and then centrifuged to obtain extracts. In most cases, methanol extraction is recommended as the initial extraction procedure. The Bligh–Dyer method for total lipid extraction [5] is not recommended because of poor recovery of polar eicosanoids.

The methanol extracts are subjected to solid-phase extraction (SPE). Sep-Pak C18 (Waters, Milford, MA, USA) or similar reversed-phase SPE cartridges have been widely used for this purpose, but recent products such as Oasis HLB (Waters) or similar polymer-based reversed-phase sorbents with polar functional groups provide better recovery and reproducibility. An Oasis HLB cartridge 1 ml/10 mg (or 1 ml/30 mg) can be used for 1 ml crude methanol extract (Fig. 24.1). Methanol extracts are diluted four- or fivefold with 0.03 % formic acid:water and loaded onto preconditioned cartridges with vacuum. Once diluted, samples are immediately loaded to avoid losing hydrophobic eicosanoids by adsorption to any contacting surfaces. The cartridges are then washed serially with 0.1 % formic acid:water, 15 % ethanol:0.03 % formic acid:water, and petroleum ether. Washing with aqueous solutions removes polar components such as salts, sugars, and polar peptides. Petroleum ether removes neutral lipids (i.e., triglycerides). After briefly drying the cartridges by passing air, eicosanoids are eluted with 200 μl 0.2 % formic acid:methanol either by low-speed centrifugation (800 *g*, 2 min) or by vacuum to collection vials. If needed, samples can be evaporated and reconstituted to smaller volumes for analysis.

24.4 Reversed-Phase Liquid Chromatography of Eicosanoids

Reversed-phase liquid chromatography (RPLC) with a C8 or C18 column is widely used for separation of eicosanoids. Columns from many manufacturers, including XBridge C8/C18 (Waters), Acquity BEH C8/C18 (Waters), Capcell Pak C18 MG

(Shiseido, Tokyo, Japan), Kinetex C8 (Phenomenex, Torrance, CA, USA), or similar, with 1.0–3.0 mm inner diameter and 50–100 mm length can be used. An acetonitrile:water solvent system with formic acid (~0.03–0.1 %) is recommended for its chromatographic performance.

The diverse hydrophobicity of eicosanoids requires gradient elution for a comprehensive analysis. Two gradient types, a continuous gradient or step gradient can be used, depending on the target panel of eicosanoids. A continuous gradient is used for a comprehensive analysis of large number of eicosanoid species, whereas a step gradient is preferred to measure a limited number of eicosanoids in a short time. Because eicosanoids are small molecules, isocratic or optimized gradient elution is needed for isomer separations. For example, separation of PGE_2 and PGD_2 is best resolved by isocratic condition, typically with 35–37 % acetonitrile, but inappropriate gradient conditions (too shallow or too steep gradients) do not provide acceptable resolutions of these eicosanoids. Depending on the column parameters (inner diameter, length, and particle type), flow rate and gradient program should be optimized.

There are several tips for major eicosanoid separation. TXB_2 does not give a single peak, resulting in a broad, odd-shaped peak (Fig. 24.2a), because of interconversion between two hemiacetal isomers. The peak shape is affected by gradient

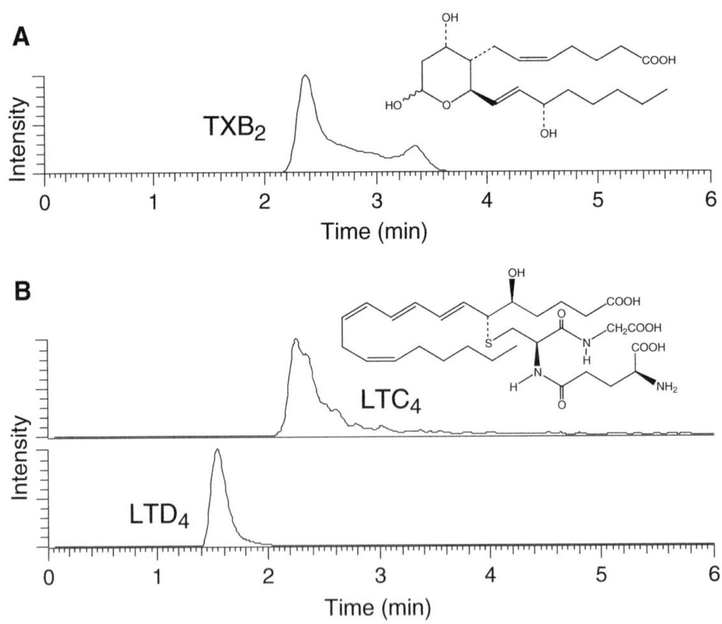

Fig. 24.2 Chromatography of TXB_2 and LTC_4. (**a**) Resolution of TXB_2 by reversed-phase LC (Capcell Pak MGS3, 1 × 100 mm, 0.1 % formic acid:37 % acetonitrile/water, isocratic). Characteristic peak shape is observed. (**b**) Resolution of LTC_4 and LTD_4 (Capcell Pak MGS3, 1 × 100 mm, 0.1 % formic acid:45 % acetonitrile:water, isocratic). Peak tailing of LTC_4 is obvious even under optimized condition (Modified from Kita et al. [6])

condition and should be optimized carefully when monitoring TXB_2. Analysis of LTC_4 needs special attention (Fig. 24.2b). The metal-interacting property of LTC_4 causes a severe peak tailing or even a peak disappearance for metal-contaminated LC columns. Any possible metal ion contamination should be eliminated from LC systems, including rust from stainless steel. Carryover is relatively high for LTC_4 quantification, requiring careful data reviewing and validation [6].

Sample solvent is an important factor. An initial mobile phase is recommended as sample solvents for gradient analysis in the textbooks, but this does not apply to the eicosanoid mixtures. Use of 35 % acetonitrile in water, for example, results in adsorption of hydrophobic eicosanoids such as HETEs to any contacting surfaces, including vial walls. Instead, methanol is recommended as a sample solvent for general purposes. However, upon analysis, injection of methanol solution results in peak deterioration (broadening, fronting, or splitting) because methanol is a stronger eluent than the initial mobile-phase composition. Minimizing injection volume improves peak deterioration to a certain extent, but the sample should be highly concentrated during preparation. The initial mobile phase can be changed to more aqueous composition (i.e., 5 % acetonitrile), in which injected samples are rapidly diluted by the mobile phase. Alternatively, coinjection of sample with water also improves peak shape by dilution effects.

24.5 Detection of Eicosanoids

Output from LC is connected to the electrospray interface of MS for detection. Eicosanoids have one or more carboxy group(s) in common, and can be ionized as a deprotonated species in the negative-ion mode electrospray ionization [6]. Some eicosanoids including cysteinyl LTs can be also ionized in the positive ion mode. Ionization efficiencies, an important factor for detection sensitivity, are affected by pH of the solvent. Higher pH facilitates deprotonation and results in better ionization in the negative ESI mode. However, a basic LC mobile phase is not suitable for practical use; deprotonation makes eicosanoids more hydrophilic, causing poor chromatographic retention. An acidic mobile phase provides better chromatographic resolution and acceptable (moderate) ionization for most of the fatty acid metabolites, excepting precursor fatty acids whose ionizations are severely compromised under acidic condition.

The MS/MS fragmentation pattern of eicosanoids can be used for their selective detection. Figure 24.3 shows collision-induced dissociation (CID) spectra for HETE structural isomers. Their deprotonated ions share the same m/z of 319, but give different MS/MS fragments that can be used for isomer differentiation. However, isomers are sometimes not well distinguished by MS/MS. For example, PGD_2 and PGE_2 (m/z 351 in the negative ESI) do not give specific MS/MS peak(s), requiring chromatographic separation for differentiation (Fig. 24.4).

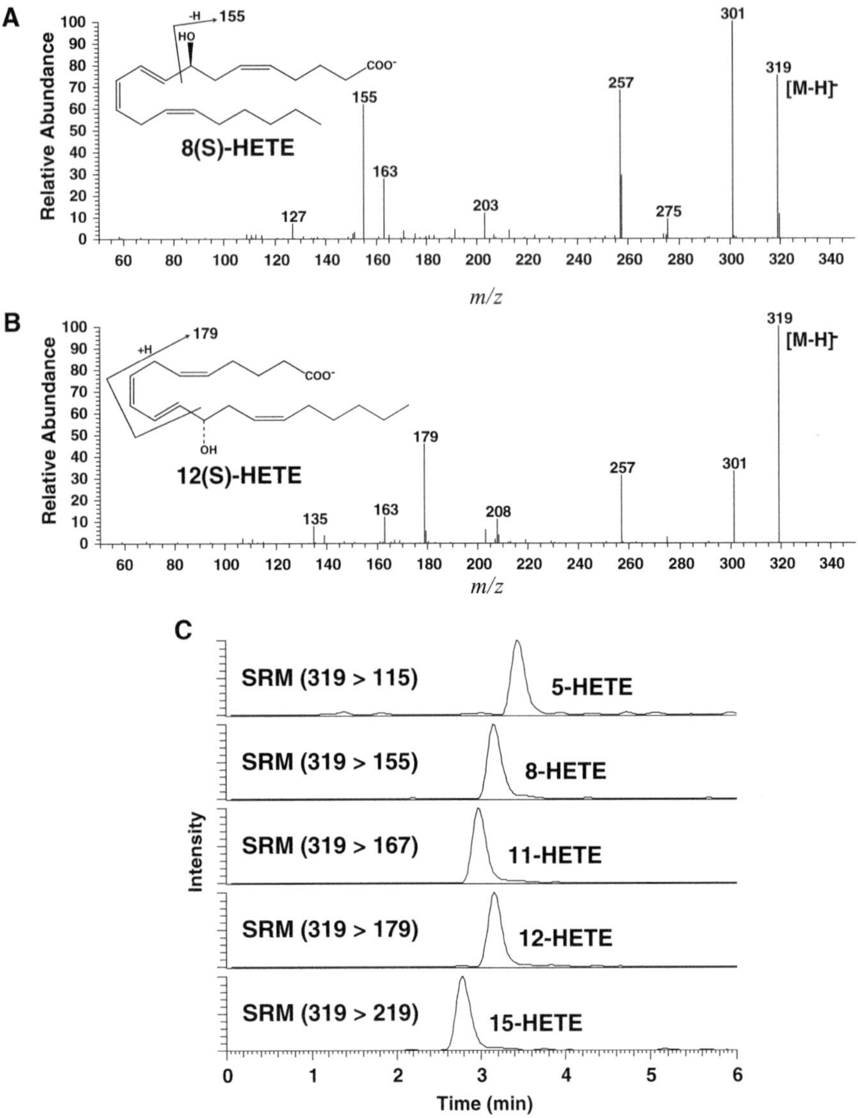

Fig. 24.3 Fragmentation and chromatography of HETE isomers. (**a**, **b**) MS/MS spectra for 8-HETE and 12-HETE. (**c**) Resolution of HETE isomers (Capcell Pak MGS3, 1 × 100 mm, 0.1 % formic acid:65 % acetonitrile:water, isocratic). 8-HETE and 12-HETE are not chromatographically resolved, but all HETE isomers are differentiated by isomer-specific selected reaction monitoring (SRM) transitions (Modified from Kita et al. [6])

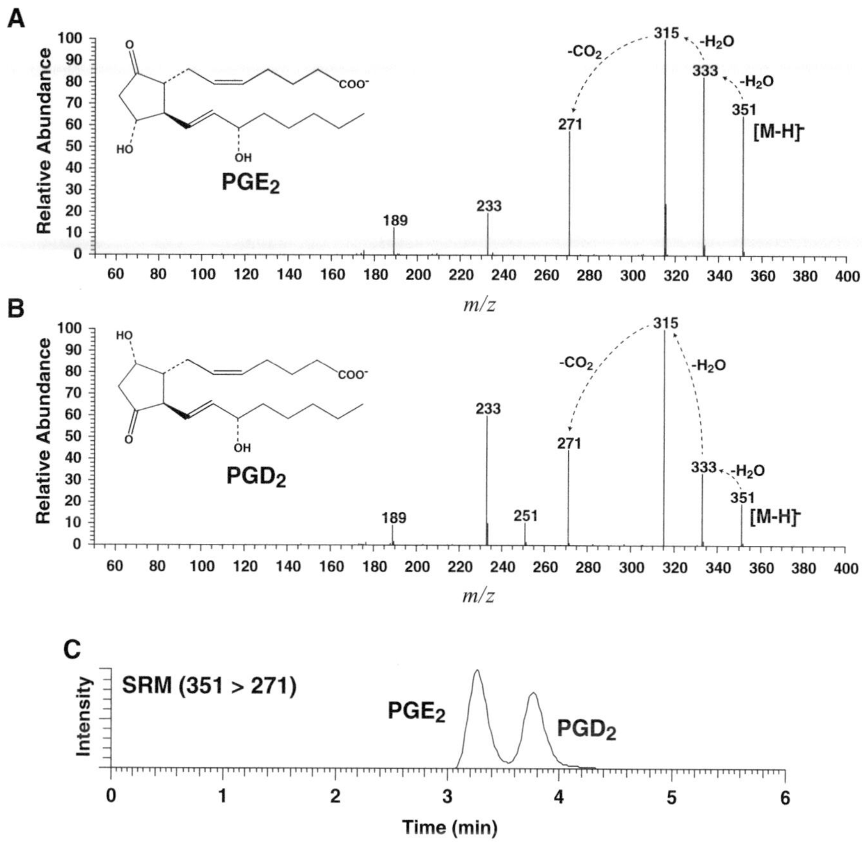

Fig. 24.4 Fragmentation and chromatography of PGE_2 and PGD_2. (**a**, **b**) MS/MS spectra for PGE_2 (**a**) and PGD_2 (**b**). Common fragment peak (*m/z* 271) is used for SRM. (**c**) Resolution of PGE_2 and PGD_2 (Capcell Pak MGS3, 1 × 100 mm, 0.1 % formic acid:37 % acetonitrile:water, isocratic) (Modified from Kita et al. [6])

The selected reaction monitoring (SRM; also referred to as multiple reaction monitoring, MRM) mode is used for sensitive detection and quantification by a triple quadrupole MS. In SRM mode, the first (Q1) and the third (Q3) quadrupoles do not scan, and are fixed to pass user-specified precursor and product ions, respectively. The precursor–product ion pair is referred to as transition, and MS instruments can be programmed to simultaneously monitor multiple SRM transitions. Table 24.1 lists SRM transitions for major eicosanoids and internal standards. Typical collision energies (CE) used in the TSQ 7000 (Thermo Fischer Scientific, Waltham, MA, USA) are also shown. Parameters for ion source including electrospray probe, heaters, gas flows, and ion guides, as well as collision cell gas density and CE, are major factors that affect sensitivity of SRM detection; on-site optimization using authentic compounds is recommended for best results.

Table 24.1 Selected reaction monitoring (SRM) transitions for major eicosanoids and internal standards

Compounds	Q1 (m/z)	Q3 (m/z)	Collision energy (eV)
6-Keto-$PGF_{1\alpha}$	369	245	32
TXB_2	369	195	18
$PGF_{2\alpha}$	353	193	25
PGE_2	351	271	18
PGD_2	351	271	18
LTB_4	335	195	20
LTC_4	624	272	26
LTD_4	495	177	20
5-HETE	319	115	20
8-HETE	319	155	18
11-HETE	319	167	20
12-HETE	319	179	16
15-HETE	319	219	16
6-Keto-$PGF_{1\alpha}$–d4	373	249	32
TxB_2-d4	373	199	18
$PGF_{2\alpha}$-d4	357	197	25
PGE_2-d4	355	275	18
PGD_2-d4	355	275	18
PGB_2-d4	337	239	22
LTB_4-d4	339	197	20
5-HETE-d8	327	116	20
12-HETE-d8	327	184	16
15-HETE-d8	327	226	16

Source: Modified from Kita et al. [6]

Deuterated eicosanoids are commercially available and can be used for quantification by internal standard method. For comprehensive analysis, a mixture of internal standards is added to the samples before eicosanoid extraction to correct recoveries throughout the sample preparation and LC-MS procedures. The amount of internal standards is determined depending on the system sensitivity and calibration range; generally, the signal of internal standards should be high enough to achieve good reproducibility in repeated measurements (%CV less than 5 % is recommended). For example, deuterated PGs can be spiked to the samples at amounts that give signals corresponding to approximately 100 pg on ordinary triple quadrupole MS instruments. Injection of an excess amount of deuterated compound that contains a small number of deuterium often disturbs the detection of target signals. For example, injection of PGE_2-d4 (*m/z* 355 > 275) can cause a small, but significant, PGE_2 signal (*m/z* 351 > 271), caused by the existence of PGE_2 as an impurity for PGE_2-d4 preparation, which becomes more significant when quadrupoles are operated at low-resolution settings where less deuterated compounds such as PGE_2-d1 or PGE_2-d2 also cause PGE_2 signals. Thus, the use of excess internal standards that mask the true signals in the samples should be avoided, although mathematical correction can be performed either manually or by software. For deuterated compounds containing more numbers of deuterium, such as HETE-d8, such interfering signals are rarely observed.

24.6 Column-Switching LC-MS Method for Eicosanoid Quantification

We have developed a column-switching LC-MS method that monitors major eicosanoids including PGs, LTs, HETEs, and platelet-activating factor (PAF) in a short analytical cycle time of less than 15 min [6]. The system has two sets of binary gradient pump systems (Fig. 24.5, pumps A–D) and two columns (trap column and separating column) connected to a six-port switching valve. Samples injected at the downstream of pump D are diluted on-line with water:0.1 % formic acid and then concentrated at the top of a short trap column. After 3 min of trapping period, the valve switches to introduce the concentrated sample to the separating column. Eicosanoids are separated in 10 min by a three-step gradient (37 %–45 %–65 % of mobile phase B) elution and detected by SRM (Fig. 24.6). The LC time program is shown in Table 24.2.

The benefit in adopting an online-dilution column-switching configuration is that SPE eluent can be efficiently introduced to the LC-MS system. The method allows an injection volume larger than 50 μl, which corresponds to more than 25 % of 200 μl SPE eluent. To perform similarly in ordinary LC-MS systems, SPE eluent should be off-line concentrated before analysis. Also, the column-switching method eliminates a peak deterioration problem for polar eicosanoids as already described because the sample solvent is replaced by an aqueous solution on the trap column.

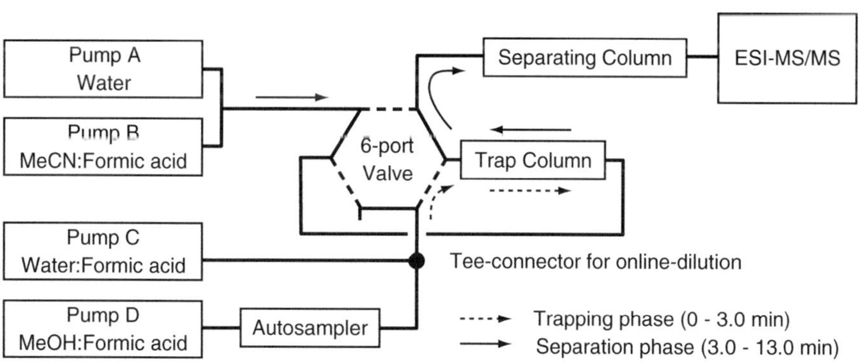

Fig. 24.5 Online-dilution column-switching LC-MS system. Sample introduced from autosampler is fivefold diluted and concentrated on a trapping column (Opti-guard mini, 1 × 10 mm; Optimize Technologies, Oregon City, OR, USA). After valve switches, sample is introduced to separating column (Capcell Pak MGS3, 1 × 100 mm) (Modified from Kita et al. [6])

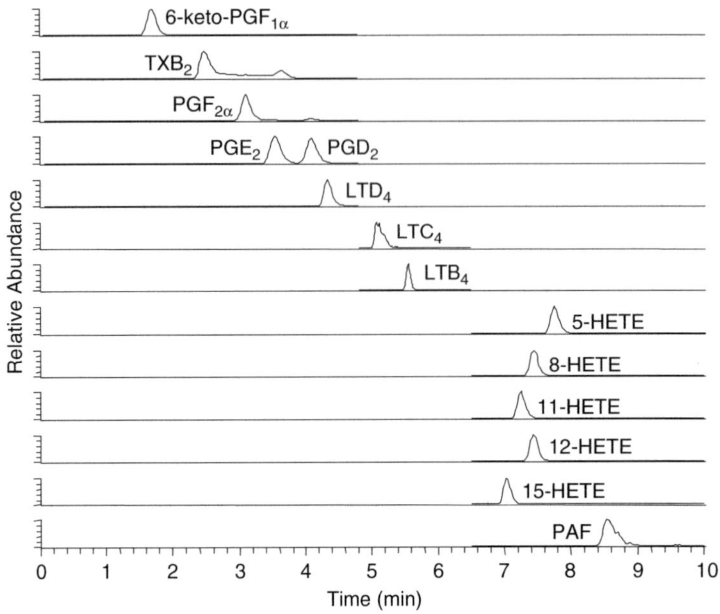

Fig. 24.6 Separation of eicosanoids in online-dilution column-switching method (Adopted from Kita et al. [6])

Table 24.2 Liquid chromatography (LC) time program

LC time (min)[a]	B % (A–B)	D % (C–D)	Valve position
0.00	65	20	Trap
0.01	37	|	|
3.00	|	|	Separation
3.10	|	90	|
6.00	37	|	|
6.01	45	|	|
7.00	45	|	|
7.01	65	|	|
12.00	|	|	Trap
12.50	|	90	|
12.51	|	20	|
13.00	END	END	

[a]Flow rate: pumps A–B, 120 μl/min; pumps C–D, 500 μl/min

24.7 Concluding Remarks

We described LC-MS-based quantitative methodologies for eicosanoids. In the future, as instruments improve, a more sensitive analysis using very small amounts of biological samples will be realized. In such cases, human skills for handling a trace amount of lipid samples become more critical, because optimal sampling and pretreatment procedures vary depending on the nature of samples. Also, assays will be more comprehensive, as ultrahigh performance LC (UHPLC) system and MS instruments capable of performing very fast SRM (>500 transitions/s) enable quantification of hundreds of targets in a short LC-MS run. An important concern here is the quality control of datasets. Although peak processing and quantitative calculation can be automatically done by software, researchers should review all the raw data. To minimize human errors in the data reviewing process, we strongly recommend using the software quality control functions when available. Currently available LC-MS systems have easy-to-use interfaces so that anyone can operate them, but the quality of data still depends on individual researchers. Improvement of software is desired for quality control of large-scale assays.

References

1. Samuelsson B (2012) Role of basic science in the development of new medicines: examples from the eicosanoid field. J Biol Chem 287:10070–10080
2. Shimizu T (2009) Lipid mediators in health and disease: enzymes and receptors as therapeutic targets for the regulation of immunity and inflammation. Annu Rev Pharmacol Toxicol 49:123–150
3. Serhan CN (2005) Mediator lipidomics. Prostaglandins Other Lipid Mediat 77:4–14
4. Rago B, Fu C (2013) Development of a high-throughput ultra performance liquid chromatography-mass spectrometry assay to profile 18 eicosanoids as exploratory biomarkers for atherosclerotic diseases. J Chromatogr B Anal Technol Biomed Life Sci 936:25–32
5. Bligh EG, Dyer WJ (1959) A rapid method of total lipid extraction and purification. Can J Biochem Physiol 37:911–917
6. Kita Y, Takahashi T, Uozumi N, Shimizu T (2005) A multiplex quantitation method for eicosanoids and platelet-activating factor using column-switching reversed-phase liquid chromatography-tandem mass spectrometry. Anal Biochem 342:134–143

Chapter 25
Mass Spectrometric Analysis of Phospholipids by Target Discovery Approach

Kazutaka Ikeda

Abstract Phospholipids (PLs), the most abundant lipids in the cell membrane, are composed of various types of polar head and fatty acid. Their abundance and distribution vary among organs, and some are oxidized by oxidant stress or inflammation. For this structural diversity, the optimal analytical method for profiling is needed.

Lipidomics is an approach well suited to obtain structural information about PL molecular species and to acquire quantitative profiles with a combination of liquid chromatography (LC) and mass spectrometry (MS). As for the conventional analysis method, a non-target, focusing, or target approach is usually selected, but these approaches have a disadvantage in sensitivity or completeness. To solve these conventional problems, this chapter introduce the target discovery approach, with which it is possible to obtain MS/MS data sensitively at high speeds and to combine a non-target or target approach. This approach has a strong potential as an improved lipidomics strategy to search new lipids of interest or to identify unknown lipid species in a nonbiased manner.

Keywords Phospholipid • Lipidomics • LC-MS • TOF-MS • Target discovery approach • Non-biased

25.1 Introduction

Phospholipids (PLs), the most abundant lipids in the cell membrane, are composed of a polar head and a nonpolar tail region. The nonpolar tail has two fatty acids of various lengths and saturation. The polar head region has a phosphate group such as choline (PC), ethanolamine (PE), inositol (PI), serine (PS), glycerol (PG), or phosphatidate (PA). The abundance and distribution of PLs vary among organs, and

K. Ikeda (✉)
Laboratory for Metabolomics, RIKEN Center for Integrative Medical Sciences, 1-7-22 Suehiro-cho, Tsurumi, Yokohama, Kanagawa 230-0045, Japan

Core Research for Evolutional Science and Technology (CREST), Japan Science and Technology Agency (JST), 7 Goban-cho, Chiyoda-ku, Tokyo 102-0076, Japan
e-mail: kazutaka.ikeda@riken.jp

T. Yokomizo, M. Murakami (eds.), *Bioactive Lipid Mediators*,
DOI 10.1007/978-4-431-55669-5_25

some are oxidized by oxidant stress or inflammation. For this structural diversity, the optimal analytical method for profiling is needed.

Lipidomics is an approach well suited to obtain structural information about PL molecular species and to acquire quantitative profiles with a combination of liquid chromatography (LC) and mass spectrometry (MS). Electrospray ionization (ESI) MS is considered to be a soft ionization method and is used widely in lipidomics. As for the conventional analysis method, a non-target, focusing, or target approach is usually selected [1]. High-resolution MS such as the time-of-flight (TOF) or fourier transform (FT) type is applied for the non-target approach in many cases, and PL molecular species are globally detected by MS scanning. Precursor ion scanning or neutral loss scanning of the polar head group or the fatty acid is performed in the focusing approach, with which it is possible to detect the same class of PL molecular species widely and sensitively [2, 3]. However, the non-target approach and focusing approach have a disadvantage in the detection of minor PLs. Multiple reaction monitoring (MRM) (also known as selected reaction monitoring, SRM) is applied for the target approach, and individual molecular species are detectable selectively with high sensitively [1, 4], whereas the target approach has a disadvantage in completeness of search range.

To solve these conventional problems, the target discovery approach, with which it is possible to obtain MS/MS data sensitively at high speeds and to combine with the non-target or target approach, is effective. The following protocols demonstrate two types of PL screening methods using the target discovery approach.

25.2 Analytical Conditions for Target Discovery Analysis

25.2.1 Materials

Acetonitrile (LC-MS grade/Sigma)
Methanol (LC-MS grade/Sigma)
Chloroform (HPLC grade/Sigma)
Ammonium formate (LC-MS grade/Sigma)
Milli-Q water (Millipore)
Internal standard (I.S.) mixture (Avanti Polar Lipids)
  14:0/14:0 PC, 14:0/14:0 PE, 14:0/14:0 PS, 16:0/16:0 PI, 14:0/14:0 PG, 14:0/14:0 PA
Bead pulverizing machine (Bms)
Glass tube (2 ml/FCR & Bio)
LC System: 1290 Infinity (Agilent)
LC column: ACQUITY UPLC HSS T3 (2.1 × 50 mm, 1.8 μm/Waters)
MS: Triple TOF 5600+ System (AB SCIEX)

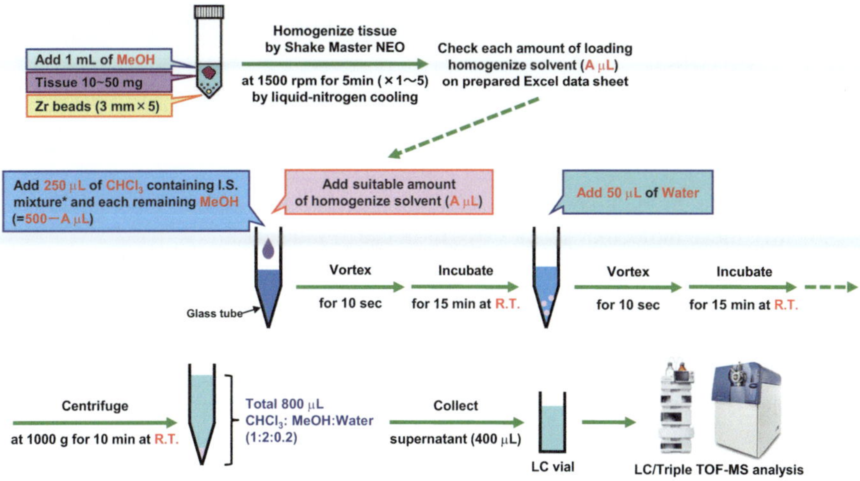

Fig. 25.1 Lipid extraction procedure

25.2.2 *Single-Phase Extraction of Muscle PL by the Modified Folch Method*

1. Pulverize and homogenize mouse skeletal muscle (10 mg) in cold methanol by a bead pulverizing machine (1500 rpm for 5 min) (Fig. 25.1)
2. Put the homogenized solvent in a 2-ml glass tube [5]
3. Add 250 μl chloroform containing internal standard (I.S.) mixture* and remain ing methanol (reaching total methanol volume, 500 μl)
4. Vortex for 10 s and incubate for 15 min at room temperature
5. Add 50 μl Milli-Q water, vortex for 10 s, and incubate for 15 min at room temperature
6. Centrifuge at 1000 *g* for 10 min at room temperature
7. Collect the supernatant in an LC-MS vial
8. Analyze the supernatant by LC-MS using the target discovery approach as follows

25.2.3 *Reversed-Phase LC Separation*

Refer to column conditions in Table 25.1

25.3 Procedure for PLs Analysis by Target Discovery Approach

25.3.1 Scanning Modes of Target Discovery Approach

1. Analyze PLs by data-dependent MS/MS scanning (Fig. 25.2; Table 25.2) and identify these molecular species by "in-house" lipid search
2. Analyze PLs by data "in-dependent" scanning (Fig. 25.3; Table 25.3)

Table 25.1 Column conditions for reversed-phase liquid chromatography (LC) separation

LC condition	Flow rate	0.3 ml/min
	Temp	45 °C
	Solvent	A → ACN:MeOH:water (=2:2:6) + 5 mM ammonium formate
		B → IPA + 5 mM ammonium formate
	Gradient (B %)	0 min → 0 %
		5 min → 40 %
		7.5 min → 64 %
		12 min → 64 %
		12.5 min → 82.5 %
		19 min → 85 %
		20 min → 95 %
		※Equilibration time: 5 min

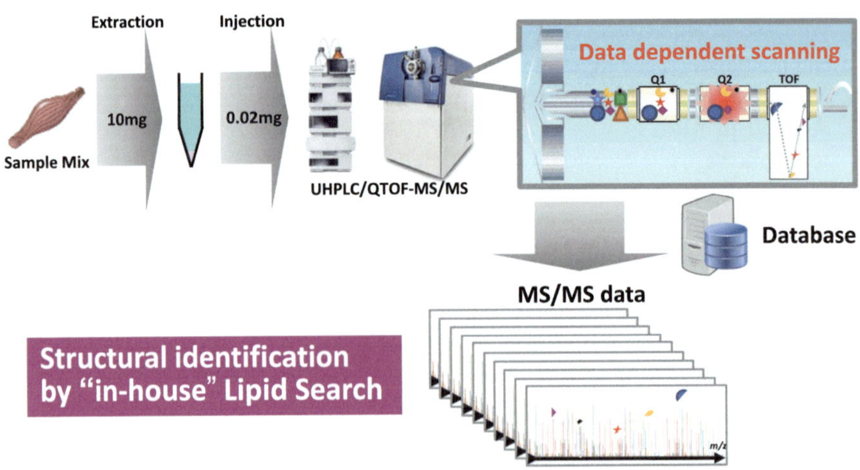

Fig. 25.2 Target discovery approach by data-dependent MS/MS scanning in combination with "in-house" lipid search

Table 25.2 Mass spectroscopy (MS) conditions for data-dependent scanning

MS condition	Duration			20 min
	Cycle			1.151 s
	Source/gas	Curtain gas		30 V
		Ion spray voltage		–4500 V
		Temperature		500 °C
		Ion source gas-1		50 V
		Ion source gas-2		50 V
	MS scan	Declustering potential (DP)		–80 V
		Collision energy (CE)		–10 V
	Data-dependent scan	Declustering potential (DP)		–80 V
		Collision energy (CE)		–45 V
		Collision energy spread (CES)		15 V
		Exclude former target ions	For	10 s
			After	2 occurrences

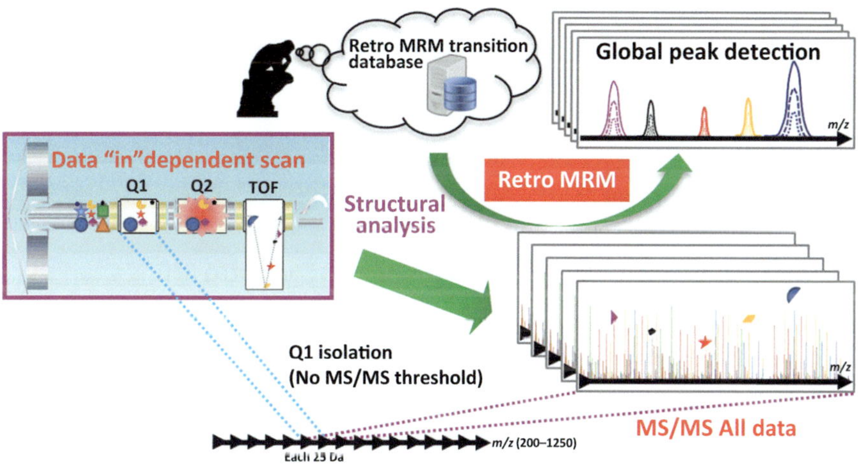

Fig. 25.3 Target discovery approach by data-"independent" scanning for MS/MS: all analysis

25.3.2 Screening of PL Molecular Species by Retrospective MRM (Retro-MRM)

Retro-MRM transitions to identify individual molecular species of PLs and lyso-phospholipids (lyso-PLs) were mainly composed of theoretical *m/z* data sets of predictive fragment ions that were derived from the polar head groups and fatty acids (Fig. 25.4). The MRM transition database was applied to the data from "in-dependent" scan of the lipid extraction of mouse muscle. As a result, a total of 134 molecular species, which consisted of various types of the class and fatty acids, were correctly identified even in just 0.02 mg mouse muscle (Table 25.4).

Table 25.3 MS conditions for data-"in"dependent scanning

MS condition	Duration		20 min
	Cycle		2.506 s
	Source/gas	Curtain gas	30 V
		Collision gas	−10 V
		Ion spray voltage	−4500 V
		Temperature	500 °C
		Ion source gas-1	50 V
		Ion source gas-2	50 V
	MS scan	Declustering potential (DP)	−80 V
		Collision energy (CE)	−10 V
	Data "in"dependent scan	Declustering potential (DP)	−80 V
		Collision energy (CE)	−14.4 to −80.4 V
		Collision energy spread (CES)	15 V
		Q1 step	25 amu

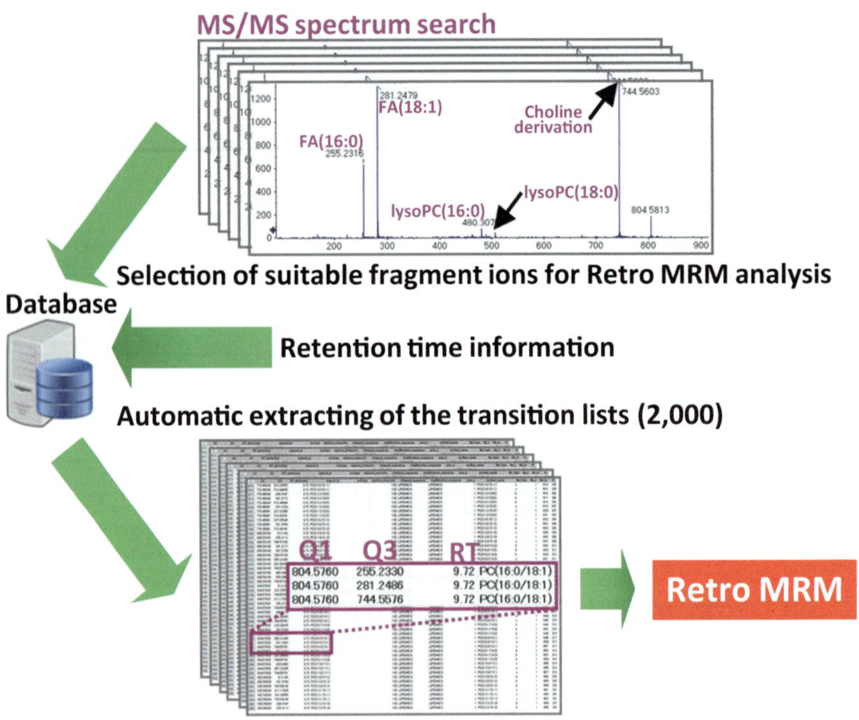

Fig. 25.4 Retrospective MRM analysis of PL molecular species from MS/MS: all data

Table 25.4 Result of identified phospholipids (PLs) and lyso-PLs by retro MRM analysis

Class	Identified molecular species	
PC	Acyl/acyl	44
lysoPC	Acyl	7
PE	Acyl/acyl	19
	Alkyl/acyl	21
lysoPE	Acyl	6
	Alkyl	3
PI	Acyl/acyl	12
lysoPI	Acyl	2
PS	Acyl/acyl	12
PG	Acyl/acyl	8

25.4 Commentary

25.4.1 MS/MS Analysis by Data-Dependent Scan

In the data-dependent scan, precursor ions of PL and lyso-PL molecular species are globally scanned by high-resolution TOF MS mode, and if some of them are over the setup threshold, they are secondarily scanned by the sensitive TOF MS/MS mode, and it is possible then to acquire the structural information (Fig. 25.2). These MS/MS datasets are applicable to "in-house" lipid search, and these molecular species are identified globally and accurately (manuscript in preparation). The lipid search is a search engine for the identification of lipid molecular species from MS raw data that was originally developed by Dr. Taguchi [6] collaborating with Mitsui Knowledge Industry (MKI).

Our "in-house" lipid search made some improvements in the accuracy of identification by new algorithms as compared with conventional searching techniques. For instance, "knockout selection," which sets the order of priority of MS/MS peaks for high-precision identification of the individual molecular species by reference to many raw MS/MS data patterns, was adopted in our group and with MKI.

25.4.2 MS/MS Analysis by "In"-Dependent Scan

For data "in-dependent" scan, all precursor ions of PL and lyso-PL molecular species are scanned without threshold by sequential window acquisition of all theoretical fragment ion spectra (SWATH) mode [7]. With this method, it is possible to obtain all the MS/MS data with high sensitively. In the SWATH mode, the Q1 quadrupole is typically stepped at 12.5–25 amu increments across the 200–1250 *m/z* precursor range (Fig. 25.3). For data mining, multiple reaction monitoring (MRM)-like analysis, our termed retro-MRM, is used to accurately identify the molecular species. Suitable candidates for fragment ions to identify individual molecular species of PLs and lyso-PLs are automatically selected from our MS/MS database, and the combinations of retro-MRM transitions were assembled from theoretical *m/z* data sets of the precursor ion and predictive fragment ions. Individual molecular species were automatically recognized by the references to the number of MRM overlapping chromatograms, retention time accuracy, and the intensity order of each MRM chromatogram. Structural isomers or isotopes, which frequently cause annotation errors in the conventional data processing tools because of chromatogram complexity, are distinguishable by these chromatogram references. This approach has strong potential as an improved lipidomics strategy to search new lipids of interest or to identify unknown lipid species retroactively in a non-biased manner.

References

1. Taguchi R, Nishijima M, Shimizu T (2007) Basic analytical systems for lipidomics by mass spectrometry in Japan. Methods Enzymol 432:185–211
2. Ekroos K, Chernushevich IV, Simons K, Shevchenko A (2002) Quantitative profiling of phospholipids by multiple precursor ion scanning on a hybrid quadrupole time-of-flight mass spectrometer. Anal Chem 74:941–949
3. Ikeda K, Kubo A, Akahoshi N, Yamada H, Miura N, Hishiki T, Nagahata Y, Matsuura T, Suematsu M, Taguchi R, Ishii I (2011) Triacylglycerol/phospholipid molecular species profiling of fatty livers and regenerated non-fatty livers in cystathionine beta-synthase-deficient mice, an animal model for homocysteinemia/homocystinuria. Anal Bioanal Chem 400:1853–1863
4. Nakanishi H, Iida Y, Shimizu T, Taguchi R (2009) Analysis of oxidized phosphatidylcholines as markers for oxidative stress, using multiple reaction monitoring with theoretically expanded data sets with reversed-phase liquid chromatography/tandem mass spectrometry. J Chromatogr B 877:1366–1374
5. Folch J, Lees M, Sloane Stanley GH (1957) A simple method for the isolation and purification of total lipids from animal tissues. J Biol Chem 226:497–509
6. Taguchi R, Ishikawa M (2010) Precise and global identification of phospholipid molecular species by an Orbitrap mass spectrometer and automated search engine Lipid Search. J Chromatogr A 1217:4229–4239
7. Gillet LC, Navarro P, Tate S, Röst H, Selevsek N, Reiter L, Bonner R, Aebersold R (2012) Targeted data extraction of the MS/MS spectra generated by data-independent acquisition: a new concept for consistent and accurate proteome analysis. Mol Cell Proteomics 11:O111.016717

Chapter 26
Determination of Sphingolipids by LC-MS/MS

Tomohiro Takahashi, Daisuke Saigusa, Chihiro Takeda, Kohei Saito, Naoto Suzuki, Hiroki Tsukamoto, and Yoshihisa Tomioka

Abstract To investigate the biological functions and roles of sphingolipids, sensitive and compound-specific methods are required to measure their levels in biological samples. Liquid chromatography–mass spectrometry (LC-MS) using electrospray ionization (ESI) is suitable for the reliable simultaneous analysis of multiple compounds. In addition, the selected reaction monitoring (SRM) mode of tandem mass spectrometry (MS/MS) is effective to quantify with high sensitivity and selectivity. Therefore, LC-MS/MS came to be utilized for simultaneous analysis of the sphingolipids in vivo. Useful methods for the sphingolipids and related features are also summarized. The following protocol demonstrates information on determination of sphingolipids, especially sphingosine and sphingosine-1-phosphate, by LC-ESI-MS/MS in biological samples such as cell lysates, plasma, serum, or urine.

Keywords Sphingolipids • Sphingosine • Sphingosine-1-phosphate • Dihydrosphingosine • Phytosphingosine • Dihydrosphingosine-1-phosphate • Phytosphingosine-1-phosphate • LC-MS/MS

T. Takahashi • C. Takeda • K. Saito • N. Suzuki • H. Tsukamoto
Graduate School of Pharmaceutical Sciences, Tohoku University,
6-3, Aoba, Aramaki, Aoba-ku, Sendai, Miyagi 980-8578, Japan

D. Saigusa
Department of Integrative Genomics, Tohoku Medical Megabank, Tohoku University,
2-1, Seiryo-machi, Aoba-ku, Sendai 980-8573, Japan

Graduate School of Medicine, Tohoku University,
2-1, Seiryo-machi, Aoba-ku, Sendai, Miyagi 980-8573, Japan

Y. Tomioka (✉)
Graduate School of Pharmaceutical Sciences, Tohoku University,
6-3, Aoba, Aramaki, Aoba-ku, Sendai, Miyagi 980-8578, Japan

Graduate School of Medicine, Tohoku University,
2-1, Seiryo-machi, Aoba-ku, Sendai, Miyagi 980-8573, Japan
e-mail: ytomioka@m.tohoku.ac.jp

T. Yokomizo, M. Murakami (eds.), *Bioactive Lipid Mediators*,
DOI 10.1007/978-4-431-55669-5_26

26.1 Reagents

Sphingolipids (Avanti Polar Lipids)
C18 sphingosine (Sph)
C18 dihydrosphingosine (dhSph)
C18 phytosphingosine (pSph)
C18 sphingosine-1-phosphate (S1P)
C18 dihydrosphingosine-1-phosphate (dhS1P)
C18 phytosphingosine-1-phosphate (pS1P)
C17 sphingosine (C17-Sph)
C17 sphingosine-1-phosphate (C17-S1P)
Methanol for LC/MS
Chloroform for HPLC
Acetonitrile for LC/MS
Formic acid for LC/MS
Ammonium formate 1 mol/l solution for HPLC
Polytyrosine-1,3,6 calibration solution (Thermo Fisher Scientific)
Water (ultrapure grade)

26.2 Equipments

High performance liquid chromatography system
ESI-triple quadrupole mass spectrometer with analytical software
Polyether ether ketone (PEEK) tubing
HPLC column (ODS):
Example: CAPCELL PAK C18 ACR (3 μm, 1.0 mm i.d. ×150 mm, Shiseido) for analytical column and CAPCELL PAK ACR (3 μm, 1.0 mm i.d. ×35 mm, Shiseido) for trap column
Syringe (500 μl, Hamilton Company)
Centrifuge
Siliconized sample tube (2.0 ml, WATOSON)
YMC Duo Filter (0.2 μm, pore size, 4 mm inner diameter, YMC)
Micro tube mixer
Ultrasonic bath
Ultrasonic homogenizer

26.3 Prepare Stock Solutions and Working Solutions of Sphingolipids

Siliconized sample tubes and pipette tips are used. For quantitative analysis, stable isotope analogue internal standards are added to biological samples. The ratio between the internal standard and the compound of interest can be measured by LC/MS.

Table 26.1 Preparation of seven working solutions

Solution	Concentration/ nmol l^{-1}	Dilution of solutions
1	1000	100 µl of solution as diluted to 1 ml
2	200	200 µl of 1 diluted to 1 ml
3	20.0	100 µl of 2 diluted to 1 ml
4	10.0	500 µl of 3 diluted to 1 ml
5	2.0	200 µl of 4 diluted to 1 ml
6	1.0	500 µl of 5 diluted to 1 ml
7	0.2	200 µl of 6 diluted to 1 ml

However, many compounds in which acyl groups differ exist in sphingolipids. The separation and selective detection by MS will be difficult if the stable isotope-labeled compound overlaps the same *m/z* as the target compounds. Therefore, C17 (C17-Sph and C17-S1P) are used for sphingolipid analysis as the internal standard.

1. Dissolve each powder of S1P, pS1P, C17-S1P, Sph, dhSph, and pSph in an appropriate volume of solution to prepare a stock solution; that is, Sph, C17-Sph, dhSph, and pSph are dissolved in methanol (1 mg/ml); S1P, C17-S1P, and pS1P are dissolved in methanol (0.04 mg/ml); dhS1P is dissolved in chloroform:methanol:H_2O = 1.5:1.4:0.2 (0.04 mg/ml). Store the solutions at −80°C.
2. Dilute the stock solutions with methanol to a final concentration of 10.0 µmol/l (solution AS1P, pS1P, dhS1P, Sph, dhSph, pSph), and then use it (them) for the optimization of a mass spectrometer.
3. Prepare seven working solutions (1000, 200, 20.0, 10.0, 2.0, 1.0, and 0.2 nmol/l) using each stock solution (A_{S1P}, pS1P, dhS1P, Sph, dhSph, pSph) as described in Table 26.1. Store at −80 °C.
4. Prepare IS mixture 1 (each 1.0 µmol/l) and IS mixture 2 (each 200 nmol/l) of the internal standard containing C17-S1P and C17-Sph. Store at −80 °C.
5. For each calibration curve, mix equal amounts of working solutions and IS mixture 2 (final concentration of sphingolipids are 500, 100, 10, 5.0, 1.0, 0.5, 0.1 nmol/l, and C17-S1P and C17-Sph are 100 nmol/l) to determine the concentration–response relationship.
6. Prepare quality control (QC) samples by spiking blank plasma or biological samples with stock solution. QC samples should be prepared at four levels: high-level QC (HQC), middle-level QC (MQC), low-level QC (LQC), and lower low-level QC (LLQC).
7. Prepare a deproteinization solution, 0.1 % formic acid in methanol.

26.4 Protocol for Sample Preparation from Plasma

Before analysis, all frozen samples (−80 °C) should be thawed and allow to equilibrate at 4 °C (or on ice). Do not leave the blood samples on ice before preparing plasma because red blood cells are a source of the plasma S1P. Thus, it is necessary

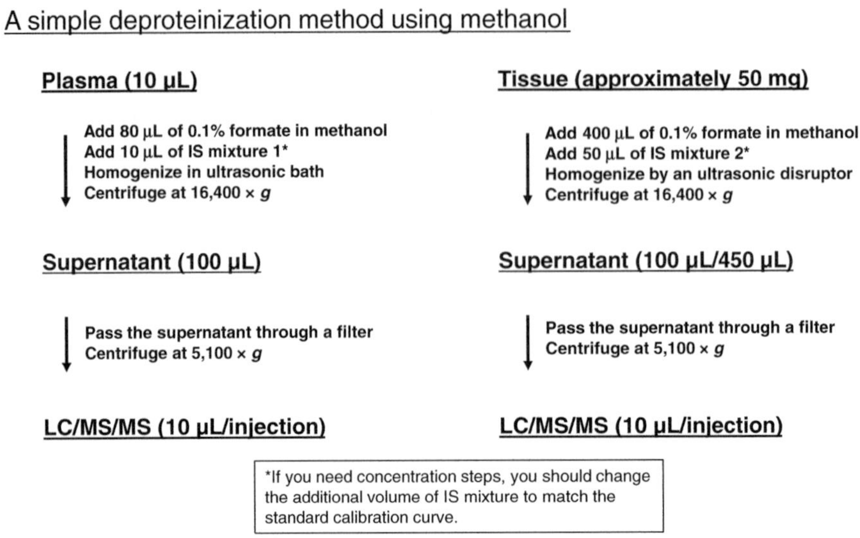

Fig. 26.1 Deproteinization method for sample preparation

to be prepared for plasma fraction from whole blood as soon as possible. Sample preparation for MS is recommended to utilize deproteinization of an organic solvent such as methanol (Fig. 26.1) as follows:

1. Transfer 10 µl plasma into a siliconized sample tube.
2. Add 80 µl deproteinization solution (0.1 % formic acid in methanol).
3. Add 10 µl IS mixture 1 (final concentration of IS is 100 nmol/l).
   Tthe additional volume of IS mixture should be changed to match the standard calibration curve if you need concentration steps for sample preparation.
4. Close the tube and vortex for 30 s.
5. Then, homogenize for 5 min in an ultrasonic bath on a tube floater.
6. Centrifuge at 16,400 *g* for 10 min at 4 °C and separate supernatant.
7. The supernatant (approximately 100 µl) should be passed through a filter (0.2 µm pore size, 4 mm inner diameter).
8. Inject 10 µl filtered solution into the chromatographic system using an automatic injector. That is, each internal standard is 1 pmol/on column.

26.5 Protocol for Sample Preparation from Tissue

The sample preparation for tissue is also shown in Fig. 26.1 as follows:

1. Tissue (approximately 50 mg) is placed into a siliconized sample tube (2.0 ml).
2. Add 400 µl deproteinization solution.
3. Add 50 µl IS mixture 1 (final concentration of IS is 100 nmol/l).
   The additional volume of IS mixture should be changed to match the standard calibration curve if you need concentration steps for sample preparation.

4. Close the tube and vortex for 30 s.
5. Then, homogenize for 30 s by an ultrasonic cell disruptor.
6. Centrifuge at 16,400 *g* for 10 min at 4 °C.
7. Transfer 100 μl supernatant to another siliconized tube (2.0 ml).
8. Pass the supernatant through a filter (0.2 μm pore size, 4 mm inner diameter, YMC).
9. Inject 10 μl filtered solution into the chromatographic system using an automatic injector.

26.6 Optimization of H-ESI-MS/MS by Infusion of Sphingolipids

A triple quadrupole mass spectrometer equipped with an ESI source is used for sphingolipids analysis such as TSQ Quantum Ultra with heated electrospray ionization (H-ESI) probe (Thermo Fisher Scientific). Because carryover of samples based on interactions with metal components and the phosphate group of lipids is a concern, PEEK tubing and connector are useful and are recommended. To optimize the sensitivity of sphingolipids, several parameters such as spray voltage, sheath gas pressure, auxiliary gas flow rate, capillary temperature, tube lens offset voltage, and vaporizer temperature need to be optimized.

1. Infuse a polytyrosine-1,3,6 calibration solution directly into the ESI source with a syringe pump to tune and calibrate.
2. Save the tune method and calibration files.
3. Fill a clean syringe with 10 μg/ml sphingolipid solution and place in the syringe holder of the syringe pump of MS.
4. Set up the MS detector to tune in H-ESI/MS mode and start infusion. The MS/MS transitions will be detected in the full scan mode (*m/z* 50–500). Optimal conditions of tube lens offset and collision energy for SRM analysis employ the transition of the [M+H]+ precursor ions to their product ions of S1P, dhS1P, pS1P, Sph, dhSph, and pSph. The parameters of ionization, especially vaporizer temperature, sheath gas pressure, and auxiliary gas pressure, should be retuned after LC condition is fixed because they depend on the flow rate of LC. Example of defined parameters of ionization and MS/MS transitions are described in Table 26.2.

26.7 Separation Condition of Liquid Chromatography (LC)-MS/MS

The basic LC instrumentation for MS consists of degassing equipment, pumps, an automatic injector, columns, and a column oven with PEEK tubing and connector. To check carryover, make a blank injection during the course of multiple runs as

Table 26.2 Parameters of ionization and MS/MS transition for SRM

Parameters of ionization		
MS/MS system	TSQ quantum ultra (Thermo Fisher Scientific)	
Ionization	ESI (+)	
Spray voltage	3.5 kV	
Vaporizer temperature	50 °C	
Sheath gas pressure	10 psi	
Ion sweep gas pressure	0 psi	
Auxiliary gas pressure	0 psi	
Capillary temperature	270 °C	
Collision gas pressure	1.5 mTorr	
Parameters of MS/MS transition for SRM		
Tube lens offset/collision energy	S1P	−95 V/18 eV (*m/z* 380.3 > 264.2)
	dhS1P	−105 V/13 eV (*m/z* 382.3 > 284.1)
	pS1P	−90 eV/14 eV (*m/z* 398.3 > 300.2)
	C17-S1P	−85 V/16 eV (*m/z* 366.2 > 250.1)
	Sph	−80 V/11 eV (*m/z* 300.3 > 282.2)
	dhSph	−100 V/13 eV (*m/z* 302.3 > 284.2)
	pSph	−80 V/13 eV (*m/z* 318.3 > 300.2)
	C17-Sph	−75 V/11 eV (*m/z* 286.3 > 268.2)

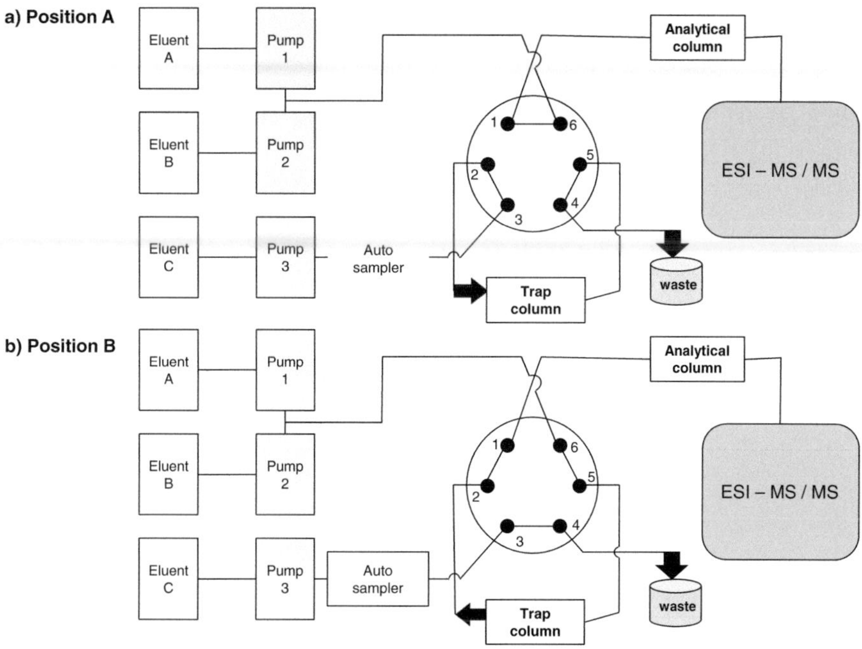

Fig. 26.2 **(a)** The sample is injected by autosampler and located to atrap column by eluent C at position A. **(b)** The traped compounds to a trap column are eluted and loaded to the analytical column at position B

needed. For low carryover, washing and cleaning of the needle attached to an automatic injector may be effective with more than one washing/cleaning solvent. An online solid-phase extraction may be valid for ultramicroanalysis. Our LC system for sphingolipids analysis is described in Fig. 26.2. The sample is injected by an autosampler and located to a trap column by eluent C for 7 min at position A (1–6). After changing from position A to B of a six-port valve, trapped compounds are eluted from the trap column and loaded to the analytical column by eluent A and B for 10 min at position B (1–2).

Isocratic elution or gradient elution may be chosen. The column selection (phase material, inner diameter, length, etc.) and optimization of the mobile phase are very important for stable analysis. On sphingolipid analysis using reverse-phase chromatography, pH 4.0 of the mobile phase is especially important (Table 26.3).

Generally, an ODS column is selected for sphingolipids analysis. However, a peak-tailing problem may be detected because of an interaction between a phosphate group of sphingolipids and the residual silanol group (imperfect hydration state) or a metal ion inside a ODS column. Therefore, column that has as reduced a residual silanol group as possible to improve the peak-tailing problem should be selected for sphingolipids analysis. To conduct good analysis, the improvement of peak shape should be aimed at evaluating a number of theoretical stages and a symmetry factor carefully, and this leads to the rise of sensitivity.

Table 26.3 HPLC system and run condition

HPLC system	Nanospace SI-2 system (Shiseido)
Analytical column	Capcell Pak C18 ACR (3 μm, 1.0 mm i.d. × 150 mm, Shiseido)
Trap column	Capcell Pak C18 ACR (3 μm, 1.0 mm i.d. × 35 mm, Shiseido)
Oven temperature	40 °C
Mobile phase	A: 5 mM $HCOONH_4$ in H_2O, pH 4.0 (HCOOH)
	B: 5 mM $HCOONH_4$ in $H_2O/CH_3CN = 5/95$, pH 4.0 (HCOOH)
	C: A/B = 70/30
	Gradient: A/B = 70/30, 0 min; A/B = 70/30, 9 min; A/B = 0/100, 14 min; A/B = 0/100, 17 min; A/B = 70/30, 17 min; A/B = 70/30, 22 min
Divert valve	Waste. 0–13.5 min; load, 13.5–15.7 min; waste, 15.7–22 min
Flow rate	A/B, 60 μl/min; C, 60 μl/min
Run time	22 min
Switching valve	Position A (1–6), 0–7 min; position B (1–2), 7–17 min; position A (1–6), 17–22 min
Injection volume	10 μl

26.8 Preparation of the Mobile Phase

1. Mobile phase A [5 mM $HCOONH_4$ in H_2O, pH 4.0 (HCOOH)] is prepared by mixing 99.5 ml water and 0.5 ml 1 mol/l ammonium formate solution, and then adjusting to pH 4.0 with formic acid (~9.6 μl).
2. Mobile phase B [5 mM $HCOONH_4$ in $H_2O/CH_3CN = 5/95$, pH 4.0 (HCOOH)] is prepared by mixing 95 ml acetonitrile, 4.5 ml water, and 0.5 ml 1 mol/l ammonium formate solution, and then adjusting to apparent pH 4.0 with formic acid (~1160 μl).

26.9 Condition of Ionization by LC-MS/MS

The parameters of ionization using LC change depending on the flow rate, especially vaporizer temperature, sheath gas pressure, and auxiliary gas pressure. Our optimal parameters for Nanospace SI 2-Quantum Ultra system are described in Table 26.4. SRM chromatograms are shown in Fig. 26.3.

26.10 Preparation of Calibration Curve and Determination of Sphingolipids by LC-MS/MS

1. All peaks were integrated automatically by Xcalibur software (Thermo Fisher Scientific).

Table 26.4 Parameters of ionization

MS/MS system	TSQ quantum ultra (Thermo Fisher Scientific)
Ionization	ESI (+)
Spray voltage	3.5 kV
Vaporizer temperature	50 °C
Sheath gas pressure	10 psi
Ion sweep gas pressure	0 psi
Auxiliary gas pressure	0 psi
Capillary temperature	300 °C
Collision gas pressure	1.5 mTorr

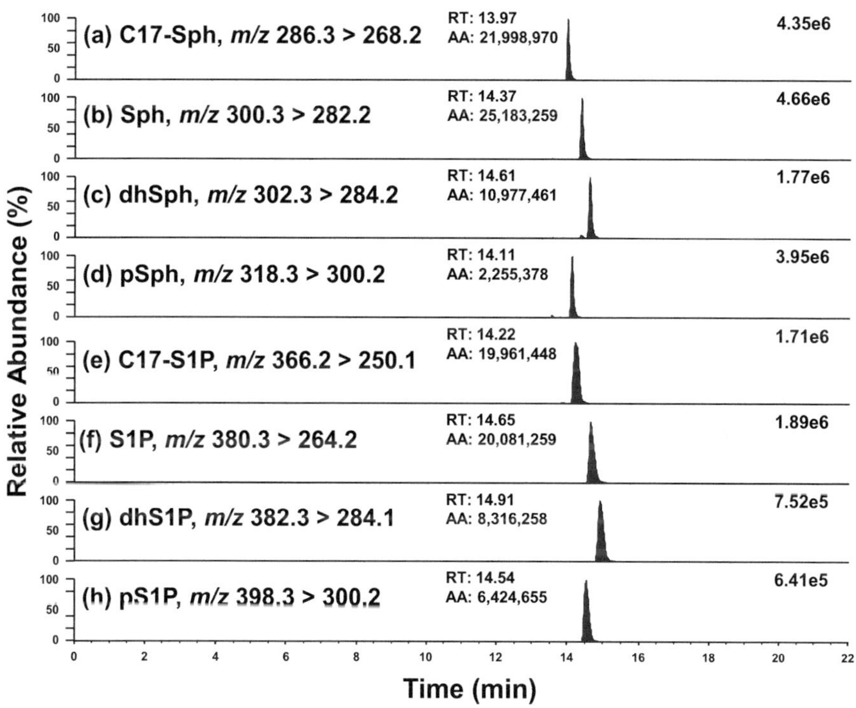

Fig. 26.3 Selected reaction monitoring (SRM) chromatograms of sphingolipids by LC-MS/MS

2. The unknown sphingolipid amount is estimated from the calibration curves by the ratios of their peak areas to that of IS, which regression coefficients are given with appropriate software. For example, a calibration curve fitting a straight line with nonlinear regression for S1P and Sph is shown in Fig. 26.4a, b, which was drawn using GraphPad Prism (GraphPad Software, Inc.) from the experimental data with 1/x weighting method, i.e., $y = mx + c$, where y is area ratio, x is analyte level, m is the gradient of the line, and c is its intercept with the y-axis.
3. Estimate the concentration of unknown sample using the line equation.

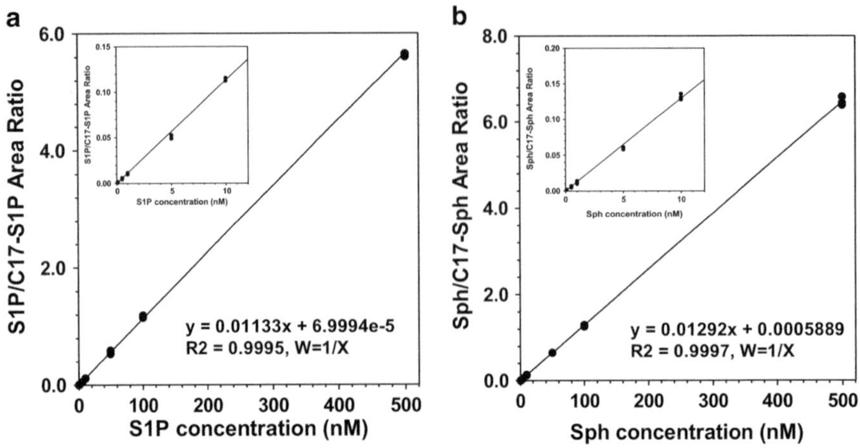

Fig. 26.4 **(a)** Caliburation curve for S1P. **(b)** Caliburation curve for Sph

For the analysis of sphingolipids by MS in various biological samples, ionization of matrix effects are taken into account. A matrix effect is the phenomenon that increases or decreases in the sensitivity of the target compound in a tested sample. To perform an exact determination quantity, a matrix effect has to be reduced as much as possible. Therefore, the validation test needs to be carried out using a practical biological sample, that is, plasma, liver, or brain, for developing the determination method, and evaluated accuracy and precision of reproducibility from intraday and interday assay. The analytical method validation is referenced from generally published guidance, i.e., Bioanalytical Method Validation; Food and Drug Administration.

This section describes a protocol for analysis of sphingolipids by LC-MS/MS in a biological sample. However, the present method was specific for 1-acyl sphingolipids, and there are other sphingo base compounds and many species of diacyl sphingolipids of glycosphingolipids, sphingomyerine, ceramide, and their phosphates. We recommend replacing the analytical column with a C8-based column, or using ultrahigh performance liquid chromatography, because the simultaneous analysis for these compounds takes 1–2 h using an ODS (C18-based) column [1]. The useful methods for sphingolipids and related features are summarized in Table 26.5. Although the performance of sphingolipids profiling depends on the improvement of equipment, this basic protocol is utilized for analysis in future studies.

Table 26.5 Methods of sphingolipids and related features

Sphingolipids	Method (column)	Sample preparation[a]	References
Cer	Q-TOF MS/MS	LLE	[2]
Cer	ESI-MS/MS (direct-injection)	LLE	[3, 4]
Cer	GC/EI-MS (OV-1)	LLE, derivatization, TLC	[5]
Cer	LC/ESI-MS/MS (XPER-CHROM 100 C_8)	LLE, SPE (silica gel)	[1]
Cer	LC/ESI-MS/MS (RP_8)	LLE	[6]
Cer	LC/APCI-MS (ultrasphere C_{18})	LLE, derivatization	[7]
Cer, S1P, dhS1P, Sph, dhSph	LC/ESI-MS/MS (Spectra C8SR)	LLE	[8, 9]
Cer, S1P, Sph	LC/ESI-MS/MS (LC-NH_2 or Discovery C18)	Hydrolysis, LLE	[10]
Cer, Sph, S1P. pSph	LC/ESI-MS/MS (HILIC silica)	LLE	[11, 12]
GSL	SFC/CI-MS (SB-cyanopropyl-50 and SB-phenyl-5)	Derivatization	[13]
GSL	FAB-MS	LLE	[14]
Sph	ESI-MS/MS	LLE (dried blood spots)	[15]
Sph, pSph	LC/ESI-MS/MS (Symmetry C_{18})	Deproteinization	[16]
Sph, dhSph	LC/ESI-MS/MS (LiChrospher Si60)	LLE	[17]
Sph, dhSph	LC/ESI-MS/MS (LiChrospher Si60)	LLE	[18]
Sph, S1P	FAB-MS	LLE, SPE (C18), derivatization	[19]
Sph, S1P	LC/ESI-MS/MS (Develosil ODS HG-5)	LLE	[20]
Sph, S1P	LC/ESI-MS/MS (Hypersil-Keystone Beta Basic CYANO)	LLE	[21]
Sph, S1P	LC/ESI-MS/MS (Luna-RP)	Deproteinization	[22]
S1P	LC/ESI-MS/MS (Phenomenex Luna C5 or Phenomenex Luna C18-HC)	LLE	[23]
Sph, S1P, dhS1P	LC/ESI-MS/MS (Luna-RP)	Deproteinization	[24]
S1P, dhS1P	LC/ESI-MS/MS (Zorbax Eclipse XDB-C8)	LLE, acetylation	[25]
S1P, dhS1P	LC/ESI-MS/MS (Discovery C18)	LLE	[26]
S1P, dhS1P	ESI-MS/MS (direct infusion)	LLE	[27]

(continued)

Table 26.5 (continued)

Sphingolipids	Method (column)	Sample preparation[a]	References
S1P, dhS1P	LC/ESI-MS/MS (Acquity UPLC BEH Hilic)	LLE	[28]
pSph	LC/MS/MS (Hypersil RP-C_{18})	LLE	[29]
S1P, pS1P, dhS1P	LC/ESI-MS/MS (Luna-RP)	LLE, IMAC	[30]
Sph, S1P, dhSph, dhS1P, pSph, pS1P	LC/ESI-MS/MS (Capcell Pak ACR)	Deproteinization	[31]
Metabolomics	LC/Q-TOF (HSS T3)	Dilution	[32]
Lipidomics	LC/ESI-MS/MS (LC-NH_2 or Discovery C18)	Hydrolysis, LLE	[33]
Lipidomics	Nano-ESI-MS (direct infusion)	LLE, derivatization	[34]

[a]*LLE* liquid–liquid extraction, *TLC* thin-layer chromatography, *SPE* solid-phase extraction, *IMAC* immobilized metal ion-adsorption chromatography

References

1. Kasumov T, Huang H, Chung YM, Zhang R, McCullough AJ, Kirwan JP (2010) Quantification of ceramide species in biological samples by liquid chromatography electrospray ionization tandem mass spectrometry. Anal Biochem 401:154
2. Wewer V, Brands M, Dormann P (2014) Fatty acid synthesis and lipid metabolism in the obligate biotrophic fungus *Rhizophagus irregularis* during mycorrhization of *Lotus japonicus*. Plant J 79:398
3. Liebisch G, Drobnik W, Reil M, Trumbach B, Arnecke R, Olgemoller B, Roscher A, Schmitz G (1999) Quantitative measurement of different ceramide species from crude cellular extracts by electrospray ionization tandem mass spectrometry (ESI-MS/MS). J Lipid Res 40:1539
4. Han X (2002) Characterization and direct quantitation of ceramide molecular species from lipid extracts of biological samples by electrospray ionization tandem mass spectrometry. Anal Biochem 302:199
5. Gaskell SJ, Edmonds CG, Brooks CJ (1976) Applications of boronate derivatives in the study of ceramides by gas-liquid chromatography-mass spectrometry. J Chromatogr 126:591
6. Camera E, Picardo M, Presutti C, Catarcini P, Fanali S (2004) Separation and characterisation of sphingoceramides by high-performance liquid chromatography-electrospray ionisation mass spectrometry. J Sep Sci 27:971
7. Couch LH, Churchwell MI, Doerge DR, Tolleson WH, Howard PC (1997) Identification of ceramides in human cells using liquid chromatography with detection by atmospheric pressure chemical ionization-mass spectrometry. Rapid Commun Mass Spectrom 11:504
8. Hammad SM, Pierce JS, Soodavar F, Smith KJ, Al Gadban MM, Rembiesa B, Klein RL, Hannun YA, Bielawski J, Bielawska A (2010) Blood sphingolipidomics in healthy humans: impact of sample collection methodology. J Lipid Res 51:3074
9. Bielawski J, Pierce JS, Snider J, Rembiesa B, Szulc ZM, Bielawska A (2009) Comprehensive quantitative analysis of bioactive sphingolipids by high-performance liquid chromatography-tandem mass spectrometry. Methods Mol Biol 579:443
10. Shaner RL, Allegood JC, Park H, Wang E, Kelly S, Haynes CA, Sullards MC, Merrill AH Jr (2009) Quantitative analysis of sphingolipids for lipidomics using triple quadrupole and quadrupole linear ion trap mass spectrometers. J Lipid Res 50:1692

11. Scherer M, Leuthäuser-Jaschinski K, Ecker J, Schmitz G, Liebisch G (2010) A rapid and quantitative LC-MS/MS method to profile sphingolipids. J Lipid Res 51:2001
12. Scherer M, Bottcher A, Schmitz G, Liebisch G (2011) Sphingolipid profiling of human plasma and FPLC-separated lipoprotein fractions by hydrophilic interaction chromatography tandem mass spectrometry. Biochim Biophys Acta 1811:68
13. Merritt MV, Sheeley DM, Reinhold VN (1991) Characterization of glycosphingolipids by supercritical fluid chromatography-mass spectrometry. Anal Biochem 193:24
14. Norberg P, Månsson JE, Liljenberg C (1991) Characterization of glucosylceramide from plasma membranes of plant root cells. Biochim Biophys Acta 1066:257
15. Legnini E, Orsini JJ, Mühl A, Johnson B, Dajnoki A, Bodamer OA (2012) Analysis of acid sphingomyelinase activity in dried blood spots using tandem mass spectrometry. Ann Lab Med 32:319
16. Seefelder W, Schwerdt G, Freudinger R, Gekle M, Humpf HU (2002) Liquid chromatography/electrospray ionisation-mass spectrometry method for the quantification of sphingosine and sphinganine in cell cultures exposed to fumonisins. J Chromatogr B Anal Technol Biomed Life Sci 780:137
17. Blaas N, Schüürmann C, Bartke N, Stahl B, Humpf HU (2011) Structural profiling and quantification of sphingomyelin in human breast milk by HPLC-MS/MS. J Agric Food Chem 59:6018
18. Fischbeck A, Krüger M, Blaas N, Humpf HU (2009) Analysis of sphingomyelin in meat based on hydrophilic interaction liquid chromatography coupled to electrospray ionization-tandem mass spectrometry (HILIC-HPLC-ESI-MS/MS). J Agric Food Chem 57:9469
19. Van Veldhoven PP, De Ceuster P, Rozenberg R, Mannaerts GP, de Hoffmann E (1994) On the presence of phosphorylated sphingoid bases in rat tissues. A mass-spectrometric approach. FEBS Lett 350:91
20. Mano N, Oda Y, Yamada K, Asakawa N, Katayama K (1997) Simultaneous quantitative determination method for sphingolipid metabolites by liquid chromatography/ionspray ionization tandem mass spectrometry. Anal Biochem 244:291
21. Lieser B, Liebisch G, Drobnik W, Schmitz G (2003) Quantification of sphingosine and sphinganine from crude lipid extracts by HPLC electrospray ionization tandem mass spectrometry. J Lipid Res 44:2209
22. Lan T, Bi H, Liu W, Xie X, Xu S, Huang H (2011) Simultaneous determination of sphingosine and sphingosine 1-phosphate in biological samples by liquid chromatography-tandem mass spectrometry. J Chromatogr B Anal Technol Biomed Life Sci 879:520
23. Murph M, Tanaka T, Pang J, Felix E, Liu S, Trost R, Godwin AK, Newman R, Mills G (2007) Liquid chromatography mass spectrometry for quantifying plasma lysophospholipids: potential biomarkers for cancer diagnosis. Methods Enzymol 433:1
24. Schmidt H, Schmidt R, Geisslinger G (2006) LC-MS/MS-analysis of sphingosine-1-phosphate and related compounds in plasma samples. Prostaglandins Other Lipid Mediat 81:162
25. Berdyshev EV, Gorshkova IA, Garcia JG, Natarajan V, Hubbard WC (2005) Quantitative analysis of sphingoid base-1-phosphates as bisacetylated derivatives by liquid chromatography-tandem mass spectrometry. Anal Biochem 339:129
26. Berdyshev EV, Gorshkova IA, Usatyuk P, Zhao Y, Saatian B, Hubbard W, Natarajan V (2006) De novo biosynthesis of dihydrosphingosine-1-phosphate by sphingosine kinase 1 in mammalian cells. Cell Signal 18:1779
27. Jiang X, Han X (2006) Characterization and direct quantitation of sphingoid base-1-phosphates from lipid extracts: a shotgun lipidomics approach. J Lipid Res 47:1865
28. Cutignano A, Chiuminatto U, Petruzziello F, Vella FM, Fontana A (2010) UPLC–MS/MS method for analysis of sphingosine 1-phosphate in biological samples. Prostaglandins Other Lipid Mediat 93:25
29. Sun D, Froman BE, Orth RG, MacIsaac SA, Larosa T, Dong F, Valentin HE (2009) Identification of plant sphingolipid desaturases using chromatography and mass spectrometry. J Chromatogr Sci 47:895

30. Jin YX, Cui XH, Paek KY, Yim YH (2012) A strategy for enrichment of the bioactive sphingoid base-1-phosphates produced by Hypericum perforatum L. in a balloon type airlift reactor. Bioresour Technol 123:284
31. Saigusa D, Shiba K, Inoue A, Hama K, Okutani M, Iida N, Saito M, Suzuki K, Kaneko T, Suzuki N, Yamaguchi H, Mano N, Goto J, Hishinuma T, Aoki J, Tomioka Y (2012) Simultaneous quantitation of sphingoid bases and their phosphates in biological samples by liquid chromatography/electrospray ionization tandem mass spectrometry. Anal Bioanal Chem 403:1897
32. Zhao YY, Liu J, Cheng XL, Bai X, Lin RC (2012) Urinary metabonomics study on biochemical changes in an experimental model of chronic renal failure by adenine based on UPLC Q-TOF/MS. Clin Chim Acta 413:642
33. Quehenberger O, Armando AM, Brown AH, Milne SB, Myers DS, Merrill AH, Bandyopadhyay S, Jones KN, Kelly S, Shaner RL, Sullards CM, Wang E, Murphy RC, Barkley RM, Leiker TJ, Raetz CR, Guan Z, Laird GM, Six DA, Russell DW, McDonald JG, Subramaniam S, Fahy E, Dennis EA (2010) Lipidomics reveals a remarkable diversity of lipids in human plasma. J Lipid Res 51:3299
34. Lydic TA, Busik JV, Reid GE (2014) A monophasic extraction strategy for the simultaneous lipidome analysis of polar and nonpolar retina lipids. J Lipid Res 55:1797

Chapter 27
Lipid Machinery Investigation Using MALDI Imaging Mass Spectrometry

Ikuko Yao, Gustavo A. Romero-Pérez, Dan Nicolaescu, and Mitsutoshi Setou

Abstract Lipids have essential roles in several biological processes, including normal cell functioning as well as in the onset of diseases. Renewed interest in lipid research has been largely fueled by emerging evidence linking the augmenting of corporal adipose tissue in humans with metabolic diseases and cancer. Imaging mass spectrometry techniques including matrix-assisted laser desorption/ionization time-of-flight mass spectrometry (MALDI-TOF MS) are analytical methods quite suitable for the investigation of compounds in biological samples, providing an enormous amount of information regarding their chemical composition and spatial distribution with no need for prior labeling. Particularly, MALDI-TOF imaging mass spectrometry (IMS) has evolved into a very versatile technique capable of analyzing, at molecular level, a wide range of compounds, including lipids. This review presents a survey of recent work conducted mainly in our laboratory on full protocol development for lipid analysis by MALDI-TOF IMS of biological samples, with emphasis on sample collection and preparation, matrix selection and deposition, and typical analysis condition settings.

Keywords Lipids • Imaging mass spectrometry • Tissue collection • Tissue sections • Matrix deposition • Spraying • Sublimation • TLC-blot-MALDI-TOF MS

I. Yao • G.A. Romero-Pérez
Department of Optical Imaging, Medical Photonics Research Center, Hamamatsu University School of Medicine, Hamamatsu, Shizuoka 431-3192, Japan
e-mail: yaoik@hama-med.ac.jp; gromprz@hama-med.ac.jp

D. Nicolaescu • M. Setou (✉)
Department of Cell Biology and Anatomy, Hamamatsu University School of Medicine, Hamamatsu, Shizuoka 431-3192, Japan
e-mail: nicol168@hama-med.ac.jp; setou@hama-med.ac.jp

T. Yokomizo, M. Murakami (eds.), *Bioactive Lipid Mediators*,
DOI 10.1007/978-4-431-55669-5_27

27.1 Introduction

The functions of lipids in energy storage in the body have been well studied and documented [1]. Nevertheless, although there have been speculations and some evidence about other functions that lipids may have in homeostatic processes and the onset of disease, until now it has been a greatly neglected area of research. Recently, scientists have shown renewed interest in lipid research largely fueled by mounting evidence linking an increase of corporal adipose tissue in humans with metabolic diseases [2] and cancer [3]. Indeed, new studies are evidencing that a wide range of lipids, including fatty acids, glycerophospholipids, and sphingolipids, are critical in a vast number of biological processes [4]. Several mass spectrometry (MS) techniques have been developed for the characterization of biological samples, including electrospray ionization (ESI) [5], secondary ion mass spectrometry (SIMS) [6], and matrix-assisted laser desorption/ionization time-of-flight mass spectrometry (MALDI-TOF MS) [7, 8]. Initially, MALDI-TOF MS was limited to the analysis of proteins, but since then it has steadily evolved into a very versatile technique capable of analyzing a wider range of compounds, including lipids.

Lipid analysis by MALDI-TOF MS is potentially a better alternative to more traditional techniques such as gas chromatography and high performance liquid chromatography. For example, it requires minimum sample preparation and has high tolerance to impurities, no cumbersome extractions are needed, and measurements can be completed in a relatively short time. Several factors concur for this outcome, such as the high abundance of lipids in tissue, inherent ionization of many lipids forming the bilayer cell membrane, which allows positive or negative ions to be generated in great number upon laser pulse impact, and efficient ion collection by modern mass spectrometers for ions having a molecular mass less than 1000 Da [9]. Nevertheless, as peak intensities of lipids can widely vary during MALDI-TOF MS analysis, certain identified species may be overestimated, such as in the case of lipids with ammonia groups, while others are underestimated. Moreover, in the case of free fatty acids, signals of standard matrices often interfere with those of targeted fatty acids as they both occupy the same low mass range [10].

To overcome the aforementioned difficulties and to enhance the identification and characterization of lipids, a number of protocols, including sample collection and preparation techniques as well as MALDI-TOF MS analysis methods, have been developed and used worldwide in the past few years to study a wide number of lipid species at molecular level. Specifically, our laboratory has made a great effort to develop new methods and modify existing techniques for the study of lipid species in tissues showing indication of diseases, including cancer. In the following sections, we describe some of these achievements in the form of protocols that we have used for the investigation of the lipid machinery.

27.2 Tissue Collection

Biological material, especially tissues rich in lipids, is very susceptible to compositional changes caused by rapid degradation of its constituents. Nonetheless, it has been found that very low temperature conditions during sample preparation and subsequent storage dramatically inhibit degradation [11]. In our premises, we have developed a number of cryopreservation methods to safely remove and sample tissues while minimizing autolytic damage. In addition, focused microwave irradiation has been hailed in recent years as a promising method for the stabilization of tissues for MALDI-TOF MS analysis [12]. Focused microwave irradiation inactivates enzymes within 0.5–2.5 s [13–15], effectively stopping metabolism and further tissue degradation. We have used focused microwave irradiation for the stabilization of brain tissues preceding analysis of glucose-derived metabolites by imaging MS [12], although our group has yet to use this irradiation technique to treat tissues before lipid analysis by MALDI-TOF MS. Nonetheless, Bazan and colleagues conducted a series of elegant experiments using focused microwave irradiation to rapidly kill rats, which also inhibited postmortem compositional changes in the brain [15–19]. These workers successfully analyzed the content of fatty acids, phospholipids, and diacylglycerols in rat brain tissues, with no reported tissue or metabolite damage.

In the following section, we list the required materials and describe protocols developed by our group [20–22] that are recommended for sampling and handling brain tissues of experimental rodent models.

27.2.1 Tissue Collection Protocols

Materials

- Anesthetic agents (pentobarbital, ketamine, etc.)
- Diethyl ether
- Surgical blades and scalpel
- Scissors
- Dry ice
- Liquid nitrogen
- Cryogenic protective gloves
- Wooden mallet
- Sturdy poly-mesh bags
- Sieves
- Polystyrene foam boxes

Postmortem Freezing

1. Pulverize the dry ice with the mallet to produce a very fine powder. If large chunks still remain, use the sieve to retrieve them. Transfer the powdered dry ice to a polystyrene foam box. Keep the lid on until further use.
2. To euthanize, deeply anesthesize the animal as required, and cut the head off. With the scissors, cut open the skin across the top of the head and toward the nasal region.
3. Insert the tip of a small scissor blade straight into the foramen magnum, lift and cut the interparietal bone, and push the parietal bones to the sides, like a swing door.
4. Without damaging the surface of the brain, insert the tip of the scissor blade underneath the frontal bones and force open the fissure in between them. Do the same with the nasal bones. Push all bones to the sides as it was done with the parietal bones.
5. Scoop out the brain with a spatula. It should be removed from the base of the skull, either from the olfactory nerve side or the cerebellum side.
6. With a gentle but quick motion, place the brain onto the powdered dry ice and cover completely with it. Let it flash-freeze for approximately 1 min. Remove the frozen brain and place it inside a precooled Ziploc plastic bag or 50-ml conical tube. Store the brain at −80 °C until further use.

In Situ Freezing

1. In a well-ventilated room, fill a Dewar flask to the brim with liquid nitrogen.
2. Deeply anesthetize the animal with diethyl ether and skin the head.
3. To euthanize the animal and inhibit tissue degradation, dip the top of the head with care so that the nose is not immersed.
4. Remove the brain and store it as described in Protocol 1.

27.3 Tissue Sectioning

MALDI-TOF MS analysis can be carried out on a wide array of clinical samples, from heart, brain, and liver to skin, hair, and bone. During tissue sectioning it must be ensured that the original morphology of the tissue is retained while maintaining its chemical composition unchanged. For example, although embedding in Optimal Cutting Temperature (OCT) compound enables tissues to retain their morphology, this product shows a high ionization efficiency that reduces the detection sensitivity of other molecules in biological samples, especially in the *m/z* range of 1000–2000 [23]. Thus, OCT compound should only be used for anchoring the tissues to the object holder. A precooled semiliquid gel prepared with 2 % sodium carboxymethylcellulose [$C_6H_7O_2(OH)_2OCH_2COONa$] is a good alternative embedding compound that does not interfere with MS analysis [23]. Other embedding media may be used for embedding, including gelatine [24] and agarose [25].

With regard to the thickness of sections, thin slices are very frail and difficult to handle [26]. Conversely, thick slices are not recommended because they take a longer time to dry and behave as insulating material, which may adversely affect MS analysis. Thus, slices of 10–20 μm have been found to be optimal for handling and analyzing by MS [26]. Once tissues are sectioned, they should be mounted onto a surface with good electric conductivity. Metal plates are the best conductive material for MALDI-TOF MS analysis, but the need to thoroughly clean these plates after analysis makes their use impractical for multiple measurements. Alternatively, there are commercially available plastic sheets and glass slides coated with an excellent conductive layer such as indium tin oxide (ITO) [23], which are also disposable.

27.4 Selection of Compounds as Matrices for Lipid Analysis

Because lipids are relatively small molecular weight compounds, signals from the matrix should be sufficiently small to avoid overlap with the target lipids [27]. The matrix 2,5-dihydroxybenzoic acid (DHB) has been widely used for the analysis of lipids, as it generally produces few matrix-related ions [28]. Goto-Inoue et al. successfully employed DHB to visualize, by MALDI-TOF IMS, the distribution of ceramides in skin tissue samples from patients suffering from Dorfman–Chanarin syndrome [29]. Similarly, He et al. studied the distribution of phophatidylcholines in Warthin tumor tissues using DHB as matrix [30]. However, the use of traditional matrices such as DHB and α-cyano-4-hydroxycinnamic acid (CHCA) for the MALDI-TOF MS analysis of lipids such as cholesterol and fatty acids sometimes results in poor-quality data because an important overlapping occurs between lipid and matrix ion signals [31], especially in negative-ion mode [28]. The matrix 9-aminoacridine (9AA) is a moderately strong basic compound that has been found to be a better alternative for the analysis of free fatty acids in negative-ion mode [32]. Matrix 9AA has also been found to produce better results than newer compounds such as 2,4,6-trihydroxyacetophenone (THAP) [33]. In addition, although DHB has been used as matrix to detect lipid oxidation products such as peroxides and reactive oxygen species (occurring in tissues from patients with inflammatory diseases), chlorinated phosphatidyl ethanolamine species are not easily detected if analyses are conducted by MALDI-TOF MS with DHB as matrix [31]. Nonetheless, if 4-chloro-α-cyanocinnamic acid (Cl-CCA), a chlorinated analogue of cinnamic acid, is used as matrix, these chloraminic species can be easily detected by MALDI-TOF MS [34].

27.4.1 Alternative Compounds Used as Matrix

As imaging MS of lipids is rapidly gaining popularity among lipid scientists, the number of compounds of interest is also increasing. Thus, in addition to the aforementioned matrices, research is actively being conducted to find new matrices or to produce combinations of existing MALDI matrices that would improve performance. For example, to enhance imaging, lithium salts can be added to the matrix solution [33], and fixation of tissues before sectioning can be performed using formal lithium [35]. If formal lithium solutions (0.154 M LiCl or $LiNO_3$ salt in formalin) are sprayed on tissue sections, lipid detection is improved in terms of image contrast because simplified mass spectra contained only peaks related to lithium-cationized molecules instead of a mixture of sodium- and potassium-cationized species. In addition, an optimized protocol for rat brain MS imaging has been proposed using lithium trifluoroacetate salt, which slightly increases desorption of phosphatidylcholines [33]. Furthermore, compatibility between formal lithium fixation and histological staining proved to be useful for human liver tissue analysis [35]. In a comprehensive evaluation, Thomas et al. [36] tested 12 different matrices including DHB, CHCA, and 9AA as well as other new ones such as THAP and 1,5-diaminonaphtalene (DAN). These workers concluded that DAN had a high vacuum stability and was particularly efficient in both positive- and negative-ion modes, offering rich lipid signatures.

27.4.1.1 Flavonoids

A new family of matrices for MALDI tissue imaging, represented by flavonoid compounds, was recently developed [37]. The comparative study included flavone and nine of its other mono- or polyhydroxyl-substituted analogues. The flavonoids compounds were sprayed on tissue using a high pH (0.1–0.5 % ammonia hydroxide) matrix solvent. The best results were obtained with quercetin and morin, which are penta-OH flavones. Quercetin showed characteristics superior to DHB, CHCA, and 2-mercaptobenzothiazole (2-MBT), allowing simultaneous visualization of more than 200 lipid species in rat brain experiments.

27.4.1.2 Ionic Matrices

Two common problems when using DHB as the sole compound in a matrix solution for MS imaging are (1) the production of nonhomogeneous, needle-shaped crystallization, which appears on the outer rim of the deposited solution, and (2) delocalization of tissue lipids by the solvent used to dissolve this matrix. Ionic matrices are organic salts used as matrices that are formed by an acid–base reaction, usually combining a conventional MALDI matrix with an organic base. Ionic matrices may be either liquid or solid, depending on the molar ratio of components and other factors. Liquid ionic matrices based on 2,5-DHB in combination with aniline (ANI), pyridine (Pyr), and 3-acetylpyridine (3-AP) were investigated to overcome the drawbacks of crystalline 2,5-DHB [38]. An automatic micro-spotter (Chemical Ink-Jet Printer CHIP-1000; Shimadzu Biotech, Kyoto, Japan) was used for precise spotting of freshly prepared matrices. Improved imaging of lipids was validated on human ovarian cancer biopsies. Liquid ionic matrices for *r*-cyano-4-hydroxycinnamic acid butylamine (CHCAB) and 2,5-dihydroxybenzoic acid butylamine (DHBB) were also validated against DHB for enhanced visualization of phospholipids in mouse liver and brain tissue sections [39].

27.4.1.3 Nanomaterials

MALDI matrix layers allow laser energy absorption and energy transfer to analyte molecules with minimal fragmentation, achieving otherwise unlikely MS measurements. Nonetheless, matrix materials usually have small molecular weights, which hampers the detection of weak signals of chemical compounds in the small mass range. Protocols for MS analysis using compounds other than traditional matrices for enhanced sensitivity in the small mass range rely on nanostructured surfaces, the techniques being generally called surface-assisted laser desorption/ionization (SALDI) [40]. Different nanomaterials have been used as substrate for SALDI IMS lipid measurements, including porous silicon [41], graphene layers [42–45], nanodots [46], and nanometals [47].

Recently, we developed a technique we termed nanoparticle (nano-P)ALDI-IMS using extremely small ($d = 3.7$ nm) nanoparticles for the analysis of lipids. Indeed, this technique employs ferrous nanoparticles as matrix to detect the lipid distribution in rat brain tissues [48]. In addition, we used alkylcarboxylate- and alkylamine-modified silver nanoparticles for the lipid analysis of mouse retinal and liver sections to identify fatty acids such as palmitic, stearic, oleic, linoleic, arachidonic, and eicosapentaenoic acids, which were not detected with matrix DHB [49]. Similarly, mapping of the distribution and localization of glycosphingolipids in brain sections was possible using alkylamine modified gold nanoparticles as matrix [50]. In this analysis, detection of glycosphingolipids was 20 fold higher than by DHB.

27.5 Matrix Deposition

As it is the step whereby cocrystallization of matrices with sample analytes takes place on the tissue surface, matrix application is generally considered the most critical procedure for MALDI-TOF MS analyses [51]. There are three major procedures to apply matrices onto samples, namely, manual spraying with an artist's airbrush, (manual and automated) droplet deposition, and sublimation under reduced pressure and elevated temperature [51]. The spraying method has been largely the preferred procedure for lipid analyses by MALDI-TOF MS in our laboratories [2, 29,

30, 50, 52–60]. Nonetheless, more recently sublimation methods for matrix deposition have also been applied by our group using an automated deposition device [61]. In the following sections, protocols for both methods are described, but readers are encouraged to search for information on alternative methods.

27.5.1 Airbrush Spraying

An artist's airbrush is generally used for matrix spraying. Two minimum features are required for the airbrush: (1) the operator should be able to adjust the droplet size and control the mist volume, and (2) the device should be resistant to organic solvents to avoid contamination of samples with residues derived from parts of the instrument [51]. In addition, the airbrush should be equipped with a 0.2-mm nozzle. In our laboratory, the preferred device is Proton Boy FWA Platinum (Mr. Hobby, Tokyo, Japan) for its simplicity and the ease of handling of its design. Representative parameters of spraying operation with an airbrush are shown in Fig. 27.1. Important points to keep in mind to master the airbrush's hand operation follow.

1. Minimize droplet size during spraying.
2. Maintain invariable the distance and angle between the nozzle and the sample surface.
3. Move the airbrush horizontally in a gradual and smooth manner.

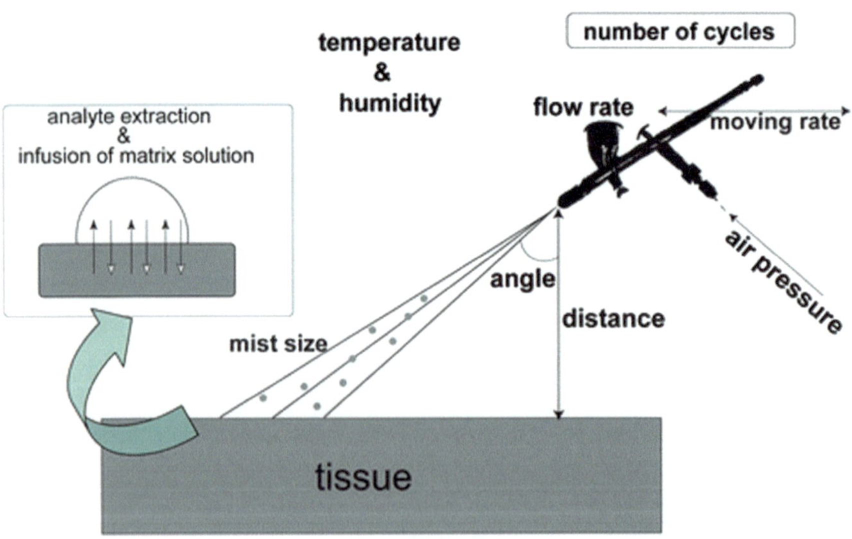

Fig. 27.1 Representative parameters of the spraying operation with an air-brush that must be maintained between trials for reproducible imaging mass spectrometry (IMS) experiment results (Reproduced from Sugiura et al. [51])

4. Maintain equilibrium between tissue moistening and solvent evaporation rates.
5. Minimize fluctuation of droplet size, amount of mist, and surrounding temperature and humidity.

Protocol

1. Pour 1 ml matrix-free solvent into the paint cup. Test and optimize droplet size, amount of mist, angle, and distance. When done, eject all solvent residues.
2. Cover the ITO glass slide surface with moisture-resistant material (e.g., tape, aluminium foil), except the area where tissues are mounted.
3. Firmly place the ITO glass slide onto a perpendicular board.
4. Pour the desired volume of matrix solution into the paint cup and start spraying on the tissue.
5. Remove the cover from the ITO glass slide and dry at room temperature (RT), or place it in an airtight container with silica beads. Soon afterward, analyze by IMS.

27.5.1.1 Preparation and Spray Deposition of 2,5-Dihydroxybenzoic Acid as Matrix

A typical protocol of matrix DHB preparation and deposition by spraying [62] is found next.

Materials

- 2,5-Dihydroxybenzoic acid (DHB)
- MeOH

Protocol

1. Dissolve 50 mg DHB in 1 ml 70 % MeOH
2. Pour 0.5–1.0 ml of DHB matrix solution into the paint cup of an artist's airbrush and spray onto the sample.
3. Place the sample in a cooling dryer.

27.5.1.2 Preparation and Spray Deposition of Ferrous Nanoparticles as Matrix

The material and protocol for the production and spraying deposition of ferrous nanoparticle matrix on samples [48] is described next.

General Materials

- Alkylcarboxylic acids
- Alkylamines
- γ-Aminopropyltriethoxysilane
- NaOH
- MeOH

- Sodium acetate (CH_3COONa)
- Acetone
- Tetrahydrate iron chloride ($FeCl_24H_2O$)
- Silver nitrate ($AgNO_3$)
- Chloro (dimethyl sulfide) gold [$AuCl(SMe_2)$]

Protocol

1. Mix an aqueous solution of 100 mM $FeCl_2•4H_2O$ with 20 ml γ-aminopropyltriethoxysilane and stir for 1 h.
2. Wash the precipitates several times with distilled water, dry at 80 °C, and pulverize in a mortar.
3. Disperse and centrifuge 10 mg ferrous nanoparticles in 1 ml 10 mM methanolic solution of sodium. Retrieve the supernatant fluid.
4. Thaw-mount the tissue section onto an ITO-coated glass slide. Spray the supernatant fluid. If spraying is carried out with an artist's airbrush, ensure maintenance of a 15-cm distance between the nozzle and the tissue surface. Let dry at RT.

27.5.1.3 Preparation and Spray Deposition of Silver Nanoparticles as Matrix

The protocol for the production and spraying deposition of silver nanoparticle as matrix onto samples [49, 63] is described next. Please note that some materials are the same as in the previous section.

Protocol

1. Add an aqueous solution of NaOH (0.150 M) to a 0.150 M suspension of the required alkylcarboxylic acid in 1.0 l hot water.
2. To the resulting solution, add an aqueous solution of $AgNO_3$ (0.165 M) to obtain a white precipitate (silver alkylcarboxylate).
3. Collect and dry under reduced pressure at 60 °C to produce alkylcarboxylates.
4. Place the silver alkylcarboxylate (1.0 mM) and an equal amount of alkylamines (1.0 mM) into a three-necked flask. Heat the compounds for 5 h to 120 or 180 °C to cause the reaction mixture to pass from liquid to a brown dispersion with a metallic luster.
5. Cool the mixture to 80 °C and add MeOH. Collect the precipates (silver nanoparticles) by filtration, wash with MeOH, and dry under vacuum.
6. Thaw-mount the tissue section onto an ITO-coated glass slide. Spray 500 μl silver nanoparticle solution (50 mg/ml in 100 % hexane). If spraying with an artist's airbrush, ensure 15-cm distance between nozzle and tissue surface. Let dry at RT.

27.5.1.4 Preparation and Spray Deposition of Gold Nanoparticles as Matrix

The protocol for the production and spraying deposition of gold nanoparticles as matrix onto samples [50, 64] is described next. Please note that some materials are the same as in the previous section.

Protocol

1. Place $AuCl(SMe_2)$ (295 mg, 1.0 mM) and the alkylamine (10.0 mM) in a 10-ml flask accessorized with a magnetic stirrer. Depending on the alkylamine, gradually heat the mixture to 100 or 120 °C. Maintain required temperature for 1 h. Afterward, let solution cool to RT.
2. Add acetone (5 ml) and MeOH (1 ml).
3. Centrifuge at 2000 rpm for 5 min. Decant the solvents; collect and dry the gold nanoparticles under vacuum.
4. Thaw-mount the tissue section onto an ITO-coated glass slide. Spray 1000 μl of gold nanoparticle solution (50 mg/ml in 100 % hexane). If spraying with artist's airbrush, ensure 15-cm distance between nozzle and tissue surface. Let dry at RT.

27.5.2 *Sublimation*

Matrix deposition through sublimation was first proposed by Hankin et al. and subseuqently has been extensively used [65]. The sublimation process, that is, the transfer of matrix molecules from solid to gas phase, is a process that depends on temperature and vacuum conditions. High purity of the matrix material is desirable for avoiding unknown and unreproducible interactions and ion suppression effects (described later) with the analytes of the biological sample. The layer quality, that is, the size of the crystals and thickness uniformity that is suitable for MALDI measurements, depends on several factors. For example, a deposition time that is too short may produce unwanted sputtering of matrix grains. In contrast, depositions at slower speed produce a sublimated layer with small-sized crystals that are uniformly distributed. Nonetheless, uniformity of the layer thickness across the sample is also affected by geometric factors such as the distance between the crucible and the substrate and the crucible lid aperture (i.e., its shape and size). Therefore, ideally the larger the distance between the source and the substrate, and the aperture size of the crucible lid, the better the uniformity of the deposited layer. In reality, however, these settings also cause the sublimated matrix to spread more in space, and thus an optimal layer thickness for a given matrix quantity in the crucible may be difficult to achieve.

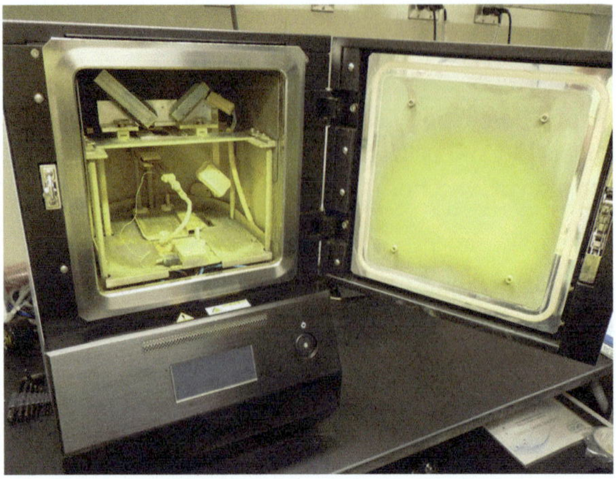

Fig. 27.2 Picture of iMLayer (Shimadzu, Kyoto, Japan), which is a desktop-size device used for automatic matrix deposition using sublimation. The vacuum chamber includes the crucible and the substrate with holders, as well as a laser light source and a detector for measuring the matrix layer thickness using transmission light signal change and an internal calibration curve. The vacuum pump and the controlling unit are separate devices and thus not shown in the picture (Photograph courtesy of Keigo Sano)

27.5.2.1 Intelligent Matrix Layer

A prototype vacuum vapor deposition device (RK27-4069) was first developed by Shimadzu (Kyoto, Japan) and used in our laboratory for analysis of lipid distribution in breast cancer tissues [3]. This equipment was further improved, and a commercial automatic device named intelligent Matrix Layer (iMLayer) was produced [61] (Fig. 27.2). A brief description of iMLayer operation is as follows. iMLayer works by passing a high electric current through a crucible or “boat” that is made from a refractory metal sheet, such as tungsten or molybdenum. As a result, the crucible temperature increases by the Joule effect. The temperature is read by a thermocouple attached to the bottom of the crucible. Information about the vacuum conditions (pressure) is also available to the automatic control system. The deposition system contains a laser light source and a detector that can measure the thickness of the deposited layer using changes in light transmission intensity. The temperature and the type of material are system settings that are required for measuring the layer thickness using built-in calibration curves. The maximal deposition time is also selectable.

The following protocol applies to iMLayer, although the steps may also have relevance for other situations.

Materials

- iMLayer
- Precision weighing scale
- High-purity matrix material for deposition using sublimation
- Mortar and pestle

- Microspatula
- Laboratory sieves for sorting the matrix material by grain size (optional)

Protocol

1. Prepare the matrix material by grinding it manually using a mortar and a pestle.
2. Select the matrix grain size using a sieve with a predetermined hole size (optional). Only grains smaller than the hole size shall be used for sublimation (i.e., the distribution in grain size has an upper limit).
3. Weigh the amount of matrix powder for deposition.
4. Place the matrix powder in the crucible.
5. Gently hit the crucible several times with a microspatula to let the powder rearrange itself at the surface and in volume.
6. Select the lid (cover) for the crucible. The lid may have a round hole or a slit aperture.
7. Place the lid over the crucible, or alternatively, maintain the crucible uncovered.
8. Set the desired distance between crucible and substrate.
9. Place the crucible in the special holder electrode.
10. Place the slide with the biological samples in the special slide holder of iMLayer.
11. Close the device door and set the conditions for temperature and maximal deposition time, depending on the matrix material used.
12. Start the sublimation.
13. After the matrix sublimation is completed, immediately analyze by MALDI TOF IMS.

Typical iMLayer conditions for the sublimation of 9-AA matrix

- Matrix grinding time: approximately 5 min
- Matrix weight for one sublimation: 500 mg
- Temperature: 230 °C
- Distance from crucible to substrate: 5–10 cm (better layer uniformity is achieved when the distance between the crucible and the slide is larger)
- Maximum deposition time: 90 min
- Expected results: 1-μm matrix layer thickness (for MALDI IMS analysis of biological tissue)

27.6 Matrix-Assisted Laser Desorption/Ionization Time-of-Flight Imaging Mass Spectrometry

Technologically, lipid analysis by MALDI-TOF MS offers several advantages when compared with other analytical approaches. Perhaps the most critical advantage is that both matrices and lipid are readily soluble in organic solvents, eliminating the need for water addition [66]. This dual solubility increases the homogeneous crystallization of the matrix–sample complex, and thus measurements are enhanced and good-quality spectra obtained [66]. Summarized sets of MALDI-TOF MS operating conditions as published by our laboratory for the analysis and identification of phospholipids ([55], Sect. 27.6.1), fatty acids ([49], Sect. 27.6.2), ceramides ([29], Sect. 27.6.3), triacylglycerides ([67], Sect. 27.6.4), and sulfatides ([60], Sect. 27.6.5) are shown next.

27.6.1 Phospholipids in Rat Brain Tissues

- Device: MALDI-TOF/TOF-type MS (Mass Microscope; Shimadzu)
- Ion mode: positive or negative
- Laser frequency: 1000-Hz repetition rate
- Laser shots: 200 at each spot
- Data point interval: 10 μm

27.6.2 Fatty Acids in Mouse Hepatic and Retinal Tissues

- Device: MALDI-TOF/TOF-type MS (ultraflex II; Bruker Daltonics)
- Ion mode: negative
- Laser frequency: 200-Hz repetition rate
- Laser shots: 200 at each spot
- Scan pitch: 100 μm (liver) and 10 μm (retina)

27.6.3 Ceramides in Dorfman–Chanarin Syndrome Patient Skin Tissues

- Device: MALDI-hybrid quadrupole TOF-type MS (QSTAR Elite; Applied Biosystems)
- Ion mode: positive
- Laser frequency: 200-Hz repetition rate

*27.6.4 Triacylglycerides in Banana Shrimp (*Penaeus merguiensis*) Ovary Tissues*

- Device: MALDI-TOF/TOF-type MS (ultraflex II; Bruker Daltonics)
- Ion mode: positive
- Laser frequency: 200-Hz repetition rate

- Laser shots: 200 at each spot
- Raster width: 35 μm per laser spot

27.6.5 Sulfatides in Human Brain Tissues

- Device: MALDI-TOF/TOF-type MS (ultraflex II, Bruker Daltonics)
- Ion mode: negative
- Raster scan: automatic
- Laser shots: 100 at each spot

27.7 Thin-Layer Chromatography-Blot-Matrix-Assisted Laser Desorption/Ionization Imaging Mass Spectrometry

Thin-layer chromatography-blot-matrix-assisted laser desorption/ionization imaging mass spectrometry (TLC-blot-MALDI-TOF MS) is a promising technique developed by our research group that combines existing lipid analytical methods, namely, thin layer and IMS. This method achieves precise separation of complex lipids into individual constituents that largely avoids preferential detection of certain lipid classes over others, a phenomenon known as ion suppression effect [68, 69]. TLC-blot-MALDI-TOF IMS has been successfully used in our laboratory to visualize and identify lipid classes such as glycosphingolipids and phospholipids from various types of tissue including mouse epididymis, human brain, and bluefin tuna flesh [68, 70–73]. A flowchart of the TLC blot MALDI-TOF IMS (Fig. 27.3) shows a schematic of the typical procedure for analysis of human brain tissues, as described next.

TLC Analyses

1. Homogenize the tissue sample (~0.1 g) with a 20-fold volume of chloroform: methanol (2:1, v/v). Sonicate for 5 min.
2. Extract the lipids and store at 4 °C.
3. Apply the samples and the corresponding lipid standards onto a silica gel 60 HPTLC plate.
4. Develop the plate to 6 cm with a solvent mixture of chloroform:methanol:0.2 % $CaCl_2$ (60:40: 9, v/v).
5. After air-drying, further develop the plate to 8 cm with a solvent mixture of methyl acetate:propanol:chloroform:methanol:0.25 % KCl (25:25:25:10:9, v/v).
6. To visualize the lipids, spray the plates with primuline reagent (1 ml 0.1 % primuline aqueous solution in 100 ml in acetone:water; 80:20, v/v). Air dry thoroughly. Place the HPTLC plate under UV light at 315 nm.

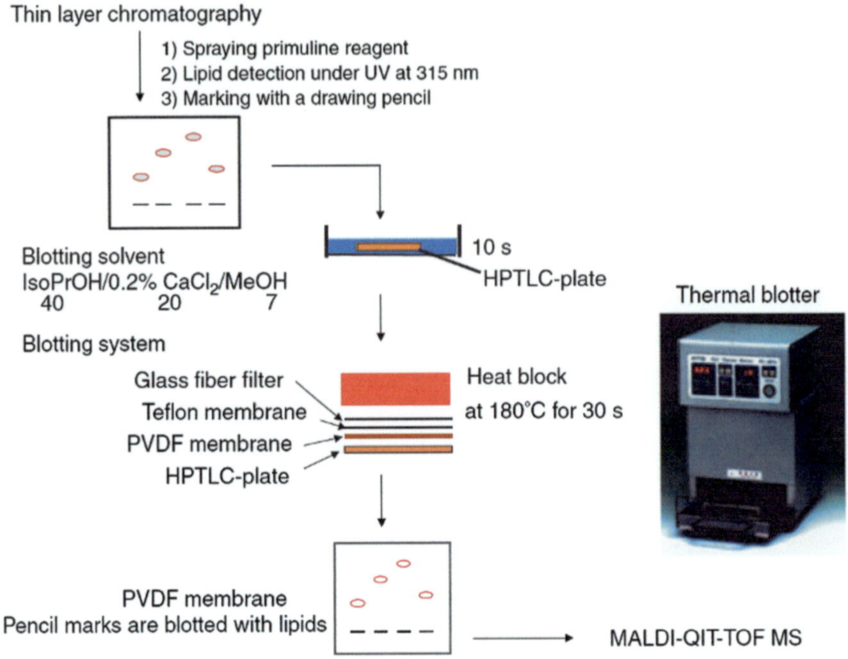

Fig. 27.3 Scheme of our approach for glyco- and lipidomics analyses by TLC-Blot-MALDI-QIT-TOF MS (Reproduced from © 2011 Valdes-Gonzalez T, Goto-Inoue N, Hirano W, Ishiyama H, Hayasaka T, Setou M, Taki T. *Journal of Neurochemistry* © 2011 International Society for Neurochemistry)

TLC-Blot

1. Dip the HPTLC plate into a blotting solvent (isopropanol:0.2 % $CaCl_2$:methanol; 40:20:7, v/v). Remove immediately.
2. Mount the plate with the PVDF membrane, the Teflon membrane, and glass fiber filter paper.
3. Press this assembly for 30 s at 180 °C by a thermal blotter. Remove the PVDF membrane and air dry.
4. Analyze the PVDF membrane with the lipids attached on the reverse side by MALDI-QIT-TOF MS.

MALDI-QIT-TOF MS

- Device: MALDI-QIT-TOF MS (AXIMA-QIT MS; Shimadzu)
- Ion mode: positive
- Raster scan: automatic
- Laser shots: 5 at each spot
- Data point interval: 100 μm

27.8 Conclusion

Recent research has been increasingly unveiling the importance of lipids in various biological processes and the onset of metabolic and degenerative diseases. Upon its introduction in the late 1980s as a relatively limited tool for the analysis of proteins, MALDI-TOF MS has become a very versatile technique capable of analyzing a much wider range of compounds including lipids. Several characteristics including a low mass range similar to those of matrices commonly used for coating samples, as well as a preferential identification of certain lipid classes over others, make lipid analysis by MALDI-TOF MS still a difficult analytical technique to master. Nonetheless, rapid, breathtaking developments in imaging MS technology are signaling that a breakthrough in subcellular research of the lipid machinery is about to occur. Thus, we foresee that future work using MALDI-TOF MS as the preferred tool will be able to further elucidate the importance of individual lipid species and precisely connect their functions with biological events taking place in tissues, as well to provide information on new markers for more accurate clinical diagnosis.

References

1. Rosen ED, Spiegelman BM (2006) Adipocytes as regulators of energy balance and glucose homeostasis. Nature 444(7121):847–853. doi:10.1038/nature05483
2. Goto-Inoue N, Manabe Y, Miyatake S, Ogino S, Morishita A, Hayasaka T, Masaki N, Setou M, Fujii N (2012) Visualization of dynamic change in contraction-induced lipid composition in mouse skeletal muscle by matrix-assisted laser desorption/ionization imaging mass spectrometry. Anal Bioanal Chem 403(7):1863–1871. doi:10.1007/s00216-012-5809-x
3. Ide Y, Waki M, Hayasaka T, Nishio T, Morita Y, Tanaka H, Sasaki T, Koizumi K, Matsunuma R, Hosokawa Y, Ogura H, Shiiya N, Setou M (2013) Human breast cancer tissues contain abundant phosphatidylcholine (36:1) with high stearoyl-CoA desaturase-1 expression. PLoS One 8(4):e61204. doi:10.1371/journal.pone.0061204
4. Goto-Inoue N, Hayasaka T, Zaima N, Setou M (2011) Imaging mass spectrometry for lipidomics. Biochim Biophys Acta 1811(11):961–969. doi:10.1016/j.bbalip.2011.03.004
5. Fenn J, Mann M, Meng C, Wong S, Whitehouse C (1989) Electrospray ionization for mass spectrometry of large biomolecules. Science 246(4926):64–71. doi:10.1126/science.2675315
6. Heeren RMA, McDonnell LA, Amstalden E, Luxembourg SL, Altelaar AFM, Piersma SR (2006) Why don't biologists use SIMS? A critical evaluation of imaging MS. Appl Surf Sci 252(19):6827–6835. doi:10.1016/j.apsusc.2006.02.134
7. Karas M, Hillenkamp F (1988) Laser desorption ionization of proteins with molecular masses exceeding 10,000 daltons. Anal Chem 60(20):2299–2301. doi:10.1021/ac00171a028
8. Tanaka K, Waki H, Ido Y, Akita S, Yoshida Y, Yoshida T, Matsuo T (1988) Protein and polymer analyses up to *m/z* 100 000 by laser ionization time-of-flight mass spectrometry. Rapid Commun Mass Spectrom 2(8):151–153. doi:10.1002/rcm.1290020802
9. Berry KAZ, Hankin JA, Barkley RM, Spraggins JM, Caprioli RM, Murphy RC (2011) MALDI imaging of lipid biochemistry in tissues by mass spectrometry. Chem Rev 111(10):6491–6512. doi:10.1021/cr200280p
10. Fuchs B, Süß R, Schiller J (2010) An update of MALDI-TOF mass spectrometry in lipid research. Prog Lipid Res 49(4):450–475. doi:10.1016/j.plipres.2010.07.001

11. Patterson NH, Thomas A, Chaurand P (2014) Monitoring time-dependent degradation of phospholipids in sectioned tissues by MALDI imaging mass spectrometry. J Mass Spectrom 49(7):622–627. doi:10.1002/jms.3382
12. Sugiura Y, Honda K, Kajimura M, Suematsu M (2014) Visualization and quantification of cerebral metabolic fluxes of glucose in awake mice. Proteomics 14(7-8):829–838. doi:10.1002/pmic.201300047
13. Galli C, Spagnuolo C (1976) The release of brain free fatty acids during ischaemia in essential fatty acid-deficient rats. J Neurochem 26(2):401–404. doi:10.1111/j.1471-4159.1976.tb04493.x
14. Guidotti A, Cheney DL, Trabucchi M, Doteuchi M, Wang C, Hawkins RA (1974) Focussed microwave radiation: a technique to minimize post mortem changes of cyclic nucleotides, dopa and choline and to preserve brain morphology. Neuropharmacology 13(12):1115–1122. doi:10.1016/0028-3908(74)90061-6
15. Visioli F, Rodriguez de Turco EB, Kreisman NR, Bazan NG (1994) Membrane lipid degradation is related to interictal cortical activity in a series of seizures. Metab Brain Dis 9(2):161–170. doi:10.1007/BF01999769
16. Birkle DL, Bazan NG (1988) Cerebral perfusion of metabolic inactivators: a new method for rapid fixation of labile lipid pools in brain. Neurochem Res 13(9):849–852. doi:10.1007/BF00970752
17. Birkle DL, Kurian P, Braquet P, Bazan NG (1988) Platelet-activating factor antagonist BN52021 decreases accumulation of free polyunsaturated fatty acid in mouse brain during ischemia and electroconvulsive shock. J Neurochem 51(6):1900–1905. doi:10.1111/j.1471-4159.1988.tb01175.x
18. Reddy TS, Bazan NG (1987) Arachidonic acid, stearic acid, and diacylglycerol accumulation correlates with the loss of phosphatidylinositol 4,5-bisphosphate in cerebrum 2 seconds after electroconvulsive shock: complete reversion of changes 5 minutes after stimulation. J Neurosci Res 18(3):449–455. doi:10.1002/jnr.490180311
19. Van Rooijen LAA, Vadnal R, Dobard P, Bazan NG (1986) Enhanced inositide turnover in brain during bicuculline-induced status epilepticus. Biochem Biophys Res Commun 136(2):827–834. doi:10.1016/0006-291X(86)90515-2
20. Hattori K, Kajimura M, Hishiki T, Nakanishi T, Kubo A, Nagahata Y, Ohmura M, Yachie-Kinoshita A, Matsuura T, Morikawa T, Nakamura T, Setou M, Suematsu M (2010) Paradoxical ATP elevation in ischemic penumbra revealed by quantitative imaging mass spectrometry. Antioxid Redox Signal 13(8):1157–1167. doi:10.1089/ars.2010.3290
21. Sugiura Y, Zaima N, Setou M, Ito S, Yao I (2012) Visualization of acetylcholine distribution in central nervous system tissue sections by tandem imaging mass spectrometry. Anal Bioanal Chem 403(7):1851–1861. doi:10.1007/s00216-012-5988-5
22. Yao I, Setou M (2010) Animal care and tissue sample extraction for IMS. In: Setou M (ed) Imaging mass spectrometry. Protocols for mass microscopy. Springer, Tokyo, pp 33–39
23. Sugiura Y, Setou M (2010) Preparing biological tissue sections for imaging mass spectrometry. In: Setou M (ed) Imaging mass spectrometry. Protocol for mass microscopy. Springer, Tokyo, pp 41–54
24. Altelaar AFM, van Minnen J, Jiménez CR, Heeren RMA, Piersma SR (2004) Direct molecular imaging of *Lymnaea stagnalis* nervous tissue at subcellular spatial resolution by mass spectrometry. Anal Chem 77(3):735–741. doi:10.1021/ac048329g
25. Kruse R, Sweedler JV (2003) Spatial profiling invertebrate ganglia using MALDI MS. J Am Soc Mass Spectrom 14(7):752–759. doi:10.1016/S1044-0305(03)00288-5
26. Walch A, Rauser S, Deininger S-O, Höfler H (2008) MALDI imaging mass spectrometry for direct tissue analysis: a new frontier for molecular histology. Histochem Cell Biol 130(3):421–434. doi:10.1007/s00418-008-0469-9
27. Schiller J, Süß R, Fuchs B, Müller M, Petković M, Zschörnig O, Waschipky H (2007) The suitability of different DHB isomers as matrices for the MALDI-TOF MS analysis of phospholipids: which isomer for what purpose? Eur Biophys J 36(4-5):517–527. doi:10.1007/s00249-006-0090-6

28. Calvano CD, Monopoli A, Ditaranto N, Palmisano F (2013) 1,8-Bis(dimethylamino) naphthalene/9-aminoacridine: a new binary matrix for lipid fingerprinting of intact bacteria by matrix assisted laser desorption ionization mass spectrometry. Anal Chim Acta 798(0):56–63. doi:10.1016/j.aca.2013.08.050
29. Goto-Inoue N, Hayasaka T, Zaima N, Nakajima K, Holleran WM, Sano S, Uchida Y, Setou M (2012) Imaging mass spectrometry visualizes ceramides and the pathogenesis of Dorfman-Chanarin Syndrome due to ceramide metabolic abnormality in the skin. PLoS One 7(11):e49519. doi:10.1371/journal.pone.0049519
30. He Q, Takizawa Y, Hayasaka T, Masaki N, Kusama Y, Su J, Mineta H, Setou M (2014) Increased phosphatidylcholine (16:0/16:0) in the folliculus lymphaticus of Warthin tumor. Anal Bioanal Chem 406(24): 5815–5825 doi:10.1007/s00216-014-7890-9. Erratum in Anal Bioanal Chem 406(28):7409-7410. doi:10.1007/s00216-014-8113-0
31. Fuchs B, Schiller J (2009) Recent developments of useful MALDI matrices for the mass spectrometric characterization of apolar compounds. Curr Org Chem 13(16):1664–1681. doi:10.2174/138527209789578108
32. Vermillion-Salsbury RL, Hercules DM (2002) 9-Aminoacridine as a matrix for negative mode matrix-assisted laser desorption/ionization. Rapid Commun Mass Spectrom 16(16):1575–1581. doi:10.1002/rcm.750
33. Cerruti CD, Touboul D, Guérineau V, Petit VW, Laprévote O, Brunelle A (2011) MALDI imaging mass spectrometry of lipids by adding lithium salts to the matrix solution. Anal Bioanal Chem 401(1):75–87. doi:10.1007/s00216-011-4814-9
34. Jaskolla T, Karas M, Roth U, Steinert K, Menzel C, Reihs K (2009) Comparison between vacuum sublimed matrices and conventional dried droplet preparation in MALDI-TOF mass spectrometry. J Am Soc Mass Spectrom 20(6):1104–1114. doi:10.1016/j.jasms.2009.02.010
35. Griffiths RL, Sarsby J, Guggenheim EJ, Race AM, Steven RT, Fear J, Lalor PF, Bunch J (2013) Formal lithium fixation improves direct analysis of lipids in tissue by mass spectrometry. Anal Chem 85(15):7146–7153. doi:10.1021/ac400737z
36. Thomas A, Charbonneau JL, Fournaise E, Chaurand P (2012) Sublimation of new matrix candidates for high spatial resolution imaging mass spectrometry of lipids: enhanced information in both positive and negative polarities after 1,5-diaminonapthalene deposition. Anal Chem 84(4):2048–2054. doi:10.1021/ac2033547
37. Wang X, Han J, Chou A, Yang J, Pan J, Borchers CH (2013) Hydroxyflavones as a new family of matrices for MALDI tissue imaging. Anal Chem 85(15):7566–7573. doi:10.1021/ac401595a
38. Meriaux C, Franck J, Wisztorski M, Salzet M, Fournier I (2010) Liquid ionic matrixes for MALDI mass spectrometry imaging of lipids. J Proteomics 73(6):1204–1218. doi:10.1016/j.jprot.2010.02.010
39. Shrivas K, Hayasaka T, Goto-Inoue N, Sugiura Y, Zaima N, Setou M (2010) Ionic matrix for enhanced MALDI imaging mass spectrometry for identification of phospholipids in mouse liver and cerebellum tissue sections. Anal Chem 82(21):8800–8806. doi:10.1021/ac102422b
40. Silina YE, Volmer DA (2013) Nanostructured solid substrates for efficient laser desorption/ionization mass spectrometry (LDI-MS) of low molecular weight compounds. Analyst 138(23):7053–7065. doi:10.1039/C3AN01120H
41. Ronci M, Rudd D, Guinan T, Benkendorff K, Voelcker NH (2012) Mass spectrometry imaging on porous silicon: investigating the distribution of bioactives in marine mollusc tissues. Anal Chem 84(21):8996–9001. doi:10.1021/ac3027433
42. Kim Y-K, Na H-K, Kwack S-J, Ryoo S-R, Lee Y, Hong S, Hong S, Jeong Y, Min D-H (2011) Synergistic effect of graphene oxide/MWCNT films in laser desorption/ionization mass spectrometry of small molecules and tissue imaging. ACS Nano 5(6):4550–4561. doi:10.1021/nn200245v
43. Kong X, Huang Y (2014) Applications of graphene in mass spectrometry. J Nanosci Nanotechnol 14(7):4719–4732. doi:10.1166/jnn.2014.9503
44. Lu M, Lai Y, Chen G, Cai Z (2011) Matrix interference-free method for the analysis of small molecules by using negative ion laser desorption/ionization on graphene flakes. Anal Chem 83(8):3161–3169. doi:10.1021/ac2002559

45. Qian K, Zhou L, Liu J, Yang J, Xu H, Yu M, Nouwens A, Zou J, Monteiro MJ, Yu C (2013) Laser engineered graphene paper for mass spectrometry imaging. Sci Rep 3:1415. doi:10.1038/srep01415
46. Chen W-Y, Chen L-Y, Ou C-M, Huang C-C, Wei S-C, Chang H-T (2013) Synthesis of fluorescent gold nanodot–liposome hybrids for detection of phospholipase C and its inhibitor. Anal Chem 85(18):8834–8840. doi:10.1021/ac402043t
47. Arakawa R, Kawasaki H (2010) Functionalized nanoparticles and nanostructured surfaces for surface-assisted laser desorption/ionization mass spectrometry. Anal Sci 26(12):1229–1240. doi:10.2116/analsci.26.1229
48. Taira S, Sugiura Y, Moritake S, Shimma S, Ichiyanagi Y, Setou M (2008) Nanoparticle-assisted laser desorption/ionization based mass imaging with cellular resolution. Anal Chem 80(12):4761–4766. doi:10.1021/ac800081z
49. Hayasaka T, Goto-Inoue N, Zaima N, Shrivas K, Kashiwagi Y, Yamamoto M, Nakamoto M, Setou M (2010) Imaging mass spectrometry with silver nanoparticles reveals the distribution of fatty acids in mouse retinal sections. J Am Soc Mass Spectrom 21(8):1446–1454. doi:10.1016/j.jasms.2010.04.005
50. Goto-Inoue N, Hayasaka T, Zaima N, Kashiwagi Y, Yamamoto M, Nakamoto M, Setou M (2010) The detection of glycosphingolipids in brain tissue sections by imaging mass spectrometry using gold nanoparticles. J Am Soc Mass Spectrom 21(11):1940–1943. doi:10.1016/j.jasms.2010.08.002
51. Sugiura Y, Setou M, Horigome D (2010) Methods of matrix application. In: Setou M (ed) Imaging mass spectrometry. Protocols for mass microscopy. Springer, Tokyo, pp 71–85
52. Ishikawa S, Tateya I, Hayasaka T, Masaki N, Takizawa Y, Ohno S, Kojima T, Kitani Y, Kitamura M, Hirano S, Setou M, Ito J (2012) Increased expression of phosphatidylcholine (16:0/18:1) and (16:0/18:2) in thyroid papillary cancer. PLoS One 7(11):e48873. doi:10.1371/journal.pone.0048873
53. Kaneko Y, Obata Y, Nishino T, Kakeya H, Miyazaki Y, Hayasaka T, Setou M, Furusu A, Kohno S (2011) Imaging mass spectrometry analysis reveals an altered lipid distribution pattern in the tubular areas of hyper-IgA murine kidneys. Exp Mol Pathol 91(2):614–621. doi:10.1016/j.yexmp.2011.07.002
54. Kobayashi Y, Hayasaka T, Setou M, Itoh H, Kanayama N (2010) Comparison of phospholipid molecular species between terminal and stem villi of human term placenta by imaging mass spectrometry. Placenta 31(3):245–248. doi:10.1016/j.placenta.2009.12.026
55. Hayasaka T, Goto-Inoue N, Masaki N, Ikegami K, Setou M (2014) Application of 2,5-dihydroxyacetophenone with sublimation provides more efficient ionization of lipid species by atmospheric pressure matrix-assisted laser desorption/ionization imaging mass spectrometry. Surf Interface Anal 46(12–13):1219–1222. doi:10.1002/sia.5592
56. Miyamura N, Nakamura T, Goto-Inoue N, Zaima N, Hayasaka T, Yamasaki T, Terai S, Sakaida I, Setou M, Nishina H (2011) Imaging mass spectrometry reveals characteristic changes in triglyceride and phospholipid species in regenerating mouse liver. Biochem Biophys Res Commun 408(1):120–125. doi:10.1016/j.bbrc.2011.03.133
57. Nakajima K, Terao M, Takaishi M, Kataoka S, Goto-Inoue N, Setou M, Horie K, Sakamoto F, Ito M, Azukizawa H, Kitaba S, Murota H, Itami S, Katayama I, Takeda J, Sano S (2013) Barrier abnormality due to ceramide deficiency leads to psoriasiform inflammation in a mouse model. J Invest Dermatol 133(11):2555–2565. doi:10.1038/jid.2013.199
58. Onoue K, Zaima N, Sugiura Y, Isojima T, Okayama S, Horii M, Akai Y, Uemura S, Takemura G, Sakuraba H, Sakaguchi Y, Setou M, Saito Y (2011) Using imaging mass spectrometry to accurately diagnose Fabry's disease. Circ J 75(1):221–223. doi:10.1253/circj.CJ-10-0767
59. Tanaka H, Zaima N, Yamamoto N, Sagara D, Suzuki M, Nishiyama M, Mano Y, Sano M, Hayasaka T, Goto-Inoue N, Sasaki T, Konno H, Unno N, Setou M (2010) Imaging mass spectrometry reveals unique lipid distribution in primary varicose veins. Eur J Vasc Endovasc Surg 40(5):657–663. doi:10.1016/j.ejvs.2010.08.001

60. Yuki D, Sugiura Y, Zaima N, Akatsu H, Hashizume Y, Yamamoto T, Fujiwara M, Sugiyama K, Setou M (2011) Hydroxylated and non-hydroxylated sulfatide are distinctly distributed in the human cerebral cortex. J Neurosci 193(0):44–53. doi:10.1016/j.neuroscience.2011.07.045
61. Takahashi K, Ogawa K (2013) Sample preparation device for MALDI and sample preparation method. United States Patent 20130171349, 7 Apr 2013
62. Goto-Inoue N, Hayasaka T, Setou M (2010) Chapter fourteen – imaging mass spectrometry of glycolipids. In: Minoru F (ed) Methods in enzymology, vol 478. Academic Press, New York, pp 287–301. doi:10.1016/S0076-6879(10)78014-9
63. Kashiwagi Y, Yamamoto M, Nakamoto M (2006) Facile size-regulated synthesis of silver nanoparticles by controlled thermolysis of silver alkylcarboxylates in the presence of alkylamines with different chain lengths. J Colloid Interface Sci 300(1):169–175. doi:10.1016/j.jcis.2006.03.041
64. Yamamoto M, Kashiwagi Y, Nakamoto M (2009) Size-controlled synthesis of gold nanoparticles by thermolysis of a gold(I)-sulfide complex in the presence of alkylamines. Z Naturforsch B 64(11-12):1305–1311. doi:10.1515/znb-2009-11-1208
65. Hankin J, Barkley R, Murphy R (2007) Sublimation as a method of matrix application for mass spectrometric imaging. J Am Soc Mass Spectrom 18(9):1646–1652. doi:10.1016/j.jasms.2007.06.010
66. Schiller J, Süß R, Arnhold J, Fuchs B, Leßig J, Müller M, Petković M, Spalteholz H, Zschörnig O, Arnold K (2004) Matrix-assisted laser desorption and ionization time-of-flight (MALDI-TOF) mass spectrometry in lipid and phospholipid research. Prog Lipid Res 43(5):449–488. doi:10.1016/j.plipres.2004.08.001
67. Chansela P, Goto-Inoue N, Zaima N, Hayasaka T, Sroyraya M, Kornthong N, Engsusophon A, Tamtin M, Chaisri C, Sobhon P, Setou M (2012) Composition and localization of lipids in *Penaeus merguiensis* ovaries during the ovarian maturation cycle as revealed by imaging mass spectrometry. PLoS One 7(3):e33154. doi:10.1371/journal.pone.0033154
68. Goto-Inoue N, Hayasaka T, Taki T, Gonzalez TV, Setou M (2009) A new lipidomics approach by thin-layer chromatography-blot-matrix-assisted laser desorption/ionization imaging mass spectrometry for analyzing detailed patterns of phospholipid molecular species. J Chromatogr A 1216(42):7096–7101. doi:10.1016/j.chroma.2009.08.056
69. Ellis S, Bruinen A, Heeren RA (2014) A critical evaluation of the current state-of-the-art in quantitative imaging mass spectrometry. Anal Bioanal Chem 406(5):1275–1289. doi:10.1007/s00216-013-7478-9
70. Katagiri YU, Sato B, Yamatoya K, Taki T, Goto-Inoue N, Setou M, Okita H, Fujimoto J, Ito C, Toshimori K, Kiyokawa N (2011) GalNAcβ1,3-linked paragloboside carries the epitope of a sperm maturation-related glycoprotein that is recognized by the monoclonal antibody MC121. Biochem Biophys Res Commun 406(3):326–331. doi:10.1016/j.bbrc.2011.02.019
71. Valdes-Gonzalez T, Goto-Inoue N, Hirano W, Ishiyama H, Hayasaka T, Setou M, Taki T (2011) New approach for glyco- and lipidomics: molecular scanning of human brain gangliosides by TLC-Blot and MALDI-QIT-TOF MS. J Neurochem 116(5):678–683. doi:10.1111/j.1471-4159.2010.07152.x
72. Zaima N, Goto-Inoue N, Adachi K, Setou M (2011) Selective analysis of lipids by thin-layer chromatography blot matrix-assisted laser desorption/ionization imaging mass spectrometry. J Oleo Sci 60(2):93–98. doi:10.5650/jos.60.93
73. Goto-Inoue N, Hayasaka T, Sugiura Y, Taki T, Li Y-T, Matsumoto M, Setou M (2008) High-sensitivity analysis of glycosphingolipids by matrix-assisted laser desorption/ionization quadrupole ion trap time-of-flight imaging mass spectrometry on transfer membranes. J Chromatogr B 870(1):74–83. doi:10.1016/j.jchromb.2008.06.001

Chapter 28
Measuring Activation of Lipid G Protein-Coupled Receptors Using the TGF-α Shedding Assay

Asuka Inoue and Junken Aoki

Abstract Lipid mediators are intercellular bioactive lipid molecules that induce various cellular responses mostly through G protein-coupled receptors (GPCRs). According to the IUPHAR database and recent publications, there are now approximately 50 lipid-recognizing GPCRs. Monitoring activation of GPCRs by lipid mediators or synthetic ligands is crucial for studying the actions of bioactive lipids as well as lipid GPCRs. Because each GPCR differentially couples with heterotrimeric G proteins (G_s, $G_{i/o}$, $G_{q/11}$, $G_{12/13}$), in general it is necessary to prepare multiple GPCR assays to detect distinct G-protein signaling and individual optimization. In this protocol section, we describe a recently developed transforming growth factor-α (TGF-α) shedding assay, which is a simple and accurate method for measuring activation of lipid GPCRs. By utilizing coexpression of chimeric Gα subunits, the TGF-α shedding assay can detect activation of a wide range of GPCRs that are coupled with any of the four G proteins. Indeed, 39 of 45 (87 %) lipid GPCRs examined were detectable in the single format of the TGF-α shedding assay, demonstrating, to the best of our knowledge, the greatest coverage of a lipid GPCR assay. Thus, the TGF-α shedding assay provides a useful platform for analyzing lipid GPCRs.

Keywords Alkaline phosphatase (AP) • Bioactive lipid • Chimeric Gα subunit • G protein-coupled receptor (GPCR) • HEK293 cells • Lipid mediator • p-Nitrophenylphosphate (*p*-NPP) • TGF-α shedding assay • TNF-α-converting enzyme (TACE)

A. Inoue (✉)
Laboratory of Molecular and Cellular Biochemistry, Graduate School of Pharmaceutical Sciences, Tohoku University, Sendai 980-8578, Miyagi, Japan

PRESTO, Japan Science and Technology Agency (JST), Kawaguchi 332-0012, Saitama, Japan
e-mail: iaska@m.tohoku.ac.jp

J. Aoki
Laboratory of Molecular and Cellular Biochemistry, Graduate School of Pharmaceutical Sciences, Tohoku University, Sendai 980-8578, Miyagi, Japan

CREST, Japan Science and Technology Agency (JST), Kawaguchi 332-0012, Saitama, Japan

T. Yokomizo, M. Murakami (eds.), *Bioactive Lipid Mediators*,
DOI 10.1007/978-4-431-55669-5_28

28.1 Background Information

G protein-coupled receptors (GPCRs), also known as seven-transmembrane receptors (7TMs), represent the most divergent and important class of lipid receptors. According to the International Union of Basic and Clinical Pharmacology (IUPHAR) database, there are approximately 50 lipid-recognizing GPCRs such as platelet-activating factor (PAF) receptor, prostanoid receptors, leukotriene receptors, lysophospholipid receptors, and free fatty acid receptors. Numerous studies including human and mouse genetics have established lipid GPCRs as important components for development, physiology, and pathophysiology. Importantly, some of them are validated as therapeutic drug targets. For example, cysteinyl leukotriene 1 receptor (CysLT1) antagonist is used for treating asthma, prostacyclin receptor (IP) agonist is used for treating pulmonary hypertension, and the functional antagonist sphingosine 1-phosphate receptor 1 ($S1P_1$) is used for treating multiple sclerosis. In addition, several lipid GPCR-targeting drugs such as lysophosphatidic acid receptor 1 (LPA_1) antagonist for treating pulmonary fibrosis are under clinical trials. Thus, lipid GPCRs attract interest from both biological and drug discovery aspects.

Conventionally, GPCR activity (or ligand-induced GPCR activation/inhibition) is measured by monitoring G protein-evoked downstream cellular events. For instances, accumulation or inhibition of cAMP, Ca^{2+} mobilization, and transcriptional activation of reporter genes are widely used events. However, because a given GPCR is coupled with one or a few G proteins among the four classes of heterotrimeric G proteins (G_s, $G_{i/o}$, $G_{q/11}$, and $G_{12/13}$) and GPCR assays detect a limited G-protein pathway, it is necessary to prepare a series of GPCR assays to detect activation of many GPCRs without bias. GPCR assays often require costly reagents/kits and expensive, specialized equipment. Furthermore, although many lipid GPCRs, especially lysophospholipid GPCRs, are coupled with $G_{12/13}$, current methods are limited in efficient detection of $G_{12/13}$ signaling.

We have recently developed a novel GPCR assay [1], which is based on cleavage of a transmembrane reporter protein, alkaline phosphatase-fused transforming growth factor-α (AP-TGF-α), and its release into conditioned media (Fig. 28.1). AP-TGF-α is initially expressed as a membrane proform. When GPCRs are activated and induce $G_{q/11}$ signaling or $G_{12/13}$ signaling, a membrane protease named TNF-α-converting enzyme (TACE, also known as ADAM17) is activated and cleaves an extracellular site near the transmembrane region of the AP-TGF-α. This process is called ectodomain shedding, from which we named the method "TGF-α shedding assay." The amount of shedding of the AP-TGF-α can be determined by measuring AP activity in conditioned media using an inexpensive reagent (p-nitrophenylphosphate, a substrate for AP). Importantly, chimeric Gα subunits and a promiscuous $G\alpha_{16}$ subunit (Fig. 28.2) are applicable to the assay, and coexpression of these Gα subunits can induce AP-TGF-α shedding responses in G_s- or $G_{i/o}$-coupled receptors. Among 45 lipid GPCRs examined, 39 GPCRs induced AP-TGF-α shedding responses (Fig. 28.3). This percentage of coverage (87 % detection rate) in a single assay format is, to the best of our knowledge, the highest

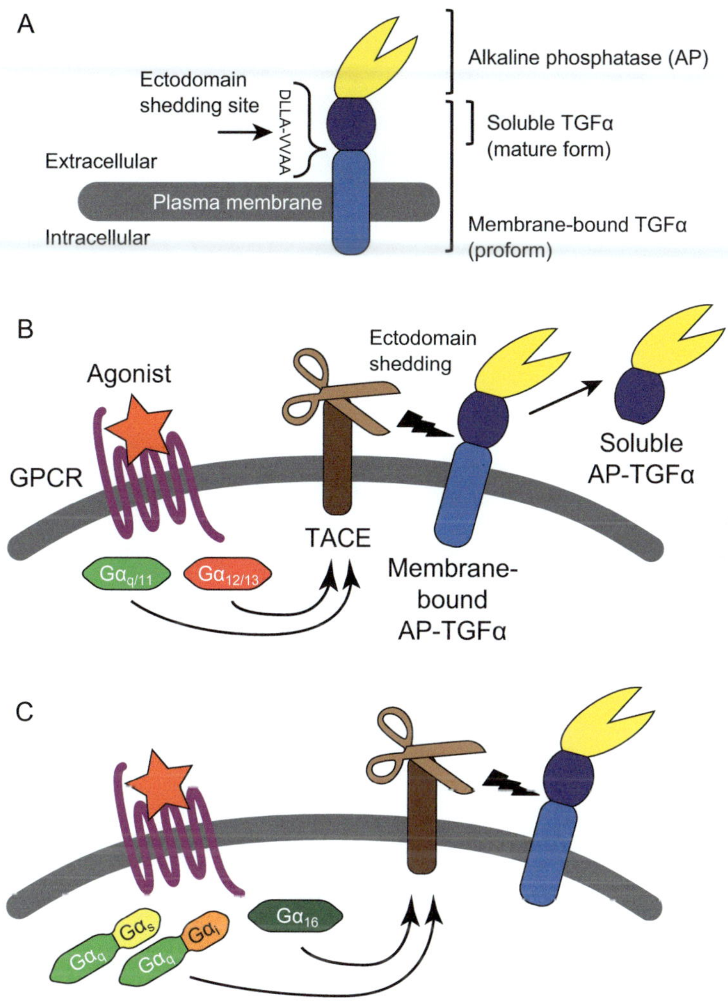

Fig. 28.1 Mechanism of the transforming growth factor (TGF)-α shedding assay. (**a**) Schematic structure of the AP-TGF-α construct. AP-TGF-α is a fusion protein consisting of N-terminal placental alkaline phosphatase (AP) and C-terminal TGF-α with a signal peptide. AP-TGF-α is initially expressed as membrane-bound proform and its juxtamembrane region is cleaved (ectodomain shedding). In HEK293 cells, TACE is responsible for the ectodomain shedding of AP-TGF-α. Amount of shedding of AP-TGF-α is determined by measuring AP activity in conditioned media (soluble AP-TGF-α). (**b**) $G_{q/11}$- and/or $G_{12/13}$-coupled receptors induce TACE-dependent AP-TGF-α shedding responses upon agonist stimulation. G-protein effectors such as phospholipase C, protein kinase C, RhoA, and ROCK mediate the process (not shown). (**c**) In G_s- or G_i-coupled receptors, chimeric $G\alpha_q$ subunits harboring C-terminal $G\alpha_s$-derived sequences ($G\alpha_{q/s}$; also see Fig. 28.2) or $G\alpha_i$-derived sequences ($G\alpha_{q/i}$) and/or promiscuous $G\alpha_{16}$ subunit are coexpressed. When agonist-bound GPCRs interact with and activate the introduced Gα subunit(s), TACE-dependent AP-TGF-α shedding responses are induced

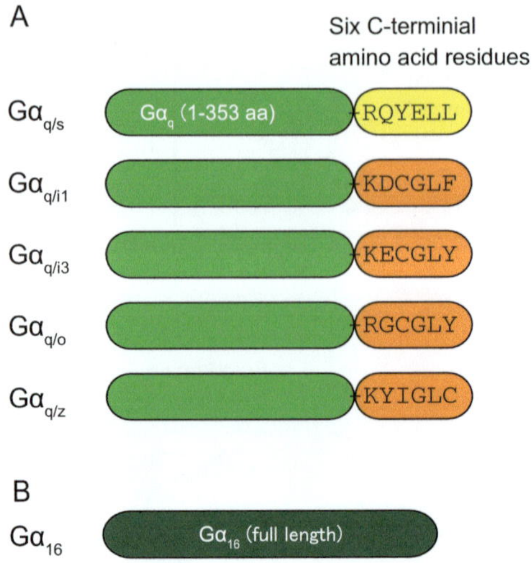

Fig. 28.2 Schematic structures of the Gα subunits. (**a**) Chimeric Gα subunits used in the TGF-α shedding assay consist of the $G\alpha_q$ backbone with substitutions of six C-terminal amino acids. Among the eight $G\alpha_i$ members ($G\alpha_{i1}$, $G\alpha_{i2}$, $G\alpha_{i3}$, $G\alpha_o$, $G\alpha_z$, $G\alpha_{t1}$, $G\alpha_{t2}$, $G\alpha_{t3}$), there are four unique C-terminal sequences because the six amino acid sequences of $G\alpha_{i1}$, $G\alpha_{i2}$, $G\alpha_{t1}$, $G\alpha_{t2}$, and $G\alpha_{t3}$ are identical. (**b**) $G\alpha_{16}$ belongs to the $G\alpha_q$ family, yet is capable of coupling with GPCRs in a relatively unspecific manner. (Note that C-terminal residues and the backbones of Gα subunits are critical for interaction with GPCRs and effectors such as phospholipase C, respectively, and that chimeric Gα subunits serve as a "converter" that switches G_s- or $G_{i/o}$-coupled receptor activation to G_q signaling). aa, amino acids

among GPCR assays developed so far. In the following protocol, a standard TGF-α shedding assay (24-well format in a 96-well plate; transfection performed in a 12-well plate) is described.

28.2 TGF-α Shedding Assay

28.2.1 Materials

Expression plasmid vector encoding AP-TGF-α (alkaline phosphatase-fused transforming growth factor-alpha).

Expression plasmid vectors encoding chimeric Gα subunits ($G\alpha_{q/s}$, $G\alpha_{q/i1}$, $G\alpha_{q/i3}$, $G\alpha_{q/o}$, $G\alpha_{q/z}$) and promiscuous $G\alpha_{16}$ subunit.

Expression plasmid vector encoding histamine H1 receptor (H1R) with N-terminal FLAG epitope tag (a positive control GPCR)

pCAGGS expression vector (an empty vector; a negative control)

We use the pCAGGS expression vector and pcDNA3 vector (or its derivative pcDNA3.1 vector) because expression of these vectors is high in HEK293 cells.

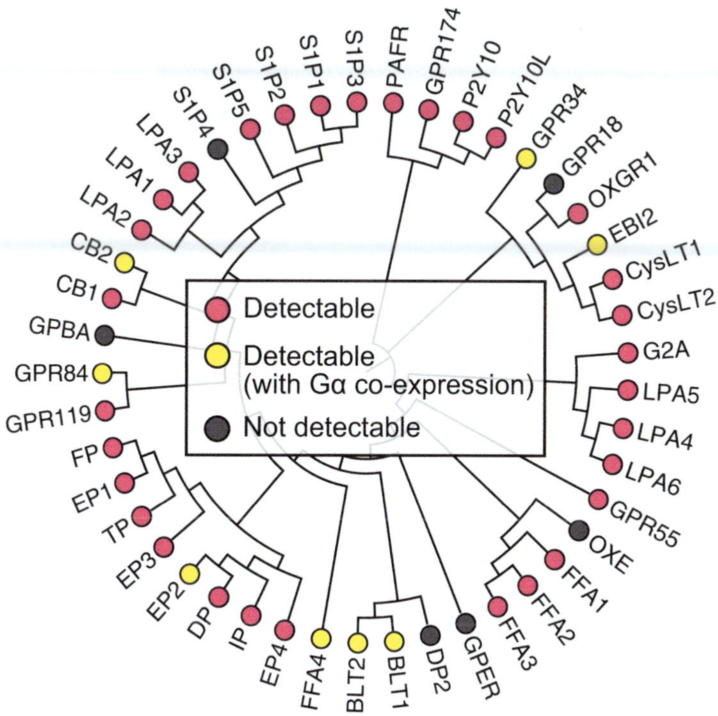

Fig. 28.3 Detection of lipid GPCRs using the TGF-α shedding assay. Forty-five lipid-recognizing GPCRs were subjected to the TGF-α shedding assay with or without coexpression of Gα subunits. In the absence of Gα coexpression, 31 GPCRs (69 %) were detectable (shown in *red*). In the presence of Gα coexpression, 8 GPCRs were additionally detectable (*yellow*). In total, 39 GPCRs (87 %) were detectable in the TGF-α shedding assay. We set 3 % AP-TGF-α release as a threshold to judge "detectable" or "not detectable." A phylogenetic tree is obtained from the Clustal W algorithm and visualized by the FigTree software

Plasmids are available upon request to the authors (iaska@m.tohoku.ac.jp to Asuka Inoue and jaoki@m.tohoku.ac.jp to Junken Aoki).

HEK293 cells

293A (Life Technologies, cat. no. R705-07) is preferable.

293FT (Life Technologies, cat. no. R700-07) and 293T (ATCC, cat. no. CRL-3216) can be used.

12-well tissue culture plate

CELLSTAR, 12W Multiwell Plate (Greiner Bio One, cat. no. 665165)

96-well tissue culture plate (flat bottom, clear)

CELLSTAR, 96W Microplate (Greiner Bio One, cat. no. 655180)

96-well assay plate (flat bottom, clear)

Assay plate 96 well (Iwaki, cat. no. 3881-096)

96-well assay plate (round bottom or V bottom, nonbinding)

96 well, PS, V bottom (Greiner Bio One, cat. no. 651901)

Absorbance microplate reader (for measurement of optical density at 405 nm)

VersaMax ELISA Microplate Reader (Molecular Devices)

12-channel or 8-channel pipette (electric pipettes are preferred)

Picus 12-channel electronic pipette, range 5–120 μl (Sartorius, cat. no. 735441)
Picus 12-channel electronic pipette, range 50–1200 μl (Sartorius, cat. no. 735491)

5-ml manual pipette (single channel)

PIPETMAN P5000 (Gilson, cat. no. F123603)

Centrifuge tubes (15 ml, 50 ml)

Corning 15-ml PP centrifuge tubes (Corning, cat. no. 430791)
Corning 50-ml PP centrifuge tubes (Corning, cat. no. 430829)

Tube centrifuge (Kubota, cat. no. 5910)
Microplate centrifuge (Kubota, cat. code Plate Spin II), optional
37 °C humidified incubator with 5 % CO_2 gas (for cell culture)
37 °C incubator without CO_2 gas (for AP reaction), optional

Incubator (AS-One, cat. no. IW-450)

37 °C hot plate, optional

Hot plate (NISSIN, cat. no. NHP-45N)

Complete Dulbecco's modified Eagle medium (DMEM) (see recipe)
Dulbecco's phosphate-buffered saline (PBS) (D-PBS)
HBSS (Hank's balanced salt solution) (see recipe)
0.05 % Trypsin/0.53 mM EDTA (see recipe)
Lipofectamine 2000 (Life Technologies, cat. no. 11668-019)
Opti-MEM (Life Technologies, cat. no. 31985-070)
1 M *p*-NPP (see recipe)
1 mM TPA (12-*O*-tetradecanoylphorbol 13-acetate, also known as PMA) (phorbol 12-myristate 13-acetate) (see recipe)
AP solution (see recipe)

28.2.2 *Three-Day Protocol (Also See Fig. 28.4)*

28.2.2.1 Seed Seeding (Day 1)

1. Prepare semi-confluent HEK293 cells grown in a culture dish or flask.
2. Rinse cells once with D-PBS.

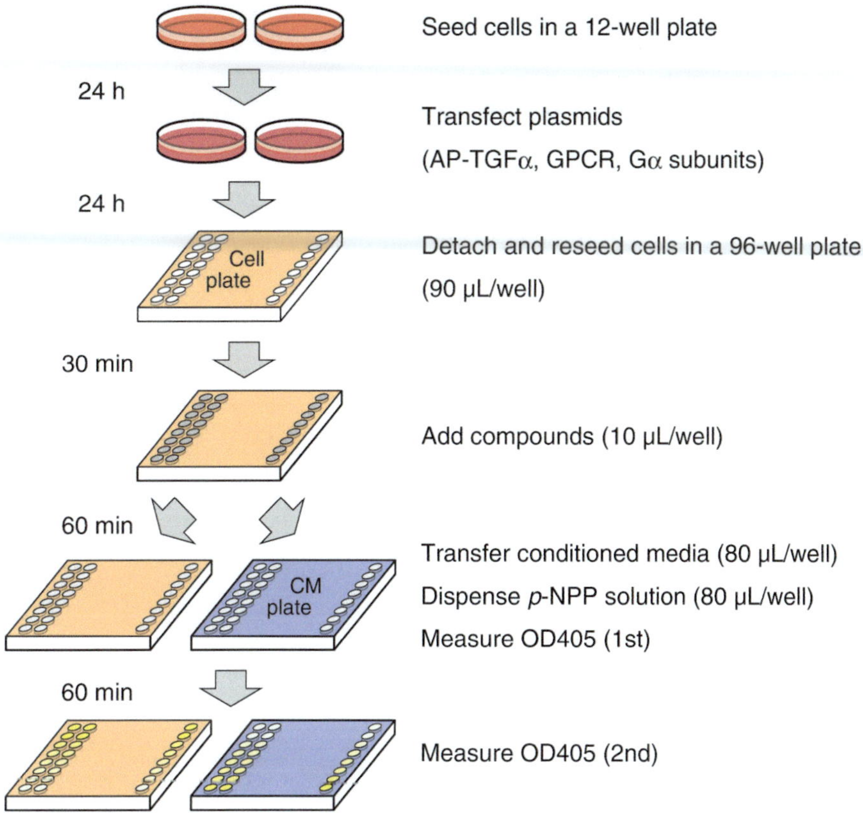

Fig. 28.4 Procedure of the TGF-α shedding assay. HEK293 cells are seeded in a 12-well plate and cultured for 24 h. Cells are transfected with a mixture of plasmids containing the AP-TGF-α, GPCR, and/or Gα subunit(s) and cultured for 24 h. Transfected cells are detached, suspended in HBSS, and reseeded in a 96-well plate (cell plate). After 30-min incubation, cells are stimulated with compounds and incubated for 60 min. Conditioned media are transferred to a blank plate (CM plate), and *p*-NPP solution is dispensed in both the cell plate and the CM plate. OD_{405} is measured before (first) and after (second) 60-min incubation with *p* NPP

3. Add 0.05 % trypsin/0.53 mM EDTA and incubate cells at room temperature until cells start to detach from bottom of a dish.
4. Add complete DMEM to neutralize trypsin and quench EDTA.
5. Suspend cells several times and transfer them into a 15-ml tube.
6. Transfer small amount of cell suspension (~20 μl) and count cell numbers using a hematocytometer.
7. While counting cells, centrifuge the cell-containing tube at 190 *g* for 5 min.

8. Remove supernatant.
9. Suspend cells in complete DMEM at a cell density of 2×10^5 cells/ml.
   Option: cells can be seeded 2 days before transfection (1×10^5 cells/ml) or same day as transfection (4×10^5 cells/ml).
10. Seed cell suspension in a 12-well plate (1 ml per well; hereafter, volumes refer to a single well of a 12-well plate, unless otherwise noted).
    If larger amounts of transfected cells are desired, use larger plates, dishes, or flasks (e.g., 6-well plate, 6-cm dish, 10-cm dish, T75).
11. Place a plate in a CO_2 incubator and incubate 24 h until plasmid transfection

28.2.2.2 Plasmid Transfection (Day 2)

12. Prepare following plasmid mixture in 125 μl Opti-MEM:

    10a. $G_{q/11}$- and/or $G_{12/13}$-coupled GPCRs

    250 ng AP-TGF-α
    100 ng GPCR

    10b. G_s-coupled GPCRs

    250 ng AP-TGF-α
    100 ng GPCR
    50 ng $G\alpha_{q/s}$

    10c. $G_{i/o}$-coupled GPCRs

    250 ng AP-TGF-α
    100 ng GPCR
    50 ng $G\alpha_{q/i1}$
    or
    250 ng AP-TGF-α
    100 ng GPCR
    12.5 ng $G\alpha_{q/i1}$
    12.5 ng $G\alpha_{q/i3}$
    12.5 ng $G\alpha_{q/o}$
    12.5 ng $G\alpha_{q/z}$

    10d. G-protein coupling is unknown

    250 ng AP-TGF-α
    100 ng GPCR
    10 ng $G\alpha_{q/s}$
    10 ng $G\alpha_{q/i1}$
    10 ng $G\alpha_{q/i3}$
    10 ng $G\alpha_{q/o}$
    10 ng $G\alpha_{q/z}$
    10 ng $G\alpha_{16}$

Always prepare a negative control for GPCR-encoding plasmid (i.e., an empty plasmid vector plus AP-TGF-α-encoding plasmid) to confirm that compound-induced response is dependent on transfected GPCR.

As an initial experiment, we recommend the following setup (also see Fig. 28.5).

(i) 350 ng pCAGGS

(ii) 250 ng pCAGGS
100 ng H1R

(iii) 250 ng AP-TGF-α
100 ng pCAGGS

(iv) 250 ng AP-TGF-α
100 ng H1R

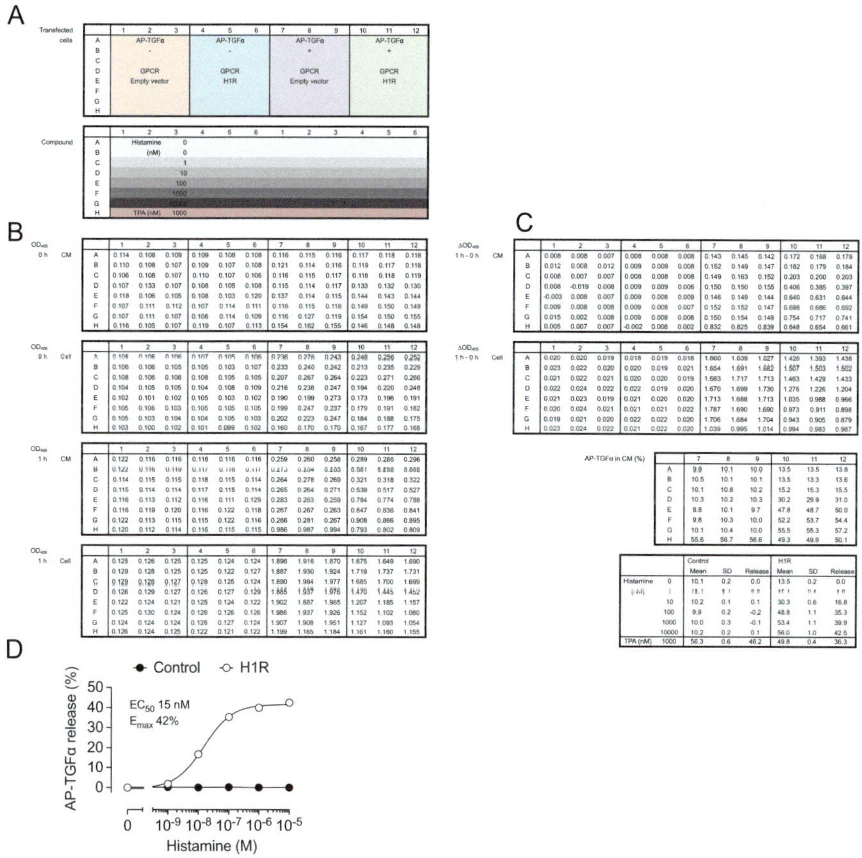

Fig. 28.5 Calculation processes. (**a**) A 96-well plate format showing reseeding of transfected cells and compound treatment. Cells were transfected with the AP-TGF-α-encoding plasmid (or an empty vector) and H1R-encoding plasmid (or an empty vector) and treated with vehicle (row *A* and *B*), histamine (row *C–G*, 1 nM to 10 μM), or 1 μM TPA (row *H*). (**b**) Raw OD values. In total, four measurements for a single 96-well plate assay were performed. (**c**) Calculation of AP-TGF-α release. (**d**) Plotting of AP-TGF-α release and fitting to four-sigmoidal curves, from which E_{max} and EC_{50} values were obtained

Conditions (i) and (ii) are negative controls for AP-TGF-α expression and AP activity measurements. Expression of endogenous AP activity in HEK293 cells should be more than 20 fold as low as AP-TGF-α-transfected cells. Histamine H1 receptor (H1R) serves as one of excellent positive controls for the following reasons: H1R potently induces shedding response of AP-TGF-α; HEK293 do not endogenously express H1R; variety of H1R ligands including agonists, antagonist, and inverse agonists are available and testable in the TGF-α shedding assay.

13. Dilute 1.25 μl Lipofectamine 2000 with 125 μl Opti-MEM and incubate 30 min at room temperature.
14. Add 125 μl Lipofectamine 2000-containing Opti-MEM to plasmid-containing Opti-MEM, mix well, and incubate 20 min at room temperature (transfection reagent).
15. Gently add 250 μl transfection reagent into cell culture plate.
16. Place a plate in a CO_2 incubator and incubate 24 h until shedding assay.

28.2.2.3 Reseeding and Stimulation of Transfected Cells (Day 3)

17. Remove conditioned media by aspiration.
18. Gently add D-PBS (~1 ml).
19. Remove D-PBS by aspiration.
20. Add warm 0.05 % Trypsin/0.53 mM EDTA (~0.5 mL).
21. Incubate briefly (~30 s) at room temperature.
22. Add complete DMEM (~1 ml) and suspend
23. Add complete DMEM to neutralize trypsin and quench EDTA.
24. Detach cells by pipetting up and down several times and transfer cells into a 15-ml tube.
25. Centrifuge tubes at 190 *g* for 3–5 min.
26. Remove supernatant by aspiration.
27. Break cell pellets by tapping bottom of tube a few times, and add D-PBS (~3 ml).

    If cell pellets appear in solutions, pipette cell suspension up and down several times.
28. Incubate 10 min.

    This incubation step is important for reducing spontaneous release of AP-TGF-α. During this incubation, detachment-induced shedding of AP-TGF-α is accumulated in media. Following the next centrifugation and removal of media, the spontaneously released AP-TGF-α is removed. Incubation time can be extended up to 30 min.
29. Centrifuge tubes at 190 *g* for 3–5 min.
30. Remove supernatant by aspiration.
31. Break cell pellets by tapping bottom of a tube a few times, and add warm HBSS (~3.5 ml).

    Volume of HBSS can be varied from 3 to 8 ml. Note that response in some GPCRs depends on cell density.

32. Suspend cells by pipetting up and down several times. Use a 5-ml liquid pipette (P5000), if possible.
33. Transfer cell suspension to a reservoir and dispense 90 μl cell suspension per well (hereafter, volume refers to a single well of a 96-well plate, unless otherwise noted) in a 96-well plate (cell plate). Use 24 wells for each transfected condition (also see Fig. 28.5).

    When using an electric multichannel pipette, select "slow" dispense speed.
34. Place the cell plate in a CO_2 incubator.
35. Incubate 30 min at 37 °C.
36. Add 10 μl 10× test compounds diluted in HBSS. Use an electric multichannel pipette, if possible.

    If test compounds are not readily soluble in water or have "sticky" properties, use bovine serum albumin (BSA) as a carrier. We recommend a concentration of 0.01 % (w/v) BSA in HBSS as solvent to dilute 10× compounds. In this condition, a final BSA concentration is 0.001 % (w/v).
37. Incubate 60 min at 37 °C in CO_2 incubator.

28.2.2.4 Measuring AP Activity in Cells and Conditioned Media (Day 3, Continued)

38. Centrifuge the cell plate at 190 *g* for 2 min at room temperature (optional).

    Set "slow" deceleration speed, if possible. We recommend including this step to prevent possible contamination of cells in transferred conditioned media in the following procedure. After centrifugation, quickly start to transfer conditioned media. Do not warm the cell plate as this causes convection within wells and the cells might be stirred up.
39. Transfer 80 μl conditioned media to a new 96-well plate (CM plate) using an electric multichannel pipette.

    Set "slow" pipetting-up speed. When starting a transfer from row A to B, to C, using a 12-channel pipette, there is no need to change or wash tips. In between plates, wash tips once with water. If bubbles appear in the CM plate (likely to occur in high (≥0.01 %) concentration of BSA), briefly centrifuge plates.
40. Dispense 60 μl HBSS in cell plate to adjust liquid volumes of 80 μl in both cell plate and CM plate (optional).
41. Leave the cell plate and the CM plate for 10 min at room temperature and cool them down.
42. Dispense 80 μl *p*-NPP solution to both the cell plate and the CM plate.

    Prepare *p*-NPP solution 1 h before the assay and leave it at room temperature. Be sure to use room-temperature *p*-NPP solution as AP reaction depends on temperature.
43. Measure optical density of a wavelength at 405 nm (OD_{405}) for both the cell plate and the CM plate.
44. Leave the cell plate and the CM plate in dark for 60 min at room temperature.

45. Measure OD_{405} for both the cell plate and the CM plate.

    If OD values are small [typically, total OD values ($OD_{405\,cell} + OD_{405\,CM}$) ≤ 0.5], additionally incubate the cell plate and the CM plate for 60 min and measure OD_{405} thereafter.

28.2.2.5 Alternate Method to Measure AP Activity Using a 37 °C Incubator (Day 3)

For steps 41–45, a temperature of 37 °C can be applied for measuring AP activity. AP reaction is faster at 37 °C than at room temperature (approximately 1.4-fold faster than at 25 °C), thus allowing assay time to be shortened.

41b. Place the cell plate and the CM plate on a 37 °C hot plate.

42b. Dispense 80 μl *p*-NPP solution to both the cell plate and the CM plate. Leave the cell plate and the CM plate on a 37 °C hot plate for 5 min.

Prepare *p*-NPP solution 1 h before the assay and warm it to 37 °C. Because AP reaction depends on temperature, it is strongly recommended to leave plates on a hot plate until the solution of the plates is entirely warmed.

43b. Measure OD_{405} for both the cell plate and the CM plate.

44b. Immediately place the cell plate and the CM plate in a 37 °C incubator.

Use an incubator without CO_2 gas supplementation as CO_2 gas acidifies solution and slows down AP reaction.

45b. Measure OD_{405} for both the cell plate and the CM plate.

If OD values are small [typically, total OD values ($OD_{405\,cell} + OD_{405\,CM}$) ≤ 0.5], additionally incubate the cell plate and the CM plate for 60 min and measure OD_{405} thereafter.

28.2.2.6 Calculation of AP-TGF-α Release (Also See Fig. 28.5)

46. For each well, calculate AP activity in CM using the following formula.

    $\Delta OD_{405\,Cell} = OD_{405\,Cell}$ (at 1 h reaction) – $OD_{405\,Cell}$ (at 0 h)

    $\Delta OD_{405\,CM} = OD_{405\,CM}$ (at 1 h reaction) – $OD_{405\,CM}$ (at 0 h)

    AP-TGF-α in CM (% of total AP-TGF-α) = $\Delta OD_{405\,CM} / (\Delta OD_{405\,Cell} + \Delta OD_{405\,CM}) \times 100 \times 1.25$

    A factor of 1.25 (100/80) in the formula is used to normalize transferred (measured) CM volume to the total CM volume (80 μl of 100 μl in total).

47. Calculate mean and standard deviation (SD) values from multiple measurements.
48. Subtract mean values of vehicle-treated AP-TGF-α in CM from those of compound-treated ones.

AP-TGF-α release (compound-treated group) (% of total AP-TGF-α) = AP-TGF-α in CM (compound-treated group) (%) − AP-TGF-α release (vehicle-treated group) (%)

49. Plot mean values of AP-TGF-α release with corresponding SD values and fit a four-sigmoidal dose–response curve using a graph-drawing software such as GraphPad Prism.

28.2.3 Critical Parameters and Optimization

There are a few critical steps to obtain accurate results: (1) uniform resuspension of cell pellet in HBSS, (2) cautious transfer of CM, and (3) temperature control, especially during AP reaction.

Development of the TGF-α shedding assay for a specific GPCR requires optimization of the following parameters:

1. Gα subunit
   A mixture of the six Gα subunits is recommended as a default condition. However, it is strongly advisable to examine coexpression of Gα subunits separately (seven conditions: no Gα, $G\alpha_{q/s}$, $G\alpha_{q/i1}$, $G\alpha_{q/i3}$, $G\alpha_{q/o}$, $G\alpha_{q/z}$, $G\alpha_{16}$) and to select one that induces the most potent shedding response.
2. GPCR plasmid
   In some GPCRs, GPCR transfection reduces AP-TGF-α expression levels (or total OD_{405} value $OD_{405\ cell} + OD_{405\ CM}$) as compared with mock transfection. We speculate that the decreased AP-TGF-α results from constitutive activity of GPCRs and/or induction of ER stress by unstable, hydrophobic GPCRs. GPCRs that have high constitutive activity spontaneously cleave AP-TGF-α during 24-h incubation after transfection, resulting in a smaller amount of AP-TGF-α on the cell surface at the time of the assay.

28.2.4 Reagents and Solutions

Use deionized, distilled water or equivalent ultrapure (e.g., Milli-Q from Merck Millipore) water in all recipes and protocol steps except for *p*-NPP reagents (deionized water).

Complete DMEM [DMEM supplemented with 10 % fetal calf serum (FCS), 100 U/ml penicillin, 100 μg/ml streptomycin, 2 mM L-glutamine].

10 % (w/v) $NaHCO_3$

$NaHCO_3$, 10 g
Water to 100 ml
Close cap tightly, autoclave at 121 °C for 15 min, and store at room temperature

100× PSG

Penicillin, 1,000,000 U
Streptomycin, 1 g
L-glutamine, 2.92 g
Water to 100 ml
Filtrate through a 0.45-μm-pore filter
Make aliquots (10 ml solution in 15-ml tubes) and store at −20 °C.

Complete DMEM

DMEM (low glucose, Nissui Pharmaceuticals, cat. no. DMEM2) 4.75 g
Water to 500 ml
Autoclave at 121 °C for 15 min; store at 4 °C
10 % $NaHCO_3$, 7.5 ml
Heat-inactivated FCS (Life Technologies, cat. no. 26140-079), 50 ml
100× PSG, 5 ml

1 M *p*-NPP (10 ml)

3.71 g *p*-nitrophenylphosphate disodium salt hexahydrate (Wako Pure Chemical, cat. no. 145-02344)
Water to 10 ml (~7.8 ml)
Make aliquots (1 ml solution in 1.5-ml tubes) and store at −20 °C.

HBSS (Hank's balanced salt solution) (with Ca^{2+} and Mg^{2+} and with 5 mM HEPES (pH 7.4), without phenol red) (*note 1)

10× HBSS solution 1

KCl, 4 g
KH_2PO_4, 0.6 g
NaCl, 80 g
Na_2HPO_4•$12H_2O$, 1.2 g
D-Glucose, 10 g
Water to 1 l
Autoclave at 121 °C for 15 min and store at 4 °C

10× HBSS solution 2

$CaCl_2$•$2H_2O$, 1.85 g
$MgCl_2$•$6H_2O$, 1 g
$MgSO_4$•$7H_2O$, 1 g
Water to 1 l
Autoclave at 121 °C for 15 min and store at 4 °C

1× HBSS

Water, 790 ml
0.5 M HEPES (pH 7.4), 10 ml
1 M KOH (*note 2)

Autoclave at 121 °C for 15 min
Add 100 ml 10× HBSS solution 1 and 100 ml 10× HBSS solution 2
Store at 4 °C

Note 1

- Ensure use of Ca^{2+}- and Mg^{2+}-containing solution. TGF-α shedding response is decreased in Ca^{2+}- and Mg^{2+}-free solution.

Note 2

- When preparing HBSS for the first time, check pH after adding 10× HBSS solution 1 and 2 and warming them at 37 °C. Adjust pH to 7.4 using 1 M KOH and record volume of added KOH solution. For the next time, add recorded volume of 1 M KOH (typically 0.5–1 ml).

0.05 % Trypsin/0.53 mM EDTA

100× Trypsin/EDTA

Trypsin (Life Technologies, cat. no. 27250-018), 2 g
D-PBS, to 38 ml
500 mM EDTA (pH 8.0), 2.15 ml
Dissolve trypsin powder in a 50-ml tube and filtrate through a 0.45-μm-pore filter.
Make aliquots (~0.5 ml) and store at −20 °C.

1× Trypsin/EDTA

100× Trypsin/EDTA, 0.5 ml
D-PBS, 49.5 ml
Store at 4 °C.

1 mM TPA

TPA (12-*O*-tetradecanoylphorbol 13-acetate, also known as PMA, phorbol 12-myristate 13-acetate); Wako Pure Chemical, cat. no. 162-23591) 1 mg
DMSO, 1.62 ml
Make aliquots (5 μl per tube) and store at −20 °C. Avoid freeze–thaw cycles.

AP reaction solution

Water, 92 ml
2 M Tris–HCl (pH 9.5), 6 ml
4 M NaCl, 1 ml
1 M $MgSO_4$, 1 ml
1 M *p*-NPP, 1 ml
For one set of 96-well plate assay (CM plate and cell plate), prepare 16 ml AP reaction solution.

Reference

1. Inoue A et al (2012) TGFα shedding assay: an accurate and versatile method for detecting GPCR activation. Nat Methods 9:1021

Chapter 29
A Novel Anti-FLAG Monoclonal Antibody Is Useful to Study GPCRs

Fumiyuki Sasaki and Takehiko Yokomizo

Abstract Epitope tagging is a technique widely used in molecular and cellular biology. FLAG (DYKDDDDK), influenza virus hemagglutinin (YPYDVPDYA), and c-*myc* (EQKLISEEDL) tags are famous and are frequently used because high-affinity antibodies against these tags are commercially available. This technique is extremely useful for G protein-coupled receptor (GPCR) research because it is generally difficult to establish specific monoclonal antibodies (mAb) against GPCRs. We unexpectedly established a novel anti-FLAG mAb (2H8) during an attempt to generate an anti-mouse leukotriene B_4 receptor 1 mAb. This mAb is a powerful tool to analyze various FLAG-fusion proteins, particularly GPCRs, both in vitro and in vivo. In this chapter, we describe experimental protocols to utilize the 2H8 mAb for flow cytometric, immunofluorescence staining, and immunoprecipitation analyses of various FLAG-tagged GPCRs.

Keywords Epitope tag • G protein-coupled receptor • Flow cytometry • Immunofluorescence staining • Western blotting • Immuno3precipitation

Common Materials

2H8 monoclonal antibody (mAb) (TransGenic or Gentaur Molecular Products, available as "anti-DYKDDDDK antibody 2H8")

Note: The 2H8 mAb only recognizes an amino-terminal FLAG sequence and does not recognize carboxy-terminal FLAG or 3× FLAG sequences [8]

Cells or tissues expressing N-terminally FLAG-tagged G protein-coupled receptors (GPCRs)

Note: We frequently use pcDNA3 or pCXN2 vectors for expression studies of GPCRs [5, 7]

F. Sasaki • T. Yokomizo, M.D., Ph.D. (✉)
Department of Biochemistry, Juntendo University Graduate School of Medicine, 2-1-1 Hongo, Bunkyo-ku, Tokyo 113-8421, Japan
e-mail: yokomizo-tky@umin.ac.jp

T. Yokomizo, M. Murakami (eds.), *Bioactive Lipid Mediators*,
DOI 10.1007/978-4-431-55669-5_29

29.1 Flow Cytometric Analysis of N-Terminally FLAG-Tagged GPCRs Using the 2H8 mAb

Materials

Phosphate-buffered saline (PBS) (ice-cold)
PBS/EDTA [PBS, pH 7.4, containing 2 mM ethylenediamine-*N,N,N',N'*-tetraacetic acid, disodium salt (EDTA-2Na)] (ice-cold)
FACS buffer (PBS, pH 7.4, containing 2 % fetal calf serum and 2 mM EDTA-2Na) (ice-cold)
Fluorescent dye-labeled anti-mouse IgG antibody [e.g., goat anti-mouse IgG-phycoerythrin (PE), Becton Dickinson]
7-Amino-actinomycin D (7-AAD, Becton Dickinson)

1. In the case of adherent cells, detach the cells transiently or stably expressing FLAG-GPCRs using PBS/EDTA after washing with PBS. (*Note*: Avoid using trypsin because trypsinization often results in digestion of GPCRs and FLAG tags.)
2. Transfer the cells to a 96-well V-bottom plate (Nunc) (e.g., $1–5 \times 10^5$ cells/well).
3. Stain the cells with 1 μg/ml 2H8 mAb prepared in FACS buffer for 30 min at 4 °C.
4. Wash the cells three times with 150 μl PBS/EDTA.
5. Stain the cells with 0.5 μg/ml anti-mouse IgG-PE prepared in FACS buffer for 30 min at 4 °C.
6. Wash the cells three times with 150 μl PBS/EDTA.
7. Suspend the cells in 300 μl FACS buffer.
8. To exclude dead cells, stain the cells with 5 μl 7-AAD per million cells and incubate for 10 min in the dark.
9. Analyze the cells using a flow cytometer (e.g., FACSCalibur, Becton Dickinson).

29.2 Immunofluorescence (IF) Staining of FLAG-GPCRs Expressed in Cells or Tissues Using the 2H8 mAb

29.2.1 *IF Staining of Cultured Cells*

Materials

10 mM hydrogen chloride (HCl) (pH 3.0)
Collagen (Cellmatrix Type I-P, Nitta Gelatin)
Fix solution [4 % paraformaldehyde (PFA) prepared in PBS containing 10 mM glycine] (ice-cold)
PBS-G solution (PBS containing 10 mM glycine)
Blocking solution [PBS containing 3 % bovine serum albumin (BSA)]
Staining solution (PBS containing 1 % BSA)

Washing solution (PBS containing 0.1 % BSA)
Fluorescent dye-labeled anti-mouse IgG antibody (e.g., goat anti-mouse IgG-Alexa Fluor 488; Life Technologies)
4′6-Diamidino-2-phenylindole (DAPI, Sigma)
Mounting medium [Mowiol containing 2.5 % 1,4-diazobicyclo-[2.2.2]-octane (DABCO)]

1. Coat glass-bottomed dishes (Matsunami) with collagen diluted tenfold in 10 mM HCl for 30 min at room temperature.
2. Seed the cells onto collagen-coated glass-bottom dishes and culture overnight in a CO_2 incubator.
3. Fix the cells with fix solution for 5 min at room temperature.
4. Wash the cells twice with PBS-G solution.
5. Incubate the cells with blocking solution for 30 min at room temperature.
6. Stain the cells with 1 μg/ml 2H8 mAb prepared in staining solution for 30 min at room temperature.
7. Wash the cells five times with wash solution.
8. Stain the cells with 1 μg/ml anti-mouse IgG-Alexa Fluor 488 prepared in staining solution for 30 min at room temperature.
9. Wash the cells five times with wash solution.
10. Incubate the cells with 1 μg/ml DAPI prepared in PBS for 30 min at room temperature.
11. Wash the cells twice with PBS.
12. Mount with mounting medium. (*Note*: Avoid air bubbles.)
13. Visualize using a confocal laser scanning microscope (e.g., A1Rsi, Nikon or LSM780, Carl Zeiss).

29.2.2 IF Staining of Tissue Sections

Materials

Fix solution (PBS containing 4 % PFA) (ice-cold)
10 mM citrate buffer solution (pH 6.0)
Blocking solution (5 % BSA prepared in PBS containing 0.5 % Triton X-100) (ice-cold)
Wash solution (PBS containing 1 % BSA and 0.1 % Tween 20)
Horseradish peroxidase (HRP)-labeled anti-mouse IgG antibody (e.g., rat anti-mouse IgG-HRP (Trueblot), Rockland)
Fluorescent dye-labeled tyramide (Tyramide-Alexa Fluor 488, Life Technologies)
DAPI (Sigma)
Mounting medium (Permafluor, Thermo)

1. Prepare the frozen sections of tissues on Matsunami adhesive silane (MAS)-coated slide glasses (Matsunami).
2. Fix the sections with fix solution for 30 min at room temperature.

3. Boil the sections in 10 mM citrate buffer solution for 30 min for antigen retrieval.
4. Incubate the sections in blocking solution for 30 min at room temperature.
5. Stain the sections with 5 μg/ml 2H8 mAb prepared in blocking solution overnight at 4 °C.
6. Wash the sections five times with wash solution.
7. Stain the sections with anti-mouse IgG-HRP diluted 100 fold in blocking solution for 60 min at 4 °C.
8. Wash the sections five times with wash solution.
9. Incubate the sections with 1 μg/ml DAPI for 30 min at room temperature.
10. Wash the sections twice with PBS.
11. Mount with mounting medium. (*Note*: Avoid air bubbles.)
12. Visualize the samples using a confocal laser scanning microscope.

29.3 Immunoprecipitation and Western Blotting of FLAG-GPCRs Using the 2H8 mAb

29.3.1 Immunoprecipitation

Materials

Lysis buffer [20 mM Tris-HCl, pH 7.4, 150 mM NaCl, 2 mM EDTA, 10 % glycerol, 0.3 % sodium deoxycholate, 1 % NP-40, and a protease inhibitor cocktail (Nacalai)]
Wash buffer (lysis buffer lacking the protease inhibitor cocktail)
Protein A/G-agarose (Santa Cruz)
2× SDS sample buffer (25 mM Tris-HCl, pH 6.8, 0.8 % SDS, 10 % glycerol, 0.1 % bromophenol blue, 2 % 2-mercaptoethanol)

1. Lyse the cells with 500 μl lysis buffer for 15 min at 4 °C.
2. Centrifuge at 10,000 *g* for 30 min at 4 °C.
3. Transfer the supernatant to new tubes.
4. Add 10 μl protein A/G-agarose and 10–100 ng 2H8 mAb.
5. Rotate the tube overnight at 4 °C using a rotator.
6. Centrifuge at 1,000 *g* for 5 min at 4 °C.
7. Discard the supernatant.
8. Add 1 ml wash buffer to the pellets.
9. Repeat the procedure (steps 6–8) four times.
10. Dissolve the immunoprecipitates in 60 μl 2× SDS sample buffer.
11. Denature the samples for 30 min at 60 °C. (*Note*: Some GPCRs aggregate after heat denaturation at 100 °C.)

29.3.2 Western Blotting

Materials

30 % Acrylamide (AA)/Bis-acrylamide (Bis) solution (29:1)
1.5 M Tris-HCl, pH 8.8
0.5 M Tris-HCl, pH 6.8
10 % SDS
10 % Ammonium persulfate (AMPS)
N,N,N',N'-Tetramethylethylenediamine (TEMED)

12 % SDS-polyacrylamide gel (upper, 6 % stacking gel; lower, 12 % separating gel)
12 % separating gel (8 ml AA/Bis, 5 ml 1.5 M Tris-HCl, 0.2 ml 10 % SDS, 0.2 ml 10 % AMPS, 0.02 ml TEMED, 6.58 ml H_2O)
6 % stacking gel (2 ml AA/Bis, 2.5 ml 0.5 M Tris-HCl, 0.1 ml 10 % SDS, 0.1 ml 10 % AMPS, 0.01 ml TEMED, 5.29 ml H_2O)

1× running buffer (25 mM Tris, 192 mM glycine, 0.1 % SDS, pH 8.3)
1× transfer buffer (25 mM Tris, 192 mM glycine)
TBS-T solution (20 mM Tris-HCl, pH 8.0, 150 mM NaCl, 0.1 % Tween 20)
Blocking solution (TBS-T containing 5 % skimmed milk)
HRP-labeled anti-mouse IgG polyclonal antibody (PoAb) (e.g., sheep anti-mouse IgG-HRP, GE Healthcare)
Immobilon Western chemiluminescence HRP substrate solution (Millipore)

1. Separate the proteins (e.g., 10–20 μl immunoprecipitated samples) by SDS-PAGE (12 % separating gel).
2. Transfer to a polyvinylidene fluoride (Millipore) membrane in 1× transfer buffer using a wet transfer system (Bio Craft).
3. Incubate the membrane with blocking solution overnight at 4 °C.
4. Incubate the membrane with 0.1 μg/ml 2H8 mAb prepared in TBS-T containing 0.5 % skimmed milk for 60 min at room temperature.
5. Wash the membrane five times with TBS-T solution.
6. Incubate the membrane with anti-mouse IgG-HRP diluted 1000 fold in TBS-T containing 0.5 % skimmed milk for 60 min at room temperature.
7. Wash the membrane five times with TBS-T solution.
8. Incubate the membrane with HRP substrate (e.g., ImmunoStar, Wako) and visualize the signal using an image analyzer (e.g., LAS4000, Fujifilm).

29.4 Results

To evaluate the usefulness of the 2H8 mAb for flow cytometric analysis, we established CHO cells expressing various GPCRs with an N-terminal FLAG tag, namely, mouse and human leukotriene B_4 receptor 1 (mBLT1 [3] and hBLT1 [10],

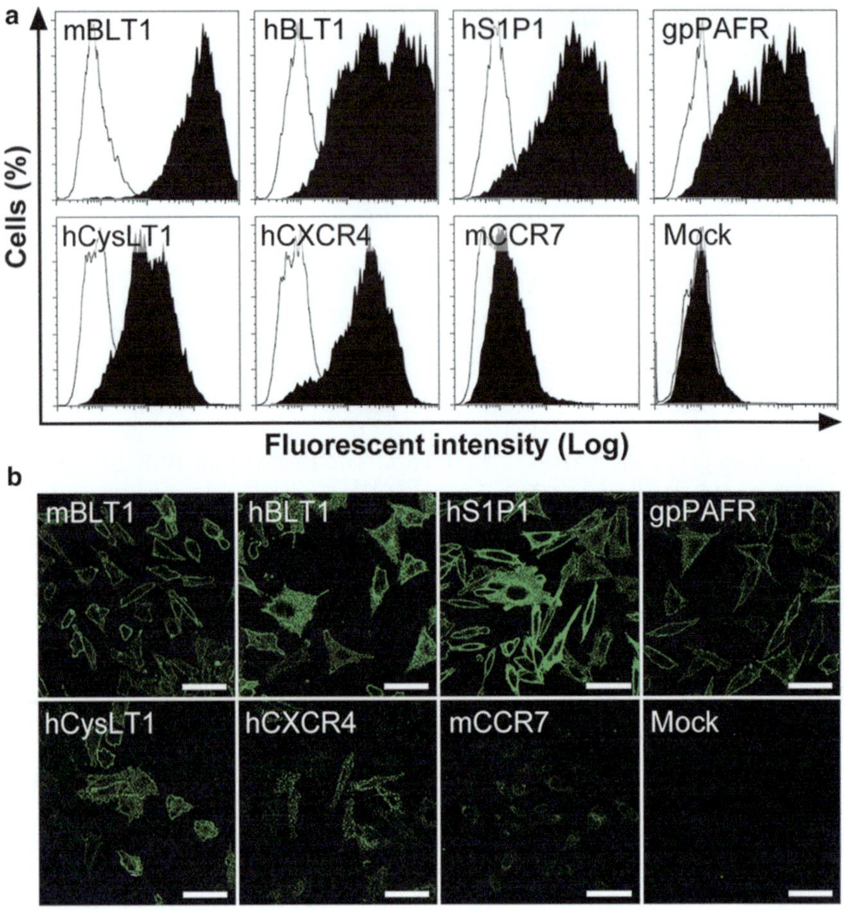

Fig. 29.1 The 2H8 mAb is sufficiently sensitive to detect N-terminally FLAG-tagged G protein-coupled receptors (GPCRs). (**a**) CHO cells stably expressing FLAG-GPCRs and mock cells were stained with 1 μg/ml 2H8 mAb (*black*) or isotype control (mouse IgG_1, *white*) and then with 1 μg/ml anti-mouse IgG-Alexa Fluor 488 and analyzed with a flow cytometer. (**b**) CHO cells expressing FLAG-GPCRs and mock cells were similarly stained in a glass-bottomed dish and visualized by confocal microscopy. Bar = 50 μm

respectively), human sphingosine-1-phosphate receptor 1 (hS1P1, [4]), guinea pig platelet-activating factor receptor (gpPAFR, [2]), human cysteinyl leukotriene receptor 1 (hCysLT1, [6]), human CXC chemokine receptor 4 (hCXCR4, [1]), and mouse CC chemokine receptor 7 (mCCR7, [9]). These cells were stained with the 2H8 mAb (1 μg/ml) followed by anti-mouse IgG-Alexa Fluor 488, and then analyzed by flow cytometry (Fig. 29.1a). Two log shifts were observed when cells expressing FLAG-tagged mBLT1, hBLT1, hS1P1, or gpPAFR were stained with the 2H8 mAb. Of note, the 2H8 mAb also detected GPCRs that were expressed at intermediate or extremely low levels (hCysLT1, hCXCR4 or mCCR7), which were not detected by commercially available anti-FLAG mAbs (M2 and M5, Sigma) [8].

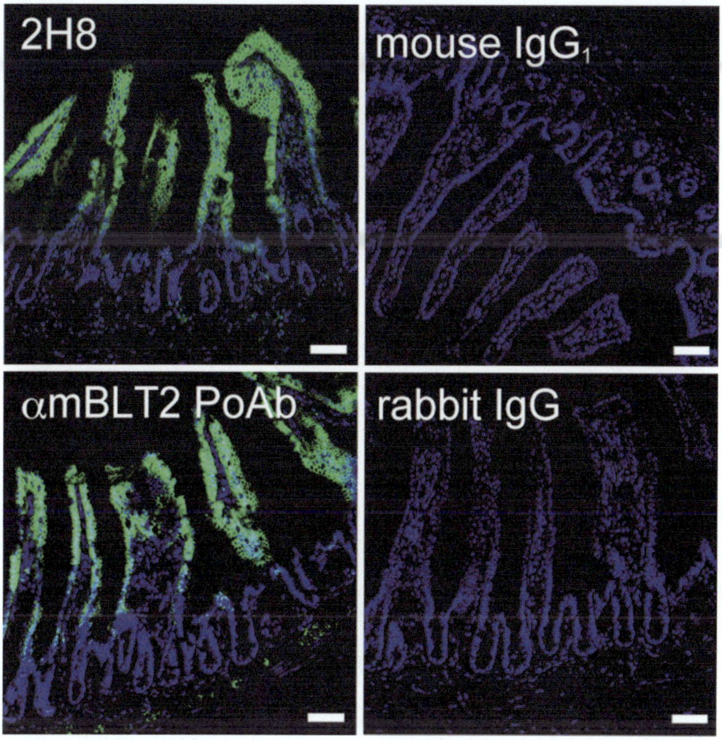

Fig. 29.2 Immunofluorescence staining in the small intestine from villin FLAG-mBLT2-Tg mice. Frozen sections of the small intestine were stained with 5 μg/ml 2H8 mAb followed by anti-mouse IgG-HRP (*upper left panel*), or with 10 μg/ml rabbit anti-mouse BLT2 PoAb followed by anti-rabbit IgG-HRP (*lower left panel*), and then reacted with Alexa Fluor 488-conjugated tyramide. *Right panels* show staining using control primary antibodies (mouse IgG_1 and rabbit IgG). Nuclei were visualized by incubation with 1 μg/ml DAPI. Bar = 50 μm

Some GPCRs are poorly expressed on the plasma membrane, possibly because of trafficking problems, even when cells overexpress these proteins. Given that it is generally difficult to generate good mAbs against GPCRs, the 2H8 mAb will be a powerful tool to detect FLAG-tagged GPCRs with modest or low expression levels that are currently undetectable using the M2 and M5 mAbs. In addition, we stained CHO cells expressing FLAG-tagged GPCRs with the 2H8 mAb and detected a bright signal on the plasma membrane (Fig. 29.1b).

Because of high background staining, especially in the nucleus, conventional anti-FLAG mAbs (e.g., M2 and M5) are not amenable to immunohistochemical staining. Therefore, sections of small intestine from villin FLAG-mBLT2 transgenic mice were stained with the 2H8 mAb or an anti-mBLT2 polyclonal antibodies (PoAb) (unpublished). We generated villin promoter-driven FLAG-mBLT2 transgenic mice, which express FLAG-mBLT2 only in intestinal epithelial cells because of the specificity of the villin promoter. Both the 2H8 mAb and the anti-mBLT2 PoAb positively stained intestinal epithelial cells (Fig. 29.2, left panels). The signal generated by the 2H8 mAb was specific because control antibodies did not produce

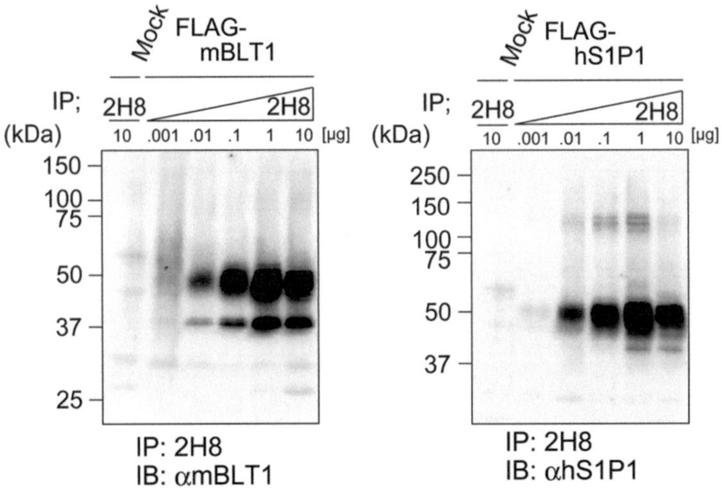

Fig. 29.3 The 2H8 mAb efficiently immunoprecipitates FLAG-tagged GPCRs. FLAG-mBLT1 (*left*) and FLAG-hS1P1 (*right*) were immunoprecipitated with increasing amounts (0.001–10 μg) of the 2H8 mAb. Immunoprecipitates were immunoblotted with 1 μg/ml rabbit anti-mBLT1 PoAb or 1 μg/ml mouse anti-hS1P1 mAb and detected with anti-rabbit or anti-mouse PoAb-HRP (1:1000) as a secondary antibody

any signals (Fig. 29.2, right panels). These data clearly demonstrate that the 2H8 mAb specifically recognizes FLAG sequences fused to GPCRs in vivo and that it is suitable for immunohistochemical analysis.

Furthermore, we examined whether the 2H8 mAb could immunoprecipitate FLAG-tagged GPCRs. The 2H8 mAb efficiently immunoprecipitated both FLAG-mBLT1 and FLAG-hS1P1 (Fig. 29.3). One hundred nanograms of 2H8 mAb was sufficient for successful immunoprecipitation, and a substantial level of immunoprecipitation was obtained using only 10 ng 2H8.

29.5 Conclusion

The 2H8 anti-FLAG mAb is a powerful tool to detect and immunoprecipitate GPCRs fused to FLAG at their amino termini.

References

1. Feng Y, Broder CC, Kennedy PE, Berger EA (1996) HIV-1 entry cofactor: functional cDNA cloning of a seven-transmembrane, G protein-coupled receptor. Science 272:872–877
2. Honda Z, Nakamura M, Miki I, Minami M, Watanabe T, Seyama Y, Okado H, Toh H, Ito K, Miyamoto T et al (1991) Cloning by functional expression of platelet-activating factor receptor from guinea-pig lung. Nature 349:342–346

3. Huang WW, Garcia-Zepeda EA, Sauty A, Oettgen HC, Rothenberg ME, Luster AD (1998) Molecular and biological characterization of the murine leukotriene B4 receptor expressed on eosinophils. J Exp Med 188:1063–1074
4. Lee MJ, Van Brocklyn JR, Thangada S, Liu CH, Hand AR, Menzeleev R, Spiegel S, Hla T (1998) Sphingosine-1-phosphate as a ligand for the G protein-coupled receptor EDG-1. Science 279:1552–1555
5. Liu M, Saeki K, Matsunobu T, Okuno T, Koga T, Sugimoto Y, Yokoyama C, Nakamizo S, Kabashima K, Narumiya S, Shimizu T, Yokomizo T (2014) 12-Hydroxyheptadecatrienoic acid promotes epidermal wound healing by accelerating keratinocyte migration via the BLT2 receptor. J Exp Med 211:1063–1078
6. Lynch KR, O'Neill GP, Liu Q, Im DS, Sawyer N, Metters KM, Coulombe N, Abramovitz M, Figueroa DJ, Zeng Z, Connolly BM, Bai C, Austin CP, Chateauneuf A, Stocco R, Greig GM, Kargman S, Hooks SB, Hosfield E, Williams DL Jr, Ford-Hutchinson AW, Caskey CT, Evans JF (1999) Characterization of the human cysteinyl leukotriene CysLT1 receptor. Nature 399:789–793
7. Okuno T, Iizuka Y, Okazaki H, Yokomizo T, Taguchi R, Shimizu T (2008) 12(*S*)-Hydroxyheptadeca-5*Z*,8*E*,10*E*-trienoic acid is a natural ligand for leukotriene B4 receptor 2. J Exp Med 205:759–766
8. Sasaki F, Okuno T, Saeki K, Min L, Onohara N, Kato H, Shimizu T, Yokomizo T (2012) A high-affinity monoclonal antibody against the FLAG tag useful for G-protein-coupled receptor study. Anal Biochem 425:157–165
9. Schweickart VL, Raport CJ, Godiska R, Byers MG, Eddy RL Jr, Shows TB, Gray PW (1994) Cloning of human and mouse EBI1, a lymphoid-specific G-protein-coupled receptor encoded on human chromosome 17q12-q21.2. Genomics 23:643–650
10. Yokomizo T, Izumi T, Chang K, Takuwa Y, Shimizu T (1997) A G-protein-coupled receptor for leukotriene B4 that mediates chemotaxis. Nature 387:620–624

Index

T. Yokomizo, M. Murakami (eds.), *Bioactive Lipid Mediators*,
DOI 10.1007/978-4-431-55669-5

M

N

FSC
www.fsc.org
MIX
Papier aus verantwortungsvollen Quellen
Paper from responsible sources
FSC® C105338

If you have any concerns about our products,
you can contact us on
ProductSafety@springernature.com

In case Publisher is established outside the EU,
the EU authorized representative is:
Springer Nature Customer Service Center GmbH
Europaplatz 3, 69115 Heidelberg, Germany

Printed by Libri Plureos GmbH
in Hamburg, Germany